Gene Flow between Crops and Their Wild Relatives

Gene Flow between Crops and Their Wild Relatives

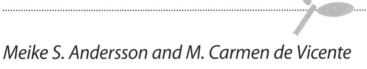

Meike S. Andersson and M. Carmen de Vicente

Foreword by Norman C. Ellstrand

The Johns Hopkins University Press

Baltimore

The Johns Hopkins University Press
2715 North Charles Street
Baltimore, Maryland 21218-4363
www.press.jhu.edu

Library of Congress Cataloging-in-Publication Data

Andersson, Meike S.
    Gene flow between crops and their wild relatives / Meike S. Andersson and M. Carmen de Vicente ; foreword by Norman C. Ellstrand.
        p. cm.
    Includes bibliographical references and index.
    ISBN-13: 978-0-8018-9314-8 (hardcover : alk. paper)
    ISBN-10: 0-8018-9314-3 (hardcover : alk. paper)
        1. Crops—Genetic engineering—Environmental aspects.   2. Transgenic plants—Risk assessment.   3. Crops—Germplasm resources.   4. Plant genetic transformation.
    I. Vicente, M. Carmen de.   II. Title.
    SB123.57.A55 2009
    631.5'23—dc22

                                                                    2008048590

A catalog record for this book is available from the British Library.

Special discounts are available for bulk purchases of this book. For more information, please contact Special Sales at 410-516-6936 or specialsales@press.jhu.edu.

Contents

Foreword

During the past decade there has been what one can only describe as a dramatic increase in interest related to the exchange of genetic material between crops and their wild relatives. This interest has been motivated by concern about whether transgenic (genetically modified) crops will mate with related free-living (wild) populations. If such mating does occur, the next logical concerns are whether the crop transgenes will persist in wild populations and have unanticipated effects in those populations. These issues, while important, are not the only reasons that we need information on the geography and reproductive biology of domesticated plants and their relatives.

Perhaps equally important is the need to have geographic and reproductive biology information for use when employing traditional methods of plant breeding. Traditional breeding has often utilized wide crosses to transfer useful alleles from wild plants into cultivars. Identifying cross-compatible relatives and finding them has always been a priority for this technique. Such data are critically important for the new, sophisticated marker-assisted breeding techniques that may rival transgenic approaches for expeditious crop improvement.

Andersson and de Vicente's masterpiece of synthesis and analysis will be welcomed as a critical book for breeders, regulators, plant population geneticists, and evolutionists, particularly those who work on crops and weeds. What an effort it must have been to find, critique, and assemble the information! An earlier, but simple, attempt, my own Chapter 7 in *Dangerous Liaisons? When Cultivated Plants Mate with Their Wild Relatives,* was the culmination of years of research. Although my task seemed Sisyphean, its product is a mere shadow of what

Andersson and de Vicente present here. Despite the ever-increasing literature on the topic, they had the courage and energy to create the first advanced crop-to-wild gene flow "field guide." For the twenty crops reviewed, they provide everything a biosafety regulator, a breeder, or a crop evolutionist would need to start on a project—including identifying research gaps. I, for one, am very grateful for their willingness to tackle this project!

Norman C. Ellstrand

Acknowledgments

....

This book represents an important undertaking for the authors. However, it would not have been possible without the support and motivation from many people and institutions. To them, we owe our gratitude.

Also, the concurrence of several circumstances is worth mentioning. First, we realized that there was valuable information available regarding the existence of gene flow among cultivated crops and assessments of its consequences. However, similar data concerning the relationships between cultivated crops and their wild relatives appeared to be rather scarce. Thus the main objective of our work was to compile this additional set of essential information in order to promote the conservation and utilization of crop genetic resources. This undertaking was the result of many interactions with research partners while working at Bioversity International.

Second, we benefited from an encounter and then subsequent discussions with Ana Mercedes Espinoza and Griselda Arrieta, researchers from the Centro de Investigación en Biología Celular y Molecular (CIBCM) of the Universidad de Costa Rica who were engaged in a study dealing with this same topic—an environmental risk assessment of gene flow from a genetically modified rice cultivar to a native wild relative, *Oryza glumaepatula* Steud.—as a model for the deployment of genetically modified rice in a tropical center of diversity.

Third, the GTZ (Deutsche Gesellschaft für Technische Zusammenarbeit, or German Technical Cooperation) provided extraordinary financial support, initially in the form of a postdoctoral grant to Andersson, because the officers understood the relevance of the subject matter and its fit with one of their main interests—biodiversity—and later in funding for the completion of this book and its publication.

Once started, the preparation of the book required the assistance of many people in different capacities. Diego F. Álvarez Sánchez (Universidad del Valle, Cali, Colombia), a geographical information systems student, helped prepare a very attractive component of this work, the world maps showing the presence of cultivated crops and their wild relatives. Andy Jarvis (Bioversity International and the International Center for Tropical Agriculture [CIAT], Cali, Colombia) and Glenn Hyman (CIAT) brainstormed with us to propose ways to complement our compendium of information with graphical tools and presentations. Stanley Wood, an economist at the International Food Policy Research Institute (IFPRI) in Washington, D.C., shared its extremely valuable worldwide crop production dataset with us. Without their help, producing the maps that are an essential part of this work would not have been possible. We are grateful to many colleagues at Bioversity International who helped make this book a reality. Ehsan Dulloo, in particular, provided moral support all along the way and continued to serve as a liaison with Bioversity once the authors had moved on to other jobs.

We thank the Johns Hopkins University Press and our editor, Vince Burke, who worked with us, patiently, during the final process in such a way that made us think that publishing a book was indeed trouble free. We also thank Kathleen Capels, who edited the manuscript, taking special care not just of details but also of the content. She definitely helped mold it into the book we had envisioned.

All the scientists who took the time to preview the crop chapters and provide us with invaluable and thoughtful feedback, corrections, and suggestions are also deserving of our gratitude. Their comments were obviously the result of knowledge gathered from years of experience with a particular crop and were full with details that could never be found in the literature. In addition, their contributions were often accompanied by praise and support for the work we had undertaken. It is fair to say that this encouragement laid the foundation for our decision to publish this work. In addition, we have benefited greatly from the exchange of ideas with many of them. Thank you for teaching us. These scientists are (in alphabetical order): Elizabeth Arnaud, Fréderic

Bakry, Alberto Duarte Vilarinhos, Charlotte Lusty, Sebastião de Oliveira E., and Franklin Rosales (bananas and plantains); Michael Baum, Roland von Bothmer, and Helmut Knüpffer (barley); Henri Darmency, Keith Downey, and Rikke Bagger Jørgensen (canola); Hernan Ceballos, Clair Hershey, Doyle McKey, and Nagib Nassar (cassava); Faiz Ahmad, Pooran Gaur, and Sripada Udupa (chickpeas); Jean-Pierre Baudoin, Daniel Debouck, Valérie Geffroy, Paul Gepts, and Eduardo Vallejos (common beans); Curt Brubaker, Paul Fryxell, and Andrew Paterson (cotton); Ndiaga Cissé, Jeff Ehlers, and Christian Fatokun (cowpeas); Katrien Devos, Mathews Dida, and Chrispus Oduori (finger millet); Julien Berthaud, Jerry Kermicle, Jesús Sánchez González, Jose Antonio Serratos-Hernández, Suketoshi Taba, and Garrison Wilkes (maize); Igor Loskutov, Jennifer Mitchell Fetch, Howard Rines, and Deon Stuthman (oats); David Bertioli, Charles Simpson, and Tom Stalker (peanuts); Bettina Haussmann and Jeffrey Wilson (pearl millet); Kul Bhushan Saxena and Said Silim (pigeonpeas); Marc Ghislain and David Spooner (potatoes); Zaida Lentini, Mathias Lorieux, Bao-Rong Lu, and César Martínez (rice); William Rooney and Allison Snow (sorghum); Ted Hymowitz and Reid Palmer (soybeans); Daniel Austin and Swee Lian Tan (sweetpotatoes); and Michel Bernard, Thomas Payne, and Wolfgang Pfeiffer (wheat).

The crop photos used as chapter openers are courtesy of the following individuals: Eduardo Marquetti (bananas), Helmut Knüpffer (barley), Claus Rebler (canola), Meike S. Andersson (cassava, common beans, rice, sorghum), Pooran Gaur of the International Crops Research Institute for the Semi-Arid Tropics (ICRISAT) (chickpeas), Linda de Volder (cotton), Jeff Ehlers (cowpeas), Mark Laing (finger millet), Carl E. Lewis (maize), Paulo J. Cardoso Marques (oats), Nancy L. Foote (peanuts), Bettina Haussmann of ICRISAT (pearl millet), André Benedito (pigeonpeas), Pamela Bevins (potatoes), Rosanne Healy (soybeans), Malaka Chessher (sweetpotatoes), and Karim Ammar of the Centro Internacional de Mejoramiento de Maíz y Trigo (CIMMYT, or International Maize and Wheat Improvement Center) (wheat).

Last but not least, in an endeavor of this type, it is common for the planned time frame to be much more optimistic than the actual amount of time needed to complete it. The period we spent compiling the information, having it revised by experts, implementing improvements, and giving proper attention to details before submitting it to the editor greatly surpassed our original expectation. All of this took much more of our personal time than we had foreseen, and while we enjoyed what we were doing, we were aware that we were depriving the people around us of our participation in other important activities. Because of that we especially want to thank Franklin Cruz Rodríguez and Humberto Gomez-Paniagua, our husbands, and promise that we will now do our best to catch up on all that we missed with them.

Abbreviations and Acronyms

....

ACRE	Advisory Committee on Releases to the Environment (UK), www.defra.gov.uk/Environment/acre/.
AOSCA	Association of Official Seed Certifying Agencies (USA), www.aosca.org.
ARS	Agricultural Research Service of the **USDA** (USA), www.ars.usda.gov.
BC	backcross(es)
Bt	*Bacillus thuringiensis* Berliner
C	central
CCIA	California Crop Improvement Association (USA), http://ccia.ucdavis.edu/.
CEC	Commission for Environmental Cooperation (Canada, Mexico, United States), www.ccemtl.org.
CIAT	Centro Internacional de Agricultura Tropical, or International Center for Tropical Agriculture (Colombia), www.ciat.cgiar.org.
CIMMYT	Centro Internacional de Mejoramiento de Maíz y Trigo, or International Maize and Wheat Improvement Center (Mexico), www.cimmyt.org.
CSGA	Canadian Seed Growers Association (Canada), www.seedgrowers.ca.
CWR	crop wild relative(s)
DEFRA	Department of the Environment, Food, and Rural Affairs (UK), www.defra.gov.uk.

DNA	deoxyribonucleic acid
E	east
EC	European Commission, http://ec.europa.eu/.
EMBRAPA	Empresa Brasileira de Pesquisa Agropecuária (Brazil), www.embrapa.br.
EU	European Union, http://europa.eu/.
FAO	Food and Agriculture Organization of the United Nations (Italy), www.fao.org.
FSE	farm-scale evaluations
GIS	geographic information system
GISD	Global Invasive Species Database, www.issg.org/database/welcome/.
GM	genetically modified
GMO	genetically modified organism(s)
GP	gene pool (GP-1: primary, GP-2: secondary, and GP-3: tertiary gene pool)
GRIN	Germplasm Resources Information Network of the USDA (USA), www.ars-grin.gov.
IBPGR	International Board for Plant Genetic Resources (now Bioversity International; Italy), www.bioversityinternational.org.
ICRISAT	International Crops Research Institute for the Semi-Arid Tropics (India), www.icrisat.org.
IITA	International Institute of Tropical Agriculture (Nigeria), www.iita.org.
INIBAP	International Network for the Improvement of Banana and Plantain (now Bioversity International; Italy), www.bioversityinternational.org.
IPGRI	International Plant Genetic Resources Institute (now Bioversity International; Italy), www.bioversityinternational.org.
ISAAA	International Service for the Acquisition of Agri-biotech Applications (Africa, SE Asia, USA), www.isaaa.org.
ITS	internal transcribed spacer

m	meter
max.	maximum
MS	male sterile
N	north
n.a.	not available
n.r.	not reported
OECD	Organization for Economic Cooperation and Development (France), www.oecd.org.
PBIP	Plantain and Banana Improvement Program of the IITA (Nigeria), www.iita.org.
PCR	polymerase chain reaction
S	south
SCIMAC	Supply Chain Initiative on Modified Agricultural Crops (UK), www.scimac.org.uk.
SSR	simple sequence repeats
USDA	United States Department of Agriculture (USA), www.usda.gov/wps/portal/usdahome/.
W	west
WHO	World Health Organization (Switzerland), www.who.int/en/.

Gene Flow between Crops and Their Wild Relatives

ONE

Introduction

The centers of origin and diversity of many crop species are located in developing countries (Damania et al. 1998; Gepts 2001). The conservation of crop genetic resources in these regions is of key importance to protect both the environment and the functioning of ecological processes. Likewise, the existence of diverse crop genetic resources forms the basis not only of their sustainable use but also of long-term food security. This is especially important, as developing regions throughout the world are those that are most threatened by famines and persistent poverty. This very argument is also used by proponents of modern agricultural advances based on genetically modified (GM) crops. Indeed, GM crops offer an array of solutions that are highly suitable for confronting the constraints of agriculture found in developing countries, including environmental issues, herbicide tolerance, and insect resistance.

Similarly, and in parallel with developments in the genetic modification of crops, numerous studies have been performed to measure the occurrence of gene flow. Often, the avalanche of gene-flow measurements has overshadowed the fact that gene flow and introgression are both natural phenomena, that is, they occur spontaneously and play a very important role in crop evolution. Gene flow and introgression have actively contributed to the forms and behavior types characterizing the crops we know today.

Even so, less research and fewer assessments have been carried out on the actual impact of GM crops on wild populations of crop relatives and their landraces. Because gene flow and introgression will occur under most circumstances where a crop and its wild relatives coexist, pertinent information needs to be available so that experts and authorities can make well-informed decisions on whether measures should be taken to prevent potentially negative effects of these natural phenomena. Our objective, then, is neither to provide judgments nor to arrive at conclusions on the type of crop that can be or should not be planted in the vicinity of its wild relatives or landraces. Instead, our goal is to guide basic and scientifically sound decision-making by taking into account the facts that gene flow and introgression exist, and that the preservation of crop genetic resources in their habitat is a requisite for the sustainable development of modern crops.

Because of the need for continued improvement in productivity and other gains, since 1996 the global area planted with GM crops has been constantly increasing in both industrialized and developing countries (James 2007). In 2008, for the first time, the latter outnumbered the former by 15 to 10. James (2008) estimates that 90% of the world's 13.3 million farmers planting GM crops are small and resource-poor farmers in developing countries. The increase in global GM crop production has led to the emergence of numerous issues related to, among others, ecological, agricultural, ethical, policy, legal, socioeconomic, health, and food safety concerns (Commandeur et al. 1996; NCB 1999; Moeller 2001; NRC 2002). An integrated risk assessment, addressing these concerns, is mandatory and usually legally required before GM crops are released.

This book focuses on one particular concern that is of utmost importance for the release of GM crops in or around areas with concentrated crop diversity—the likelihood of gene flow and introgression to crop wild relatives (CWR) and other domesticated species. Our perspective is not that gene flow and introgression necessarily result in negative impacts, but more one of discovering and understanding facts (e.g., biological information on crop plants, existing CWR, the likelihood of gene flow between crops and their relatives, crop production areas, and the geographical range of wild relatives) so that informed decisions can be made on whether or not a risk exists. This book does not make judgments, although we do indicate potential effects, depending on the area of spread and on coexistence (sympatry).

Nor is it intended to be a comprehensive review of all research pertaining to the crops treated here. Because new information from field, greenhouse, and laboratory studies is continuously becoming available, our book does not pretend to be an exclusive tool for addressing issues related to field testing, deregulation, or the commercialization of GM plants. Indeed, we recommend consulting the online resources mentioned in the respective crop sections for updated information, particularly for GM crops of major economic importance, such as canola, rice, cotton, wheat, maize, and potatoes. For those crops of lesser economic value globally, the information presented here may be valid for a longer period, as scientific advances on gene flow in these crops are slower because not as much is invested in research on them.

The following chapters summarize both state-of-the-art knowledge and research gaps related to gene flow and introgression between crops and their wild relatives. The information is based on a comprehensive review of the relevant scientific and technical literature. It emphasizes gene transfer from GM crops to wild species, rather than between different cultivars (i.e., from GM to conventional varieties) of the same crop.

In this context, we consider gene flow to be the transfer of genetic information between different individuals, populations, or species. Gene flow occurs naturally via spontaneous hybridization, which can be either seed or pollen mediated. It can also be artificial, whether through hand pollination or more sophisticated hybridization techniques such as ovule culture, embryo rescue, or protoplast fusion. At first glance, only spontaneous hybridization seems to be a serious issue, as it does indeed increase the likelihood of gene flow in the wild.

However, if artificial hybridization is relatively easy (e.g., using hand pollination), then it, too, offers an opportunity for gene flow to occur unexpectedly. Hence such situations may be of as much concern as spontaneous hybridization. In contrast, if artificial hybridization is possible only with considerable human intervention (i.e., using sophisticated hybridization techniques), then spontaneous hybridization in the wild is highly unlikely.

We consider introgression to be the permanent incorporation of genetic information from one set of differentiated populations (species, subspecies, races, and so on) into another. Hybridization is a prerequisite for gene flow, and gene flow a prerequisite for introgression. However, not all hybridization events lead to gene flow and not all gene-flow events lead to introgression.

Any assessment of the likelihood of gene flow and introgression between plant species or populations deals with the following:

1   the emission, dispersal, and deposition of pollen and/or seed
2   the likelihood and extent of gene flow from crops to wild relatives
3   stabilization and the spread of traits in wild species
4   ecological effects caused by the introduction of traits into the new host population

This book targets the first three issues by analyzing the biological framework for protecting the genetic integrity of indigenous wild relatives in centers of crop origin and diversity. However, to conduct such an analysis, the distribution of CWR must be known. Very few maps are published on the geographical distribution of CWR; these usually provide information on a local or regional scale and rarely show global distributions. Thus a handbook supplying comprehensive information on how to assess the likelihood of gene flow in the world's most important crops is both timely and valuable.

Gene Flow

The flow of crop genes or novel GM traits can be transmitted either through seeds or pollen to populations of (1) other forms of the same crop, (2) weeds, and (3) wild relatives.

Seed-Mediated Gene Flow

Gene flow via seeds can be easily overlooked. For example, seed banks of GM crops with dormant seeds can build up naturally in the soil. In subsequent cultivation cycles, volunteers arise from the seeds in the seed bank and then outcross with the newly planted crop and sexually compatible wild relatives. Gene flow, and potential introgression, could thus occur. This is a natural process that has been documented for several crops, particularly sugar beets (Arnaud et al. 2003) and canola (Hails et al. 1997; Simard et al. 2002). The importance of controlling volunteers in subsequent crop generations is well established, and a range of management options is now available, such as mechanical and chemical roguing, tillage, and crop rotation.

However, the actual production area of GM crops, as well as the locations where the crops are grown, are not the only significant vehicles for unwanted gene flow. Humans can also be a vector for seed-mediated gene escape, for example, during the transport, processing, and exchange of GM seeds. This has been documented in Japan, where several GM canola plants (also known as oilseed rape, or *Brassica napus* L.) were detected at major ports and along roadsides (Saji et al. 2005; Aono et al. 2006). No GM traits, however, were identified in seeds collected from canola plants growing along riverbanks or in seeds from closely related species (*B. rapa* L. and *B. juncea* (L.) Czern.). Because GM canola is not commercially cultivated in Japan, the GM plants most probably resulted from imported GM canola seeds that had spilled during transport to oilseed-processing facilities.

These reports were the first published example of feral GM plants occurring in a country where the GM crop is not and has not been cultivated commercially. Shortly thereafter, a similar case was reported in Korea (Kim et al. 2006), where a feral GM maize plant was detected along a roadside near a Korean seaport.

Gene flow is also facilitated by farmers informally exchanging seeds or other planting materials—such as potato tubers and cassava rootstocks or stem cuttings—a frequent practice in many traditional cropping systems in Central and South America, Africa, and Asia (Louette 1997; Álvarez et al. 2005; Nagarajan and Smale 2007). These autochthonous farming practices stimulate extensive gene flow, not only to landraces but also eventually to sexually compatible wild relatives.

Pollen-Mediated Gene Flow

For interspecific gene flow to occur via pollen exchange (outcrossing), hybridization needs to be differentiated from actual gene introgression (Jenczewski et al. 2003; Stewart et al. 2003). Introgression involves the permanent transfer of genetic information from one plant to another, such as through repeated backcrossing. First, not every movement of pollen grains (potential gene flow) will result in outcrossing. Pollen grains must fertilize the ovules for gene flow to actually occur. If pollination does result in fertilization, then only the seeds of the pollen-receiving plant will contain the GM trait—the recipient plant itself does not become a GM plant. However, if the resulting hybrid seeds are viable and germinate, then their progeny (i.e., the entirety of the new plants arising from these seeds) are affected. Nonetheless, the resulting plants or hybrids may be sterile or less viable than the parent plants.

Hybridization

Artificial methods such as ovule culture, embryo rescue, and protoplast fusion have extended the range of possibilities for intra- and interspecific crosses that involve crop species. Most of the resulting new artificial hybrid combinations would not occur naturally, because of hybridization barriers that prevent normal embryo or endosperm development. While these techniques provide an important method for plant improvement by allowing the transfer of genes that would not otherwise be accessible, the ease with which hybrids are formed when *in vitro* methods are used is not indicative of the probability of similar hybrids occurring by cross-pollination under natural outdoor conditions. Such artificially obtained hybrids can, however, provide valuable information on the chromosome pairing behavior between weakly homologous genomes. They can also facilitate evaluating hybrids between crops and related species in order to assess their potential to survive and persist under natural conditions. The extent of gene flow and the actual frequency of hybridization depend on the four factors described below.

Pre- and Post-Zygotic Barriers

In plants, cross-incompatibility after pollination is caused by two types of fertilization barriers (Stebbins 1958; Hadley and Openshaw 1980; MacNair 1989). One occurs before fertilization and results mainly from pollen-pistil interactions, whereas the other happens when the development of young zygotes is arrested.

Pollen Competition

In general, when plants of two different species exchange pollen, the pollen of the other species (alien pollen) is slow to effect fertilization and is therefore at a disadvantage with respect to pollen from another plant of the same species (Grant 1975). Several examples exist, such as in cotton, where pollen competition clearly occurs when alien pollen reaches cultivated cotton (Kearney 1924; Kearney and Harrison 1932).

Direction of the Cross

At times, hybridization is more likely to occur in one direction rather than the other. For example, hybrids between domesticated maize and its wild relative, the Mexican teosinte (*Zea mays* subsp. *mexicana* (Schrad.) H.H. Iltis), are produced much more easily when the crop plant is the female parent (pollen receptor) rather than vice versa (Kermicle and Allen 1990; Kermicle 1997; Baltazar et al. 2005).

In general, when male-sterile crop varieties are used, hybrids from crosses in which the crop plant is the female are again much more likely to occur than the reverse. This has been observed, for example, in canola (Darmency et al. 1998; Chèvre et al. 2000). Hybrids resulting from crosses of male-sterile canola with wild relatives (e.g., wild radish, *Raphanus raphanistrum* L.) are usually partially sterile. However, hybrid fertility can be restored by backcrossing with the wild relative, which would then result in GM traits moving towards wild relatives that act as the female plant. Even unfit and ephemeral hybrid plants could therefore serve as a bridge to the evolution of fitter GM wild relatives. This example illustrates the fact that the direction of hybridization does not necessarily allow for conclusions about the possible containment of GM traits, as backcrossing can reverse this effect within only a few generations.

Role of Other, Often Local, Conditions

In addition to the above-mentioned crop-specific parameters, gene flow also depends on several variable local parameters that must be assessed on a case-by-case basis, according to the given geographic location. These factors include:

- the layout of the experimental fields, such as the size and orientation of the plots/fields, since smaller fields are subject to a higher relative degree of pollen importation than larger fields

- topographic features, such as valleys or water surfaces, which influence air currents and wind strength

- other surrounding structures, such as hedges, forests, and uncultivated areas, and the spatial arrangements of the fields
- the incidence and/or overlap of flowering times
- the type of vector that facilitates pollination
- the species composition, variation in activity and behavior, and population size of the pollinators present in the area or region
- the distances between pollen-emitting and pollen-receiving populations and the length of time that the pollen remains viable
- local climatic conditions that influence the activity of pollinators and the transport of airborne pollen (e.g., rainfall distribution, humidity, wind direction and strength, and air and soil temperatures)
- regional crop management practices
- regional crop rotation systems and cropping patterns, taking into account crop-specific seed longevity and other factors

The combination of relevant variables differs from crop to crop and is also influenced by their respective mating systems.

Hybridization is therefore context dependent and may vary in time and space (Harlan 1992; Ellstrand 2003). In this book, however, we assume that hybridization is homogeneous, that is, if hybridization has been experimentally demonstrated in one region or location, then we suppose that it may potentially occur at any other place or moment.[1]

Introgression

Successful hybridization does not necessarily lead to gene introgression. The permanent introgression of genetic material into the receiving population or species depends mainly on the viability and fitness of the offspring and on the nature of the transferred gene.

Hybrid Viability and Fitness

Fitness-related factors such as fertility, dormancy, seed persistence in the soil, germination rate, abiotic stress tolerance, and seedling mortality may confer selective advantages and lead to increased weediness. These traits must therefore be considered together with the package of legal requirements involved in the risk-assessment procedures carried out before a GM crop variety is released.

However, these characteristics vary considerably, depending on the variety and the GM trait or traits under consideration, as well as on environmental factors (environment x genotype interactions). For example, Linder (1998) has shown that the seed dormancy of some GM canola varieties is significantly higher than that of their non-GM controls under certain field conditions (darkness, high nutrients). These site-specific differences in hybrid fitness emphasize the need for appropriate risk assessments at various sites, thereby expanding the evaluations to cover the actual range where there are plans to market a GM crop.

Nature of a GM Trait

The likelihood and impact of gene flow and introgression are also influenced by factors related to the nature of the GM trait(s). These usually refer to the stability of the introgressed trait, its location in the genome, its inheritance pattern (dominant or recessive), and its potential to increase or reduce the fitness of the receptor plant (see, e.g., Linder and Schmitt 1994; Jenczewski et al. 2003; Lu 2008).

In theory, traits with negative effects on the receptor plant's fitness (e.g., male sterility, altered fiber quality, changes in lignin biosynthesis) are expected to be selected against in nature, and thus be unlikely to spread. In contrast, traits such as resistance to pests, diseases, or environmental stresses are likely to be beneficial for the receptor plant in areas where these constraints affect its performance or survival. This may apply to cropping systems as well as to natural habitats. Introgression of an advantageous trait into a wild or weedy species can then pose an ecological problem, as it is likely to result in increased overall fitness and competitiveness. A prominent example is the increased fitness of wild sunflowers after hybridization between cultivated Bt sunflowers (*Helianthus annuus* L.) and wild sunflowers. The wild plants had higher seed production because of reduced herbivory (Snow et al. 2003).

A third group are the so-called neutral traits, such as enhanced yield or nutritional content, which neither improve nor reduce fitness. In practice, however, even GM traits that appear neutral can have unexpected secondary effects. For example, seed-protein and seed-oil modifications can affect seed dormancy and survival in the soil, the timing of emergence, and seedling vigor and performance (Levin 1974; Linder and Schmitt 1995).

Ecological Effects of Introgression

The ecological effects of introgression between crops and their wild relatives depend mainly on three aspects: the nature of the introgressed trait (see above); the ecology of the organism receiving the trait; and the local context, that is, the prevalent agricultural management practices, location in cultivated fields versus wild habitat, and other environmental conditions. Depending on the site-specific combination of these factors, possible ecological impacts include (1) the evolution of various resistances (hence the importance of refuges), (2) increased weediness or invasiveness, and (3) alterations in genetic diversity.

We will not go into detail here, as extensive discussions on the probability and scope of these events have been published elsewhere (Ellstrand et al. 1999; NAS 2000; Ellstrand 2001; Dale et al. 2002; Conner et al. 2003; Stewart et al. 2003). Specific studies on the likelihood and consequences of evolved resistance (point 1 above) were made by Gould (1998), Traynor and Westwood (1999), Andow (2001) and Tabashnik et al. (2000). Increased weediness and the development of invasive characteristics (point 2) were reviewed by Parker and Kareiva (1996), Ellstrand et al. (1999), Ellstrand and Schierenbeck (2000), Snow (2002), and Novak (2007).

Altered genetic diversity (point 3) is of particular concern with regard to centers of crop origin and diversity. Depending on the nature of the introgressed trait and the predominant direction of gene flow, the ecological implications for landraces and wild species can be considerable. Hybridization can lead to an increase or decrease of genetic diversity in the receiving population, or even to its extinction (Wolf et al. 2001; Gepts and Papa 2003). For example, in Asia, Kiang et al. (1979) found that the natural hybridization of wild rice species with domesticated rice has led to the near extinction of the endemic wild Taiwanese rice *Oryza perennis formosana* (= *Oryza sativa* var. *formosana* (Masam. & Suzuki) Yeh & Henderson, a synonym of *O. sativa* L.). Similarly, the disappearance of several other wild rice populations throughout Asia is attributed to hybridization with the crop (chapter 18).

The impact of changes in genetic diversity due to hybridization is further worsened when landrace or wild populations are small. We know this is frequently the case in many of the centers of crop origin and diversity, where CWR are often threatened by habitat destruction and anthropogenic activities (see chapter 6 on cassava, chapter 8 on common beans, chapter 12 on maize, chapter 18 on rice, and chapter 22 on wheat). The loss of genetic diversity, or even extinction, through hybridization with domesticated species is an additional factor that threatens the sustained existence of diverse crop genetic resources in their natural habitats.

Organization of the Book

Before the likelihood and potential impact of gene flow can be adequately assessed, studies must be carried out on a crop-by-crop basis. A good understanding of the cropping system, together with that of the biotic and abiotic conditions at the specific geographic location being considered, is essential. As we have seen above, the GM trait itself, and the ecology and phenology of the modified and receiving plants can also influence the likelihood and potential impact of gene flow. The information presented in this book is meant to provide initial clues as to which crops and variables could benefit from closer scrutiny.

The next chapter is an overview of the choices we have made and our methods of presentation. The subsequent crop chapters are divided into the following major sections. First, general biological information is discussed, including the crop's scientific name, its center(s) of origin and diversity, flowering, pollen dispersal and longevity, sexual and vegetative reproduction, seed dispersal and dormancy, the existence and persistence of volunteers and feral plants, and the existence of weeds and their invasiveness potential. Second, the most important CWR are presented, and information about their ploidy levels, diverse genomes, centers of origin, and geographic distribution is given. The list generally includes all other species of the same genus as the crop, but it may also contain species from closely related genera. Third, the crop's potential for hybridization with its wild relatives is described, and detailed information is provided on relevant studies and literature. Fourth, pollen flow studies related to pollen dispersal distances and hybridization rates are analyzed, in conjunction with the separation distances recommended by regulatory authorities for maintaining seed purity standards. Fifth, an overview summarizes the state of development of GM technology regarding that crop, including global GM crop area, GM traits currently researched for the crop, and countries where GM varieties are grown commercially or in field trials. For some crops, information is provided on the extent of gene flow via seed exchange, the conservation status of CWR, and management recommendations to minimize gene flow in the field.

At the end of each crop chapter, research gaps are identified and conclusions are drawn about the likelihood of gene flow and introgression. These conclusions contain an innovation; for the first time, world (rather than localized) maps are presented that show regions where crops are likely to occur in sympatry with their sexually compatible wild relatives, based on the modeled distribution of CWR. These maps illustrate the likelihood of introgression between each crop and its CWR, as identified in the conclusion for each crop chapter, where we use a classification system with four categories of likelihood: low, moderate, high, and very high. This visual presentation thus immediately signals areas where gene flow and introgression may be an issue and hence could benefit from closer scrutiny.

NOTE
....

1. Two further assumptions we make regarding hybridization are (1) that crop and CWR populations are sufficiently close to each other to allow outcrossing, and (2) that flowering times overlap.

REFERENCES
....

Álvarez N, Garine E, Khasah C, Dounias E, Hossaert-McKey M, and McKey D (2005) Farmers' practices, metapopulation dynamics, and conservation of agricultural biodiversity on-farm: A case study of sorghum among the Duupa in sub-Sahelian Cameroon. *Biol. Cons.* 121: 533–543.

Andow DA (2001) Resisting resistance in *Bt* corn. In: Letourneau DK and Burrows

BE (eds.) *Genetically Engineered Organisms: Assessing Environmental and Human Health Effects* (CRC Press: Boca Raton, FL, USA), pp. 99–124.

Aono M, Wakiyama S, Nagatsu M, Nakajima N, Tamaoki M, Kubo A, and Saji H (2006) Detection of feral transgenic oilseed rape with multiple-herbicide resistance in Japan. *Environ. Biosafety Res.* 5: 77–87.

Arnaud JF, Viard F, Delescluse M, and Cuguen J (2003) Evidence for gene flow via seed dispersal from crop to wild relatives in *Beta vulgaris* (Chenopodiaceae): Consequences for the release of genetically modified crop species with weedy lineages. *Proc. Roy. Soc. London, Ser. B, Biol. Sci.* 270: 1565–1571.

Baltazar BM, Sánchez JJ, de la Cruz L, and Schoper JB (2005) Pollination between maize and teosinte: An important determinant of gene flow in Mexico. *Theor. Appl. Genet.* 110: 519–526.

Chèvre AM, Eber F, Darmency H, Fleury A, Picault H, Letanneur JC, and Renard M (2000) Assessment of interspecific hybridization between transgenic oilseed rape and wild radish under normal agronomic conditions. *Theor. Appl. Genet.* 100: 1233–1239.

Commandeur P, Joly PB, Levidow L, Tappeser B, and Terragni F (1996) Public debate and regulation of biotechnology in Europe. *Biotech. Dev. Monit.* 26: 2–9.

Conner AJ, Glare TR, and Nap JP (2003) The release of genetically modified crops into the environment: Part II. Overview of ecological risk assessment. *Plant J.* 33: 19–46.

Dale PJ, Clarke B, and Fontes EM (2002) Potential for the environmental impact of transgenic crops. *Nat. Biotech.* 20: 567–574.

Damania A, Valkoun J, Willcox G, and Qualset CO (eds.) (1998) *The Origins of Agriculture and the Domestication of Crop Plants: Proceedings of the Harlan Symposium, Aleppo, Syria, May 10–14, 1997* (International Center for Agricultural Research in Dry Areas [ICARDA], International Plant Genetic Resources Institute [IPGRI], Food and Agriculture Organization of the United Nations [FAO], and Genetic Resources Conservation Program [GRCP]: Aleppo, Syria). www.bioversityinternational.org/publications/Web_version/47/begin.htm.

Darmency H, Lefol E, and Fleury A (1998) Spontaneous hybridizations between oilseed rape and wild radish. *Mol. Ecol.* 7: 1467–1473.

Ellstrand NC (2001) When transgenes wander, should we worry? *Plant Physiol.* 125: 1543–1545.

——— (2003) *Dangerous Liaisons? When Cultivated Plants Mate with Their Wild Relatives* (Johns Hopkins University Press: Baltimore, MD, USA). 244 p.

Ellstrand NC and Schierenbeck K (2000) Hybridization as a stimulus for the evolution of invasiveness in plants? *Proc. Natl. Acad. Sci. USA* 97: 7043–7050.

Ellstrand NC, Prentice HC, and Hancock JF (1999) Gene flow and introgression from domesticated plants into their wild relatives. *Ann. Rev. Ecol. Syst.* 30: 539–563.

Gepts P (2001) Origins of plant agriculture and major crop plants. In: Tolba M (ed.) *Our Fragile World: Challenges and Opportunities for Sustainable Development* (EOLSS Publishers: Oxford, UK), pp. 629–637.

Gepts P and Papa R (2003) Possible effects of (trans) gene flow from crops on the genetic diversity from landraces and wild relatives. *Environ. Biosafety Res.* 2: 89–103.

Gould F (1998) Sustainability of transgenic insecticidal cultivars: Integrating pest genetics and ecology. *Ann. Rev. Entomol.* 43: 701–726.

Grant M (1975) *Genetics of Flowering Plants* (Columbia University Press: New York, NY, USA). 514 p.

Hadley HH and Openshaw SJ (1980) Interspecific and intergeneric hybridization. In: Fehr W and Hadley HH (eds.) *Hybridization of Crop Plants* (American Society of Agronomy: Madison, WI, USA), pp. 133–159.

Hails RS, Rees M, Kohn DD, and Crawley MJ (1997) Burial and seed survival in *Brassica napus* subsp. *oleifera* and *Sinapis arvensis* including a comparison of transgenic and non-transgenic lines of the crop. *Proc. Roy. Soc. London, Ser. B, Biol. Sci.* 264: 1–7.

Harlan JR (1992) *Crops and Man.* 2nd ed. (American Society of Agronomy: Madison, WI, USA). 284 p.

James C (2007) *Global Status of Commercialized Biotech/GM Crops 2007.* ISAAA Brief 37 (International Service for the Acquisition of Agri-biotech Applications [ISAAA]: Ithaca, NY, USA). 143 p.

——— (2008) *Global Status of Commercialized Biotech/GM Crops 2008.* ISAAA Brief 39 (International Service for the Acquisition of Agri-biotech Applications [ISAAA]: Ithaca, NY, USA). 243 p.

Jenczewski E, Ronfort J, and Chèvre AM (2003) Crop-to-wild gene flow, introgression, and possible fitness effects of transgenes. *Environ. Biosafety Res.* 2: 9–24.

Kearney TH (1924) Selective fertilization in cotton. *J. Agric. Res.* 27: 329–340.

Kearney TH and Harrison GJ (1932) Pollen antagonism in cotton. *J. Agric. Res.* 44: 191–226.

Kermicle JL (1997) Cross-compatibility within the genus *Zea.* In: Serratos JA, Willcox MC, and Castillo F (eds.) *Proceedings of a Forum: Gene Flow among Maize Landraces, Improved Maize Varieties, and Teosinte; Implications for Transgenic Maize* (International Maize and Wheat Improvement Center [CIMMYT, or Centro Internacional de Mejoramiento de Maíz y Trigo]: México, DF, Mexico), pp. 40–43.

Kermicle JL and Allen JO (1990) Cross-incompatibility between maize and teosinte. *Maydica* 35: 399–408.

Kiang YT, Antonovics J, and Wu L (1979) The extinction of wild rice (*Oryza perennis formosana*) in Taiwan. *J. Asian. Ecol.* 1: 1–9.

Kim CG, Yi H, Park S, Yeon JE, Kim DY, Kim DI, Lee KH, Lee TC, Paek IS, Yoon WK, Jeong SC, and Mook H (2006) Monitoring the occurrence of genetically modified soybean and maize around cultivated fields and at a grain receiving port in Korea. *J. Plant Biol.* 49: 218–223 [English abstract].

Levin DA (1974) The oil content of seeds: An ecological perspective. *Am. Naturalist* 108: 193–206.

Linder CR (1998) Potential persistence of transgenes: Seed performance of transgenic canola and wild relatives × canola hybrids. *Ecol. Appl.* 8: 1180–1195.

Linder CR and Schmitt J (1994) Assessing the risks of transgene escape through time and crop-wild hybrid persistence. *Mol. Ecol.* 3: 23–30.

——— (1995) Potential persistence of escaped transgenes: Performance of transgenic oil-modified *Brassica* seeds and seedlings. *Ecol. Appl.* 5: 1056–1068.

Louette D (1997) Seed exchange among farmers and gene flow among maize varieties in traditional agricultural systems. In: Serratos JA, Willcox MC, and Castillo F (eds.) *Proceedings of a Forum: Gene Flow among Maize Landraces, Improved Maize Varieties, and Teosinte; Implications for Transgenic Maize* (International Maize and Wheat Improvement Center [CIMMYT]: México, DF, Mexico), pp. 56–66.

Lu B-R (2008) Transgene escape from GM crops and potential biosafety consequences: An environmental perspective. In: Italian Ministry for Environment and Territory (ed.) *ICGEB Collection of Biosafety Reviews*, vol. 4 (International Centre for Genetic Engineering and Biotechnology [ICGEB], Biosafety Unit: Trieste, Italy), pp. 66–141.

MacNair MR (1989) The potential for rapid speciation in plants. *Genome* 31: 203–210.

Moeller DR (2001) *GMO Liability Threats for Farmers: Legal Issues Surrounding the Planting of Genetically Modified Crops* (Institute for Agriculture and Trade Policy [IATP]: Minneapolis, MN, USA). 8 p. www.gefoodalert.org/library/admin/uploadedfiles/GMO_Liability_Threats_for_Farmers_PDF_Ver .pdf.

Nagarajan L and Smale M (2007) Village seed systems and the biological diversity of millet crops in marginal environments of India. *Euphytica* 155: 167–182.

NAS [National Academy of Sciences] (2000) *Transgenic Plants and World Agriculture* (National Academy Press: Washington, DC, USA). 46 p. www.nap.edu/catalog.php?record_id=9889#toc/.

NCB [Nuffield Council on Bioethics] (1999) *Genetically Modified Crops: The Ethical and Social Issues* (Latimer Trend: Plymouth, UK). 171 p. www.nuffieldbioethics.org/go/ourwork/gmcrops/publication_301.html.

Novak SJ (2007) The role of evolution in the invasion process. *Proc. Natl. Acad. Sci. USA* 104: 3671–3672.

NRC [National Research Council] (2002) *Environmental Effects of Transgenic Plants: The Scope and Adequacy of Regulation* (National Academies Press: Washington, DC, USA). 342 p. http://books.nap.edu/books/0309082633/html/index.html.

Parker IM and Kareiva P (1996) Assessing risks of invasion for genetically engineered plants: Acceptable evidence and reasonable doubt. *Biol. Cons.* 78: 193–203.

Saji H, Nakajima N, Aono M, Tamaoki M, Kubo A, Wakiyama S, Hatase Y, and Nagatsu M (2005) Monitoring the escape of transgenic oilseed rape around Japanese ports and roadsides. *Environ. Biosafety Res.* 4: 217–222.

Simard MJ, Légère A, Pageau D, Lajeunnesse J, and Warwick S (2002) The frequency and persistence of canola (*Brassica napus*) volunteers in Québec cropping systems. *Weed Technol.* 16: 433–439.

Snow AA (2002) Transgenic crops—why gene flow matters. *Nat. Biotechnol.* 20: 542.

Snow AA, Pilson D, Rieseberg LH, Paulsen M, Pleskac N, Reagon MR, Wolf DE, and Selbo SM (2003) A *Bt* transgene reduces herbivory and enhances fecundity in wild sunflowers. *Ecol. Appl.* 13: 279–286.

Stebbins GL (1958) The inviability, weakness, and sterility of interspecific hybrids. *Adv. Genet.* 9: 147–215.

Stewart CN Jr, Halfhill MD, and Warwick SI (2003) Transgene introgression from genetically modified crops to their wild relatives. *Nature* 4: 806–817.

Tabashnik BE, Patin AL, Dennehy TJ, Liu YB, Carrière Y, Sims MA, and Antilla L (2000) Frequency of resistance to *Bacillus thuringiensis* in field populations of pink bollworm. *Proc. Natl. Acad. Sci. USA* 97: 12980–12984.

Traynor PL and Westwood JH (eds.) (1999) *Proceedings of a Workshop on Ecological Effects of Pest Resistance Genes in Managed Ecosystems, Bethesda, Maryland, USA, January 31–February 3, 1999* (Information Systems for Biotechnology: Blacksburg, VA, USA). www.isb.vt.edu/proceedings99/proceedings.intro.html.

Wolf DE, Takebayashi N, and Rieseberg LH (2001) Predicting the risk of extinction through hybridization. *Conserv. Biol.* 15: 1039–1053.

TWO

Methodology

There are several aspects related to our choices in the assemblage of information we used to write the crop chapters and assess the likelihood of gene flow and introgression between crops and their wild relatives, which are enumerated here.

Crop Species

Crop species were selected according to their global importance, in terms of the following criteria:

- harvested world production area (FAO 2009)
- advances in GM technology (only those crops for which GM technology is available or being developed were considered)
- their relative contribution to food security (FAO 2004)

As a result, the following 20 crops were chosen:

1 bananas and plantains (*Musa* spp.)

2 barley (*Hordeum vulgare* L.)

3 canola / oilseed rape (*Brassica napus* L.)

4 cassava / yuca / manioc (*Manihot esculenta* Crantz)

5 chickpeas (*Cicer arietinum* L.)

6 common beans (*Phaseolus vulgaris* L.)

7 cotton (*Gossypium hirsutum* L.)

8 cowpeas (*Vigna unguiculata* (L.) Walp.)

9   finger millet (*Eleusine coracana* (L.) Gaertn.)

10 maize / corn (*Zea mays* L.)

11 oats (*Avena sativa* L.)

12 peanuts / groundnuts (*Arachis hypogaea* L.)

13 pearl millet (*Pennisetum glaucum* (L.) R. Br.)

14 pigeonpeas (*Cajanus cajan* (L.) Millsp.)

15 potatoes (*Solanum tuberosum* L.)

16 rice (*Oryza sativa* L.)

17 sorghum (*Sorghum bicolor* (L.) Moench)

18 soybeans (*Glycine max* (L.) Merr.)

19 sweetpotatoes / batata / camote (*Ipomoea batatas* (L.) Lam.)

20 wheat / bread wheat (*Triticum aestivum* L.)

Taxonomy and Nomenclature

The taxonomic classification of the 20 selected crops and their wild relatives agrees with the most recent taxonomic treatments and are cited as such in the respective crop chapters. Where no specific references or sources are provided, the nomenclature follows that of the GRIN taxonomy for plants, a part of the Germplasm Resources Information Network (GRIN), a U.S. Department of Agriculture (USDA) database on plant species (USDA-GRIN 2009).

Identifying Wild Relatives of Crops

A list of all sexually compatible crop wild relatives (CWR) was assembled for each of the selected crops, following a three-step approach.

1 All *congeneric* species were identified, initially using the GRIN database (USDA-GRIN 2009). As a rule, only wild relatives were included. However, for canola, common beans, cotton, rice, and wheat, other domesticated species in the same genus were also integrated when they were known to hybridize with the crop. In families where *intergeneric* hybridization is known to be important, members of other genera were also considered.

2 A literature and Internet search was conducted to assemble articles that identified or defined the primary, secondary, and tertiary gene pools of each crop. The previously compiled list of CWR was reduced or extended accordingly.

3 Relevant scientific articles, books, book chapters, technical reports, and conference proceedings that would help in deducing the extent of sexual compatibility between the crops and their CWR were consulted. They covered a wide range of research topics, including traditional and modern crop breeding, phylogeny, molecular biology, and genetic diversity. The information was used to further refine the CWR list by adding species that were identified as being sexually compatible with the listed crops or excluding those that were not. This step resulted in the final list of CWR for each crop.

State of GM Technology

Since the release of the first GM crops in 1996, the worldwide area planted with GM crops for commercial production has increased on a yearly basis, with similar growth rates in both developing and industrialized countries (fig. 2.1; James 1997, 1999, 2004, 2005, 2007, 2008). Today, GM crops are commercially planted on more than 125 million ha in more than 25 countries (James 2008).

Each crop chapter contains a list of the countries in which field trials have been conducted with GM plants. This list includes experimental field trials, as well as field trials of new GM varieties before their commercial release. These data are publicly available for many countries, and have been obtained by consulting the following sources:

- Argentina: www.sagpya.mecon.gov.ar
- Australia: www.maps.ogtr.gov.au/jsp/index.jsp
- Canada: www.inspection.gc.ca/english/plaveg/bio/confine.shtml
- Europe: http://gmoinfo.jrc.ec.europa.eu/, www.gmo-compass .org/eng/agri_biotechnology/field_trials/, and www.coextra.eu/ country_reports/
- India: http://igmoris.nic.in/TrailGMPro.asp

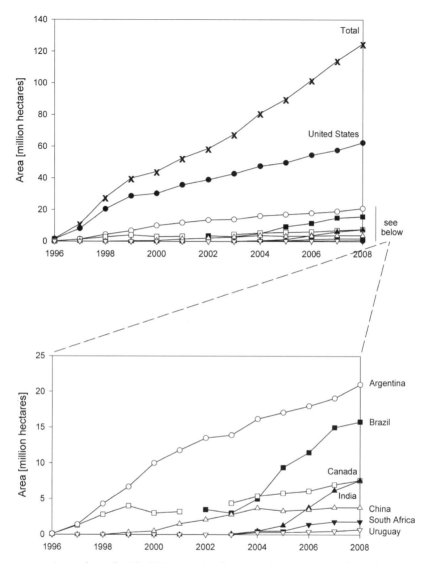

Fig. 2.1   Area planted with GM crops (soybean, maize, cotton, and canola), 1996 to 2008

- Japan: www.bch.biodic.go.jp/english/lmo.html
- New Zealand: www.ermanz.govt.nz/no/
- USA: www.nbiap.vt.edu/cfdocs/fieldtests1.cfm or www.isb .vt.edu/cfdocs/fieldtests1.cfm

Field trial data often provide information regarding the types of GM traits being tested. A further source of information on GM traits that are used or being experimented with to genetically transform crops is the AGBIOS online database, found at www.agbios.com/dbase.php?action=ShowForm.

Likelihood of Gene Flow and Introgression

The CWR were grouped into five categories, with each one indicating a particular likelihood of gene flow and introgression, based on the extent of sexual compatibility between the CWR and their respective crops[1]. This was assessed from both literature searches and the opinions of crop experts (see the acknowledgments) and is delimited as follows (adapted from Armstrong et al. 2005):

1 *Very high*: The CWR is fully interfertile with the crop, and spontaneous hybrids are frequent under natural conditions. Introgression may occur at relatively high rates.

2 *High*: Strong evidence exists for substantial reproductive compatibility between the CWR and the crop (e.g., manual or open pollination results in the production of viable, fertile F1 hybrids, and later generations are also known to be fertile). Spontaneous hybrids are often reported from the wild. Natural introgression has been reported, but it occurs at relatively low rates.

3 *Moderate*: Cross-compatibility of the CWR with the crop is reduced. Spontaneous hybridization is possible under field conditions or in the wild, but the resulting F1 hybrids are completely sterile, have significantly reduced fertility, or show hybrid weakness (reduced viability). Introgression is possible in theory, but rather unlikely.

4 *Low*: The CWR is reproductively isolated from the crop by pre- or post-zygotic hybridization barriers. Viable or fertile F1 hybrids

usually cannot be produced through manual pollination; they require considerable human intervention, such as the use of embryo rescue or protoplast fusion methods. In some cases, spontaneous hybridization may be possible, but strong pre- or post-zygotic hybridization barriers prevent the production of viable or fertile F1 hybrids.

5 *Very low*: In theory, very low levels of interspecific crosses between the CWR and the crop may occur under exceptional circumstances. However, the likelihood of introgression is extremely remote, because of genetic incompatibility.

World Maps Showing the Modeled Likelihood of Introgression

Another important aspect in determining the likelihood of gene flow is the distribution of CWR that are sexually compatible with the crop. However, very few maps showing this distribution are available, and most provide information only on a local or regional scale; they rarely indicate global distributions.

The map section describes the method we developed to produce maps that link modeled CWR distributions with (1) global data on crop production areas and (2) the above-mentioned five categories of likelihood of introgression between crops and sexually compatible CWR. These maps show, for the first time, global crop production areas and the modeled distribution of sexually compatible CWR. They allow immediate identification of those regions where gene flow and introgression may be an issue[1] because of the distribution of crops and their overlap with sexually compatible CWR. The approach we developed for gathering information and modeling the geographic distributions of crops and their wild relatives constitutes an innovation that can be applied to the geographic mapping of any other plant species, and a detailed description of the methodology we used will be published shortly (Andersson et al., in prep.).

NOTES
....

1. Three further assumptions we make regarding introgression are (1) that crop and CWR populations are sufficiently close to each other to allow outcrossing,

(2) that flowering times overlap, and (3) that the nature of the respective gene or trait (i.e., its stability, location in the genome, inheritance pattern, and selective value) favors introgression.

REFERENCES

Andersson MS, Alvarez DF, Jarvis A, Wood S, Hyman G, and de Vicente MC (forthcoming) Mapping the global likelihood of gene flow using crop production information and distribution data of wild relatives.

Armstrong TT, Fitzjohn RJ, Newstrom LE, Wilton AD, and Lee WG (2005) Transgene escape: What potential for crop-wild hybridization? *Mol. Ecol.* 14: 2111–2132.

FAO [Food and Agriculture Organization of the United Nations] (2004) Annex 1 of the International Treaty on Plant Genetic Resources for Food and Agriculture (FAO: Rome, Italy). 25 p. www.fao.org/Ag/Magazine/ITPGRe.pdf.

——— (2009) FAOSTAT data. http://faostat.fao.org/site/567/default.aspx.

James C (1997) *Global Status of Transgenic Crops in 1997.* ISAAA Brief 5 (International Service for the Acquisition of Agri-biotech Applications [ISAAA]: Ithaca, NY, USA). 31 p.

——— (1999) *Preview: Global Review of Commercialized Transgenic Crops 1999.* ISAAA Brief 12 (International Service for the Acquisition of Agri-biotech Applications [ISAAA]: Ithaca, NY, USA). 14 p.

——— (2004) *Preview: Global Status of Commercialized Biotech/GM Crops 2004.* ISAAA Brief 32 (International Service for the Acquisition of Agri-biotech Applications [ISAAA]: Ithaca, NY, USA). 43 p.

——— (2005) *Global Status of Commercialized Biotech/GM Crops 2005.* ISAAA Brief 34 (International Service for the Acquisition of Agri-biotech Applications [ISAAA]: Ithaca, NY, USA). 46 p.

——— (2007) *Global Status of Commercialized Biotech/GM Crops 2007.* ISAAA Brief 37 (International Service for the Acquisition of Agri-biotech Applications [ISAAA]: Ithaca, NY, USA). 143 p.

——— (2008) *Global Status of Commercialized Biotech/GM Crops 2008.* ISAAA Brief 39 (International Service for the Acquisition of Agri-biotech Applications [ISAAA]: Ithaca, NY, USA). 243 p.

USDA-GRIN [United States Department of Agriculture, Germplasm Resources Information Network] (2009) GRIN Taxonomy for Plants: An Online Plant Species Database (National Germplasm Resources Laboratory: Beltsville, MD, USA). www.ars-grin.gov/cgi-bin/npgs/html/index.pl?language=en.

THREE

Banana and Plantain
(*Musa* spp.)

Center(s) of Origin and Diversity

Bananas and plantains are native to the Indo-Malayan, SE Asian, and Australian tropics. It is commonly accepted that the two diploid ancestors of cultivated bananas and plantains—*Musa acuminata* Colla and *M. balbisiana* Colla—originated in the regions stretching from Papua New Guinea to the Indian subcontinent, the former in Malaysia and the latter in India (Simmonds 1962, 1995). Another cultivated group—also known as the Fe'i cultivar group—originated in the Pacific and is of hybrid origin. Little is known about its possible wild ancestors (Cheesman 1950; Simmonds 1956).

The primary center of diversity for both diploid bananas and plantains is SE Asia, either in Malaysia or Indonesia (Daniells et al. 2001). After their introduction to Africa, a secondary diversification of triploid plantains in W Africa and of triploid cooking and beer bananas in E Africa occurred, based mainly on mutations (Champion 1967; Pillay et al. 2004; Nelson et al. 2006). Today, the greatest variability of banana cultivars in the world is found in India and SE Asia, and of plantains in W and central Africa.

Biological Information

Edible cultivated bananas and plantains and wild *Musa* species are perennial herbs. Most cultivated bananas are triploids ($2n = 3x = 33$) and account for the major portion of world banana production, but diploid ($2n = 2x = 22$) and tetraploid ($2n = 4x = 44$, mostly artificial hybrids) landraces and cultivars are also grown. All wild *Musa* species are diploids ($2n = 2x$).

Flowering

Banana plants flower only once, at the end of their life cycle. They have unisexual flowers, with successive female and male phases. The inflorescence—also called the bud, bell, or heart—first displays semicircular clusters of large female flowers whose ovaries later develop into the fruit. As the bud elongates, bisexual flowers and then small, generally nonfunctional male flowers develop (Nelson et al. 2006). Each cluster consists of 12 to 20 flowers. Pollen production and the pollen fertility of cultivated bananas and plantains depend mainly on genetic factors, although environmental influences have also been recognized (Landa et al. 1999). After flowering and fruiting, the mother plant dies and is replaced by new suckers that develop from underground buds.

Pollen Dispersal and Viability

Bananas and plantains are pollinated by insects (honeybees), birds (e.g., sunbirds), and bats (Start and Marshall 1976; Fahn and Benovaiche 1979; Mutsaers 1993; Ortiz and Crouch 1997; M. Liu et al. 2004).

Pollen viability varies between and within genome groups. Fortescue and Turner (2004) noted that Australian seeded diploid *Musa* species have three times (mean 88%) more viable pollen than edible tetraploids (29%) and nine times more than edible triploids (6%–10%). Pollen fertility decreases as follows: *M. balbisiana* > *M. ornata* > *M. acuminata* > **AAAB** > **AAA** > **ABB** > **AAB**. The **AAA** triploid cultivar 'Gros Michel' has 13% viable pollen.

Some Indian *Musa* species have slightly lower pollen viability: cultivated and wild diploids at 50%–66%, tetraploids at 28%, and triploid cultivars at 21%–29% (Sathiamoorthy 1994).

Sexual Reproduction

Cultivated and wild *Musa* species have an allogamous reproductive system. The majority of edible banana and plantain cultivars are partially or completely sterile. For example, in diploid edible bananas, many female-sterile and male-fertile cultivars are known. Sterility in diploids can be due to (1) structural heterozygosity (translocations, inversions), (2) a lack of total homology between **A** and **B** chromosomes in **AB** varieties, or (3) genic effects (F. Bakry, pers. comm. 2008).

Triploidy reinforces the sterility already observed at the diploid level. The reproductive cells of triploid cultivars can contain anywhere from one to three sets of chromosomes. The plants produce little or no pollen and are partially sterile. They develop fruit pulp without pollination (parthenocarpy); the fruits are seedless and are propagated vegetatively (Shepherd et al. 1986; Silva et al. 2001). Crossing the triploid cultivar (as the female parent) with a diploid (as the male) will produce descendants that have two, three, or four sets of chromosomes. Fertility is restored in those descendents with an even set of chromosomes, which may lead to the production of seeded fruits when the flowers are pollinated. Some sterile triploid cultivars can be stimulated to produce seeds when fertilized with the pollen of wild diploid bananas (Shepherd 1992; Pillay 2005).

Inedible ornamental *Musa* species such as *M. velutina* H. Wendl. & Drude ('Fuzzy Pink' or 'Pink Velvet' bananas), and fiber species such as *M. textilis* Née (abaca, or Manila hemp), are commonly grown from seed.

All wild bananas are fully fertile diploids, and some species flower and fruit regularly every year. Seeds produced in breeding programs often have very low (about 1%) germination rates (Pillay 2005), but the viability and germination rates of seeds produced in the wild are usually much higher. For instance, wild *M. balbisiana* populations in China showed very low levels of inbreeding, due to the species' high fertility and fecundity. Yet in spite of good seed set, banana plants in the wild often propagate asexually and expand via budding (Ge et al. 2005).

Vegetative Reproduction

Cultivated bananas and plantains are mainly propagated by means of suckers, or basal corms, which develop from underground buds, and far more rarely by seeds (as a rule this occurs only for ornamental types and in banana breeding). Each sucker is genetically identical to its parent plant. This means that each banana cultivar consists of a line of genetically identical plants (a clone). Also, many wild banana species—although fully fertile—propagate asexually via budding (Ge et al. 2005).

Seed Dispersal and Dormancy

Birds, small mammals such as flying foxes (a bat species in the western Pacific), foraging rodents, squirrels, and monkeys are important disseminators of wild species and of seed-producing cultivars (Ge et al. 2005; Nelson et al. 2006).

Simmonds (1962) found that dry *Musa* seeds can be safely stored for several years when placed over calcium chloride, and cryopreservation studies indicate that dry *M. balbisiana* seeds can survive freezing by liquid nitrogen, showing high germination rates after thawing (Bhat et al. 1994).

To our knowledge, no information is available on banana seed survival in the wild, nor is there any published documentation that a persistent seed bank can be formed. However, Simmonds (1959) and Häkkinen (2004) state that dormant seeds of some wild *Musa* species remain viable in the soil for years and will germinate rapidly when an opening is created in the canopy.

Volunteers, Ferals, and Their Persistence

Due to the fact that most cultivated bananas and plantains are seedless, and thus sterile, the probability of volunteers or ferals establishing themselves spontaneously is relatively low. Such populations can, however, result from human expansion, and—once established—can be very persistent, due to their ability to propagate vegetatively.

Weediness and Invasiveness Potential

Cultivated *Musa* varieties are not considered to be invasive, nor do they compete well with forest species under natural conditions. Wild *Musa* species and seeded varieties, however, have the potential to spread (Nelson et al. 2006) and are locally considered to be weeds, such as wild *M. balbisiana* in Costa Rica (F. Bakry, pers. comm. 2008).

Crop Wild Relatives

The genus *Musa* L. belongs to the family Musaceae Juss., which also comprises some eight species of *Ensete* Horan. and possibly a third genus, the monospecific *Musella* (Franch.) H.W. Li, which is closely related to *Musa*. *Musa* species are mainly wild forest-dwellers that can be found between India and the Pacific, reaching as far north as Nepal and extending south to the northern tip of Australia (INIBAP 2006).

Here, we will distinguish between the better-known cultivated, edible bananas and plantains and the lesser-known wild *Musa* species.

Edible or Cultivated Bananas and Plantains

The vast majority of cultivated, edible varieties—also referred to as bananas and plantains—belong to sect. *Eumusa* (Baker) Cheesman and originated from the two diploid ($2n = 22$) wild *Musa* species, *M. acuminata* (**A** genome) and *M. balbisiana* (**B** genome), that are native to SE Asia (Simmonds and Shepherd 1955; Purseglove 1972; Stover and Simmonds 1987). In spite of their ability to hybridize, these two wild seeded diploids are considered to be two distinct biological species, due to clear differences in genomic constitution (**A** versus **B** genomes) and morphological characteristics (Cheesman 1948).

A third species in sect. *Eumusa*, *M. schizocarpa* N.W. Simmonds (**S** genome), may have contributed to the origin of some of the diploids cultivated in Papua New Guinea (Shepherd and Ferreira 1982; Sharrock 1990). Another group of cultivated edible bananas (Fe'i) evolved from the diploid wild *M. textilis* (**T** genome), and these cultivars are placed in sect. *Australimusa* Cheesman (Nelson et al. 2006). The evolution of diploid ($2n = 2x = 22$), triploid ($2n = 3x = 33$), and tetraploid ($2n = 4x = 44$) *Musa* cultivars from these wild species, their dispersal, and their human domestication have been documented in detail (Simmonds 1962; Price 1995).

Today, a wide range of different types of edible bananas and plantains exists for different end uses, including dessert, cooking, roasting, and beer bananas (Acland 1971; Karamura 1999). These different edible cultivars can be classified into genome groups, based on their ploidy levels and their relative genome proportion of *M. acuminata* (**AA**) and *M. balbisiana* (**BB**) (Simmonds and Shepherd 1955; Stover and Simmonds 1987). Table 3.1 gives some examples of these cultivars, along with their genome groups and uses.

Wild *Musa* Species

While farmers domesticated bananas by selecting plants that were sterile and seedless, the fertile ones continued propagating, hybridizing among themselves and consequently generating further diversity. These wild

Table 3.1 Classification of cultivated bananas and plantains (*Musa* spp.) according to genome type

Cultivar and genome group	Cultivar subgroups*	Banana type	Origin
M. acuminata cultivars			
AA	Sucrier, Pisang Mas, Amas	dessert	Malaysia
	Pisang Jari Buaya, Niukin	dessert	Papua New Guinea
	Pisang Lilin	dessert	Philippines
	Inarnibal	dessert	
AAA	Gros Michel (Highgate)	dessert	Guadeloupe, Honduras
	Cavendish (Petite Naine, Dwarf Cavendish, Kiri Tia Mwin, Grande Naine, Williams)	dessert	
	Red (Mossi, Figue Rose, Morato, Red Dacca, Morado)	dessert	Comoro Islands
	Ambon (Pisang Ambon)		
	Ibota (Yangambi Km5)		Congo Democratic Republic
East African highland bananas (AAA-EA)	Mutika/Lujugira (Mpologoma, Muvubo, Namunwe, Musakala, Nakabululu, Enyoya, Entente, etc.)	cooking	Uganda, Tanzania
	Mutika/Lujugira (Nalukira, Kabula, Nsowe, Ntukura)	beer	Uganda
AAAA	Pisang Ustrali		Papua New Guinea
M. acuminata × *M. balbisiana* cultivars			
AB	Ney Poovan (Lady's Finger, Kisubi, Safet Velchi)	dessert	
AAB	Iholena (Uzakan)	dessert	Papua New Guinea
	Laknau	cooking	
	Mysore (Pisang Ceylan)	dessert	
	Silk (Malbhog, Manzana, Latundan, Pisang Rasthali, Silk Fig)	dessert	Malaysia

Genome group	Cultivars*	Uses	Country
	Pome (Prata Añá, Lady's Finger, Santa Catarina Prata)	dessert	
	Maia Maoli/Popoulu (Ainu, Mangaro Torotea, Hua Moa, Pacific Plantain)	cooking and dessert	Cook Islands
	Pisang Nangka	cooking	Indonesia
	Pisang Raja	dessert	Indonesia
Plantains (AAB)	Plantain (French Plantain, French Clair, Rouge de Loum, Nyombé 1, Rose d'Ekona, Red Yade, 3 Vert, Corn Type, Esang)	cooking, frying, pounding, roasting	Cameroon, Congo, Ivory Coast, Gabon, Ghana, Nigeria, Rwanda
ABB	Bluggoe (Silver Bluggoe)	cooking	
	Pisang Awak (Ducasse)	cooking and dessert	
	Monthan (Pacha Bontha Bathees)	cooking	
	Kalapua	cooking	Papua New Guinea
	Klue Teparod	cooking	Thailand
	Saba (Benedetta)	cooking	Philippines
	Pelipita	cooking	
AABB AAAB ABBB	Laknau der (Ngoen, Ngern)		
M. acuminata × *M. schizocarpa* cultivars			
AS	(Ato, Ungota, Vunamami, Kokor)	cooking and dessert	Papua New Guinea
Australimusa cultivars			
Fe'i	Menei, Rimina, Wain	cooking and dessert	Papua New Guinea
Eumusa × *Australimusa* cultivars			
AT	Umbubu	cooking	Papua New Guinea
AAT	Karoina, Mayalopa, Sar	cooking	Papua New Guinea
ABBT	Yawa 2, Giant Kalapua	cooking, dessert, beer	Papua New Guinea

Source: Daniells et al. 2001.

*Examples of common and local cultivars are in parentheses.

Table 3.2 Wild relatives of bananas and plantains (wild *Musa* species)

Species	Common name	Genome	$2n$*	Origin and distribution
sect. *Eumusa* (Baker) Cheesman				
M. acuminata[†] Colla		AA	22	native to New Guinea, Samoa; distributed from SE Asia to Australia (see subspecies below); naturalized in some areas of tropical America (e.g., Costa Rica)
M. balbisiana Colla	Wild banana	BB		native to SE Asia and E Indies (India, S China to Japan and Philippines); naturalized in Madagascar
M. basjoo Siebold	Japanese fiber banana	AA		cultivated; native to Japan and introduced to China
M. cheesmanii N.W. Simmonds				NE India, Bhutan, Myanmar
M. flaviflora N.W. Simmonds				NE India, Bhutan
M. halabanensis Meijer				Indonesia (Sumatra)
M. itinerans Cheesman				native to and widely distributed in China, India, Myanmar, Thailand, Vietnam, and Philippines
M. nagensium Prain				SE India (Naga Mountains, Assam), SW China
M. ochracea Sheph.				India
M. schizocarpa N.W. Simmonds		SS		Papua New Guinea
M. sikkimensis Kurz				NE India, Bhutan
sect. *Rhodochlamys* (S. Schauer) Cheesman				India, Myanmar, Bangladesh, NW Thailand
M. laterita Cheesman	Indian dwarf banana, ornamental ('Bronce,' 'Red Salmon')	AA	22	native to NE India, Myanmar, N Thailand; cultivated worldwide
M. ornata Roxb. (probably = *M. velutina* × *M. flaviflora*)	ornamental ('Ornata,' 'Standard Lavender,' 'Dwarf Blue')			native from Pakistan to Myanmar (Bangladesh, N India, Thailand); cultivated worldwide
M. sanguinea Hook.	ornamental			NE India, Tibet, SW China, Indonesia
M. velutina H. Wendl. & Drude	ornamental ('Fuzzy Pink,' 'Pink Velvet')			native to NE India; cultivated worldwide

	Common name	Genome	$2n$	Distribution
sect. _Australimusa_ Cheesman		TT	20	
M. bukensis Argent				Papua New Guinea
M. fitzalanii F. Muell. (extinct)				native to Australia (Queensland); now extinct
M. jackeyi W. Hill	Johnstone River banana			Papua New Guinea, Australia (Queensland)
M. lolodensis Cheesman				Indonesia, Papua New Guinea
M. maclayi F. Muell				endemic in Papua New Guinea
M. peekelii Lauterb.				endemic in Papua New Guinea; introduced to
Hawaii				
M. textilis Née	abaca, Manila or Taiwan hemp			native to Philippines, Borneo, Papua New Guinea; introduced in China; seriously declining as a crop
sect. _Callimusa_ Cheesman			20	
M. beccarii N.W. Simmonds			18	Malaysia, Indonesia (Borneo)
M. borneënsis Becc.				Vietnam, Indonesia (Borneo)
M. coccinea Andrews	Red banana, ornamental ('Okinawa Torch')			native to China, Indonesia, Thailand; cultivated worldwide
M. gracilis Holttum				Malaysia, Thailand
M. salaccensis Zoll.	Javanese wild banana			Indonesia (Java, Sumatra)
M. violascens Ridl.				Malaysia
Incertae sedis				
M. boman Argent				endemic in Papua New Guinea
M. ingens N.W. Simmonds			14	Papua New Guinea
M. lasiocarpa Franch.	Chinese yellow banana, ground lotus		18	China, Vietnam, Laos

Sources: de Langhe et al. 2000; Daniells et al. 2001; Sharrock 2001; Häkkinen and Sharrock 2002; A. Liu et al. 2002a; Uma et al. 2006.

*$2n$ = chromosome number.

†Within _M. acuminata_ there are several wild subspecies: _M. acuminata_ subsp. _banksii_ N.W. Simmonds (New Guinea, Samoa, Australia); subsp. _burmannica_ N.W. Simmonds (= subsp. _burmannicoides_ DeLanghe) (Myanmar, Thailand); subsp. _errans_ (Blanco) R.V. Valmayor (Philippines); subsp. _malaccensis_ (Ridl.) N.W. Simmonds (Malaysia); subsp. _microcarpa_ (Becc.) N.W. Simmonds (Indonesia, Malaysia, Thailand); subsp. _siamea_ N.W. Simmonds (Indochina, Malaysia, Thailand); subsp. _truncata_ (Ridl.) Kiew (Malaysia); and subsp. _zebrina_ (Van Houtte) Nasution ('Blood/Rojo/Zebrina' or 'Variegated Red,' ornamental; Indonesia).

relatives are native to SE Asia and are found primarily in tropical regions, from India to Polynesia (Stover and Simmonds 1987). Only *Ensete* species are native to Africa (Constantine 2003).

The total number of species and the taxonomic status of many wild species contained in the genus *Musa* are still unknown, and further research is needed. All wild *Musa* species are diploids ($2n = 2x$), and, as of now, the genus *Musa* is grouped into four sections (sects. *Eumusa*, *Rhodochlamys* (S. Schauer) Cheesman, *Callimusa* Cheesman, and *Australimusa*), based mainly on morphological differences, geographical distribution, and hybridization studies (Cheesman 1947; table 3.2). Two of the sections contain species with a chromosome number of $2n = 22$ (*Eumusa* and *Rhodochlamys*). The chromosome number for species in the other two sections (*Callimusa* and *Australimusa*) is $2n = 20$, and three species for which the relevant section has yet to be determined are meanwhile classified as *incertae sedis* (Horry et al. 1997). Here we list only the most commonly accepted *Musa* species, but a number of additional species of dubious taxonomic status exist (see, for example, Govaerts and Häkkinen 2007). The grouping of *Musa* species into four sections is controversial, and genetic evidence has been provided supporting (Ude et al. 2002a) as well as refuting (Wong et al. 2001, 2002; Nwakanma et al. 2003) this classification. A new global initiative, including a Taxonomic Advisory Group (TAG) of *Musa* taxonomists, breeders, germplasm managers, and molecular experts, has been launched to characterize and regularize the taxonomy in collections and derive an agreed-upon taxonomy for use by the *Musa* community (INIBAP 2006; TAG 2006).

Hybridization

The majority of edible banana and plantain cultivars are derived from the two wild *Musa* species, *M. acuminata* and *M. balbisiana* (table 3.1). Natural hybridization between *M. acuminata* (**AA**) and wild-seeded *M. balbisiana* (**BB**) occurred—and still occurs—in regions where their natural distributions overlap (Valmayor et al. 2000). Historically, hybridization may have taken place repeatedly when derivatives (**AA, AAA, AAAA**) of domesticated *M. acuminata* spread into the native distributional range of *M. balbisiana*, resulting in several (**AB, AAB,** and **ABB**) genome combinations (Karamura 1999). We could therefore refer to a banana and plantain "crop-wild complex" that would include the wild progenitors *M. acuminata* and *M. balbisiana*, and all diploid, triploid, and tetraploid cultivars derived from them (containing either the **A** or **B** genome, or any combination of them). All these forms are cross-fertile and can, in theory, hybridize with each other.

In practice, most edible cultivars are partially or completely sterile. However, a considerable number of fertile cultivars and landraces are known (see, e.g., Simmonds 1960; Vuylsteke et al. 1993a, b; Ortiz et al. 1995; Oselebe et al. 2006). Examples include the following:

- male-fertile diploids (e.g., Landa et al. 1999)
- partially or fully fertile triploids, such as 'Pisang Awak' (**ABB**) cultivars in the Philippines (Simmonds 1959), and 'Cavendish' (**AAA**) cultivars in colder climates (F. Bakry, pers. comm. 2008)
- male-fertile tetraploid landraces (Vuylsteke et al. 1993a)

Thus some cultivated bananas and plantains produce viable, fertile pollen and can serve as the male parent in crosses with compatible wild *Musa* species or female-fertile cultivars. Also, some triploid cultivars produce seeds when fertilized with pollen from wild diploid bananas—an approach frequently used in conventional banana breeding (Ortiz 1997; Pillay 2005).

Musa acuminata (AA) × *M. balbisiana* (BB)

In general terms, the **A** and **B** genomes show relatively high cross-compatibility. To improve the success of hybridizations, artificial techniques such as ploidy manipulation and embryo rescue are used (see, e.g., Ortiz and Vuylsteke 1996; Rowe and Rosales 1996; Vuylsteke et al. 1997; Shepherd 1999 and the references cited therein; Silva et al. 2001). The resulting F1 hybrids (**AB**), however, tend to be sterile (F. Bakry, pers. comm. 2008).

Hybrids between *M. acuminata* and *M. balbisiana* can be obtained quite easily in both directions under natural field conditions, as well as by hand pollination (Simmonds 1954, 1962; Shepherd 1999; Ude et al. 2002b). Depending on the combination of genotypes and environmental conditions, the F1 hybrids are moderate to highly vigorous and female fertile (Shepherd 1999). Backcrossing results in progeny with varying chromosome numbers, including diploid, triploid, pentaploid, and aneuploid plants (Shepherd 1999).

Hybridization between *Musa* Species with 2n = 22 (sects. *Eumusa* and *Rhodochlamys*)

In spite of having different genome constitutions, both *M. acuminata* (**A** genome) and *M. balbisiana* (**B** genome) are cross-compatible with a number of other wild *Musa* species that have the same chromosome number ($2n = 22$, sects. *Eumusa* and *Rhodochlamys*).

For example, crosses of *M. acuminata* and *M. balbisiana* with *M. laterita* Cheesman, *M. ornata* Roxb. and *M. velutina* (all in sect. *Rhodochlamys*) can be made with relative ease in the field under natural conditions. Hybridization yields viable F1 and backcross (BC) hybrids, some of them fertile (Simmonds 1954, 1962; Shepherd 1988, 1999). In Thailand there is evidence of natural *M. acuminata* × *M. laterita* hybrid populations in the wild (Simmonds 1956; de Langhe et al. 2000).

Furthermore, the occurrence of natural male-fertile hybrids between *M. acuminata* and *M. schizocarpa* has been reported in Papua New Guinea (Argent 1976; Shepherd and Ferreira 1982; Tezenas du Montcel et al. 1995). In field experiments under natural conditions, Shepherd (1999) obtained F1, F2, and BC progeny, including a substantial number of either moderately or highly vigorous plants with varying levels of pollen production, pollen fertility (ranging from nearly male sterile to highly fertile), and seed production. Although *M. schizocarpa* carries the **S** genome, which shows a somewhat reduced compatibility with the **A** and **B** genomes, Shepherd concluded that "it seems very likely that no critical isolation barrier exists between *M. schizocarpa* and [sympatric] *M. acuminata* subsp. *banksii*" (1999, p. 78).

Hybridization between *Musa* Species with Different Basic Chromosome Numbers

There is very limited phenotypic information available on intersectional hybrids between species with 22 and 20 chromosomes. For the most part, species from sect. *Australimusa* (T genome) have been reported to have a very limited compatibility with A genome species, although the T genome was involved in the evolution of some cultivars (d'Hont et al. 2000).

Bernardo (1957) describes the characteristics of a hybrid between *M. balbisiana* (2*n* = 22, B genome) and Manila hemp, or *M. textilis* (2*n* = 20, T genome), in the Philippines. Shepherd (1999) obtained some surprisingly vigorous F1 plants from different combinations of species with 20 and 22 chromosomes when the former were used as the male parent, while crosses always failed in the reverse direction. All F1 hybrids were completely male and female sterile.

Hybridization with Species from Other Genera

To our knowledge, no studies have been carried out investigating the hybridization between *Musa* species and species belonging to other genera (e.g., *Ensete, Heliconia* L., *Musella*).

Outcrossing Rates and Distances

No information has been found about experimental trials carried out to study pollen dispersal distances and outcrossing rates in *Musa*.

Gene flow via pollen, estimated from molecular data, is 3.65 times greater among *Musa* populations in China than gene flow via seed (Ge et al. 2005). Varying levels of pollen flow between *Musa* populations are strongly related to the presence of vector species for pollen dispersal (M. Liu et al. 2004). Honeybees migrate within a limited range, so most pollination take place locally, whereas fruit bats are responsible for longer-distance dispersal (A. Liu et al. 2002b). No effective dispersal distances have been reported thus far.

State of Development of GM Technology

Work on the genetic transformation of bananas and plantains is recent. The first successful transformation events for bananas were reported in the 1990s (May et al. 1995; Sági et al. 1995; Remy et al. 1998). Today, stable protocols are available using *Agrobacterium*-mediated transformation of meristems or particle bombardment of cell suspensions (see, e.g., Becker et al. 2000; Dugdale et al. 2000; Sági et al. 2000; Ganapathi et al. 2001; Khanna et al. 2004).

As of now, no commercial transgenic *Musa* varieties have been released. However, field trials are being conducted in Australia, Belgium, Brazil, Costa Rica, Honduras, India, Israel, Mexico, and Saint Lucia. GM traits that have been researched include virus resistance, fungal disease resistance, nematode resistance, extended shelf life, and enhanced carotenoid content.

Agronomic Management Recommendations

There are no specific recommendations.

Crop Production Area

Nowadays, the distribution of *Musa* species is pantropical; they are widely cultivated in more than 80 countries throughout the tropics. Bananas and plantains are also successfully grown in the Mediterranean region (Cyprus, Egypt, Israel, Morocco, Turkey) and in certain temperate, somewhat frost-free climates (e.g., N Argentina, California).

The total crop production area of bananas and plantains is 9.8 million ha, with 4.4 million in bananas and 5.4 million in plantains (FAO 2009). More than 50% of the world's production takes place in Africa (Cameroon, Ghana, Nigeria, Rwanda, Uganda), 20% in SE Asia (China, India, Indonesia, the Philippines, Thailand), 20% in Latin America (Brazil, Colombia, Costa Rica, Ecuador, Peru), and 3% in the Caribbean (Lescot 2006; FAO 2009). Over 85% of the bananas harvested in the world are grown by smallholders (Heslop-Harrison 2005).

Of the 105 million tons currently being produced, some 16 million are accounted for by the export trade, which is represented by just a few, genetically very similar cultivars belonging to the 'Cavendish' group. The remainder are eaten or sold locally, and this market is also dominated more and more by a small number of cultivars (Swennen 2005).

Research Gaps and Conclusions

Unfortunately, the information available thus far is insufficient for an in-depth assessment of gene flow between cultivated bananas and plantains and their wild relatives. Data is also lacking with respect to hybrid vigor and fertility, the extent and primary direction of introgression, and the fate of introgressed alleles in later generations. An assessment of the probability of permanent gene introgression between cultivated and wild *Musa* species is therefore not possible at this point. If sound scientific assessments of gene flow in bananas are to be made, research is needed in the following fundamental topics:

- A better understanding of flowering and reproductive biology in general, and pollen biology in particular (including pollen production, fertility, and viability), primarily for cultivated bananas and plantains, but also for wild *Musa* species.

- Greater knowledge about the distribution and outcrossing rates of cultivars and landraces in Asia, Africa, and tropical America.

- More information about the geographic distribution of wild *Eumusa* relatives outside of Asia, as a result of recent human expansion. For instance, wild *M. balbisiana* is common in tropical rainforests on the Atlantic coast of Costa Rica, and a small population of wild *M. acuminata* has been reported in a remote valley in Madagascar (F. Bakry, pers. comm. 2008). These plants, spread via seeds, are now considered to be weeds by the local farmers.

- The interspecific and intergeneric hybridization potential of *Musa*—including cytological studies to determine the genome constitution of *Musa* species (particularly those of sects. *Eumusa* and *Rhodochlamys*), as well as artificial hybridization studies in the laboratory and open-pollination studies in the field—to determine pollen dispersal distances and outcrossing rates. More research is particularly needed to assess hybridization between *Musa* cultivars and (1) species belonging to sections with the same chromosome number (i.e., sects. *Eumusa* and *Rhodochlamys*), (2) species in sects. *Callimusa* and *Australimusa*, to better understand genomic relationships and compatibility, and (3) species belonging to other genera (e.g., *Ensete, Heliconia, Musella*).

Greater knowledge of these important plant characteristics is crucial, not only for adequate gene flow assessment, but also for future breeding efforts.

Initially, edible triploid *Musa* cultivars seem to be particularly desirable for genetic transformations, since they are both seed- and pollen-sterile, which implies no likelihood of gene flow. This, however, is only partially true, since several triploid landraces and commercial hybrids are partially or even fully fertile and thus capable of yielding sexual offspring. Moreover, some sterile triploid cultivars can be stimulated to produce fertile seeds when cross-pollinated from diploid wild relatives.

Also, since wild relatives of bananas and plantains are native only to SE Asia, the lack of cross-fertile wild relatives in many other banana-producing areas is thought to reduce the likelihood of genes escaping from genetically transformed bananas (Smale and de Groote 2003; Pillay 2005). In Africa and tropical America, for example, there are no native wild *Musa* species, except for some naturalized populations of *M. balbisiana* in Costa Rica and *M. acuminata* in Madagascar (map 3.1). However, Africa constitutes an important secondary center of diversity for bananas and plantains, and E Africa in particular harbors a sizeable local diversity of triploid plantain landraces (Karamura 1999). Several triploid and tetraploid landraces and diploid hybrid cultivars are fertile and have been shown to set seed through open pollination (Vuylsteke et al. 1993a, b; PBIP 1996; Oselebe et al. 2006). In field experiments in Onne, Nigeria, germination rates of open-pollinated cultivars are higher than those of hybrid seeds derived from hand pollination (Ortiz and Crouch 1997). Hybridization can thus occur between modern cultivars and landraces, and between different landraces, thereby affecting the genetic constitution of the landraces' gene pool.

Based on the knowledge available about the relationships between edible *Musa* species and their wild relatives, the likelihood of gene introgression from cultivated bananas and plantains—assuming physical proximity and flowering overlap, and being conditional on the nature of the respective trait(s)—is *moderate* for:

- its direct progenitors *Musa acuminata* and *M. balbisiana*
- other wild *Musa* relatives that have the same basic chromosome number (i.e., species in sects. *Eumusa* and *Rhodochlamys*) in regions of Asia where wild *Musa* species are present (map 3.1)

REFERENCES

Acland JD (1971) *East African Crops: An Introduction to the Production of Field and Plantation Crops in Kenya, Tanzania, and Uganda* (Longman Scientific and Technical: London, UK). 252 p.

Argent GCG (1976) The wild bananas of Papua New Guinea. *Notes Roy. Bot. Gard. Edinburgh* 35: 77–114.

Becker DK, Dudgale B, Smith MK, Harding RM, and Dale JL (2000) Genetic transformation of Cavendish banana (*Musa* spp. AAA group) cv. 'Grand Nain' via microprojectile bombardment. *Plant Cell Rep.* 19: 229–234.

Bernardo FA (1957) Plant characters, fiber, and cytology of *Musa balbisiana* × *M. textilis* F1 hybrids. *Philipp. Agric.* 41: 117–156.

Bhat SR, Bhat KV, and Chandel KPS (1994) Studies on germination and cryopreservation of *Musa balbisiana* seed. *Seed Sci. Technol.* 22: 637–640.

Champion J (1967) *Les Bananiers et leurs cultures* (Édition Setco: Paris, France). 214 p.

Cheesman EE (1947) Classification of the bananas: II. The genus *Musa* L. *Kew Bull.* 2: 106–117.

——— (1948) On the nomenclature of edible bananas. *J. Genet.* 48: 293–296.

——— (1950) Classification of the bananas: III. Critical notes on species. *Kew Bull.* 5: 27–28.

Constantine DR (2003) *The Musaceae: An annotated List of the Species of Ensete, Musa, and Musella.* www.users.globalnet.co.uk/~drc/.

Daniells JW, Jenny C, Karamura DA, Tomekpe K, Arnaud E, and Sharrock S (comps.) (2001) Musa*logue: A Catalogue of Musa Germplasm; Diversity in the Genus* Musa. 2nd ed. (International Network for the Improvement of Banana and Plantain [INIBAP]: Montpellier, France). 207 p.

de Langhe E, Wattanachaiyingcharoen D, Volkaert H, and Piyapitchard S (2000) Biodiversity of wild Musaceae in northern Thailand. In: Molina AB and Roa VN (eds.) *Advancing Banana and Plantain R & D in Asia and the Pacific: Proceedings of the 9th INIBAP-ASPNET Regional Advisory Committee*

Meeting, South China Agricultural University, Guangzhou, China, November 2–5, 1999, pp. 71–83.

d'Hont A, Paget-Goy A, Escoute J, and Carreel F (2000) The interspecific genome structure of cultivated banana, *Musa* spp., revealed by genomic DNA *in situ* hybridization. *Theor. Appl. Genet.* 100: 177–183.

Dugdale B, Becker DK, Beetham PR, Harding RM, and Dale JL (2000) Promoters derived from banana bunchy-top virus DNA-1 to -5 direct vascular-associated expression in transgenic banana (*Musa* spp.). *Plant Cell Rep.* 19: 810–814.

Fahn A and Benovaiche P (1979) Ultrastructure, development, and secretion in the nectary of banana flowers. *Ann. Bot.* (Oxford) 44: 85–93.

FAO [Food and Agriculture Organization of the United Nations] (2009) FAOSTAT data. http://faostat.fao.org/site/567/default.aspx.

Fortescue JA and Turner DW (2004) Pollen fertility in *Musa*: Viability in cultivars grown in southern Australia. *Aust. J. Agric. Res.* 55: 1085–1091.

Ganapathi TR, Higgs NS, Balint-Kurti PJ, Arntzen C, May GD, and van Eck JM (2001) *Agrobacterium*-mediated transformation of embryogenic cell suspensions of the banana cultivar 'Rasthali' (**AAB**). *Plant Cell Rep.* 20: 157–162.

Ge XJ, Liu MH, Wang WK, Schaal BA, and Chiang TY (2005) Population structure of wild bananas, *Musa balbisiana*, in China determined by SSR fingerprinting and cpDNA PCR-RFLP. *Mol. Ecol.* 14: 933–944.

Govaerts R and Häkkinen M (2007) World Checklist of Musaceae. www.kew.org/wcsp/home.do/.

Häkkinen M (2004) Endemic *Callimusa* species of Borneo. In: *Abstract Guide, 1st International Congress on* Musa: *Harnessing Research to Improve Livelihoods, Penang, Malaysia, July 6–9, 2004* (International Network for the Improvement of Banana and Plantain [INIBAP]: Montpellier, France). 5 p. http://musalit.inibap.org/pdf/IN050503_en.pdf.

Häkkinen M and Sharrock S (2002) Diversity in the genus *Musa*: Focus on *Rhodochlamys*. In: International Network for the Improvement of Banana and Plantain [INIBAP] (ed.) *Annual Report 2001* (International Network for the Improvement of Banana and Plantain [INIBAP]: Montpellier, France), pp. 16–23.

Heslop-Harrison P (2005) Unlocking the secrets of the banana genome. In: *INIBAP Annual Report 2005: Networking Banana and Plantain* (International Network for the Improvement of Banana and Plantain [INIBAP]: Montpellier, France), pp. 14–17. www.bioversityinternational.org/nc/publications/publications/publication/?user_bioversitypublications_pi1[showUid]=3004.

Horry JP, Ortiz R, Arnaud E, Crouch JH, Ferris RSB, Jones DR, Mateo N, Picq C, and Vuylsteke D (1997) Banana and Plantain. In: Fuccillo DA, Sears L, and Stapleton P (eds.) *Biodiversity in Trust: Conservation and Use of Plant Genetic*

Resources in CGIAR [Consultative Group on International Agricultural Research] Centres (Cambridge University Press: Cambridge, UK), pp. 67–81.

INIBAP [International Network for the Improvement of Banana and Plantain] (2006) *Global Conservation Strategy for* Musa *(Banana and Plantain): A Consultative Document Prepared by INIBAP with the Collaboration of Numerous Partners in the* Musa *Research-and-Development Community* (INIBAP: Montpellier, France). 27 p. www.bioversityinternational.org/nc/publications/publications/publication/?user_bioversitypublications_pi1[showUid]=3082.

Karamura DA (1999) Numerical taxonomic studies of the East African highland bananas (*Musa* AAA-East Africa) in Uganda. PhD dissertation, University of Reading (International Network for the Improvement of Banana and Plantain [INIBAP]: Montpellier, France). 192 p.

Khanna H, Becker DK, Kleidon J, and Dale JL (2004) Centrifugation assisted *Agrobacterium tumefaciens*-mediated transformation (CAAT) of embryogenic cell suspensions of banana (*Musa* spp. 'Cavendish' AAA and 'Lady-finger' AAB). *Mol. Breed.* 14: 239–252.

Landa R, Rayas A, Ramírez T, Ventura J, Albert J, and Roca O (1999) Study of the pollen fertility of several *Musa* cultivars used in the INIVIT genetic improvement programme. *InfoMusa* 8: 27.

Lescot T (2006) La banane en chiffres: Le fruit préféré de la planète. *Fruitrop* 140: 5–9.

Liu AZ, Li DZ, and Li XW (2002a) Taxonomic notes on wild bananas (*Musa*) from China. *Bot. Bull. Acad. Sin.* 43: 77–81.

Liu AZ, Li DZ, Wang H, and Kress WJ (2002b) Ornithophilous and chiropterophilous pollination in *Musa itinerans* (Musaceae), a pioneer species in tropical rain forests of Yunnan, southwestern China. *Biotropica* 34: 254–260.

Liu MH, Ge XJ, Wang WK, Hsu TW, Schaal BA, and Chiang TY (2004) Pollen and seed dispersal of *Musa balbisiana* in South China. *Conserv. Quart.* 47: 9–24.

May G, Afza R, Mason H, Wiecko A, Novak F, and Arntzen C (1995) Generation of transgenic banana (*Musa acuminata*) plants via *Agrobacterium*-mediated transformation. *Bio/Technol.* 13: 486–492.

Mutsaers M (1993) Natural pollination of banana and plantain at Onne. *MusAfrica* 2: 2–3.

Nelson SC, Ploetz RC, and Kepler AK (2006) *Musa* Species (Bananas and Plantains), Version 2.2. In: Elevitch CR (ed.) *Species Profiles for Pacific Island Agroforestry: Ecological, Economic, and Cultural Renewal* (Permanent Agriculture Resources [PAR]: H_lualoa, Hawai'i, USA). www.traditionaltree.org.

Nwakanma DC, Pillay M, Okoli BE, and Tenkouano A (2003) Sectional relationships in the genus *Musa* L. inferred from the PCR-RFLP of organelle DNA sequences. *Theor. Appl. Genet.* 107: 850–856.

Ortiz R (1997) Secondary polyploids, heterosis, and evolutionary crop breeding for further improvement of the plantain and banana (*Musa* spp. L.) genome. *Theor. Appl. Genet.* 94: 1113–1120.

Ortiz R and Crouch JH (1997) The efficiency of natural and artificial pollinators in plantain (*Musa* spp. AAB group) hybridization and seed production. *Ann. Bot.* (Oxford) 80: 693–695.

Ortiz R and Vuylsteke D (1996) Recent advances in *Musa* genetics, breeding, and biotechnology. *Plant Breed. Abst.* 66: 1355–1363.

Ortiz R, Ferris RSB, and Vuylsteke DR (1995) Banana and plantain breeding. In: Gowen S (ed.) *Bananas and Plantains* (Chapman and Hall: London, UK), pp. 110–146.

Oselebe HO, Tenkouano A, and Pillay M (2006) Ploidy variation of *Musa* hybrids from crosses. *Afr. J. Biotech.* 5: 1048–1053.

PBIP [Plantain and Banana Improvement Program] (1996) *Annual Report 1995* (Crop Improvement Division, International Institute of Tropical Agriculture [IITA]: Ibadan, Nigeria). 160 p.

Pillay M (2005) Hungry for improvement. In: *INIBAP Annual Report 2005: Networking Banana and Plantain* (International Network for the Improvement of Banana and Plantain [INIBAP]: Montpellier, France), pp. 9–11. www.bioversityinternational.org/nc/publications/publications/publication/?user_bioversity-publications_pi1[showUid]=3004.

Pillay M, Tenkouano A, Ude G, and Ortiz R (2004) Molecular characterization of genomes in *Musa* and its applications. In: Jain SM and Swennen R (eds.) *Banana Improvement: Cellular, Molecular Biology, and Induced Mutations: Proceedings from a Meeting, Leuven, Belgium, September 24–28, 2001* (Science: Enfield, NH, USA), pp. 271–286.

Price NS (1995) The origin and development of banana and plantain cultivars. In: Gowen S (ed.) *Bananas and Plantains* (Chapman and Hall: London, UK), pp. 1–12.

Purseglove JW (1972) *Tropical Crops: Monocotyledons*, vol. 2 (Longman Scientific and Technical: Harlow, UK). 607 p.

Remy S, François I, Cammue BPA, Swennen R, and Sági L (1998) Co-transformation as a potential tool to create multiple and durable disease resistance in banana (*Musa* spp.). *Acta Hort.* 461: 361–366.

Rowe P and Rosales FE (1996) Bananas and plantains. In: Janick J and Moore J (eds.) *Fruit Breeding, Vol. 1: Tree and Tropical Fruits* (John Wiley and Sons: New York, NY, USA), pp. 167–211.

Sági L, Panis B, Remy S, Schoofs H, de Smet K, Swennen R, and Cammue B (1995) Genetic transformation of banana (*Musa* spp.) via particle bombardment. *Bio/ Technol.* 13: 481–485.

Sági L, Remy S, Cammue BPA, Maes K, Raemaekers T, Schoofs H, and Swennen R (2000) Production of transgenic banana and plantain. *Acta Hort.* 540: 203–206.

Sathiamoorthy S (1994) *Musa* improvement in India. In: Jones DR (ed.) *The Improvement and Testing of* Musa: *A Global Partnership* (International Network for the Improvement of Banana and Plantain [INIBAP]: Montpellier, France), pp. 188–200.

Sharrock S (1990) Collecting *Musa* in Papua New Guinea. In: Jarret RL (ed.) *Identification of Genetic Diversity in the Genus* Musa (International Network for the Improvement of Banana and Plantain [INIBAP]: Montpellier, France), pp. 140–157.

——— (2001) Diversity in the genus *Musa*: Focus Paper 1. Focus on *Australimusa*. In: International Network for the Improvement of Banana and Plantain [INIBAP] (ed.) *Annual Report 2000: Networking Banana and Plantain* (International Network for the Improvement of Banana and Plantain [INIBAP]: Montpellier, France), pp. 14–19.

Shepherd K (1988) Observation on *Musa* taxonomy. In: *Identification of Genetic Diversity in the Genus* Musa: *Proceedings of an International Workshop, Los Baños, Philippines, September 5–10, 1988* (International Network for the Improvement of Banana and Plantain [INIBAP]: Montpellier, France), pp. 158–165.

——— (1992) History and methods of banana breeding. In: World Bank (ed.) *Report of the First External Program and Management Review of the International Network for the Improvement of Banana and Plantain (INIBAP)* (Consultative Group on International Agricultural Research [CGIAR] Secretariat: Washington, DC, USA), pp. 108–110.

——— (1999) *Cytogenetics of the Genus* Musa (International Network for the Improvement of Banana and Plantain [INIBAP]: Montpellier, France). 154 p.

Shepherd K and Ferreira FR (1982) The Papua New Guinea Biological Foundation's banana collection at Laloki, Port Moresby, Papua New Guinea. *IBPGR/SEAN Newsl.* 8: 28–34.

Shepherd K, Dantas JLL, and Alves EJ (1986) Melhoramento genético da bananeira. *Inf. Agropec.* 12: 11–19.

Silva SO, Souza MT Jr, Alves EJ, Silveira JRS, and Lima MB (2001) Banana breeding program at Embrapa. *Crop Breed. Appl. Biotech.* 1: 399–436.

Simmonds NW (1954) Isolation in *Musa*, sections *Eumusa* and *Rhodochlamys*. *Evolution* 8: 65–74.

——— (1956) Botanical results of the banana collecting expedition, 1954–5. *Kew Bull.* 11: 463–489.

——— (1959) *Bananas* (Longman Scientific and Technical: London, UK). 466 p.

———— (1960) Megasporogenesis and female fertility in three edible triploid bananas. *J. Genet.* 57: 269–278.

———— (1962) *The Evolution of the Bananas* (Longman Scientific and Technical: Harlow, UK). 170 p.

———— (1995) Bananas. In: Simmonds S and Smartt J (eds.) *Evolution of Crop Plants.* 2nd ed. (Longman Scientific and Technical: Harlow, UK), pp. 370–375.

Simmonds NW and Shepherd K (1955) The taxonomy and origins of the cultivated bananas. *J. Linn. Soc., Bot.* 55: 302–312.

Smale M and de Groote H (2003) Diagnostic research to enable adoption of transgenic crop varieties by smallholder farmers in sub-Saharan Africa. *Afr. J. Biotech.* 2: 586–595.

Start AN and Marshall AG (1976) Nectarivorous bats as pollinators of trees in west Malaysia. In: Burley J and Styles ST (eds.) *Tropical Trees: Variation, Breeding, and Conservation* (Academic Press: San Diego, CA, USA), pp. 141–150.

Stover RH and Simmonds NW (1987) *Bananas.* 3rd ed. (Longman Scientific and Technical: Harlow, UK). 468 p.

Swennen R (2005) Banking on the future. In: *INIBAP Annual Report 2005: Networking Banana and Plantain* (International Network for the Improvement of Banana and Plantain [INIBAP]: Montpellier, France), pp. 4–7. www.bioversity international.org/nc/publications/publications/publication/?user_bioversity publications_pi1[showUid]=3004.

TAG [Taxonomic Advisory Group for *Musa*] (2006) *Launching the Taxonomic Advisory Group: Developing a Strategic Approach to the Conservation and Use of* Musa *Diversity* (International Network for the Improvement of Banana and Plantain [INIBAP]: Montpellier, France). http://bananas.bioversityinternational .org/files/files/pdf/publications/launching_tag.pdf.

Tezenas du Montcel H, Carreel F, and Bakry F (1995) Improve the diploids: The key for banana breeding. In: *Proceedings of a Workshop: New Frontiers in Resistance Breeding for Nematode, Fusarium, and Sigatoka, Kuala Lumpur, Malaysia, October 2–5, 1995* (International Network for the Improvement of Banana and Plantain [INIBAP]: Montpellier, France), pp. 119–128.

Ude G, Pillay M, Nwakanma D, and Tenkouano A (2002a) Analysis of genetic diversity and sectional relationships in *Musa* using AFLP markers. *Theor. Appl. Genet.* 104: 1239–1245.

———— (2002b) Genetic diversity in *Musa acuminata* Colla and *M. balbisiana* Colla and some of their natural hybrids using AFLP markers. *Theor. Appl. Genet.* 104: 1246–1252.

Uma S, Saraswathi MS, Durai P, and Sathiamoorthy S (2006) Diversity and distribution of section *Rhodochlamys* (Genus *Musa*, Musaceae) in India and breeding potential for banana improvement programmes. *Plant Genet. Resour. Newsl.* 146: 17–23.

Valmayor RV, Jamaluddin SH, Silayoi B, Kusumo S, Danh LD, Pascua OC, and Espino RRC (2000) *Banana Cultivar Names and Synonyms in Southeast Asia* (International Network for the Improvement of Banana and Plantain [INIBAP], Asia and the Pacific Office: Los Baños, Philippines). 28 p. www.bioversityinternational.org/publications/Pdf/713.pdf.

Vuylsteke D, Ortiz R, Ferris RSB, and Crouch JH (1997) Plantain improvement. *Plant Breed. Rev.* 14: 267–320.

Vuylsteke D, Swennen R, and Ortiz R (1993a) Development and performance of black sigatoka-resistant tetraploid hybrids. *Euphytica* 65: 33–42.

—— (1993b) Registration of 14 improved tropical *Musa* plantain hybrids with black sigatoka resistance. *Hort. Sci.* 28: 957–959.

Wong C, Kiew R, Lamb A, Ohn S, Lee SK, Gan LH, and Gan YY (2001) Sectional placement of three Bornean species of *Musa* (Musaceae) based on AFLP. *Gard. Bull. Singapore* 53: 327–341.

Wong C, Kiew R, Argent G, Set O, Lee SK, and Gan YY (2002) Assessment of the validity of the sections in *Musa* (Musaceae) using ALFP. *Ann. Bot.* (Oxford) 90: 231–238.

FOUR

Barley
(*Hordeum vulgare* L.)

Centers of Origin and Diversity

Domesticated barley (*Hordeum vulgare* L. subsp. *vulgare*) originated in the region of the Fertile Crescent in the Near East, most likely in the Israel-Jordan area (Vavilov 1926, 1951; Takahashi 1955; Zohary 1969; Zohary and Hopf 1993; Badr et al. 2000; Orabi et al. 2007). There is still some debate over whether there were other centers of barley domestication in the Himalayas, Ethiopia, and Morocco (Staudt 1961; Bekele 1983; Molina-Cano et al. 1987, 2005; Morrell and Clegg 2007; Orabi et al. 2007).

The center of diversity for barley (i.e., domesticated barley and its progenitor) is SW Asia—the Fertile Crescent where domesticated barley evolved (Vavilov 1951; Badr et al. 2000). However, the genus *Hordeum* L. has several centers of diversity—both in terms of high species numbers and high genetic diversity (for a comprehensive review see Knüpffer et al. 2003)—including SW Asia and the Near East, Middle Asia, E Asia (Himalayan region), the Mediterranean region (including N Africa), Ethiopia, Europe-Siberia, and the New World (the Americas and Oceania).

Biological Information

Domesticated barley is an annual, spring or winter, cool-season crop, adapted to climates as extreme as the Arctic and to the upper limits of cultivation in high mountains (Harlan 1995).

The cultivated and wild forms in sect. *Hordeum* are diploids ($2n = 2x = 14$) or tetraploids (some *H. bulbosum* L. cytotypes), whereas species in the other sections include diploid, tetraploid, and hexaploid forms

Table 4.1 Cultivated barley (*Hordeum vulgare* L.) and its closest wild relatives

Species	Common name	Ploidy	Genome	Origin and distribution	Life form and weediness
Primary gene pool (GP-1)					
sect. *Hordeum*					
H. vulgare L. subsp. vulgare	cultivated barley	2n = 2x = 14	I	known only from cultivation, found in temperate areas throughout the world	annual, mainly inbreeding
H. vulgare subsp. spontaneum (K. Koch) Asch. & Graebn.	wild barley	2n = 2x = 14	I	wild ancestor, native to Fertile Crescent, Greece, N Africa, SW Asia up to Pakistan	weedy annual, mainly inbreeding
Secondary gene pool (GP-2)					
H. bulbosum L.	bulbous barley	2n = 2x = 14, 2n = 4x = 28	I, II	Near East and temperate Asia, Mediterranean region, N Africa	perennial, obligate outbreeder
Tertiary gene pool (GP-3)					
H. murinum L.	false barley, wall barley	2n = 2x = 14, 2n = 4x = 28, 2n = 6x = 42	Y	native to Europe, Mediterranean region up to Afghanistan; naturalized in Australia, New Zealand, USA, and S America	crop weed, annual
Hordeum species in other sections					
sect. Sibirica Nevski		2n = 2x = 14	H	3 species; Asiatic diploids	
sect. Critesion (Raf.) Nevski		2n = 2x = 14, 2n = 4x = 28, 2n = 6x = 42	based on H	approximately 22 species; N and S America, temperate Asia	
sect. Stenostachys Nevski		2n = 2x = 14, 2n = 4x = 28, 2n = 6x = 42	X and based on H	4 species; Europe, S and N Africa, Near East, E and SW Asia	

Sources: Bothmer and Jacobsen 1985; Bothmer et al. 1995.
Note: Genome designations are according to Bothmer et al. 1986, 1995. Gene pool boundaries are according to Bothmer et al. 1995.

with a basic chromosome number of $x = 7$. The pattern of relationships between species in the genus *Hordeum* is complicated, due to the presence of several genomes. At the diploid level, four basic genomes (H, I, X, and Y) occur (Bothmer et al. 1986; table 4.1).

Flowering

The inflorescence of barley is a spike with incomplete flowers that lack sepals and petals. The stamens and pistil of individual florets are enclosed within the same floral structures (Starling 1980). The anthers often dehisce (open) before the spikes emerge from the leaf sheaths, and usually do so without the flowers opening. Most pollen is therefore shed within the spikelet; consequently, self-fertilization is usual in barley (Pope 1944). However, the degree to which a spike emerges prior to anthesis is variable and depends on environmental conditions and the cultivar type (Starling 1980). Pollen grains germinate within minutes after landing on the stigma (Pope 1937).

Pollen Dispersal and Longevity

Barley pollen, like that of most grass species, is wind dispersed (Jørgensen and Wilkinson 2005). Insect pollinators, such as bees and beetles, do visit barley flowers occasionally, but their contribution to cross-fertilization is minor.

Barley pollen remains viable for only a few hours after dehiscence (Anthony and Harlan 1920; Engledow 1924; Pope 1944). In a recent study, however, Parzies et al. (2005) found that the pollen of cultivated barley retains a sufficiently high level of viability to ensure successful cross-fertilization over a period of at least 26 hours, even at temperatures of up to 40°C. These authors also found that the pollen longevity of wild barley (*H. vulgare* subsp. *spontaneum* (K. Koch.) Thell. and *H. bulbosum*) was very similar.

Sexual Reproduction

The crop is self-fertile and highly autogamous, with outcrossing rates below 2% (Stevenson 1928; Hammer 1975; Brown et al. 1978b; Wagner and Allard 1991; Tammisola 1998; Parzies et al. 2000; Abdel-Ghani et al. 2004), although outcrossing rates of up to 7% have also been reported (Ritala et al. 2002). The barley progenitor *H. vulgare* subsp. *spontaneum* (same species as the crop) has outcrossing rates of 12% and higher (Brown et al. 1978b).

Reproductive behaviors of other *Hordeum* wild relatives range from almost exclusive inbreeding (some of the annuals, e.g., *H. murinum* L.) to intermediate figures of 40%–60% (probably *H. secalinum* Schreb.) and to exclusive allogamy caused by self-incompatibility (*H. bulbosum* and *H. brevisubulatum* (Trin.) Link) (R. von Bothmer, pers. comm. 2008).

Seed Dispersal and Dormancy

The presence of a brittle rachis in wild barley species promotes seed shattering and thus efficient seed dispersal. Most modern barley cultivars, in contrast, have lost their shattering ability, due to selection for a tough rachis during domestication (Bothmer et al. 2003; Pourkheirandish and Komatsuda 2007).

Two seed dispersal strategies operate in *Hordeum* species. Many wild species have small, light seeds that are dispersed by the wind (e.g., *H. jubatum* L.). Others—including *H. bulbosum* and both subspecies of *H. vulgare*—have large, heavy seeds that are either buried in the soil or dispersed by animals, mainly rodents and ants (Volis et al. 2004).

Animal-mediated seed dispersal—via seed predators such as rodents and ants—usually occurs on a local scale, with mean dispersal distances of less than 1 m to a maximum 40 m (Beattie 1982; Caughley et al. 1998; Gómez and Espadaler 1998). However, if they are attached to the fur of large animals such as livestock and deer, or the clothes of humans, seeds can be dispersed over considerable distances, up to several hundred kilometers (Bothmer et al. 1995; Manzano and Malo 2006).

Similar to the contrast in shattering ability mentioned above, seed dormancy is common in many wild *Hordeum* species (see, e.g., Takeda 1995; Gutterman and Gozlan 1998; Volis et al. 2004) but has been under strong selection pressure during barley domestication. Dormancy can still be found in some landraces, and a moderate level of seed dormancy is thought to be appropriate for barley cultivars (Han et al. 1999; Romagosa et al. 1999). Wild barley seeds can survive for up to three or four years in the soil under dry conditions, forming short-term seed banks (see, e.g., K. Thompson et al. 1997; Volis et al. 2002, 2004).

The survival of cereal seeds in the soil depends on the climatic and environmental conditions present during seed development, the depth of seed burial, and soil management. In general, survival is longer in deeper soil layers (> 20 cm) than at shallower depths. The seeds of domesticated barley appear to have the potential to survive longer in soil seed banks than wheat, oats, or triticale. Also, during cold winters, freezing kills moistened grains, due to starch degradation (Ritala et al. 2002).

Volunteers, Ferals, and Their Persistence

Volunteer plants are sometimes found in subsequent cropping cycles, due to barley seeds that shattered during planting and harvesting. Under continuous monocropping, seed shattering can lead to the accumulation of seeds in the ground and hence to the generation of significant populations of barley volunteers over time. Outside of cultivation, individual barley ferals and small naturalized feral populations can also be found.

Neither barley volunteers nor ferals are sufficiently vigorous to establish significant populations and compete with other species in the wild, due to their lack of efficient survival mechanisms, such as shattering ability and dormancy (Zohary 1960, 1969). Natural hybrids and backcrosses between landraces and their wild progenitor (*H. vulgare* subsp. *spontaneum*), however, have the potential to persist in the wild for several generations (R. von Bothmer, pers. comm. 2008). Winter barley is generally less hardy than the corresponding wheat or rye cultivars (Stanca et al. 2003); however, it can survive temperatures as low as –18°C (Hömmö and Pulli 1993; Tantau et al. 2004).

Weediness and Invasiveness Potential

Barley is recognized as being weedy or invasive in areas where it is cultivated, but it is not considered to be a challenging weed, either agriculturally or environmentally. However, some wild barley species do behave like weeds and are considered to be invasive in regions where they do not occur naturally, such as in S Australia or N America (Bothmer et al. 2007). "I never saw a field of any crop in Kurdistan in which wild barley [*H. vulgare* subsp. *spontaneum*] was not to be found growing as a weed" (Helbaek 1959, p. 370).

Crop Wild Relatives

The genus *Hordeum* includes approximately 30 annual and perennial species and is divided into four sections (Bothmer and Jacobsen 1985). *Hordeum* species can be found in temperate regions throughout the world, including subtropical areas in Central and S America and arctic areas in N America and central Asia (Bothmer et al. 2003).

Barley (*H. vulgare* subsp. *vulgare*) is the only domesticated species of the genus and is included in sect. *Hordeum*, together with its progenitor *H. vulgare* subsp. *spontaneum* and their closest wild relatives, *H. bulbosum* and *H. murinum* (table 4.1).

Most *Hordeum* species have quite limited geographical distributions, but some species are very widespread (e.g., *H. bulbosum*) or have become established as weeds in many parts of the world (e.g., *H. murinum*, *H. marinum* Huds., and *H. jubatum*).

Hybridization

Plant breeders have found that domesticated barley is very difficult to cross with wild barleys, aside from its wild progenitor *Hordeum vulgare* subsp. *spontaneum* (Baum et al. 1992). Most wild barley species are distantly related and sexually incompatible with the crop (Harlan 1995). Based on biosystematic and molecular studies, barley's wild relatives can be divided into primary, secondary, and tertiary gene pools according to their level of cross-compatibility with the crop (Harlan and de Wet 1971; Bothmer et al. 1992, 1995, 2003; Terzi et al. 2001; table 4.1).

Primary Gene Pool (GP-1)

This includes the *Hordeum vulgare* crop-wild-weed complex, which consists of domesticated barley (*H. vulgare* subsp. *vulgare*) and its wild ancestor (*H. vulgare* subsp. *spontaneum*). Both taxa are fully interfertile and can spontaneously outcross with each other in areas of overlap, leading to frequent introgression (Zohary 1960; Kobyljanskij 1967; Kamm 1977; Brown et al. 1978a; Allard 1988; Asfaw and Bothmer 1990; Bothmer et al. 1995; Harlan 1995).

For example, some Moroccan wild populations of subsp. *spontaneum* have been shown to have a hybrid origin (Giles and Lefkovitch 1984, 1985). In many areas where cultivated and wild barley occur in sympatry, morphologically intermediate forms can be found, including feral weedy individuals and hybrid swarms. These natural hybrids have at times been called *H. vulgare* f. *agriocrithon* (A.E. Åberg) Bowden (Zohary 1960, 1969; Tanno and Takeda 2004).

The wild progenitor has been used in breeding programs, as it has been identified as the source for several interesting agronomic traits, including disease resistance (Moseman et al. 1983; Lehmann and Bothmer 1988; Nevo 1992; Eglinton et al. 1999; Weibull et al. 2003).

Secondary Gene Pool (GP-2)

The secondary gene pool includes only one species, the perennial *H. bulbosum*, which is native to the Mediterranean region. A diploid and a tetraploid cytotype of *H. bulbosum* are known (Jørgensen 1982; Jakob et al. 2004). The species has the same I genome as *H. vulgare*, which is not present in any of the other species of *Hordeum*. The tetraploid cytotype of *H. bulbosum* is a true autotetraploid with the genomic designation II (Jørgensen 1982).

Cultivated barley and *H. bulbosum* are reproductively isolated, and hybrids at both the diploid and tetraploid levels are generally highly sterile. This is thought to be due to the existence of a dominant incompatibility gene (Pickering 1983). However, variations in incompatibility with *H. bulbosum* exist between stocks of the same cultivar (Pickering 1985).

Despite very high meiotic pairing between cultivated barley and diploid *H. bulbosum*, strong pre-zygotic hybridization barriers lead to the elimination of the *H. bulbosum* chromosomes, resulting in the formation of haploid hybrids with extremely low fertility (Kasha and Sadasivaiah 1971; Lange 1971; Bothmer et al. 1983; H. Thomas and Pickering 1988; Xu and Snape 1988; Zhang et al. 1999). However, seed set depends on genotype and can exceed 80% in some cases (Pickering and Hayes 1976).

Although the process is not understood in detail, and depends on genotype and environmental interactions, chromosome elimination at times does not occur, allowing embryos to develop into true, stable hybrids (Craig and Fedak 1985; Fedak 1985; H. Thomas and Pickering 1985; Pickering and Rennie 1990; Bothmer et al. 1995).

Crosses with tetraploid *H. bulbosum* often result in highly sterile triploids. However, partially fertile F1 plants can be obtained, although the seed set of these triploid hybrids is very low (Konzak et al. 1951; Pickering 1988; Xu and Kasha 1992; Pickering et al. 1994, 2000; Molnar-Lang et al. 2000).

In breeding programs where artificial techniques such as colchicine treatment are used, several traits of agronomic interest have been transferred from *H. bulbosum* to cultivated barley, conferring resistance to powdery mildew, leaf rust, *Septoria*, a soil-borne virus complex (BaMMV, BaYMV-1, -2), scald (*Rhynchosporium secalis* (Oudem.) Davis), and barley yellow mosaic virus (Jones and Pickering 1978; Szigat and Pohler 1982; Gustafsson and Claësson 1988; Xu and Snape 1989; Graner et al. 1999; Pickering et al. 2000, 2004; Zhang et al. 2001; Ruge et al. 2003; Toubia-Rahme et al. 2003; Weibull et al. 2003; Pickering et al. 2006). Partially fertile interspecific hybrids have been developed that are then selfed or backcrossed to barley (Pickering and Johnston 2005).

Tertiary Gene Pool (GP-3)

The tertiary gene pool contains the weedy annual *H. murinum* and the remaining *Hordeum* species in other sections of the genus. Cultivated barley is reproductively isolated from these species, as they possess dissimilar genomes (**X**, **Y**, and **H** genomes). Due to strong pre- and postzygotic hybridization barriers, interspecific hybrids can only be obtained by significant human intervention and the use of artificial hybridization techniques such as embryo rescue; in such cases, the resulting hybrids are highly sterile (see, e.g., Clark 1967; Bothmer et al. 1983, 1995; Bothmer and Linde-Laursen 1989; Harlan 1995; Ritala et al. 2002). No spontaneous hybrids have been detected in the progeny of wild GP-3 relatives growing in sympatry with the crop (Clark 1967; Savova et al. 2002). Although several of the wild GP-3 species have been shown to contain interesting resistance genes (Orton 1979; Schooler 1980; Bothmer and Hagberg 1983; Bothmer 1992), thus far no gene transfer from these wild species to barley has been achieved in breeding programs.

There has been debate about the probability of hybridization of barley with other cereals, such as wheat, rye, or oats. In general, barley does not intercross with other cereals, and—as in the case of GP-3 *Hordeum* species—any hybrid seed produced using artificial techniques is either sterile or, if fertile, results in F1 progeny with high chromosome and meiotic instability (see, e.g., W. Thompson and Johnston 1947; Kruse 1973; Fedak 1977; J. Thomas et al. 1977; Falk and Kasha 1981; Mujeeb-Kazi and Rodriguez 1983; Koba et al. 1991). However, tetraploid *H. bulbosum* has been used to produce haploids in artificial intergeneric crosses with *Triticum* (see, e.g., Fedak 1985, 1991; Sitch and Snape 1987).

Outcrossing Rates and Distances

Pollen Flow

As in most other plants, pollen-mediated gene flow in barley is influenced by several factors, including variation in flowering time, outcrossing rate, the sizes of and distance between populations, wind speed, and humidity (see, e.g., Ritala et al. 2002). Natural outcrossing rates of domesticated barley are usually below 2% (Brown et al. 1978b; Wagner and Allard 1991; Ennos 1994; Tammisola 1998; Parzies et al. 2000; Abdel-Ghani et al. 2004). In varieties with open flowers, frequencies of up to 7% have been reported under field conditions in Finland (Ritala et al. 2002). Barley varieties with closed flowers exhibit much lower rates of outcrossing (Hammer 1977).

Long-distance pollen dispersal is influenced by pollen longevity. Barley pollen grains lose viability quite quickly, within a few hours. They are also very sensitive to environmental conditions (Hammer 1977), which further reduces the possibilities of cross-pollination over long distances. This has been confirmed by a pollen flow study with transgenic barley in Australia, where very low outcrossing rates (0.005%) were measured at distances up to 12 m (Gatford et al. 2006). A pollen flow experiment at field-scale level in Finland demonstrated

Table 4.2 Isolation distances for barley (*Hordeum vulgare* L.)

Threshold*	Minimum isolation distance (m)	Country	Reference		Comment
0.3% / 0.1%	25	EU	EC	2001	foundation (basic) seed requirements (hybrids)
0.3% / 0.3%	25	EU	EC	2001	certified seed requirements (hybrids)
n.a. / 0.1%	50	OECD	OECD	2008	foundation (basic) seed requirements
n.a. / 0.3%	20	OECD	OECD	2008	certified seed requirements
0.01% / 1 seed per pound	3	USA	CCIA	2009	foundation (basic) and registered seed requirements
0.02% / 1 seed per pound	3	USA	CCIA	2009	certified seed requirements
0.03% / 0.05%	0	USA	USDA	2008	foundation (basic) seed requirements
0.05% / 0.1%	0	USA	USDA	2008	registered seed requirements
0.1% / 0.2%	0	USA	USDA	2008	certified seed requirements
0.01% / n.a.	3	Canada	CSGA	2009	foundation (basic) and registered seed requirements
0.05% / n.a.	3	Canada	CSGA	2009	certified seed requirements

*Maximum percentage of off-type plants permitted in the field / seed impurities of other varieties or off-types permitted. n.a. = not available.

that the dispersal of transgenic pollen and cross-fertilization occurred at up to 45 m in the dominant wind direction; this happened at very low rates (less than 0.04%), using male-sterile recipient plants (Ritala et al. 2002). Similar rates and distances have been reported for wild barley, *H. vulgare* subsp. *spontaneum* (Engledow 1924; Wagner and Allard 1991).

Isolation Distances

The separation distances established for barley seed production depend on the seed category, and vary between 0 to 3 m in the United States and Canada and 20 to 50 m in OECD (Organization for Economic Cooperation and Development) countries and Europe (table 4.2). The few pollen flow studies available thus far (see above) suggest that these distances may be sufficient to maintain the respective seed purity standards and remain under the maximum allowable amount (0.9%) of GM trait contamination.

State of Development of GM Technology

In comparison to wheat and other major crops, little research has been devoted thus far to the development of GM barley. However, stable and efficient transformation protocols exist for barley (reviewed by Goedeke et al. 2007; also see Hensel et al. 2008). Studied GM traits include pest and disease resistance and improved malting and brewing characteristics. GM barley field trials can be found in Australia, Canada, Finland, Germany, Hungary, Iceland, New Zealand, the United Kingdom, and the United States.

Agronomic Management Recommendations

In barley, four main paths of gene introgression to other cultivated or wild species need to be considered:

1   through pollen flow from cultivated barley

2   through volunteers in fields previously planted with barley

3   through barley seeds dispersed by livestock and seed predators, such as rodents and ants

4   through barley seeds dispersed by humans during cropping, harvesting, handling, storage, and transportation

To reduce the likelihood of gene introgression, the following management recommendations may be useful:

- Consider the land-use history of both the field to be planted and neighboring fields. This can help avoid contamination with crop-derived volunteer populations left by an earlier cropping cycle.

- Shallowly till the soil surface several days post-harvest. This encourages seed germination, and the resulting volunteer plants can then be destroyed, either by mechanical or chemical treatments. In general, volunteer cereals can be easily controlled through crop rotation, the use of herbicides, and/or by mechanical means.

- Consider establishing barriers with male-sterile bait plants (which are only fertilized by pollen from external sources) around the area planted with barley, in order to capture any escaped pollen. A 10-meter-wide continuous border should effectively contain the majority of pollen-mediated gene flow.

- If flowering overlap is likely with any of the two barley relatives identified as susceptible to gene introgression (see above), no plants should be in the area at distances shorter than those specified by legal regulations.

- The persistence of feral plants outside of fields is an important component of the assessment and management of gene flow. Special measures should be taken when transporting barley seeds to avoid seeds spilling out of harvesting vehicles. The presence of volunteer barley plants should be monitored along roads and, if present, they should be controlled by mechanical or chemical means.

Crop Production Area

Barley is now cultivated in all temperate and many warm-humid areas of the world. The major barley production areas are Canada, China, Ethiopia, Europe, India, the Near East, N Africa, the Russian Federation, and the United States. The total crop production area is estimated at 56.6 million ha (FAO 2009). To date, there is no reported commercial production of GM barley.

Research Gaps and Conclusions

More studies at field-scale level, measuring the outcrossing rates at smaller distance intervals, would be useful to better understand the dynamics of barley pollen flow and the outcrossing frequencies at different distances.

Gene flow via pollen occasionally occurs within the barley crop-wild-weed complex, which consists of domesticated barley and its conspecific wild progenitor *H. vulgare* subsp. *spontaneum*. Introgression of crop alleles to the wild species has been demonstrated. Barley can also hybridize with its wild relative *H. bulbosum*, although outcrossing frequencies are low and hybrid progeny (if any) are often completely sterile. This means that the probability of introgression between these two species is low. Given the high rates of self-pollination in barley and the lack of success in achieving hybridization with other *Hordeum* species, even with manual intervention, it is highly unlikely that spontaneous hybridization will occur with other wild relatives.

Based on the knowledge available about the relationships between barley and its wild relatives, the likelihood of gene introgression from domesticated barley—assuming physical proximity (less than 20 m) and flowering overlap, and being conditional on the nature of the respective trait(s)—is as follows:

- *high* for its conspecific wild progenitor *Hordeum vulgare* subsp. *spontaneum* (China, Middle and Near East, N Africa, SE Europe; map 4.1)

- *low* for its wild relative *H. bulbosum* (Europe, N Africa, Near and Middle East; map 4.1)

REFERENCES

Abdel-Ghani AH, Parzies HK, Omary A, and Geiger HH (2004) Estimating the outcrossing rate of barley landraces and wild barley populations collected from ecologically different regions of Jordan. *Theor. Appl. Genet.* 109: 588–595.

Allard RW (1988) Genetic changes associated with the evolution of adaptedness in cultivated plants and their wild progenitors. *J. Hered.* 79: 225–238.

Anthony S and Harlan HV (1920) Germination of barley pollen. *J. Agric. Res.* 18: 525–536.

Asfaw Z and Bothmer R von (1990) Hybridization between landrace varieties of Ethiopian barley (*Hordeum vulgare* subsp. *vulgare*) and the progenitor of barley (*H. vulgare* subsp. *spontaneum*). *Hereditas* 112: 57–64.

Badr A, Müller K, Schäfer R, El Rabey H, Ibrahim HH, Pozzi C, Rohde W, and Salamini F (2000) On the origin and domestication history of barley (*Hordeum vulgare*). *Mol. Biol. Evol.* 17: 499–510.

Baum M, Lagudah ES, and Appels R (1992) Wide crosses in cereals. *Annu. Rev. Plant Physiol.* 43: 117–143.

Beattie AJ (1982) Ants and gene dispersal in flowering plants. In: Armstrong JM, Powell JM, and Richards AJ (eds.) *Pollination and Evolution: Based on the*

Symposium on Pollination Biology, Held During the 13th International Botanical Congress (Section 6.7), at Sydney, Australia, August, 1981 (Royal Botanic Gardens: Sydney, NSW, Australia), pp. 1–8.

Bekele E (1983) A differential rate of regional distribution of barley flavonoid patterns in Ethiopia, and a view on the center of origin of barley. *Hereditas* 98: 269–280.

Bothmer R von (1992) The wild species of *Hordeum*: Relationships and potential use for improvement of cultivated barley. In: Shewry PR (ed.) *Barley: Genetics, Biochemistry, Molecular Biology, and Biotechnology* (CAB International: Wallingford, UK), pp. 3–18.

Bothmer R von, Baden C, and Jacobsen NH (2007) *Hordeum*. In: Barkworth ME, Capels KM, Long S, Anderton LK, and Piep MB (eds.) *Magnoliophyta: Commelidineae (in part); Poaceae, Part 1*. Vol. 24 in *Flora of North America North of Mexico* (Oxford University Press: New York, NY, USA). 911 p.

Bothmer R von, Flink J, Jacobsen N, Kotimäki M, and Landström T (1983) Interspecific hybridization with cultivated barley (*Hordeum vulgare* L.). *Hereditas* 99: 219–244.

Bothmer R von, Flink J, and Landström T (1986) Meiosis in interspecific *Hordeum* hybrids: I. Diploid combinations. *Can. J. Genet. Cytol.* 28: 525–535.

Bothmer R von and Hagberg A (1983) Pre-breeding and wide hybridization in barley. *Genetika* (Belgrade), Ser. F, 3 (Suppl.): 41–53.

Bothmer R von and Jacobsen N (1985) Origin, taxonomy, and related species. In: Rasmusson DC (ed.) *Barley*. American Society of Agronomy Monograph 26 (American Society of Agronomy: Madison, WI, USA), pp. 19–56.

Bothmer R von and Linde-Laursen I (1989) Backcrosses to cultivated barley (*Hordeum vulgare* L.) and partial elimination of alien chromosomes. *Hereditas* 111: 145–147.

Bothmer R von, Seberg O, and Jacobsen N (1992) Genetic resources in the Triticeae. *Hereditas* 116: 141–150.

Bothmer R von, Jacobsen N, Baden C, Jørgensen RB, and Linde-Laursen I (1995) *An Ecogeographical Study of the Genus Hordeum*. 2nd ed. Systematic and Ecogeographic Studies on Crop Genepools 7 (International Plant Genetic Resources Institute [IPGRI]: Rome, Italy). 129 p.

Bothmer R von, Hintum TJL van, Knüpffer H, and Sato H (2003) Barley diversity—an introduction. In: Bothmer R von, Hintum TJL van, Knüpffer H, and Sato H (eds.) *Diversity in Barley (*Hordeum vulgare*)* (Elsevier: Amsterdam, The Netherlands), pp. 3–8.

Brown AHD, Nevo E, Zohary D, and Dagan O (1978a) Genetic variation in natural populations of wild barley (*Hordeum spontaneum*). *Genetica* 49: 97–108.

Brown AHD, Zohary D, and Nevo E (1978b) Outcrossing rate and heterozygosity in natural populations of *Hordeum spontaneum* Koch in Israel. *Heredity* 41: 49–62.

Caughley J, Bomford M, Parker B, Sinclair R, Griffiths J, and Kelly D (1998) *Managing Vertebrate Pests: Rodents* (Bureau of Resource Sciences and Grains Research and Development Corporation: Canberra, ACT, Australia). 130 p.

CCIA [California Crop Improvement Association] (2009) *Grain Certification Standards.* http://ccia.ucdavis.edu/seed_cert/seedcert_index.htm.

Clark HH (1967) The origin and early history of the cultivated barleys: A botanical and archaeological synthesis. *Agric. Hist. Rev.* 15: 1–18.

Craig IL and Fedak G (1985) Variation in crossability of diverse genotypes of *Hordeum bulbosum* with *H. vulgare* (4x) cv. Betzes. *Cereal Res. Commun.* 13: 393–397.

CSGA [Canadian Seed Growers Association] (2009) *Canadian Regulations and Procedures for Pedigreed Seed Crop Production.* Circular 6–2005/Rev. 01.4–2009 (Canadian Seed Growers Association [CSGA]: Ottawa, ON, Canada). www.seedgrowers.ca/cropcertification/circular.asp.

EC [European Commission] (2001) European Communities (Cereal Seed) Regulations, 2001: S.I. [Statutory Instrument] No. 640/2001, Amending the Council Directive No. 66/402/EEC of 14 June 1966 on the Marketing of Cereal Seed www.irishstatutebook.ie/2001/en/si/0640.html.

Eglinton JK, Evans DE, Brown AHD, Langridge P, McDonald G, Jefferies SP, and Barr AR (1999) The use of wild barley (*Hordeum vulgare* subsp. *spontaneum*) in breeding for quality and adaptation. *Proceedings of the Ninth Australian Barley Technical Symposium, Melbourne, Australia, September 12–16, 1999* (Grains Research and Development Corporation, Australia [GRDC]: n.p.)

Engledow FL (1924) Inheritance in barley: III. The awn and lateral floret. *J. Genet.* 14: 49–87.

Ennos RA (1994) Estimating the relative rates of pollen and seed migration among plant populations. *Heredity* 72: 250–259.

Falk DE and Kasha KJ (1981) Comparison of the crossability of rye (*Secale cereale*) and *Hordeum bulbosum* onto wheat (*Triticum aestivum*). *Can. J. Genet. Cytol.* 23: 81–88.

FAO [Food and Agriculture Organization of the United Nations] (2009) FAOSTAT data. http://faostat.fao.org/site/567/default.aspx.

Fedak G (1977) Increased homoeologeous chromosome pairing in *Hordeum vulgare* × *Triticum aestivum* hybrids. *Nature* 266: 529–530.

——— (1985) Wide crosses in *Hordeum.* In: Rasmusson D (ed.) *Barley.* American Society of Agronomy Monograph 26 (American Society of Agronomy: Madison, WI, USA), pp. 155–186.

——— (1991) Intergeneric hybrids involving the genus *Hordeum.* In: Gupta PK and Tsuschiya T (eds.) *Chromosome Engineering in Plants: Genetics, Breeding, Evolution, Part A* (Elsevier: Amsterdam, The Netherlands), pp. 433–438.

Gatford K, Basri Z, Edlington J, Lloyd J, Qureshi J, Brettell R, and Fincher G (2006)

Gene flow from transgenic wheat and barley under field conditions. *Euphytica* 151: 383–391.

Giles BE and Lefkovitch LP (1984) Differential germination in *Hordeum spontaneum* from Iran and Morocco. *Z. Pflanzenzucht.* 92: 234–238.

—— (1985) Agronomic differences in *Hordeum spontaneum* from Iran and Morocco. *Z. Pflanzenzucht.* 94: 25–40.

Goedeke S, Hensel G, Kapusi E, Gahrtz M, and Kumlehn J (2007) Transgenic barley in fundamental research and biotechnology. *Transgenic Plant J.* 1: 104–117.

Gómez C and Espadaler X (1998) Myrmecochorous dispersal distances: A world survey. *J. Biogeogr.* 25: 573–580.

Graner A, Streng S, Kellermann A., Schiemann A, Bauer E, Waugh R, Pellio B, and Ordon F (1999) Molecular mapping of the Rym 5 locus encoding resistance to different strains of the barley yellow mosaic virus complex. *Theor. Appl. Genet.* 98: 285–290.

Gustafsson M and Claësson L (1988) Resistance to powdery mildew in wild species of barley. *Hereditas* 108: 231–237.

Gutterman Y and Gozlan S (1998) Amounts of winter or summer rain triggering germination and "the point of no return" of seedling desiccation tolerance of some *Hordeum spontaneum* local ecotypes in Israel. *Plant Soil* 204: 223–234.

Hammer K (1975) Die Variabilität einiger Komponenten der Allogamieneigung bei der Kulturgerste (*Hordeum vulgare* L., *s.l.*). *Kulturpflanze* 23: 167–180.

—— (1977) Fragen der Eignung des Pollens der Kulturgerste (*Hordeum vulgare* L., *s.l.*) für die Windbestäubung. *Kulturpflanze* 25: 13–23.

Han F, Ullrich SE, Clancy JA, and Romagosa I (1999) Inheritance and fine mapping of a major barley seed dormancy QTL. *Plant Sci.* 143: 113–118.

Harlan JR (1995) Barley. In: Smartt J and Simmonds NW (eds.) *Evolution of Crop Plants.* 2nd ed. (Longman Scientific and Technical: Harlow, UK), pp. 140–146.

Harlan JR and de Wet JMJ (1971) Towards a rational classification of cultivated plants. *Taxon* 20: 509–517.

Helbaek H (1959) Domestication of food plants in the Old World: Joint efforts by botanists and archeologists illuminate the obscure history of plant domestication. *Science* 130: 365–372.

Hensel G, Valkov V, Middlefell-Williams J, and Kumlehn J (2008) Efficient generation of transgenic barley: The way forward to modulate plant-microbe interactions. *J. Plant Physiol.* 165: 71–82.

Hömmö L and Pulli S (1993) Winterhardiness of some winter wheat (*Triticum aestivum*), rye (*Secale cereale*), triticale (×*Triticosecale*), and winter barley (*Hordeum vulgare*) cultivars tested at six locations in Finland. *Agric. Sci. Finl.* 2: 311–327.

Jakob SS, Meister A, and Blattner FR (2004) The considerable genome size variation of *Hordeum* species (Poaceae) is linked to phylogeny, life form, ecology, and speciation rates. *Mol. Biol. Evol.* 21: 860–869.

Jones IT and Pickering RA (1978) The mildew resistance in *Hordeum bulbosum* and its transference into *H. vulgare* genotypes. *Ann. Appl. Biol.* 88: 295–298.

Jørgensen RB (1982) Biosystematics of *Hordeum bulbosum* L. *Nord. J. Bot.* 2: 421–434.

Jørgensen RB and Wilkinson MJ (2005) Rare hybrids and methods for their detection. In: Poppy GM and Wilkinson MJ (eds.) *Gene Flow from GM Plants* (Blackwell: Oxford, UK), pp. 113–142.

Kamm A (1977) *The Range of Brittle Types of Cerealia Barleys in Israel*. Pamphlet No. 165 (Agricultural Research Organization, Volcani Center: Bet-Dagan, Israel). 43 p.

Kasha K and Sadasivaiah RS (1971) Genome relationships between *Hordeum vulgare* L. and *H. bulbosum*. *Chromosoma* 35: 264–287.

Knüpffer H, Terentyeva I, Hammer K, Kovalyova O, and Sato K (2003) Ecogeographical diversity—a Vavilovian approach. In: Bothmer R von, Hintum TJL van, Knüpffer H, and Sato H (eds.) *Diversity in Barley (*Hordeum vulgare*)* (Elsevier: Amsterdam, The Netherlands), pp. 53–67.

Koba T, Handa T, and Shimada T (1991) Efficient production of wheat-barley hybrids and preferential elimination of barley chromosomes. *Theor. Appl. Genet.* 81: 285–292.

Kobyljanskij VD (1967) Biological features of wild species of barley with a view to their utilization in breeding. *Armen. Biol. J.* 20: 41–51.

Konzak CF, Randolph LF, and Jensen NF (1951) Embryo culture of barley species hybrids: Cytological studies of *Hordeum sativum* × *Hordeum bulbosum*. *J. Hered.* 42: 124–134.

Kruse A (1973) *Hordeum × Triticum* hybrids. *Hereditas* 73: 157–161.

Lange W (1971) Crosses between *Hordeum vulgare* and *H. bulbosum*: I. Production, morphology, and meiosis of hybrids and haploids. *Euphytica* 20: 14–29.

Lehmann L and Bothmer R von (1988) *Hordeum spontaneum* and landraces as a gene resource for barley breeding. In: Jorna ML and Slootmaker LAJ (eds.) *Cereal Breeding Related to Integrated Cereal Production* (Pudoc: Wageningen, The Netherlands), pp. 190–194.

Manzano P and Malo JE (2006) Extreme long-distance seed dispersal via sheep. *Frontiers Ecol. Environ.* 4: 244–248.

Molina-Cano JL, Fra Mon P, Salcedo G, Aragoncillo C, Roca de Togores F, and García-Olmedo F (1987) Morocco as a possible domestication center for barley: Biochemical and agromorphological evidence. *Theor. Appl. Genet.* 73: 531–536.

Molina-Cano JL, Russell JR, Moralejo MA, Escacena JL, Arias G, and Powell W (2005) Chloroplast DNA microsatellite analysis supports a polyphyletic origin for barley. *Theor. Appl. Genet.* 110: 613–619.

Molnar-Lang M, Linc G, Logojan A, and Sutka J (2000) Production and meiotic pairing behaviour of new hybrids of winter wheat (*Triticum aestivum*) × winter barley (*Hordeum vulgare*). *Genome* 43: 1045–1054.

Morrell PL and Clegg MT (2007) Genetic evidence for a second domestication of barley (*Hordeum vulgare*) east of the Fertile Crescent. *Proc. Natl. Acad. Sci. USA* 104: 3289–3294.

Moseman JG, Nevo E, and Zohary D (1983) Resistance of *Hordeum spontaneum* collected in Israel to infection with *Erysiphe graminis hordei*. *Crop. Sci.* 23: 1115–1119.

Mujeeb-Kazi A and Rodriguez R (1983) Meiotic instability in *Hordeum vulgare* × *Triticum aestivum* hybrids. *J. Hered.* 74: 292–296.

Nevo E (1992) Origin, evolution, population genetics, and resources for breeding of wild barley, *Hordeum spontaneum*, in the Fertile Crescent. In: Shewry PR (ed.) *Barley Genetics, Biochemistry, Molecular Biology, and Biotechnology* (CAB International: Wallingford, UK), pp. 19–43.

OECD [Organization for Economic Cooperation and Development] (2008) OECD Scheme for the Varietal Certification of Cereal Seed Moving in International Trade: Annex VIII to the Decision. C(2000)146/FINAL. OECD Seed Schemes, February 2008. www.oecd.org/document/0/0,3343,en_2649_33905_1933504 _1_1_1_1,00.html.

Orabi J, Backes G, Wolday A, Yahyaoui A, and Jahoor A (2007) The Horn of Africa as a centre of barley diversification and a potential domestication site. *Theor. Appl. Genet.* 114: 1117–1127.

Orton TJ (1979) A quantitative analysis of growth and regeneration from tissue cultures of *Hordeum vulgare*, *H. jubatum*, and their interspecific hybrid. *Environ. Exp. Bot.* 19: 319–335.

Parzies HK, Spoor W, and Ennos RA (2000) Outcrossing rates of barley landraces from Syria. *Plant Breed.* 119: 520–522.

Parzies HK, Schnaithmann F, and Geiger HH (2005) Pollen viability of *Hordeum* spp. genotypes with different flowering characteristics. *Euphytica* 145: 229–235.

Pickering RA (1983) The location of a gene for incompatibility between *Hordeum vulgare* L. and *H. bulbosum* L. *Heredity* 51: 455–459.

——— (1985) Crossability relationships between certain species in the Hordeae. *Barley Genet. Newsl.* 14: 14–17.

——— (1988) The production of fertile triploid hybrids between *Hordeum vulgare* L. ($2n = 2x = 14$) and *H. bulbosum* L. ($2n = 4x = 28$). *Barley Genet. Newsl.* 18: 25–29.

Pickering RA and Hayes JD (1976) Partial incompatibility in crosses between *Hordeum vulgare* L. and *H. bulbosum* L. *Euphytica* 25: 671–678.

Pickering RA and Johnston PA (2005) Recent progress in barley improvement using wild species of *Hordeum. Cytogenet. Genome Res.* 109: 344–349.

Pickering RA and Rennie WF (1990) The evaluation of superior *Hordeum bulbosum* L. genotypes for use in a doubled haploid breeding programme. *Euphytica* 45: 251–255.

Pickering RA, Timmerman GM, Cromey MG, and Melz G (1994) Characterisation of progeny from backcrosses of triploid hybrid between *Hordeum vulgare* L. (2x) and *H. bulbosum* L. (4x) to *H. vulgare. Theor. Appl. Genet.* 88: 460–464.

Pickering RA, Malyshev S, Künzel G, Johnston PA, Menke M, and Schubert I (2000) Locating introgressions of *Hordeum bulbosum* chromatin within the *H. vulgare* genome. *Theor. Appl. Genet.* 100: 27–31.

Pickering RA, Niks RE, Johnston PA, and Butler RC (2004) Importance of the secondary genepool in barley genetics and breeding: II. Disease resistance, agronomic performance, and quality. *Czech J. Genet. Plant Breed.* 40: 79–850.

Pickering RA, Ruge-Wehling B, Johnston PA, Schweizer G, Ackermann P, and Wehling P (2006) The transfer of a gene conferring resistance to scald (*Rhynchosporium secalis*) from *Hordeum bulbosum* into *H. vulgare* chromosome 4HS. *Plant Breed.* 125: 576–579.

Pope MN (1937) The time factor in pollen-tube growth and fertilization in barley. *J. Agric. Res.* 54: 525–529.

——— (1944) Some notes on technique in barley breeding. *J. Hered.* 35: 99–111.

Pourkheirandish M and Komatsuda T (2007) The importance of barley genetics and domestication in a global perspective. *Ann. Bot.* (Oxford) 100: 999–1008.

Ritala A, Nuutila AM, Aikasalo R, Kauppinen V, and Tammisola J (2002) Measuring gene flow in the cultivation of transgenic barley. *Crop Sci.* 42: 278–285.

Romagosa I, Han F, Clancy JA, and Ullrich SE (1999) Individual locus effects on dormancy during seed development and after ripening in barley. *Crop Sci.* 39: 74–79.

Ruge B, Linz A, Pickering R, Proeseler G, Greif P, and Wehling P (2003) Mapping of Rym14[Hb], a gene introgressed from *Hordeum bulbosum* and conferring resistance to BaMMV and BaYMV in barley. *Theor. Appl. Genet.* 107: 965–971.

Savova Bianchi D, Keller Senften J, and Felber F (2002) Isozyme variation of *Hordeum murinum* in Switzerland and test of hybridization with cultivated barley. *Weed Res.* 42: 325–333.

Schooler AB (1980) Intergeneric and interspecific barley hybrids show tolerance to barley yellow dwarf virus. *North Dakota Farm Res.* 38: 19–21.

Sitch LA and Snape JW (1987) Factors affecting haploid production in wheat using the *Hordeum bulbosum* system: 3. Post-fertilization effects on embryo survival. *Euphytica* 36: 763–773.

Stanca AM, Romagosa I, Takeda K, Lundborg T, Terzi V, and Cattivelli L (2003)

Diversity in abiotic stress tolerances. In: Bothmer R von, Hintum TJL van, Knüpffer H, and Sato H (eds.) *Diversity in Barley (*Hordeum vulgare*)* (Elsevier: Amsterdam, The Netherlands), pp. 179–199.

Starling TM (1980) Barley. In: Fehr W and Hadley HH (eds.) *Hybridization of Crop Plants* (American Society of Agronomy: Madison, WI, USA), pp. 189–202.

Staudt G (1961) The origin of cultivated barleys: A discussion. *Econ. Bot.* 15: 205–212.

Stevenson FJ (1928) Natural crossing in barley. *J. Am. Soc. Agron.* 20: 1193–1196.

Szigat G and Pohler W (1982) *Hordeum bulbosum* × *H. vulgare* hybrids and their backcrosses with cultivated barley. *Cereal Res. Commun.* 10: 73–78.

Takahashi R (1955) The origin and evolution of cultivated barley. In: Demerec M (ed.) *Advances in Genetics 7* (Academic Press: San Diego, CA, USA), pp. 227–266.

Takeda K (1995) Varietal variation and inheritance of seed dormancy in barley. In: Noda K and Mares DJ (eds.) *Pre-Harvest Sprouting in Cereals 1995* (Center for Academic Societies: Osaka, Japan), pp. 205–212.

Tammisola J (1998) Transgene flow in barley cultivation. In: de Vries G (ed.) *Past, Present, and Future Considerations in Risk Assessment When Using GMOs: Proceedings of the 2nd International CCRO [Co-ordination Commission Risk Assessment Research] Workshop, Leeuwenhorst Congress Centre, Noordwijkerhout, The Netherlands, March 5–6, 1998* (Commission Genetic Modification: Bilthoven, The Netherlands), pp. 137–144.

Tanno K and Takeda K (2004) On the origin of six-rowed barley with brittle rachis, agriocrithon [*Hordeum vulgare* subsp. *vulgare* f. *agriocrithon* (Åberg) Bowd.], based on a DNA marker closely linked to the *vrs1* (six-row gene) locus. *Theor. Appl. Genet.* 110: 145–150.

Tantau H, Balko C, Brettschneider B, Melz G, and Dörffling K (2004) Improved frost tolerance and winter survival in winter barley (*Hordeum vulgare* L.) by *in vitro* selection of proline overaccumulating lines. *Euphytica* 139: 19–32.

Terzi V, Pecchioni N, Faccioli P, Kuera L, and Stanca AM (2001) Phyletic relationships within the genus *Hordeum* using PCR-based markers. *Genet. Resour. Crop Evol.* 48: 447–458.

Thomas HM and Pickering RA (1985) The influence of parental genotype on the chromosome behaviour of *Hordeum vulgare* × *H. bulbosum* diploid hybrids. *Theor. Appl. Genet.* 71: 437–442.

——— (1988) The cytogenetics of a triploid *Hordeum bulbosum* and of some of its hybrids and trisomic derivatives. *Theor. Appl. Genet.* 76: 93–96.

Thomas JB, Mujeeb KA, Rodríguez R, and Bates LS (1977) Barley × wheat hybrids. *Cereal Res. Comm.* 5: 181–188.

Thompson K, Bakker JP, and Bekker RM (1997) *Soil Seed Banks of North West Europe: Methodology, Density and Longevity* (Cambridge University Press: Cambridge, UK). 288 p.

Thompson WP and Johnston D (1947) The cause of incompatibility between barley and rye. *Can. J. Res.* 23(C): 1–15.

Toubia-Rahme H, Johnston PA, Pickering RA, and Steffenson BJ (2003) Inheritance and chromosomal location of *Septoria passerinii* resistance introgressed from *Hordeum bulbosum* into *Hordeum vulgare. Plant Breed.* 122: 405–409.

USDA [United States Department of Agriculture] (2008) 7 CFR [Code of Federal Regulations] § 201.76. Federal Seed Act Regulations: Minimum Land, Isolation, Field, and Seed Standards. Source: 59 FR [Federal Register] 64516, Dec. 14, 1994, as amended at 65 FR 1710, Jan. 11, 2000 (Government Printing Office: Washington, DC, USA), pp. 372–378. www.access.gpo.gov/nara/cfr/waisidx_08/7cfr201_08.html.

Vavilov NI (1926) Studies on the origin of cultivated plants. *Bull. Appl. Bot. Genet. Plant Breed.* 16: 1–245.

—— (1951) The origin, variation, immunity, and breeding of cultivated plants. *Chron. Bot.* 13: 1–366.

Volis S, Mendlinger S, and Ward D (2002) Differentiation in populations of *Hordeum spontaneum* Koch along a gradient of environmental productivity and predictability: Life history and local adaptation. *Biol. J. Linn. Soc.* 77: 479–490.

—— (2004) Demography and role of the seed bank in Mediterranean and desert populations of wild barley. *Basic Appl. Ecol.* 5: 53–64.

Wagner DB and Allard WB (1991) Pollen migration in predominantly self-fertilising plants: Barley. *Heredity* 82: 392–404.

Weibull J, Walther U, Sato K, Habekuß A, Kopahnke D, and Proeseler G (2003) Diversity in resistance to biotic stresses. In: Bothmer R von, Hintum TJL van, Knüpffer H, and Sato K (eds.) *Diversity in Barley (*Hordeum vulgare*)* (Elsevier: Amsterdam, The Netherlands), pp. 143–178.

Xu J and Kasha KJ (1992) Transfer of a dominant gene for powdery mildew resistance and DNA from *Hordeum bulbosum* into cultivated barley (*H. vulgare*). *Theor. Appl. Genet.* 84: 771–777.

Xu J and Snape JW (1988) The cytology of hybrids between *Hordeum vulgare* and *H. bulbosum* revisited. *Genome* 30: 486–494.

—— (1989) The resistance of *Hordeum bulbosum* and its hybrids with *H. vulgare* to common fungal pathogens. *Euphytica* 41: 273–276.

Zhang L, Pickering RA, and Murray BG (1999) Direct measurement of recombination frequency in interspecific hybrids between *Hordeum vulgare* and *H. bulbosum* using genomic *in situ* hybridization. *Heredity* 83: 304–309.

—— (2001) *Hordeum vulgare* × *H. bulbosum* tetraploid hybrid provides useful agronomic introgression lines for breeders. *New Zeal. J. Crop Hort. Sci.* 29: 239–246.

Zohary D (1960) Studies on the origin of cultivated barley. *Bull. Res. Counc. Isr.* 9D: 21–42.

———— (1969) The progenitors of wheat and barley in relation to domestication and agriculture dispersal in the Old World. In: Ucko PJ and Dimbley GW (eds.) *The Domestication and Exploitation of Plants and Animals* (Duckworth: London, UK), pp. 47–66.

Zohary D and Hopf M (1993) *Domestication of Plants in the Old World: The Origin and Spread of Cultivated Plants in West Asia, Europe, and the Nile Valley.* 2nd ed. (Clarendon Press: Oxford, UK). 328 p.

FIVE

Canola, Oilseed Rape
(*Brassica napus* L.)

Center(s) of Origin and Diversity

The Atlantic coast of N Europe, including the United Kingdom, is proposed as the primary center of origin of *Brassica napus* L., canola. Afghanistan is sometimes considered as a further independent center of origin of the Asian and Near Eastern types (Prakash and Hinata 1980).

The main center of diversity of the genus *Brassica* L. is the Mediterranean region (Europe and N Africa).

Biological Information

The crop commonly referred to as canola or oilseed rape actually includes two different *Brassica* species, *B. napus* var. *oleifera* Delile (canola, swede rape) and *B. rapa* L. var. *rapa* (canola, field mustard, bird rape, cultivated turnip). Also, aside from canola, there are other associated species in the genus *Brassica* that are being cultivated and for which GM varieties exist or are being developed, including *B. oleracea* L. (several crop varieties), *B. juncea* (L.) Czern. (brown mustard), and *B. carinata* A. Braun (Ethiopian mustard). In this chapter, the focus is on *B. napus*, canola; it does not analyze the likelihood of gene flow and introgression from other *Brassica* crops to allied species.

Flowering

The inflorescence of *Brassica napus* is an elongated raceme. The flowers cluster at the top and open upward from the base of the raceme (Downey et al. 1980). With compact stigmas, petals, and continuous scent and nectar production, they are well adapted to a wide range of insect pollinators (Free 1993). The flower structure also favors wind pollination.

In Canada, the flowering periods of early-, intermediate-, and late-seeded *B. napus* cultivars and volunteers in noncultivated experiments overlap with those of several early-flowering wild relatives (Simard and Légère 2004).

In central Europe, there is only a short period of flowering synchrony between cultivated *B. napus* and that of its wild relatives. The flowering period of *B. napus* volunteers, however, coincides with that of several sexually compatible wild relatives (e.g., *B. rapa* L., *B. nigra* (L.) W.D.J. Koch, *B. juncea*, *Hirschfeldia incana* (L.) Lagr.-Foss. and *Raphanus raphanistrum* L.). Flowering volunteers often branch and set new buds, thus prolonging the flowering period. Also, new volunteers emerge and flower throughout the vegetation period (Gruber and Claupein 2007).

Pollen Dispersal and Longevity

Cross-pollination among *B. napus* plants occurs primarily through physical contact with neighboring plants and, to a smaller extent, by wind (Williams 1984; Cresswell 1994; Cresswell et al. 2002). Insects (mainly honeybees and bumblebees) are responsible for long-distance pollen dispersal; they also play an important role in cross-pollination and can significantly increase the yield of rapeseed and mustard (Abrol 2007).

Canola pollen can remain viable from 24 hours to one week under controlled conditions (Mesquida and Renard 1982). Pollen viability gradually decreases after four to five days in natural circumstances (Ranito-Lehtimäki 1995), depending on environmental conditions, particularly temperature and humidity.

Sexual Reproduction

Brassica napus is an annual, predominantly autogamous, partially (12%–47%) allogamous (Rakow and Woods 1987; Becker et al. 1992; Kapteijns 1993) allotetraploid ($2n = 4x = 38$, **AACC** genome). The crop is capable of reproduction via parthenogenesis (seed production without fertilization), which is sometimes induced after foreign pollen lands on the surface of the stigma (Rieger et al. 1999).

Vegetative Reproduction

There are no reports of vegetative reproduction in *Brassica napus*.

Seed Dispersal and Dormancy

Seed loss from ripe canola, particularly in hot and windy conditions, can be up to 70% (Colton and Sykes 1992). The seeds are dispersed by wind, water, animals (ants, birds, mice, grazing animals), and humans (via clothing and vehicles) (Crawley and Brown 1995; Garnier et al. 2008; Pivard 2008). Harvest losses of 6% have been reported in Canada (Gulden et al. 2003a).

The concentration of seeds remaining on or in the soil after canola is harvested can extend from 2000 seeds/m² (in Canada; Légère et al. 2001) to 10,000 seeds/m² (in the United Kingdom; Lutman 1993). Ungerminated seeds can remain in the soil and form seed banks. In a Canadian study, the density of canola seeds in the seed bank declined 10-fold in the first year but remained almost stable thereafter, due to volunteer plants replenishing the seed bank every spring (Pekrun et al. 1997).

The survival of canola seeds in the seed bank is very low in comparison with that of wild relatives (Chadoeuf et al. 1998). However, because secondary dormancy can be induced under certain conditions (Hails et al. 1997; Pekrun et al. 1998), some seeds may survive for long periods. Canola seeds have been shown to remain viable in disturbed soils for at least 5 years, and possibly up to 10 years or more in undisturbed soils (Masden 1962; Pessel et al. 2001). The persistence of canola seeds in a seed bank depends both on seed dormancy and on their vertical distribution in the soil, as seeds are more likely to persist at deep rather than shallow depths (Pekrun et al. 1998; Simard et al. 2002). Linder (1998) found that the seed dormancy of some GM varieties was significantly higher than that of their non-GM controls under certain field conditions (darkness, high nutrients). Fitness risk assessments should thus be conducted at spaced sites covering the range where a GM crop is intended for cultivation.

Little is known about the viability of canola seeds after passing through the digestive tract of animals. They may be unaffected, as anecdotal evidence suggests that seeds of GM canola remained viable and subsequently emerged after being fed to chickens and then spread on a field through the chicken manure (Martens 2001).

Volunteers, Ferals, and Their Persistence

Volunteer canola plants can often be observed in subsequent cropping seasons, or as ferals outside of cultivated areas. The number of volunteers appearing in later plantings can be substantial, due to canola's considerable seed loss (Price et al. 1996; Pessel and Lecomte 2000; also see the seed dispersal section above). For the most part, volunteer populations are established from harvest losses of several thousand seeds/m² onto the soil. Volunteers also arise from pod shattering, which can be substantial under certain climatic conditions (strong winds close to harvest).

The frequency with which feral GM plants occurred in ruderal areas was similar to the proportion of the area planted with GM canola cultivars in preceding years. In Canada, more than 60% of the sampled feral canola plants were GM plants (Yoshimura et al. 2006).

In central Europe, canola volunteers growing in fields sown with rapeseed produced 45% of the number of seeds that were produced by the crop, and 10% if they grew in cereal fields (Gruber and Claupein 2007). In a high-risk scenario (canola field, high volunteer density, high seed persistence, no control mechanisms), volunteers would replenish the soil seed bank with 519 seeds/m² and thus exceed the maximum level (0.9%) set by European Commission (EC) regulations for the adventitious presence of genetically modified organisms (GMOs) in conventional crops. The number of seeds entering the soil seed bank in high-risk situations could be reduced by leaving the seed on the soil surface through delayed post-harvest tillage or minimum till.

As with many annual weeds, feral canola plants usually are not capable of surviving outside of cultivation and without human intervention for more than a few generations. In undisturbed habitats, they generally last one to four years, with densities of approximately one feral/ 100 m² after the second year. Still, some old canola cultivars can persist outside of cultivation for up to eight or nine years after they were last cultivated, albeit at low densities (Squire et al. 1999; Légère et al. 2001; Simard et al. 2002; Gulden et al. 2003b). In the United Kingdom, some feral populations remained stable at high densities for less than 10 years (Crawley and Brown 2004). Changes in seed bank survival appear to affect the population growth and persistence of ferals more than dormancy, fecundity, and seedling survival (Claessen et al. 2005).

Weediness and Invasiveness Potential

Although some feral canola plants can persist for several years, long-term studies in the United Kingdom show no evidence that canola is invasive in undisturbed natural habitats, nor there is evidence that GM varieties of canola are more invasive, or more persistent in disturbed habitats, than their non-GM counterparts (Crawley et al. 1993, 2001).

Crop Wild Relatives

The genus *Brassica* contains approximately 100 species, including several cultivated ones, among them *B. napus*—commonly known as canola, oilseed rape, or rapeseed. Canola shows partial sexual compatibility with several related *Brassica* species and with some other closely related species outside of the genus. Table 5.1 lists the chromosome numbers, ploidy levels, genomes, and geographical distribution of *B. napus* and its most common wild relatives.

Several other *Brassica* species (not included in table 5.1) are endemic to the Mediterranean region:

- S Europe: *B. assyriaca* Mouterde, *B. balearica* Pers., *B. cadmea* Heldr. *ex* O.E. Schulz, *B. drepanensis* (Caruel) Damanti, *B. glabrescens* Poldini, *B. hilarionis* Post, *B. incana* Ten., *B. jordanoffii* O.E. Schulz, *B. macrocarpa* Guss., *B. montana* Pourr., *B. rupestris* Raf., *B. villosa* Biv.

Table 5.1 Canola (*Brassica napus* L.), its wild relatives, and allies

Species	Common name	Ploidy level	Genome	Origin and distribution	Life form and weediness
B. napus L. (including Siberian kale, swede rape)	canola, oilseed rape, swede rape, rutabaga	allotetraploid (2*n* = 4*x* = 38)	AC	known only from cultivation; cultivated in Europe and widely naturalized elsewhere (USA, Canada, China, N India, N Africa, Australia, S America)	annual or biennial, occasional weed or volunteer in cultivated fields; derived from ancient crosses between *B. rapa* × *B. oleracea*
B. rapa L. (including Chinese cabbage, mizu-na, Chinese mustard)	oilseed rape, field mustard, bird rape, turnip	diploid (2*n* = 20)	A	native to Europe, the Russian Federation, C Asia, and Near East; weed or ruderal in Europe, N India, N Africa, USA, Canada	annual or biennial; wild form occurs along riversides, weedy forms spread widely across disturbed habitats
B. oleracea L. (including broccoli, cauliflower, brussels sprouts, kohlrabi, collards, kale)	cabbage (wild and cultivated)	diploid (2*n* = 18)	C	native to Atlantic coast of Europe, Mediterranean region, and Canary Islands; widely grown in temperate regions (Europe, USA, the Russian Federation, China, India, Australia, S America [e.g., Peru])	biennial or perennial; wild form occurs along Atlantic coast, elsewhere known only from cultivation
B. juncea (L.) Czern.	brown Oriental mustard, rai, Indian mustard	allotetraploid (2*n* = 4*x* = 36)	AB	native to C Asia, China, and Himalayan region; now widely cultivated and naturalized (reported even from Peru); large-scale production mainly in W Canada and India, also grown in China and USA	annual, occasional derived from *B. nigra* × *B. rapa*

Species	Common name	Ploidy	Code	Distribution	Notes
B. carinata A. Braun	Ethiopian or Abyssinian mustard	allotetraploid (2n = 4x = 34)	BC	known only from cultivation; widely grown in NE Africa (Sierra Leone, Ethiopia, Zambia, Guinea); occasionally grown in SE Asia, USA, E Europe, India	annual; derived from B. nigra × B. oleracea
B. nigra (L.) W.D.J. Koch	black mustard	diploid (2n = 16)	B	native to Asia Minor—Iran area; now widely naturalized in Europe, N Africa, Near East, tropical Asia (India, Nepal)	annual; weedy
B. deflexa Boiss.		diploid (2n = 14)	n.a.	native to Irano-Turanian region (Near East, S Europe)	annual
B. elongata Ehrh.		diploid (2n = 22)	n.a.	native to Europe, C and SW Asia, and Near East; naturalized in Australia and N America)	biennial to perennial; weedy
B. tournefortii Gouan	African mustard, wild turnip	diploid (2n = 20)	T	native to S Europe, N Africa, Near East, and Pakistan; naturalized in British Isles, Australia, New Zealand, and North America	annual; crop weed
Raphanus raphanistrum L.	wild radish, jointed charlock	diploid (2n = 18)	Rr	native to Europe, temperate Asia, N Africa; now widely naturalized elsewhere (reported also from Peru)	annual or biennial; crop weed
Raphanus sativus L.	garden radish	diploid (2n = 18)	R	known only from cultivation, origin unknown	annual or biennial; known only from cultivation
Hirschfeldia incana (L.) Lagr.-Foss.	hoary mustard, buchan weed	diploid (2n = 14)	Ad	native to S Europe, N Africa (Algeria, Libya, Morocco, Tunisia), Near East; widely invasive elsewhere	annual or biennial; weedy

(continued)

Table 5.1 (continued)

Species	Common name	Ploidy level	Genome	Origin and distribution	Life form and weediness
Sinapis arvensis L.	charlock, wild or field mustard	diploid ($2n = 18$)	**Sar**	probably native to Mediterranean region; now widely naturalized in Europe, temperate Asia and Pakistan, and N Africa	annual; crop weed
Sinapis alba L.	white or yellow mustard	diploid ($2n = 24$)	n.a.	native to Mediterranean region; introduced and naturalized in NW Europe, the Russian Federation, Japan, N and S America, Australia, New Zealand, N Africa, India, and China	annual
Sinapis pubescens L.		diploid ($2n = 18$)	n.a.	endemic to Mediterranean region	perennial
Erucastrum gallicum (Willd.) O.E. Schulz	dog mustard, hairy rocket	allotetraploid ($2n = 4x = 30$)	n.a.	native to Europe; invasive elsewhere	annual or biennial; weed
Erucastrum leucanthum Coss. & Durieu		diploid ($2n = 16$)	n.a.	native to Algeria and Morocco	perennial
Eruca sativa Mill.	rugula, garden rocket	diploid ($2n = 22$)	n.a.	native to Mediterranean region; also wild in N Africa (Algeria, Libya, Morocco, Tunisia), Near East; widely invasive elsewhere	annual vegetable; weedy
Diplotaxis muralis (L.) DC.	annual wall rocket	amphidiploid ($2n = 42$)	n.a.	native to S Europe, N Africa	annual or biennial; weedy
Diplotaxis tenuifolia (L.) DC.	sand rocket, Lincoln weed	diploid ($2n = 22$)	n.a.	native to Europe and W Asia; naturalized or casual elsewhere in temperate regions	perennial; weed

Species	Common name	Ploidy		Distribution	Life form
Diplotaxis erucoides (L.) DC.	white rocket, wild rocket	diploid ($2n = 14$)	n.a.	native to Mediterranean region; naturalized or casual elsewhere	annual; weed
Coincya monensis (L.) Greuter & Burdet	star mustard	diploid ($2n = 24$)	n.a.	native to Europe; naturalized or casual elsewhere	biennial; weed
Orychophragmus violaceus (L.) O.E. Schulz		diploid ($2n = 24$)		endemic to China	annual or biennial
Moricandia arvensis (L.) DC.	collejón	diploid ($2n = 28$)	n.a.	native to Mediterranean region, N Africa	biennial or perennial weed
Moricandia nitens (Viv.) E.A. Durand & Barratte		n.a.	n.a.	native to N Africa, Egypt, Israel	perennial
Rapistrum rugosum (L.) J.P. Bergeret	turnip weed, annual bastard-cabbage	diploid ($2n = 16$)	n.a.	native to Irano-Turanian and Mediterranean regions; widely distributed elsewhere (including Europe, N Africa, temperate W Asia)	annual or biennial; crop weed, considered invasive in USA (Texas)
Sisymbrium orientale L.	Oriental mustard	diploid ($2n = 14$)	n.a.	native to W Asia, Mediterranean region; widely naturalized elsewhere	annual or perennial; crop weed

Sources: Warwick et al. 2000; Salisbury 2002b.
Note: Cultivated forms are in bold. Ploidy levels are according to Warwick et al. 2000. n.a. = not available.

- N Africa: *B. desnottesii* Emb. & Maire (Morocco), *B. dimorpha* Coss. & Durieu (Tunisia, Algeria), *B. maurorum* Durieu, *B. spinescens* Pomel (Algeria, Morocco)
- Near East (Sinai): *B. deserti* Danin & Hedge
- S Europe and N Africa: *B. barrelieri* (L.) Janka, *B. fruticulosa* Cirillo, *B. gravinae* Ten., *B. insularis* Moris, *B. oxyrrhina* (Coss.) Willk., *B. procumbens* (Poir.) O.E. Schulz, *B. repanda* (Willd.) DC., *B. souliei* (Batt.) Batt.
- S Europe and Near East (Lebanon, Syria, Israel, Jordan): *B. cretica* Lam., *B. bourgeaui* (Webb. *ex* H. Christ) Kuntze (Canary Islands, Israel, Crimea)

Hybridization

Hybridization of *Brassica napus* (canola) with other species occurs with varying degrees of difficulty. It is influenced by external factors, including local climatic conditions (e.g., wind direction, wind speed, temperature, humidity, rainfall), experimental design (e.g., the size and orientation of donor and recipient fields), and insect movements (Scheffler et al. 1993), as well as the direction of the cross (i.e., which species is the pollen donor and which is the pollen recipient). Some very comprehensive overviews about hybridization between *B. napus*, its *Brassica* relatives, and associated species are available as references for further study (see, e.g., Rieger et al. 1999; Salisbury 2002a; FitzJohn et al. 2007).

The three allotetraploid *Brassica* species—*B. napus*, *B. juncea*, and *B. carinata*—undoubtedly arose from old natural crosses between diploid species, perhaps after several crosses for each species. Therefore the potential for gene exchange exists among all these species. Also, hybrids between *Brassica* species that share at least one genome can be easily obtained by manual cross-pollination (Scheffler and Dale 1994). In most other cases, however, hybridization is only possible by artificial means, such as ovary culture and embryo rescue.

There has been large-scale cultivation of GM canola in several countries (including the United States and Canada) for more than 10 years now. To date, there are no reports of problems with interspecific crosses and GM trait introgression of the herbicide-tolerant genes into cultivated or wild relatives of canola. However, spontaneous interspecific hybrids of *B. napus* with seven other species can occur naturally in the field, and there may be others that have not been detected yet. The potential for outcrossing and gene introgression between *B. napus* and these seven species is documented in more detail below.

Brassica rapa

Canola (*B. napus*, **AACC** genome) and *B. rapa* (**AA** genome) have a common set of chromosomes, which facilitates interspecific gene flow between these two species. Spontaneous hybridization and backcrossing in the field is possible, and natural *B. napus* × *B. rapa* hybrids have been reported in several countries, including Canada (Warwick et al. 2003; Yoshimura et al. 2006), the Czech Republic (Bielikova and Rakousky 2001), Denmark (Landbo et al. 1996; Hansen et al. 2001), New Zealand (Jenkins et al. 2001), the United Kingdom (Allainguillaume et al. 2006), and the United States (Halfhill et al. 2002). Subsequent introgression of alleles from *B. napus* to *B. rapa* occurs infrequently (Hansen et al. 2001, 2003; Leflon et al. 2006).

In laboratory experiments, *B. rapa* pollen has a significantly lower degree of fitness on *B. napus* than conspecific pollen, and hybrid zygote survival is markedly reduced in comparison to conspecific zygotes (Hauser et al. 1997). In field trials under natural conditions, hybridization success varies widely, depending on the experimental design and the direction of crosses (Jørgensen and Andersen 1994; Jørgensen et al. 1996; Halfhill et al. 2002; Pallett et al. 2006). Hybridization rates are significantly higher if *B. rapa* is the female parent, rather than vice versa, due to the self-incompatibility of *B. rapa*.

The triploid (**AAC** genome, $2n = 29$) F1 hybrids are viable and present rather high levels of fertility, albeit with greatly varying amounts of seed production (see, e.g., MacKay 1977; Lu and Kato 2001; Leflon et al. 2006). Hybrid fitness, both in terms of seed production and seed viability, differs between mixed and pure stands, depending on the number of parents and hybrids, which suggests that reproductive interactions and vegetative competition play a role (Pertl et al. 2002; Hauser et al. 2003). Seed production is usually much lower than that of *B. rapa* (Jørgensen et al. 1996; Ammitzbøll et al. 2005; Allainguillaume et al. 2006; Johannessen et al. 2006; Sutherland et al. 2006), but from time to time it may be higher (Hauser et al. 1998a, b). Most F1 hybrids produce small seeds with low seed dormancy, low seed viability, and reduced seedling survival (J. Brown and Brown 1996; Scott and Wilkinson 1998; Jørgensen et al. 1999; Warwick et al. 2003). F1 hybrids with high seed production result in F2 and BC progeny with improved plant fitness, seed production, and fertility, whereas F1 hybrids with low seed production produce F2 and BC progeny in which all of these characteristics are reduced.

In the greenhouse, GM F1 and BC progeny showed no significant differences in survival and fitness (growth rate, pollen fertility, seed production) with non-GM plants and performed as well as their wild *B. rapa* parents (Linder and Schmitt 1995; Linder 1998; Snow et al. 1999).

To contain the gene flow from *B. napus* to *B. rapa*, researchers have proposed positioning GM traits either in the chloroplast or in definite locations of the nuclear genome of *B. napus* (Lu et al. 2002; Li et al. 2006). However, the success of this approach is questionable, since the exchange of chloroplast DNA and the incorporation of the *B. napus* C-genome DNA into the *B. rapa* genome is possible, albeit rare (Tomiuk et al. 2000; Hansen et al. 2001, 2003).

Canola and *B. rapa* can spontaneously produce viable hybrids in the field (Harberd 1975; Landbo et al. 1996), and introgression into later generations is possible, with apparently few or no fitness costs (Mikkelsen et al. 1996a, b; Metz et al. 1997a, b; Rieger et al. 1999; Halfhill et al. 2001, 2003; B. Zhu et al. 2004; Légère 2005; Myers 2006; Jørgensen 2007). However, the long-term fate of introgressed traits is variable and depends, among other factors, on the number of surrounding parents and hybrids. In some cases, traits conferring a selective advantage—for example GM herbicide resistance—may be introgressed persistently into *B. rapa* populations, even in the absence of selection due to herbicide application (Snow et al. 1999). In others, hybridization over several generations—with or without introgression of GM traits—will result in less productive and thus less competitive populations (Halfhill et al. 2005).

Brassica juncea

Brassica juncea (**AABB** genome, $2n = 36$) has a common set of chromosomes with *B. napus* (**AACC** genome), which enhances the likelihood of interspecific hybridization and gene flow. Spontaneous hybridization and backcrossing in the field can occur and has been reported in Canada and Denmark (Bing et al. 1991, 1996; Frello et al. 1995; Jørgensen et al. 1998). Depending on the proportions of the parental species, up to 3% of hybrid offspring from *B. juncea* are reported, although hybridization is less successful when *B. napus* is the female parent (Jørgensen et al. 1998).

The pollen and seed fertility of the F1 hybrids is often less than 30% (Roy 1980; Bajaj et al. 1986; Sacristán and Gerdemann 1986; Prakash and Chopra 1988; Frello et al. 1995; Choudhary and Joshi 1999; Ghosh-Dastidar and Varma 1999), but spontaneous BC progeny with improved fertility have been documented (Bing et al. 1991, 1996; Alam et al. 1992; Jørgensen 1999). Backcrossing and subsequent gene introgression from *B. napus* to *B. juncea* could be expected, albeit infrequently.

Brassica oleracea

As with *B. rapa* and *B. juncea*, *B. oleracea* (**CC**) also has a genome in common with *B. napus* (**AACC**) and therefore introgression between the latter two species is theoretically possible (Scheffler and Dale 1994; Eastham and Sweet 2002). Wilkinson et al. (2000) examined wild *B. oleracea* populations in the United Kingdom in relation to their proximity to canola fields and the formation of natural hybrids. Because *B. oleracea* is restricted to maritime cliff habitats, only one population was found within 50 m of canola production; none of the nine new seedlings in that population were hybrids. More recently, the first spontaneous hybrids between *B. napus* and wild *B. oleracea* were reported in wild *B. oleracea* populations in the United Kingdom (Ford et al. 2006). Using flow cytometry and crop-specific microsatellite markers, the authors detected one triploid F1 hybrid, as well as nine diploid and two near-triploid introgressants.

Hybridization of *B. napus* with *B. oleracea* can also be achieved by hand pollination. In this case, the frequency of successful crosses is very low, and embryos often abort at early stages of development (Honma and Summers 1976; Chiang et al. 1977; Mattsson 1988). Thus far, no viable hybrid seeds have been obtained from crosses between *B. napus* and *B. oleracea* without the assistance of embryo rescue or ovule culture (Takeshita et al. 1980; Ayotte et al. 1987; Quazi 1988; Myers 2006).

However, if amphidiploid F1 hybrids between *B. napus* and *B. oleracea* are produced, these are fully fertile (Sundberg and Glimelius 1991; Kerlan et al. 1992). Backcrossing to either parent can produce viable and partially fertile offspring, with increasing fertility in each successive generation (Chèvre et al. 1996).

No information is available thus far on the fertility of the natural hybrids reported by Ford et al. (2006). However, because *B. oleracea* is a polycarpic perennial, a single hybrid could produce second-generation introgressants over several years.

Hirschfeldia incana

Spontaneous hybridization between male-sterile *Brassica napus* canola and *Hirschfeldia incana* in the field has been reported on several occasions and in both directions (Eber et al. 1994; Lefol et al. 1995, 1996a; Chèvre et al. 1996). In France, an average of two hybrid seeds per 100 fertilized flowers has been observed (Eber et al. 1994).

The resulting triploid F1 hybrids are often vigorous and at least as competitive as their wild parent (Eber et al. 1994; Lefol et al.1996a). However, they are usually male sterile and produce very little seed (< 1 seed per plant) under controlled conditions (Lefol et al. 1996a). In a field trial in France, where progeny from backcrosses with *H. incana* were subjected to herbicide-tolerance selection, fewer seeds were set after each successive backcross (Downey 1999b). The reproductive output decreased with each BC generation, yielding a single seed in the third BC and no viable seeds in the fourth generation. This might be due to a gene in *H. incana* that inhibits homoeologous pairing and, as such, leads to rapid expulsion of *B. napus* chromosomes in hybrids with *H. incana* (Kerlan et al. 1993; Lefol et al. 1996a). Hence hybridization between these two species is easy and recurrent, but introgression rarely occurs (Darmency and Fleury 2000).

Raphanus raphanistrum

Spontaneous hybridization between *Brassica napus* (**AACC** genome, $2n = 38$) and *Raphanus raphanistrum* (**RR** genome, $2n = 18$), or wild radish, may occur in the wild. Hybrids have been reported in standard field trials in Australia, Canada, Denmark, and France, albeit at very low frequencies (Eber et al. 1994; Baranger et al. 1995; Darmency et al. 1995; Chèvre et al. 1996, 2000; Rieger et al. 2001; Ammitzbøll and Jørgensen 2006). The frequency of hybridization varies, depending on several factors, among them the canola variety (male sterile or fully fertile) and genotype, local environmental conditions, the populations of R. raphanistrum that are used, and the numbers of parents (Eber et al. 1994; Baranger et al. 1995; Chèvre et al. 1996, 2000, 2003; Ammitzbøll and Jørgensen 2006).

Although pollen germination and ovule fertilization vary considerably within populations, post-zygotic barriers to hybridization are generally minor (Guéritaine and Darmency 2001). F1 hybrids are mainly allotriploids and show very low (< 2 seeds/plant) fertility (Kerlan et al. 1992; Baranger et al. 1995; Chèvre et al. 1996, 1998; Darmency et al. 1998; Pinder et al. 1999; Thalmann et al. 2001; Warwick et al. 2003). F1 hybrids commonly show decreased fitness in terms of reduced seedling emergence, a significant emergence delay, and a lower survival rate than both parents (Guéritaine et al. 2003). Despite this, a small number of highly fertile F1 hybrids have been reported (Salisbury 2002a).

Fertility increases after repeated BC to R. raphanistrum. However, after four generations of BC with herbicide-tolerant GM canola, neither the herbicide tolerance gene nor other B. napus marker genes were detected in the wild radish genome (Chèvre et al. 1997, 1999). After several generations, progeny having wild radish cytoplasm (RBC) were more likely to propagate in agronomic and natural conditions than offspring having canola cytoplasm derived from reciprocal crosses. The fitness of the latter was approximately 1% of that of the RBC progeny (Guéritaine et al. 2002). Under herbicide selection pressure over several generations, 18% of the progeny from resistant plants were resistant in each generation (Al Mouemar and Darmency 2004). This suggests that in the presence of herbicide selection pressure, the GM trait for herbicide resistance may be maintained despite a lack of steady introgression. In the absence of such selection, the frequency of resistance in the population is expected to decline.

Raphanus sativus

The first report on spontaneous hybridization between a male-sterile Brassica napus line and a variety of the cultivated radish Raphanus sati-

vus L. was published recently (Ammitzbøll and Jørgensen 2006). All F1 offspring were found to be hybrids with low pollen fertility (0%–15%).

Hybridization of *B. napus* with *R. sativus* can be achieved by hand pollination or through more sophisticated methods, such as ovule culture and embryo rescue (Ellerström 1978; Harberd and McArthur 1980; Takeshita et al. 1980; Paulmann and Röbbelen 1988; Lange et al. 1989; Sundberg and Glimelius 1991; Metz et al. 1995; Gupta 1997; Rhee et al. 1997; Huang et al. 2002). Thus far, all artificially produced F1 hybrids have proved to be male sterile.

Sinapis arvensis

Despite extensive studies investigating the coexistence of *Brassica napus* and *Sinapis arvensis* L. (see, e.g., Bing et al. 1996; Sweet et al. 1997; Moyes et al. 2002; Warwick et al. 2003), natural hybrids between these two species have only been observed on very rare occasions (Chèvre et al. 1996; Lefol et al. 1996a, b). In such cases, *S. arvensis* has always been the paternal parent, and only recently has a British report claimed to have identified a F1 hybrid with *S. arvensis* as the maternal parent (Daniels et al. 2005). In field trials under natural conditions in France, male-sterile *B. napus* has been shown to produce up to 0.18 hybrid seeds per plant (Chèvre et al. 1996; Lefol et al. 1996b). Lefol et al. (1996b) calculated the probability for a flower of any of these two species to spontaneously produce hybrid progeny as less than 10^{-10}. The resulting F1 progeny are completely sterile.

Artificial hybridizations using ovary culture and embryo rescue produce up to one seed per 100 pollinated flowers, but the F1 plants are weak and highly or completely sterile (Kerlan et al. 1992; Bing et al. 1995; Lefol at al. 1996b; J. Brown et al. 1997). F2 or BC seed production is extremely rare (Inomata 1988), and no gene introgression has been detected thus far (Bing et al. 1995, 1996; Chèvre et al. 1996, 2003; Lefol et al. 1996b; Rieger et al. 1999; Moyes et al. 2002; Warwick et al. 2003).

Although a very low level of interspecific crosses could occur (Leckie et al. 1993), the likelihood of gene flow between *B. napus* and *S. arvensis* is extremely remote, and there is general agreement that no gene introgression will occur (Bing et al. 1991; Eber et al. 1994; Downey 1999a, b).

Bridging

Intra- or intergeneric hybridization of *Brassica napus* relatives can indirectly act as a bridge for introgression of *B. napus* genes into other species that do not readily hybridize with *B. napus* itself.

For example, *B. nigra* (L.) W.D.J. Koch can hybridize with *B. juncea* as well as with *Sinapis arvensis*. Although it is highly unlikely in practice, it is theoretically possible for *B. nigra* to act as a genetic bridge, allowing gene introgression from *B. juncea* to *S. arvensis* via two subsequent interspecific hybridization events (Bing et al. 1991; Downey 1999b).

In the laboratory, several artificial bridge crosses have been used to transfer genes to otherwise incompatible *Brassica* species, such as *B. rapa* × *Diplotaxis siettiana* Maire as bridge-cross to *B. juncea* and *B. napus* (Nanda Kumar and Shivanna 1993) or *B. rapa* × *D. erucoides* (L.) DC. and *B. rapa* × *D. berthautii* Braun-Blanq. & Maire for gene transfer to *B. juncea* (Malik et al. 1999). The crosses employed in these latter examples require the use of artificial methods, such as embryo rescue, and do not occur naturally in the wild. The Canadian Food Inspection Agency (CFIA1994) and Warwick et al. (2000) describe more examples of *B. napus* relatives that may act as bridges.

Hybridization with Other *Brassica* Species

In addition to the species mentioned above, hybridization of *B. napus* has been attempted with other cultivated and wild *Brassica* species, including (in alphabetical order):

- *B. barrelieri* (L.) Janka (Prakash et al. 1982; Takahata and Hinata 1983; Mattsson 1988; Salisbury 2002a)
- *B. carinata* A. Braun (Harberd and McArthur 1980; Sacristán and Gerdemann 1986; Fernández-Escobar et al. 1988; Alam et al. 1992; Gupta 1997)
- *B. elongata* Ehrh. (Plümper 1995, cited in Siemens 2002)
- *B. fruticolosa* Cirillo (Heyn 1977; Prakash et al. 1982; Takahata and Hinata 1983; Plümper 1995, cited in Siemens 2002; Salisbury 2002a)

- *B. gravinae* Ten. (Prakash et al. 1982; Nanda Kumar et al. 1989)
- *B. incana* Ten. (Mithen and Herron 1991)
- *B. insularis* Moris (Mithen and Herron 1991)
- *B. macrocarpa* Guss. (Mithen and Herron 1991)
- *B. maurorum* Durieu (Prakash et al. 1982; Takahata and Hinata 1983; Bijral et al. 1995)
- *B. nigra* (L.) W.D.J. Koch (Heyn 1977; Diederichsen and Sacristán 1988; Bing et al. 1991; Kerlan et al. 1992; J. Zhu et al. 1993)
- *B. souliei* (Batt.) Batt. (Plümper 1995, cited in Siemens 2002)
- *B. spinescens* Pomel (Prakash et al. 1982)
- *B. tournefortii* Gouan (Heyn 1977; Prakash et al. 1982; Lokanadha and Sarla 1994; Liu et al. 1996; Nagpal et al. 1996; Gupta 1997; Salisbury 2002a)

In most cases, the production of F₁ hybrids requires considerable human intervention and is only successful if artificial hybridization techniques are used (usually embryo rescue). All resulting F₁ hybrids are sterile. Some species (e.g., *B. carinata*, *B. nigra*, *B. tournefortii*) are able to form hybrids with *B. napus* after hand pollination and without the use of sophisticated embryo rescue methods. However, they have significant barriers to introgression—such as pollen dehiscence prior to the flower opening, sexual incompatibility, reduced hybrid fertility, or sterility—making gene exchange with *B. napus* extremely unlikely. In theory, gene transfer from or to these species cannot be completely ruled out—e.g., via bridge species—but it is not very likely. There are, however, a number of wild *Brassica* relatives for which no studies have been conducted about their compatibility with canola (e.g., *B. assyriaca* Mouterde, *B. balearica* Pers., *B. cadmea* O.E. Schulz, etc.).

Hybridization with Species of Other Genera

Wide crosses of *B. napus* with a number of species belonging to other genera (listed in alphabetical order) have been attempted using artificial hybridization methods:

- *Arabidopsis thaliana* (L.) Heynh. (Forsberg et al. 1994, 1998; Bohman et al. 1999)

- *Barbarea vulgaris* W.T. Aiton (Fahleson et al. 1994a)

- *Coincya monensis* (L.) Greuter & Burdet (Plümper 1995, cited in Siemens 2002)

- *Diplotaxis catholica* (L.) DC., *D. cossoniana* (Reut. *ex* Boiss.) O.E. Schulz, *D. erucoides* (L.) DC., *D. harra* (Forssk.) Boiss., *D. muralis* (L.) DC., *D. siettiana* Maire, *D. siifolia* Kunze, *D. tenuifolia* (L.) DC., *D. tenuisiliqua* Delile, *D. viminea* (L.) DC., and *D. virgata* DC. (Heyn 1977; Takahata and Hinata 1983; Ringdahl et al. 1987; Klimaszewska and Keller 1988; Delourme et al. 1989; Salisbury 1989; Batra et al. 1990; Nanda Kumar and Shivanna 1993; Plümper 1995, cited in Siemens 2002; Vyas et al. 1995; Bijral and Sharma 1996b, 1998; Klewer et al. 2002; Salisbury 2002a; Klewer and Sacristán 2003)

- *Enarthrocarpus lyratus* (Forssk.) DC. (Gundimeda et al. 1992)

- *Eruca sativa* Mill. (Heyn 1977; McNaughton 1995; Bijral and Sharma 1996a; Fahleson et al. 1997)

- *Erucastrum gallicum* (Willd.) O.E. Schulz and *E. leucanthum* Coss. & Durieu (Harberd and McArthur 1980; Batra et al. 1989; Lefol et al. 1997; Warwick et al. 2003)

- *Lesquerella fendleri* (A. Gray) S. Watson (Skarzhinskaya et al. 1996)

- *Moricandia arvensis* (L.) DC. and *M. nitens* (Viv.) E.A. Durand & Barratte (Takahata and Takeda 1990; Takahata et al. 1993; O'Neill et al. 1996; Meng 1998; Rawsthorne et al. 1998; Beschorner et al. 1999; Meng et al. 1999)

- *Orychophragmus violaceus* (L.) O.E. Schulz (Hu and Hansen 1998; Hu et al. 1999)

- *Rapistrum rugosum* (L.) All. (Heyn 1977; Salisbury 2002a)

- *Sinapis alba* L. and *S. pubescens* L. (Ripley and Arnison 1990; Inomata 1991, 1994; Mathias 1991; Bijral et al. 1993; Lelivelt et al. 1993; Chèvre et al. 1994; Sridevi and Sarla 1996; J. Brown et al. 1997)

- *Sisymbrium erysimoides* Desf., *S. irio* L., *S. officinale* (L.) Scop., and *S. orientale* L. (Salisbury 2002a)

- *Thlaspi perfoliatum* L. (Fahleson et al. 1994b)

Crosses of *Brassica napus* (canola) with these species either failed or resulted in partially or completely sterile F_1 progeny. However, it should be kept in mind that, similar to the situation noted above for canola and several wild *Brassica* species, there are several wild relatives within the other genera that are known to hybridize with canola for which no hybridization studies with *B. napus* have been conducted thus far (e.g., *Diplotaxis acris* (Forssk.) Boiss., *D. assurgens* (Delile) Thell., *D. berthaultii* Braun-Blanq. & Maire, *D. brevisiliqua* (Coss.) Mart.-Laborde, *D. viminea* (L.) DC., *Sinapis flexuosa* Poir., etc.).

Outcrossing Rates and Distances

Pollen Flow

As is the case for many other plants, pollen-mediated gene flow in canola is influenced by a variety of factors, including variation in flowering time, outcrossing rates, sizes of and distance between donor and recipient populations, wind speed, and humidity. Where the pollen source and receptor plants are in close proximity (such as for adjacent plants and rows) or in small plots, spontaneous outcrossing rates vary between 3% and 21% (Rakow and Woods 1987; Förster et al. 1998; Cuthbert and McVetty 2001). Wind and insects, such as honeybees (*Apis mellifera* L.) and bumblebees (*Bombus terrestris* L.), are other important pollination vectors (Cresswell 1994; Paul et al. 1995; Picard-Nizou et al. 1995; Cresswell et al. 2002).

The majority of *Brassica* pollen travels less than 10 m (see, e.g., Thompson et al. 1999; Hüsken and Dietz-Pfeilstetter 2007), with a very steep decline over 50 m, but this dispersal can continue, albeit at low frequency, for several kilometers (Timmons et al. 1995; Rieger et al. 2002). Wind contributes to long-distance (up to 5 km) dispersal (Eisikowitch 1981), as do insects such as honeybees, bumblebees (*Bombus* spp.), hoverflies (*Simosyrphus grandicornis* Macquart), other native bees, and pollen beetles (Mesquida et al. 1988; Cresswell 1999; Ramsay et al. 1999).

While a few studies exist (Mesquida and Renard 1982; McCartney and Lacey 1991; Reboud 2003; Hayter and Cresswell 2006), more needs to be understood regarding the relative role of wind and insects in long-distance gene dispersal. Wind has been suggested as the primary vector for long-distance dispersal, transporting pollen both by upward air movements above the height of the vegetation and by local air currents created by surface features (Timmons et al. 1995; Wilkinson et al. 2003). No research has been carried out on the movement of canola pollen in atmospheric conditions such as convection currents, turbulence, and weather fronts. However, research on pollen from other plant species has demonstrated that dispersal can occur over considerable distances, such as 380 km for arboreal pollen (OGTR 2008).

Bee activity increases long-distance cross-pollination. While most foraging occurs between 1 and 5 m, bees can move pollen over several kilometers (see, e.g., Eckert 1933; Goulson and Stout 2001). Hayter and Cresswell (2006) estimate that wind and bees can contribute to the pollination of fields 1 km away to a level of up to 0.3%. Long-distance pollination by bees of up to 3.5 km was found in numerous studies at different scales (within the same field: Scheffler et al. 1995; between adjacent fields: Hall et al. 2000 and Beckie et al. 2003; at the landscape scale: Rieger et al. 2002 and Ramsay et al. 2003).

Table 5.2 presents a summary of field experiments investigating pollen dispersal distances and outcrossing rates for canola. In general, pollen flow from larger fields results in slightly higher outcrossing rates and greater pollen dispersal distances than pollen flow from small fields (Salisbury 2002b). Hence pollen flow studies in small trial plots need to take into account the fact that the capacity for actual long-range dispersal may be greater at the scale of commercial fields.

Isolation Distances

Based on experimental field studies measuring pollen dispersal and outcrossing rates, minimum crop purity thresholds and isolation distances have been established for seed production in conventional crops

Table 5.2 Some representative studies on pollen dispersal distances and outcrossing rates of canola (*Brassica napus* L.)

Pollen flow from	Pollen dispersal distance (m)	% outcrossing (max. values)	Country	Reference
B. napus	0 (mixed population)	3–12	UK	Paul et al. 1995
B. napus	0.4, 0.8, 1.2	9.5, 5.6, 3.9	Canada	Cuthbert and McVetty 2001
B. napus	5, 7.25, 9, 11.25, 13 (separated by barren zone)	5, 2.6, 1.2, 1.4, 0.9	USA	Morris et al. 1994
B. napus	5, 7.25, 9, 11.25, 13 (separated by oilseed crop as pollen trap)	5, 3.5, 1.0, 3.9, 2.5	USA	Morris et al. 1994
B. napus	0, 7.5	6.3, 0.5	USA	A. Brown et al. 1996
B. napus	0, 5, 10, 30	4, 2.5, 1.8, 0.6	France	Champolivier et al. 1999
B. napus	1.5, 4, 11.5, 21.5, 31.5	1.56, 0.68, 0.25, 0.2, 0.03	Canada	Staniland et al. 2000
B. napus	1, 16, 32	0.1, 0.001, 0.001	Hungary	Pauk et al. 1995
B. napus	25, 50	0.7, 0.4	UK	Norris and Sweet 2002
B. napus	4, 8, 20, 34, 56	2.0, 0.33, 0.16, 0.16, 0.11	UK	Sweet et al. 1999
B. napus	6, 30, 42, 50	0.05, 0.05, 0.33, 0.16	UK	Simpson et al. 1999
B. napus	10, 20, 150	0.44, 0.05, 0.22	UK	Simpson et al. 1999
B. napus	1, 3, 6, 12, 47, 70	1.5, 0.4, 0.11, 0.02, 0.0003, 0	UK	Scheffler et al. 1993
B. napus	1.5, 11.5, 26.5, 51.5, 91.5	1.0, 0.5, 0.15, 0.1, 0.05	UK	Simpson, cited by Eastham and Sweet 2002
B. napus	50, 100	0.022, 0.011	UK	Manasse and Kareiva 1991
B. napus	20, 50, 100	1.5, 0.4, 0.4	Canada	Downey 1999b
B. napus	5, 25, 40, 50, 100, 200	1.2, 0.25, 0.65, 0.10, 0.5, 0.2	UK	Norris, cited by Eastham and Sweet 2002
B. napus	47, 137, 366	2.1, 1.1, 0.6	Canada	Stringam and Downey 1982
B. napus	50, 100, 175, 200, 225	0.24, 0.21, 0.09, 0.09, 0.09	Canada	Monsanto, cited by Salisbury 2002b

Species				Reference
B. napus	50, 100, 180, 400	0.02, 0.01, 0, 0	Canada	Monsanto, cited by Salisbury 2002b
B. napus	200, 400	0.016, 0.004	UK	Scheffler et al. 1995
B. napus	0, 50, 100, 200, 400, 800	1.4, 0.2, 0.15, 0.2, 0.14, 0.07	Canada	Beckie et al. 2001, 2003
B. napus	0, 10, 50, 225, 550, 800	0.12, 0.04, 0.02, 0.02, 0.001, 0.03	UK	Ramsay et al. 2003*
B. napus	450, 800	1.0, 0.1	France	Darmency and Renard 1992
B. napus	500, 1000, 2000, 3000, 5000	0.16, 0.11, 0.2, 0.15, 0.0	Australia	Rieger et al. 2002
B. napus	> 500	16.5–92.4	France	Baranger et al. 1995
B. napus†	0, 3, 6, 9, 12, 15, 20, 30	56, 60, 6, 1, 0.3, 0.2, 0.1, 0	France	Lavigne et al. 1998
B. napus†	> 32	0.02–0.05	Scotland	Wilkinson et al. 1995
B. napus†	6, 20, 42, 54, 150	21, 0.16, 0.33, 0.11, 0.22	UK	Simpson et al. 1999
B. napus†	0, 100, 360	6.3, 0.5, 3.7	Scotland	Timmons et al. 1996
B. napus†	100, 200, 400	0.13, 0.03, 0.06	UK	Simpson, cited by Eastham and Sweet 2002
B. napus†	0, 10, 50, 225, 550, 800	0.58, 0.31, 0.33, 0.21, 0.1, 0.02	UK	Ramsay et al. 2003*
B. napus†	1500, 2500	1.2, 0.8	Scotland	Timmons et al. 1995
B. napus†	500, 1000, 2000, 3000, 4000	15–70, 25–58, 8–35, 5, 5	UK	Thompson et al. 1999

*Estimated.

†Using male-sterile varieties or emasculated "bait" plants (petals and stamens removed).

Table 5.3 Isolation distances for canola (*Brassica napus* L.)

Threshold*	Minimum isolation distance (m)	Country	Reference	Comment
0.005% / n.a.	200	Canada	CSGA 2009	foundation (basic) seed requirements (non-hybrids)
0.01% / n.a.	100			certified seed requirements (non-hybrids)
0.005% / n.a.	800			foundation (basic) seed requirements (hybrids)
0.01% / n.a.	800			certified seed requirements (hybrids)
1 plant in 30m²/ 0.1%	200	OECD	OECD 2008	foundation (basic) seed requirements (non-hybrids)
1 plant in 10m²/ 0.3%	100			certified seed requirements (non-hybrids)
0.2 / 0.1%	500			foundation (basic) seed requirements (hybrids)
0.5 / 0.3%	300			certified seed requirements (hybrids)
1 plant in 30m²/ 0.1%	200	EU	EC 2002	foundation (basic) seed requirements
1 plant in 10m²/ 0.3%	100			certified seed requirements
n.a. / 0.1%	400	UK	Ingram 2000	foundation (basic) seed requirements
n.a. / 0.3%	200			certified seed requirements
n.a. / 0.1%	100	UK	Ingram 2000	recommendations for GM canola fields > 2 ha
n.a. / 0.5%	10			recommendations for GM canola fields > 2 ha
n.a. / 1%	1.5			recommendations for GM canola fields > 2 ha
n.a. / 0.9%	50	UK	SCIMAC 1999	between GM and conventional non-GM canola
n.a. / 0.9%	200			between GM and organic non-GM canola and for certified seed production
0.05 / 0.05%	200	USA	USDA 2008	foundation (basic) seed requirements (inbred)
0.2 / 0.25%	100			certified seed requirements (inbred)
0.05 / 0.05%	400			foundation (basic) seed requirements (open-pollinated)
0.2 / 0.25%	100			certified seed requirements (open-pollinated)

*Maximum percentage of off-type plants permitted in the field / seed impurities of other varieties or off-types permitted. n.a. = not available.

(table 5.3). These can also serve as guidelines for separation distances from GM crops to minimize gene escape through pollen flow. Timmons et al. (1996) estimate that isolation distances of 4 km would be needed to prevent unwanted outcrossing from commercial-scale canola plantings.

Outcrossing rates in canola decrease exponentially with distance (Weekes et al. 2005). A comparison of various pollen dispersal studies that measured pollen flow distances and outcrossing rates for cultivated canola shows a decline in outcrossing rates to less than 0.5% beyond a distance of 50 m, less than 0.15% beyond 100 m for normal fertile varieties, and less than 0.3% for male-sterile ones (fig. 5.1). According to these data, the separation distances established for seed production by regulatory authorities (table 5.3) generally seem to be insufficient to minimize seed contamination through pollen flow. To attain seed purity levels that comply with stipulated thresholds of, for example, 0.1% impurities or less, separation distances of at least 800 m would be required.

State of Development of GM Technology

Field trials of GM canola have been reported from the following countries: Australia, Belgium, Canada, China, Denmark, Finland, France, Germany, Italy, Japan, Lithuania, Mexico, the Netherlands, New Zealand, Spain, Sweden, the United Kingdom, and the United States.

The GM traits incorporated into improved canola varieties encompass herbicide tolerance, insect and fungus resistance, increased oil content and modified oil composition in the seed, nutritional enhancement, stress tolerance, increased yield, male sterility, fertility restoration, photoperiod insensitivity, extended flower life, reduced pod shattering, reduced glucosinolate content, dwarf types, and the production of a range of proteins, enzymes, and other gene products (including pharmaceuticals).

Agronomic Management Recommendations

Although the survival and maintenance of most interspecific and intergenomic hybrids is quite unlikely in nature, three primary means of gene introgression to other cultivated or wild species need to be considered:

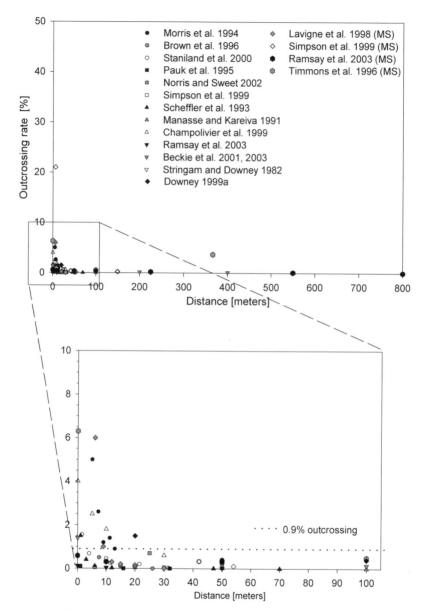

Fig. 5.1 Pollen dispersal distances and outcrossing rates of cultivated canola
(*Brassica napus* L.)
MS = male sterile

1 through pollen flow from canola fields

2 through volunteers in fields previously planted with canola and through ferals outside of cultivation

3 through canola seeds dispersed by humans during multiple stages— planting and cultivation, harvesting, post-harvest manipulation, storage, and transport of the crop

As an example, in Canada, Downey and Beckie (2002) found unexpected (non-GM) herbicide resistance traits in pedigreed canola seed lots, either through inadvertent mechanical mixing of certified seed lots during harvest or handling, or through contamination occurring in earlier generations of pedigreed seed production (i.e., Breeder or Foundation seed). Of 27 canola seed lots, 26 had detectable levels of an herbicide-resistant trait admixture. Of these, 14 had levels above 0.25%, and therefore failed to meet the 99.75% cultivar purity guideline for certified seed (Friesen et al. 2003). In France, the quality of canola seed was monitored during the harvest of a conventional variety grown in fields three to eight years after cultivation with a GM variety (Messéan et al. 2007). The quantity of GM seeds in the mixture largely exceeded the European threshold in 6 out of 18 instances, and in one case it was as high as 18%. Large differences in the GM admixture were observed, depending on the GM varieties previously planted. These were attributed to variations in seed shedding, seed dormancy, and seed survival in the soil among GM cultivars.

In Japan, where GM canola is not grown commercially, several GM canola plants have been detected at major ports and along roadsides, some of them with resistance to multiple herbicides (Saji et al. 2005; Aono et al. 2006). The feral plants almost certainly arose from imported GM seeds that were spilled during transport to oilseed-processing facilities. However, no GM traits were detected in seeds collected from B. napus plants growing along riverbanks or in seeds from closely related species (B. rapa and B. juncea). These reports are the first published examples of feral GM populations occurring in a country where the GM crop has not yet been planted commercially.

Various models have been developed to simulate seed bank dynamics (Colbach et al. 2001a, 2008; Pivard 2008) and pollen dispersal at the landscape scale (see, e.g., Colbach et al. 2001b, 2005; Cresswell et al. 2002; Walklate et al. 2004; Cresswell 2005; Damgaard and Kjellsson 2005; Cresswell and Hoyle 2006; Devaux et al. 2006; Klein et al. 2006; Hoyle and Cresswell 2009). Depending on the assumptions and conditions used for developing these models (e.g., winter or spring canola types), they can be useful in estimating the effect of management strategies in cases where there is temporal and/or spatial coexistence of GM and non-GM canola varieties.

The following management recommendations should reduce the likelihood of gene introgression as much as possible:

- Take land-use history into account to avoid contamination with crop-derived volunteer populations left by an earlier cropping cycle. This applies both to the field to be planted and to neighboring fields.

- Consider establishing border rows with male-sterile bait plants (which are only fertilized by pollen from external sources) around the area planted with canola so that these bait plants capture escaped pollen. The 10-meter-wide continuous borders recommended by some government safety regulations have been shown to effectively contain the majority of pollen-mediated gene flow.

- If flowering overlap is likely with any of the plant species identified above as being susceptible to gene introgression, no plants should be in the area at distances shorter than those specified by legal regulations.

- The persistence of volunteer plants within and feral plants outside of fields is an important component of risk assessment and management, since ferals can contribute more to crop impurities (1%) than gene flow by pollen movement (Squire et al. 2003). Special measures should be taken when transporting seeds to avoid them being blown out of harvesting vehicles. Monitor roads for feral canola plants and, if present, control them by mechanical or chemical means.

Crop Production Area

Domesticated *B. napus* is only known from cultivation. It has traditionally been grown in Europe, but it is now also widely cultivated (and naturalized) in Canada, N and S America, Asia, and Australia. The total area of *B. napus* production is about 30.2 million ha (FAO 2009), about 5.9 million ha of which are planted with GM canola (James 2008). Out of all the GM canola in commercial production, 5.3 million ha are in Canada and 0.5 million are in the United States.

Research Gaps and Conclusions

Extensive information is available about the likelihood of gene flow and introgression between canola and several of its closely related wild and weedy relatives. However, more detailed studies about the level of cross-compatibility with canola are required for some of the less closely related species, including *Brassica nigra*, *B. tournefortii*, *Diplotaxis tenuifolia*, *Sisymbrium officinale*, and *S. orientale*, to mention only a few. Within the genera already known to hybridize with *B. napus* (e.g., *Diplotaxis, Eruca, Sinapis, Sisymbrium*), there are several species for which virtually nothing is known about their interrelationship with canola. Further studies should focus on these lesser-known species.

Assuming physical proximity (less than 100 m) and flowering overlap, and being conditional on the nature of the respective trait(s), the likelihood of gene introgression from *Brassica napus* to other species, as shown in map 5.1, is as follows:

- *very high* for cultivated and wild *B. rapa* (field mustard)
- *high* for *B. juncea* (Indian mustard)
- *moderate* for cultivated and wild *B. oleracea* (cabbage)
- *low* for *Hirschfeldia incana* (Buchan weed) and *Raphanus raphanistrum* (wild radish)

REFERENCES

....

Abrol DP (2007) Honeybees and rapeseed: A pollinator-plant interaction. *Adv. Bot. Res.* 45: 337–367.

Al Mouemar A and Darmency H (2004) Lack of stable inheritance of introgressed transgene from oilseed rape in wild radish. *Environ. Biosafety Res.* 3: 209–214.

Alam M, Ahmad H, Quazi MH, and Khawaja HIT (1992) Cross-compatibility studies within the genus *Brassica*: 1. Amphidiploid combinations. *Sci. Khyber* 5: 89–92.

Allainguillaume J, Alexander M, Bullock JM, Saunders M, Allender CJ, King G, Ford CS, and Wilkinson JM (2006) Fitness of hybrids between rapeseed (*Brassica napus*) and wild *Brassica rapa* in natural habitats. *Mol. Ecol.* 15: 1175–1184.

Ammitzbøll H and Jørgensen RB (2006) Hybridization between oilseed rape (*Brassica napus*) and different populations and species of *Raphanus*. *Environ. Biosafety Res.* 5: 3–13.

Ammitzbøll H, Nørgaard Mikkelsen T, and Jørgensen RB (2005) Transgene expression and fitness of hybrids between GM oilseed rape and *Brassica rapa*. *Environ. Biosafety Res.* 4: 3–12.

Aono M, Wakiyama S, Nagatsu M, Nakajima N, Tamaoki M, Kubo A, and Saji H (2006) Detection of feral transgenic oilseed rape with multiple-herbicide resistance in Japan. *Environ. Biosafety Res.* 5: 77–87.

Ayotte R, Harney PM, and Souza Machado V (1987) The transfer of triazine resistance from *Brassica napus* L. to *B. oleracea* L.: I. Production of F1 hybrids through embryo rescue. *Euphytica* 36: 615–624.

Bajaj YPS, Mahajan SK, and Labana KS (1986) Interspecific hybridisation of *Brassica napus* and *B. juncea* through ovary, ovule, and embryo culture. *Euphytica* 35: 103–109.

Baranger A, Chèvre AM, Eber F, and Renard M (1995) Effect of oilseed rape genotype on the spontaneous hybridization rate with a weedy species: An assessment of transgene dispersal. *Theor. Appl. Genet.* 91: 956–963.

Batra V, Shivanna KR, and Prakash S (1989) Hybrids of wild species *Erucastrum gallicum* and crop *Brassicas*. In: Iyama S and Takeda G (eds.) *Breeding Research: The Key to the Survival of the Earth.* Vol. 1 in *Proceedings of the 6th International Congress SABRAO, Tsukuba, Japan, August 21–25, 1989* (Society for the Advancement of Breeding Researches in Asia and Oceania [SABRAO]: n.p.), pp. 443–446.

Batra V, Prakash S, and Shivanna KR (1990) Intergeneric hybridization between *Diplotaxis siifolia*, a wild species, and crop *Brassicas*. *Theor. Appl. Genet.* 80: 537–541.

Becker HC, Karle R, and Han SS (1992) Environmental variation for outcrossing rates in rapeseed (*Brassica napus*). *Theor. Appl. Genet.* 84: 303–306.

Beckie HJ, Hall LM, and Warwick SI (2001) Impact of herbicide-resistant crops as weeds in Canada. In: British Crop Protection Council (ed.) *Weeds 2001: The BCPC Conference Proceedings of an International Conference, Brighton, UK, November 12–15, 2001* (British Crop Protection Council [BCPC]: Farnham, UK), pp. 135–142.

Beckie HJ, Warwick SI, Nair H, and Séguin-Swartz G (2003) Gene flow in commercial fields of herbicide-resistant canola (*Brassica napus*). *Ecol. Appl.* 13: 1276–1294.

Beschorner M, Warwick SI, and Lydiate DJ (1999) Interspecies transfer of improved water use efficiency from *Moricandia* into *Brassica*. In: Wratten N and Salisbury PA (eds.) *New Horizons for an Old Crop: Proceedings of the 10th International Rapeseed Congress, Canberra, Australia, 1999* (Regional Institute: Gosford, NSW, Australia), contribution no. 83. www.regional.org.au/au/gcirc/2/83.htm.

Bielikova L and Rakousky S (2001) *Survey on Oilseed Rape Cultivation and Weed Relatives in the Czech Republic.* European Science Foundation Meeting of a Working Group on Interspecific Gene Flow from Oilseed Rape to Weedy Species, Rennes, France, June 2001. 9 p.

Bijral JS and Sharma TR (1996a) Cytogenetics of intergeneric hybrids between *Brassica napus* L. and *Eruca sativa* Lam. *Cruciferae Newsl. Eucarpia* 18: 12–13.

—— (1996b) Intergeneric hybridization between *Brassica napus* and *Diplotaxis muralis. Cruciferae Newsl. Eucarpia* 18: 10–11.

—— (1998) Production and cytology of intergeneric hybridization between *Brassica napus* and *Diplotaxis catholica. Cruciferae Newsl. Eucarpia* 20: 15.

Bijral JS, Sharma TR, and Kanwal KS (1993) Morphocytogenetics of *Brassica napus* L. × *Sinapis alba* L. sexual hybrids. *Indian J. Genet. Plant Breed.* 53: 442–444.

Bijral JS, Sharma TR, Gupta BB, and Singh K (1995) Interspecific hybrids of *Brassica maurorum* with *Brassica* crops, and their cytology. *Cruciferae Newsl. Eucarpia* 17: 18–19.

Bing DJ, Downey RK, and Rakow FW (1991) Potential of gene transfer among oilseed *Brassica* and their weedy relatives. In: McGregor DI (ed.) *Proceedings of the GCIRC 8th International Rapeseed Congress, Saskatoon, Saskatchewan, Canada, July 9–11, 1991* (Organizing Committee of the 8th International Rapeseed Congress under the auspices of the Groupe Consultatif International de Recherche sur la Colza [GCIRC] and the Canola Council of Canada: n.p.), pp. 1022–1027.

—— (1995) An evaluation of the potential of intergeneric gene transfer between *Brassica napus* and *Sinapis arvensis. Plant Breed.* 114: 481–484.

—— (1996) Hybridizations among *Brassica napus*, *B. rapa*, and *B. juncea* and their two weedy relatives *B. nigra* and *Sinapis arvensis* under open pollination conditions in the field. *Plant Breed.* 115: 470–473.

Bohman S, Forsberg J, Glimelius K, and Dixelius C (1999) Inheritance of *Arabidopsis* DNA in offspring from *Brassica napus* and *Arabidopsis thaliana* hybrids. *Theor. Appl. Genet.* 98: 99–106.

Brown AP, Brown J, Thill DC, and Brammer TA (1996) Gene transfer between canola (*Brassica napus*) and related weed species. *Cruciferae Newsl. Eucarpia* 18: 36–37.

Brown J and Brown AP (1996) Gene transfer between canola (*Brassica napus* L. and *B. campestris* L.) and related weed species. *Ann. Appl. Biol.* 129: 513–522.

Brown J, Brown AP, Davis JB, and Erickson D (1997) Intergeneric hybridization between *Sinapis alba* and *Brassica napus*. *Euphytica* 93: 163–168.

CFIA [Canadian Food Inspection Agency] (1994) Biology Document BIO1994–09: The Biology of *Brassica napus* L. (Canola/Rapeseed); A Companion Document to Directive 94–08 (Dir94–08), Assessment Criteria for Determining Environmental Safety of Plants with Novel Traits (Plant Biosafety Office, Plant Products Directorate, Canadian Food Inspection Agency: Ottawa, ON, Canada). 12 p. www.inspection.gc.ca/english/plaveg/bio/dir/biodoce.shtml.

Chadoeuf R, Darmency H, Maillet J, and Renard M (1998) Survival of buried seeds of interspecific hybrids between oilseed rape, hoary mustard, and wild radish. *Field Crop Res.* 58: 197–204.

Champolivier J, Gasquez J, Mességuer A, and Richard-Molard M (1999) Management of transgenic crops within the cropping system. In: Lutman PJW (ed.) *Gene Flow and Agriculture: Relevance for Transgenic Crops; Proceedings of a Symposium Held at the University of Keele, Staffordshire, April 12–14, 1999.* BCPC Symposium Proceedings 72 (British Crop Protection Council [BCPC]: Farnham, UK), pp. 233–240.

Chèvre AM, Eber F, Margale E, Kerlan MC, Primard C, Vedel F, Delseny M, and Pelletier G (1994) Comparison of somatic and sexual *Brassica napus-Sinapis alba* hybrids and their progeny by cytogenetical studies and molecular characterization. *Genome* 37: 367–374.

Chèvre AM, Eber F, Kerlan MC, Barret P, Festoc G, Vallee P, and Renard M (1996) Interspecific gene flow as a component of risk assessment for transgenic *Brassicas*. *Acta Hort.* 407: 169–179.

Chèvre AM, Eber F, Baranger A, and Renard M (1997) Gene flow from transgenic crops. *Nature* 389: 924.

Chèvre AM, Eber F, Baranger A, Hureau G, Barret P, Picault H, and Renard M (1998) Characterisation of backcross generations obtained under field conditions from oilseed rape-wild radish F1 interspecific hybrids: An assessment of transgene dispersal. *Theor. Appl. Genet.* 97: 90–98.

Chèvre AM, Eber F, Renard M, and Darmency H (1999) Gene flow from oilseed rape to weeds. In: Lutman PJW (ed.) *Gene Flow and Agriculture: Relevance for Transgenic Crops; Proceedings of a Symposium Held at the University of Keele, Staffordshire, April 12–14, 1999.* BCPC Symposium Proceedings 72 (British Crop Protection Council [BCPC]: Farnham, UK), pp. 125–130.

Chèvre AM, Eber F, Darmency H, Fleury A, Picault H, Letanneur JC, and Renard M (2000) Assessment of interspecific hybridization between transgenic oilseed rape and wild radish under normal agronomic conditions. *Theor. Appl. Genet.* 100: 1233–1239.

Chèvre AM, Eber F, Jenczewski E, Darmency H, and Renard M (2003) Gene flow from oilseed rape to weedy species. *Acta Agr. Scand. B.-S. P* 53 (Suppl.): 22–25.

Chiang MS, Chiang BY, and Grant WF (1977) Transfer of resistance to race 2 of *Plasmodiophora brassicae* from *Brassica napus* to cabbage (*B. oleracea* var. *capitata*): I. Interspecific hybridization between *B. napus* and *B. oleracea* var. *capitata. Euphytica* 26: 319–336.

Choudhary BR and Joshi P (1999) Interspecific Hybridization in *Brassica.* In: Wratten N and Salisbury PA (eds.) *New Horizons for an Old Crop: Proceedings of the 10th International Rapeseed Congress, Canberra, Australia, 1999* (Regional Institute: Gosford, NSW, Australia), contribution no. 516. www.regional.org.au/au/gcirc/4/516.htm.

Claessen D, Gilligan CA, Lutman PJW, and van den Bosch F (2005) Which traits promote persistence of feral GM crops? Part 1: Implications of environmental stochasticity. *Oikos* 110: 20–29.

Colbach N, Clermont Dauphin C, and Meynard JM (2001a) GENESYS: A model of the influence of cropping system on gene escape from herbicide tolerant rapeseed crops to rape volunteers: I. Temporal evolution of a population of rapeseed volunteers in a field. *Agric. Ecosys. Environ.* 83: 235–253.

——— (2001b) GENESYS: A model on the influence of cropping system on gene escape from herbicide tolerant rapeseed crops to volunteers: II. Genetic exchanges among volunteer and cropped populations in a small region. *Agric. Ecosys. Environ.* 82: 255–270.

Colbach N, Fargue A, Sausse C, and Angevin F (2005) Evaluation and use of a spatio-temporal model of cropping system effects on gene flow: Example of the GENESYS model applied to three co-existing herbicide tolerance transgenes. *Eur. J. Agron.* 22: 417–440.

Colbach N, Dürr C, Gruber S, and Pekrun C (2008) Modelling the seed bank evolution and emergence of oilseed rape volunteers for managing co-existence of GM and non-GM varieties. *Europ. J. Agron.* 28: 19–32.

Colton RT and Sykes JD (1992) *Canola.* 4th ed. Agfacts P5.2.1. (New South Wales Agriculture: Orange, NSW, Australia). 51 p.

Crawley MJ and Brown SL (1995) Seed limitation and the dynamics of feral oilseed rape on the M25 motorway. *Proc. Roy. Soc. London, Ser. B, Biol. Sci.* 259: 49–54.

——— (2004) Spatially structured population dynamics in feral oilseed rape. *Proc. Roy. Soc. London, Ser. B, Biol. Sci.* 271: 1909–1916.

Crawley MJ, Hails RS, Rees M, Kohn D, and Buxton J (1993) Ecology of transgenic oilseed rape in natural habitats. *Nature* 363: 620–623.

Crawley MJ, Brown SL, Hails RS, Kohn D, and Rees M (2001) Transgenic crops in natural habitats. *Nature* 409: 682–683.

Cresswell JE (1994) A method for quantifying the gene flow that results from a single bumblebee visit using transgenic oilseed rape, *Brassica napus* L. cv. Westar. *Transgen. Res.* 3: 134–137.

────── (1999) The influence of nectar and pollen availability on pollen transfer by individual flowers of oilseed rape (*Brassica napus*) when pollinated by bumblebees (*Bombus lapidarius*). *J. Ecol.* 87: 670–677.

────── (2005) Accurate theoretical prediction of pollinator-mediated gene dispersal. *Ecology* 86: 574–578.

Cresswell JE and Hoyle M (2006) A mathematical method for estimating patterns of flower-to-flower gene dispersal from a simple field experiment. *Funct. Ecol.* 20: 245–251.

Cresswell JE, Osborne JL, and Bell SA (2002) A model of pollinator-mediated gene flow between plant populations with numerical solutions for bumblebees pollinating oilseed rape. *Oikos* 98: 375–384.

CSGA [Canadian Seed Growers Association] (2009) Canadian Regulations and Procedures for Pedigreed Seed Crop Production. Circular 6–2005/Rev.01.4–2009 (CSGA: Ottawa, Canada). www.seedgrowers.ca/cropcertification/circular.asp.

Cuthbert JL and McVetty PBE (2001) Plot-to-plot, row-to-row, and plant-to plant outcrossing studies in oilseed rape. *Can. J. Plant Sci.* 81: 657–664.

Damgaard C and Kjellsson G (2005) Gene flow of oilseed rape (*Brassica napus*) according to isolation distance and buffer zone. *Agric. Ecosys. Environ.* 108: 291–301.

Daniels R, Boffey C, Mogg R, Bond J, and Clark R (2005) *The Potential for Dispersal of Herbicide-Tolerant Genes from Genetically-Modified, Herbicide-Tolerant Oilseed Rape Crops to Wild Relatives.* Final Report to DEFRA [Department of Environment, Food, and Rural Affairs] (Centre for Ecology and Hydrology [CEH] Dorset, Winfrith Technical Centre: Dorchester, UK). 23 p.

Darmency H and Fleury A (2000) Mating system in *Hirschfeldia incana* and hybridization to oilseed rape. *Weed Res.* 40: 231–238.

Darmency H and Renard M (1992) Efficiency of safety procedures in experiments with transgenic oilseed rape. In: Casper R and Landsmann J (eds.) *The Biosafety Results of Field Tests of Genetically Modified Plants and Microorganisms: Proceedings of the 2nd International Symposium on the Biosafety Results of Field Tests of Genetically Modified Plants and Microorganisms, Goslar, Germany, May 11–14, 1992* (Biologische Bundesanstalt für Land- und Forstwirtschaft: Braunschweig, Germany). 54 p. www.jki.bund.de/cln_045/nn_807 152/DE/Home/biolsich/gentechnik/taggoslar/Goslar02_7.html.

Darmency H, Fleury A, and Lefol E (1995) Effect of transgenic release on weed biodiversity: Oilseed rape and wild radish. *Brighton Crop Protection Conference: Weeds 1995; Proceedings of an International Conference Held at the Brighton Centre and the Brighton Metropole Hotel, Brighton, England, November 20–23, 1995*, vol. 2 (British Crop Protection Council [BCPC]: Farnham, UK), pp. 433–438.

Darmency H, Lefol E, and Fleury A (1998) Spontaneous hybridizations between oilseed rape and wild radish. *Mol. Ecol.* 7: 1467–1473.

Delourme R, Eber F, and Chèvre AM (1989) Intergeneric hybridization of *Diplotaxis erucoides* with *Brassica napus*: I. Cytogenetic analysis of F1 and BC1 progeny. *Euphytica* 41: 123–128.

Devaux C, Lavigne C, Austerlitz F, and Klein EK (2006) Modelling and estimating pollen movement in oilseed rape (*Brassica napus*) at the landscape scale using genetic markers. *Mol. Ecol.* 16: 487–499.

Diederichsen E and Sacristán MD (1988) Interspecific hybridizations in the genus *Brassica* followed by in-ovule embryo culture. *Cruciferae Newsl. Eucarpia* 13: 20–21.

Downey RK (1999a) Gene flow and rape—the Canadian experience. In: Lutman PJW (ed.) *Gene Flow and Agriculture: Relevance for Transgenic Crops; Proceedings of a Symposium Held at the University of Keele, Staffordshire, April 12–14,1999*. BCPC Symposium Proceedings 72 (British Crop Protection Council [BCPC]: Farnham, UK), pp. 109–116.

———— (1999b) Risk assessment of outcrossing of transgenic *Brassica*, with focus on *B. rapa* and *B. napus*. In: Wratten N and Salisbury PA (eds.) *New Horizons for an Old Crop: Proceedings of the 10th International Rapeseed Congress, Canberra, Australia, 1999* (Regional Institute: Gosford, NSW, Australia), contribution no. 61. www.regional.org.au/au/gcirc/4/61.htm.

Downey RK and Beckie H (2002) *Isolation Effectiveness in Canola Pedigree Seed Production*. Internal Research Report, Agriculture and Agri-Food Canada (Saskatoon Research Centre: Saskatoon, SK, Canada).

Downey RK, Klassen AJ, and Stringam GP (1980) Rapeseed and mustard. In: Fehr WR and Hadley HH (eds.) *Hybridization of Crop Plants* (American Society of Agronomy: Madison, WI, USA), pp. 495–509.

Eastham K and Sweet J (2002) *Genetically Modified Organisms (GMOs): The Significance of Gene Flow Through Pollen Transfer*. Environmental Issue Report No. 28 (European Environment Agency [EEA]: Copenhagen, Denmark). 75 p. http://reports.eea.europa.eu/environmental_issue_report_2002_28/en/GMOs%20for%20www.pdf.

Eber F, Chèvre AM, Baranger A, Vallee P, Tanguy X, and Renard M (1994) Spontaneous hybridization between a male-sterile oilseed rape and two weeds. *Theor. Appl. Genet.* 88: 362–368.

EC [European Commission] (2002) Council Directive 2002/57/EC of 13 June 2002 on the Marketing of Seed of Oil and Fibre Plants. OJ No. L 193, 20.7.2002. www.legaltext.ee/text/en/U70003.htm.

Eckert JE (1933) The flight range of the honeybee. *J. Apicult. Res.* 47: 257–285.

Eisikowitch D (1981) Some aspects of pollination of oilseed rape (*Brassica napus* L.). *J. Agric. Sci.* 96: 321–326.

Ellerström S (1978) Species crosses and sterility in *Brassica* and *Raphanus. Cruciferae Newsl. Eucarpia* 3: 16–17.

Fahleson J, Eriksson I, and Glimelius K (1994a) Intertribal somatic hybrids between *Brassica napus* and *Barbarea vulgaris*—production of *in vitro* plantlets. *Plant Cell Rep.* 13: 411–416.

Fahleson J, Eriksson I, Landgren M, Stymmne S, and Glimelius K (1994b) Intertribal somatic hybrids between *Brassica napus* and *Thlaspi perfoliatum* with high content of the *T. perfoliatum*-specific nervonic acid. *Theor. Appl. Genet.* 87: 795–804.

Fahleson J, Lagercrantz U, Mouras A, and Glimelius K (1997) Characterization of somatic hybrids between *Brassica napus* and *Eruca sativa* using species-specific repetitive sequences and genomic in situ hybridization. *Plant Sci.* 123: 133–142.

FAO [Food and Agriculture Organization of the United Nations] (2009) FAOSTAT data. http://faostat.fao.org/site/567/default.aspx.

Fernández-Escobar J, Domínguez J, Martín A, and Fernández-Martínez JM (1988) Genetics of erucic acid content in interspecific hybrids of Ethiopian mustard (*Brassica carinata* Braun) and rapeseed (*B. napus* L.). *Plant Breed.* 100: 310–315.

FitzJohn RG, Armstrong TT, Newstrom-Lloyd LE, Wilton AD, and Cochrane M (2007) Hybridisation within *Brassica* and allied genera: Evaluation of potential for transgene escape. *Euphytica* 158: 209–230.

Ford CS, Allainguillaume JI, Grilli-Chantler P, Cuccato G, Allender CJ, and Wilkinson MJ (2006) Spontaneous gene flow from rapeseed (*Brassica napus*) to wild *Brassica oleracea. Proc. Roy. Soc. London, Ser. B, Biol. Sci.* 273: 3111–3115.

Forsberg J, Landgren M, and Glimelius K (1994) Fertile somatic hybrids between *Brassica napus* and *Arabidopsis thaliana. Plant Sci.* 95: 213–223.

Forsberg J, Dixelius C, Lagercrantz U, and Glimelius K (1998) UV dose-dependent DNA elimination in asymmetric somatic hybrids between *Brassica napus* and *Arabidopsis thaliana. Plant Sci.* 131: 65–76.

Förster K, Schuster C, Belter A, and Diepenbrock W (1998) Agrarökologische Auswirkungen des Anbaus von transgenem herbizidtoleranten Raps. *Bundesgesundhbl.* 41: 547–552.

Free JB (1993) *Insect Pollination of Crops.* 2nd ed. (Academic Press: San Diego, CA, USA). 684 p.

Frello S, Hansen KR, Jensen J, and Jørgensen JF (1995) Inheritance of rapeseed

(*Brassica napus*)-specific RAPD markers and a transgene in the cross *B. juncea* × (*B. juncea* × *B. napus*). *Theor. Appl. Genet.* 91: 236–241.

Friesen LF, Nelson AG, and Acker RC van (2003) Evidence of contamination of pedigreed canola (*Brassica napus*) seed lots in western Canada with genetically engineered herbicide-resistance traits. *Agron. J.* 95: 1342–1347.

Garnier A, Pivarda S, and Lecomte J (2008) Measuring and modelling anthropogenic secondary seed dispersal along road verges for feral oilseed rape. *Basic Appl. Ecol.* 9: 533–541.

GhoshDastidar N and Varma NS (1999) A study on intercrossing between transgenic *B. juncea* and other related species. In: Wratten N and Salisbury PA (eds.) *New Horizons for an Old Crop: Proceedings of the 10th International Rapeseed Congress, Canberra, Australia, 1999* (Regional Institute: Gosford, NSW, Australia), contribution no. 244. www.regional.org.au/au/gcirc/4/244.htm.

Goulson D and Stout JC (2001) Homing ability of the bumblebee *Bombus terrestris* (Hymenoptera: Apidea). *Apidol.* 32: 105–111.

Gruber S and Claupein W (2007) Fecundity of volunteer oilseed rape and estimation of potential gene dispersal by a practice-related model. *Agric. Ecosys. Environ.* 119: 401–408.

Guéritaine G and Darmency H (2001) Polymorphism for interspecific hybridisation within a population of wild radish (*Raphanus raphanistrum*) pollinated by oilseed rape (*Brassica napus*). *Sex. Plant Reprod.* 14: 169–172.

Guéritaine G, Sester M, Eber F, Chèvre AM, and Darmency H (2002) Fitness of backcross six of hybrids between transgenic oilseed rape (*Brassica napus*) and wild radish (*Raphanus raphanistrum*). *Mol. Ecol.* 11: 1419–1426.

Guéritaine G, Bazot S, and Darmency H (2003) Emergence and growth of hybrids between *Brassica napus* and *Raphanus raphanistrum*. *New Phytol.* 158: 561–567.

Gulden RH, Shirtliffe SJ, and Thomas AG (2003a) Harvest losses of canola (*Brassica napus*) cause large seedbank inputs. *Weed Sci.* 51: 83–86.

Gulden RH, Shirtliffe SJ, and Gordon TA (2003b) Secondary seed dormancy prolongs persistence of volunteer canola in western Canada. *Weed Sci.* 51: 904–913.

Gundimeda HR, Prakash S, and Shivanna KR (1992) Intergeneric hybrids between *Enarthrocarpus lyratus*, a wild species, and crop *Brassicas*. *Theor. Appl. Genet.* 83: 655–662.

Gupta SK (1997) Production of interspecific and intergeneric hybrids in *Brassica* and *Raphanus*. *Cruciferae Newsl. Eucarpia* 19: 21–22.

Hails RS, Rees M, Kohn DD, and Crawley MJ (1997) Burial and seed survival in *Brassica napus* subsp. *oleifera* and *Sinapis arvensis* including a comparison of transgenic and non-transgenic lines of the crop. *Proc. Roy. Soc. London, Ser. B, Biol. Sci.* 264: 1–7.

Halfhill MD, Richards HA, Mabon SA, and Stewart CN Jr (2001) Expression of

GFP and *Bt* transgenes in *Brassica napus* and hybridization with *Brassica rapa*. *Theor. Appl. Genet.* 103: 659–667.

Halfhill MD, Millwood RJ, Raymer PL, and Stewart CN Jr (2002) *Bt*-transgenic oilseed rape hybridization with its weedy relative, *Brassica rapa*. *Environ. Biosafety Res.* 1: 19–28.

Halfhill MD, Millwood RJ, Weissinger AK, Warwick SI, and Stewart CN Jr (2003) Additive transgene expression and genetic introgression in multiple GFP transgenic crop × weed hybrid generations. *Theor. Appl. Genet.* 107: 1533–1540.

Halfhill MD, Sutherland JP, Moon HS, Poppy GM, Warwick SI, Weissinger AK, Rufty TW, Raymer PL, and Stewart CN Jr (2005) Growth, productivity, and competitiveness of introgressed weedy *Brassica rapa* hybrids selected for the presence of *Bt cry1*Ac and GFP transgenes. *Mol. Ecol.* 14: 3177–3189.

Hall L, Topinka K, Huffman J, Davis L, and Good A (2000) Pollen flow between herbicide-resistant *Brassica napus* is the cause of multiple-resistant *B. napus* volunteers. *Weed Sci.* 48: 688–694.

Hansen LB, Siegismund HR, and Jørgensen RB (2001) Introgression between oilseed rape (*Brassica napus* L.) and its weedy relative *B. rapa* L. in a natural population. *Genet. Resour. Crop Evol.* 48: 621–627.

——— (2003) Progressive introgression between *Brassica napus* (oilseed rape) and *B. rapa*. *Heredity* 91: 276–283.

Harberd DJ (1975) *Brassica*. In: Stace CA (ed.) *Hybridization and the Flora of the British Isles* (Academic Press: London, UK), pp. 137–139.

Harberd DJ and McArthur ED (1980) Meiotic analysis of some species and genus hybrids in the Brassiceae. In: Tsunoda S, Hinata K, and Gómez-Campo C (eds.) Brassica *Crops and Wild Allies* (Japan Scientific Societies Press: Tokyo, Japan), pp. 65–87.

Hauser TP, Jørgensen RB, and Østergård H (1997) Preferential exclusion of hybrids in mixed pollinations between oilseed rape (*Brassica napus*) and weedy *B. campestris* (Brassicaceae). *Am. J. Bot.* 84: 756–762.

——— (1998a). Fitness of backcross and F2 hybrids between weedy *Brassica rapa* and oilseed rape (*B. napus*). *Heredity* 81: 436–443.

Hauser TP, Shaw RG, and Østergård H (1998b). Fitness of F1 hybrids between weedy *Brassica rapa* and oilseed rape (*B. napus*). *Heredity* 81: 429–435.

Hauser TP, Damgaard C, and Jorgensen RB (2003) Frequency-dependent fitness of hybrids between oilseed rape (*Brassica napus*) and weedy *B. rapa* (Brassicaceae). *Am. J. Bot.* 90: 571–578.

Hayter KE and Cresswell JE (2006) The influence of pollinator abundance on the dynamics and efficiency of pollination in agricultural *Brassica napus*: Implications for landscape-scale gene dispersal. *J. Appl. Ecol.* 43: 1196–1202.

Heyn FW (1977) Analysis of unreduced gametes in the Brassiceae by crosses between species and ploidy levels. *Z. Pflanzenzucht.* 78: 13–30.

Honma S and Summers WL (1976) Interspecific hybridization between *Brassica napus* L. (Napobrassica group) and *B. oleracea* L. (Botrytis group). *J. Am. Soc. Hort. Sci.* 101: 299–302.

Hoyle M and Cresswell JE (2009) Maximum feasible distance of windborne cross-pollination in *Brassica napus*: A 'mass budget' model. *Ecol. Model.* (Online first) doi:10.1016/j.ecolmodel.2009.01.013.

Hu Q and Hansen LN (1998) Protoplast fusion between *Brassica napus* and related species for transfer of new genes for low erucic acid. *Cruciferae Newsl. Eucarpia* 20: 37–38.

Hu Q, Andersen SV, Laursen J, and Hansen LN (1999) Intergeneric hybridization by protoplast fusion aiming at modification of fatty acid composition in *Brassica napus*. In: Wratten N and Salisbury PA (eds.) *New Horizons for an Old Crop: Proceedings of the 10th International Rapeseed Congress, Canberra, Australia, 1999* (Regional Institute: Gosford, NSW, Australia), contribution no. 332. www.regional.org.au/au/gcirc/4/332.htm.

Huang B, Liu Y, Wu W, and Xue X (2002) Production and cytogenetics of intergeneric hybrids between Ogura CMS *Brassica napus* and *Raphanus sativus*. *Cruciferae Newsl. Eucarpia* 24: 25–27.

Hüsken A and Dietz-Pfeilstetter A (2007) Pollen-mediated intraspecific gene flow from herbicide-resistant oilseed rape (*Brassica napus* L.). *Transgen. Res.* 16: 557–569.

Ingram J (2000) *Report on the Separation Distances Required to Ensure Cross-Pollination Is Below Specified Limits in Non-Seed Crops of Sugar Beet, Maize, and Oilseed Rape.* Report prepared for the Ministry of Agriculture, Fisheries, and Food [MAFF], UK, Project No. RG0123. (Ministry of Agriculture, Fisheries and Food [MAFF]: London, UK). 14 p. www.defra.gov.uk/science/project_data/DocumentLibrary/RG0123/RG0123_2916_FRP.doc.

Inomata N (1988) Intergeneric hybridization between *Brassica napus* and *Sinapis arvensis* and their crossability. *Cruciferae Newsl. Eucarpia* 13: 22–23.

——— (1991) Intergeneric hybridization in *Brassica juncea* × *Sinapis pubescens* and *B. napus* × *S. pubescens*, and their cytological studies. *Cruciferae Newsl. Eucarpia* 14/15: 10–11.

——— (1994) Intergeneric hybridization between *B. napus* and *S. pubescens*, and cytology and crossability of their progenies. *Theor. Appl. Genet.* 89: 540–544.

James C (2008) *Global Status of Commercialized Biotech/GM Crops 2008.* ISAAA Brief 39 (International Service for the Acquisition of Agri-biotech Applications [ISAAA]: Ithaca, NY, USA). 243 p.

Jenkins TE, Conner AJ, and Frampton CM (2001) Investigating gene introgression from rape to wild turnip. *New Zeal. Plant Prot.* 54: 101–104.

Johannessen MM, Andersen BA, and Jørgensen RB (2006) Competition affects gene flow from oilseed rape (female) to *Brassica rapa* (male). *Heredity* 96: 360–367.

Jørgensen RB (1999) Gene flow from oilseed rape (*Brassica napus*) to related species. In: Lutman PJW (ed.) *Gene Flow and Agriculture: Relevance for Transgenic Crops; Proceedings of a Symposium Held at the University of Keele, Staffordshire, April 12–14, 1999.* BCPC Symposium Proceedings 72 (British Crop Protection Council [BCPC]: Farnham, UK), pp. 117–124.

——— (2007) Oilseed rape: Coexistence and gene flow from wild species. *Adv. Bot. Res.* 45: 451–464.

Jørgensen RB and Andersen B (1994) Spontaneous hybridization between oilseed rape (*Brassica napus*) and weedy *B. campestris* (Brassicaceae): A risk of growing genetically modified oilseed rape. *Am. J. Bot.* 8: 1620–1626.

Jørgensen RB, Andersen B, Landbo L, and Mikkelsen TR (1996) Spontaneous hybridization between oilseed rape (*Brassica napus*) and weedy relatives. *Acta Hort.* 407: 193–200.

Jørgensen RB, Andersen B, Hauser TP, Landbo L, Mikkelsen TR, and Østergård H (1998) Introgression of crop genes from oilseed rape (*Brassica napus*) to related wild species—an avenue for the escape of engineered genes. *Acta Hort.* 459: 211–217.

Jørgensen RB, Hauser T, Landbo L, Mikkelsen TR, and Østergård H (1999) *Brassica napus* and *B. campestris*: Spontaneous hybridisation, backcrossing, and fitness components of offspring plants. In: Gray AJ, Amijee F, and Gliddon CJ (eds.) *Environmental Impact of Genetically Modified Crops.* Research Report No. 10 (Department of the Environment, Transport and the Regions [DETR]: London, UK), pp. 20–27.

Kapteijns AJAM (1993) Risk assessment of genetically modified crops—potential of four arable crops to hybridize with the wild flora. *Euphytica* 66: 145–149.

Kerlan MC, Chèvre AM, Eber F, Baranger A, and Renard M (1992) Risk assessment of outcrossing of transgenic rapeseed to related species: I. Interspecific hybrid production under optimal conditions with emphasis on pollination and fertilization. *Euphytica* 62: 145–153.

Kerlan MC, Chèvre AM, and Eber F (1993) Interspecific hybrids between a transgenic rapeseed (*Brassica napus*) and related species—cytogenetical characterization and detection of the transgene. *Genome* 36: 1099–1106.

Klein EK, Lavigne C, Picault H, Renard M, and Gouyon PH (2006) Pollen dispersal of oilseed rape: Estimation of the dispersal function and effects of field dimension. *J. Appl. Ecol.* 43: 141–151.

Klewer A, Mewes S, Mai J, and Sacristán MD (2002) Alternaria-Resistenz in interspezifischen Hybriden und deren Rückkreuzungsnach kommenschaften im Tribus Brassiceae. *Vortr. Pflanzenzucht.* 54: 505–508.

Klewer A and Sacristán MD (2003) Erschließung neuer Resistenzquellen bei Raps—Schwerpunkt Alternaria-Resistenz. *Vortr. Pflanzenzucht.* 56: 63–70.

Klimaszewska K and Keller WA (1988) Regeneration and characterization of somatic hybrids between *Brassica napus* and *Diplotaxis harra. Plant Sci.* 58: 211–222.

Landbo L, Andersen B, and Jørgensen RB (1996) Natural hybridisation between oilseed rape and a wild relative: Hybrids among seeds from weedy *B. campestris. Hereditas* 125: 89–91.

Lange W, Toxopeus H, Lubberts JH, Dolstra O, and Harrewijn JL (1989) The development of raparadish (×*Brassicoraphanus*, 2n = 38), a new crop in agriculture. *Euphytica* 40: 1–4.

Lavigne C, Klein EK, Vallée P, Pierre J, Godelle B, and Renard M (1998) A pollen dispersal experiment with transgenic oilseed rape: Estimation of the average pollen dispersal of an individual plant within a field. *Theor. Appl. Genet.* 96: 886–896.

Leckie D, Smithson A, and Crute IR (1993) Gene movement from oilseed rape to weedy populations—a component of risk assessment for transgenic cultivars. *Asp. Appl. Biol.* 35: 61–66.

Leflon M, Eber F, Letanneur JC, Chelysheva L, Coriton O, Huteau V, Ryder CD, Barker G, Jenczewski E, and Chèvre AM (2006) Pairing and recombination at meiosis of *Brassica rapa* (**AA**) × *Brassica napus* (**AACC**) hybrids. *Theor. Appl. Genet.* 113: 1467–1480.

Lefol E, Danielou V, Darmency H, Boucher F, Maillet J, and Renard M (1995) Gene dispersal from transgenic crops: I. Growth of interspecific hybrids between oilseed rape and the wild hoary mustard. *J. Appl. Ecol.* 32: 803–808.

Lefol E, Fleury A, and Darmency H (1996a) Gene dispersal from transgenic crops. II. Hybridization between oilseed rape and wild hoary mustard. *Sex. Plant Reprod.* 9: 189–196.

——— (1996b) Predicting hybridization between transgenic oilseed rape and wild mustard. *Field Crop Res.* 45: 153–161.

Lefol E, Séguin-Swartz G, and Downey RK (1997) Sexual hybridization in crosses of cultivated *Brassica* species with *Erucastrum gallicum* and *Raphanus raphanistrum*: Potential for gene introgression. *Euphytica* 95: 127–139.

Légère A (2005) Risks and consequences of gene flow from herbicide-resistant crops: Canola (*Brassica napus* L.) as a case study. *Pest Managem. Sci.* 61: 292–300.

Légère A, Simard MJ, Thomas AG, Pageau D, Lajeunesse J, Warwick SI, and Derksen DA (2001) Presence and persistence of volunteer canola in Canadian cropping systems. In: British Crop Protection Council (ed.) *Weeds 2001: The BCPC*

Conference Proceedings of an International Conference, Brighton, UK, November 12–15, 2001 (British Crop Protection Council [BCPC]: Farnham, UK), pp. 143–148.

Lelivelt CLC, Leunissen EHM, Frederiks HJ, Helsper JPFG, and Krens FA (1993) Transfer of resistance to the beet cyst nematode (Heterodera schachtii Schm.) from Sinapis alba L. (white mustard) to the Brassica napus L. gene pool by means of sexual and somatic hybridization. Theor. Appl. Genet. 85: 688–696.

Li J, Fang X, Wang Z, Li J, Luo L, and Hu Q (2006) Transgene directionally integrated into C-genome of Brassica napus. Chin. Sci. Bull. 51: 1578–1585.

Linder CR (1998) Potential persistence of transgenes: Seed performance of transgenic canola and wild relatives × canola hybrids. Ecol. Appl. 8: 1180–1195.

Linder CR and Schmitt J (1995) Potential persistence of escaped transgenes: Performance of transgenic oil-modified Brassica seeds and seedlings. Ecol. Appl. 5: 1065–1068.

Liu JH, Landgren M, and Glimelius K (1996) Transfer of the Brassica tournefortii cytoplasm to B. napus for the production of cytoplasmic male sterile B. napus. Physiol. Plant. 96: 123–129.

Lokanadha RD and Sarla N (1994) Hybridization of Brassica tournefortii and cultivated Brassicas. Cruciferae Newsl. Eucarpia 16: 32–33.

Lu CM and Kato M (2001) Fertilization fitness and relation to chromosome number in interspecific progeny between Brassica napus and B. rapa: A comparative study using natural and resynthesized B. napus. Breed. Sci. 51: 73–81.

Lu CM, Kato M, and Kakihara F (2002) Destiny of a transgene escape from Brassica napus into B. rapa. Theor. Appl. Genet. 105: 78–84.

Lutman PJW (1993) The occurrence and persistence of volunteer oilseed rape (Brassica napus). Asp. Appl. Biol. 35: 29–35.

MacKay GR (1977) The introgression of S alleles into forage rape, Brassica napus L., from turnip, Brassica campestris L. subsp. rapifera. Euphytica 26: 511–519.

Malik M, Vyas P, Rangaswamy NS, and Shivanna KR (1999) Development of two new cytoplasmic male-sterile lines in Brassica juncea through wide hybridization. Plant Breed. 118: 75–78.

Manasse R and Kareiva P (1991) Quantifying the spread of recombinant genes and organisms. In: Ginzburgh L (ed.) Assessing Ecological Risks of Biotechnology (Butterworth-Heinemann: Boston, MA, USA), pp. 215–231.

Martens G (2001) From Cinderella to Cruela: Volunteer Canola; 2nd Annual Manitoba Agronomists Conference, University of Manitoba, Winnipeg, Manitoba, Canada, 2001 (University of Manitoba: Winnipeg, MB, Canada), pp. 151–154.

Masden SB (1962) Germination of buried and dry-stored seeds: III. 1934–1960. Proc. Int. Seed Test. Assoc. 27: 920–928.

Mathias R (1991) Improved embryo rescue technique for intergeneric hybridization between *Sinapis* species and *B. napus*. *Cruciferae Newsl. Eucarpia* 14/15: 90–91.

Mattsson B (1988) Interspecific crosses within the genus *Brassica* and some related genera. *Sveriges Utsadesforen. Tidskr.* 98: 187–212.

McCartney HA and Lacey ME (1991) Wind dispersal of pollen from crops of oilseed rape. *J. Aerosol. Sci.* 22: 467–477.

McNaughton IH (1995) Turnip and relatives. In: Simmonds S and Smartt J (eds.) *Evolution of Crop Plants.* 2nd ed. (Longman Scientific and Technical: Harlow, UK), pp. 62–68.

Meng J-L (1998) Studies on the relationships between *Moricandia* and *Brassica* species. *Acta Bot. Sin.* 40: 508–514.

Meng J-L, Yan Z, Tian Z, Huang R, and Huang B (1999) Somatic hybrids between *Moricandia nitens* and three *Brassica* species. In: Wratten N and Salisbury PA (eds.) *New Horizons for an Old Crop: Proceedings of the 10th International Rapeseed Congress, Canberra, Australia, 1999* (Regional Institute: Gosford, NSW, Australia), contribution no. 6. www.regional.org.au/au/gcirc/4/6.htm.

Mesquida J and Renard M (1982) Study of the pollen dispersal by wind and of the importance of wind pollination in rapeseed (*Brassica napus* var. *oleifera* Metzger). *Apidol.* 4: 353–366 [English summary].

Mesquida J, Renard M, and Pierre JS (1988) Rapeseed (*Brassica napus* L.) productivity: The effect of honeybees (*Apis mellefera* L.) and different pollination conditions in cage and field tests. *Apidol.* 19: 51–72.

Messéan A, Sausse C, Gasquez J, and Darmency H (2007) Occurrence of genetically modified oilseed rape seeds in the harvest of subsequent conventional oilseed rape over time. *Europ. J. Agron.* 27: 115–122.

Metz PLJ, Nap JP, and Stiekema WJ (1995) Hybridization of radish (*Raphanus sativus* L.) and oilseed rape (*Brassica napus* L.) through a flower-culture method. *Euphytica* 80: 159–168.

Metz PLJ, Jacobsen E, and Stiekema WJ (1997a) Aspects of the biosafety of transgenic oilseed rape. *Acta Bot. Neerland.* 46: 51–67.

Metz PLJ, Jacobsen E, Nap JP, Pereira A, and Stiekema WJ (1997b) The impact on biosafety of the phosphinothricin-tolerance transgene in inter-specific *B. rapa* × *B. napus* hybrids and their successive backcrosses. *Theor. Appl. Genet.* 95: 442–450.

Mikkelsen TR, Jensen J, and Jørgensen RB (1996a) Inheritance of oilseed rape (*Brassica napus*) RAPD markers in a backcross progeny with *Brassica campestris*. *Theor. Appl. Genet.* 92: 492–497.

Mikkelsen TR, Andersen B, and Jørgensen RB (1996b) The risk of crop transgene spread. *Nature* 380: 31.

Mithen RF and Herron C (1991) Transfer of disease resistance to oilseed rape from wild *Brassica* species. In: McGregor DI (ed.) *Proceedings of the GCIRC 8th International Rapeseed Congress, Saskatoon, Saskatchewan, Canada, July 9–11, 1991* (Organizing Committee of the Eighth International Rapeseed Congress under the auspices of the Groupe Consultatif International de Recherche sur la Colza [GCIRC] and the Canola Council of Canada: n.p.), pp. 244–249.

Morris WK, Kareiva PM, and Raymer PL (1994) Do barren zones and pollen traps reduce gene escape from transgenic crops? *Ecol. Appl.* 4: 157–165.

Moyes CL, Lilley JM, Casais CA, Cole SG, Haeger PD, and Dale PJ (2002) Barriers to gene flow from oilseed rape (*Brassica napus*) into populations of *Sinapis arvensis. Mol. Ecol.* 11: 103–112.

Myers JR (2006) *Outcrossing Potential for* Brassica *Species and Implications for Vegetable Crucifer Seed Crops of Growing Oilseed* Brassicas *in the Willamette Valley.* Oregon State University Extension Service Special Report 1064, January 2006 (Oregon State University: Corvallis, OR, USA). 7p. www.pacificbiomass.org/documents/Oilseed/BrassicaOutcrossingPotentialOR.pdf.

Nagpal R, Raina SN, Sodhi YS, Mukhopadhyay A, Arumugam N, Pradhan AK, and Pental D (1996) Transfer of *Brassica tournefortii* (**TT**) genes to allotetraploid oilseed *Brassica* species (*B. juncea* **AABB**, *B. napus* **AACC**, *B. carinata* **BBCC**): Homoeologous pairing is more pronounced in the three-genome hybrids (**TACC, TBAA, TCAA, TCBB**) as compared to allodiploids (**TA, TB, TC**). *Theor. Appl. Genet.* 92: 566–571.

Nanda Kumar PBA and Shivanna KR (1993) Intergeneric hybridization between *Diplotaxis siettiana* and crop *Brassicas* for the production of alloplasmic lines. *Theor. Appl. Genet.* 85: 770–776.

Nanda Kumar PBA, Prakash S, and Shivanna KR (1989) Wide hybridization in *Brassica*: Studies on interspecific hybrids between cultivated species (*B. napus, B. juncea*) and a wild species (*B. gravinae*). In: Iyama S and Takeda G (eds.) *Breeding Research: The Key to the Survival of the Earth.* Vol. 1 in *Proceedings of the 6th International Congress SABRAO, Tsukuba, Japan, August 21–25, 1989* (Society for the Advancement of Breeding Researches in Asia and Oceania [SABRAO]: n.p.), pp. 435–438.

Norris CE and Sweet J (2002) *Monitoring Large Scale Releases of Genetically Modified Crops (EPG 1/5/84).* Final Report of Monitoring Studies of Field Scale Releases of GM Oilseed Rape Crops in England From 1994–2000 (Department for Environment, Food. and Rural Affairs [DEFRA]: London, UK). 120 p. www.defra.gov.uk/Environment/gm/research/pdf/epg_1–5–84_screen.pdf.

OECD [Organization for Economic Cooperation and Development] (2008) *OECD Scheme for the Varietal Certification of Crucifer Seed and Other Oil or Fibre Species Seed Moving in International Trade: Annex VII to the Decision.*

C(2000)146/FINAL, OECD Seed Schemes February 2008. www.oecd.org/do cument/o/o,3343,en_2649_33905_1933504_1_1_1_1,oo.html.

OGTR [Office of the Gene Technology Regulator] (2008) *The Biology and Ecology of Canola (*Brassica napus*)* (Office of the Gene Technology Regulator [OGTR]: Canberra, ACT, Australia). 63 p. www.ogtr.gov.au/internet/ogtr/publishing. nsf/Content/riskassessments-1.

O'Neill CM, Murata T, Morgan CL, and Mathias RJ (1996) Expression of the C_3–C_4 intermediate character in somatic hybrids between *Brassica napus* and the C_3–C_4 species *Moricandia arvensis. Theor. Appl. Genet.* 93: 1234–1241.

Pallett DW, Huang L, Cooper JI, and Wang H (2006) Within-population variation in hybridisation and transgene transfer between wild *Brassica rapa* and *Brassica napus* in the UK. *Ann. Appl. Biol.* 148: 147–155.

Pauk J, Stefanov I, Fekete S, Bögre L, Karsai I, Feher A, and Dudits D (1995) A study of different (caMV 35S and mas) promoter activities and risk assessment of field use in transgenic rapeseed plants. *Euphytica* 85: 411–416.

Paul EM, Thompson C, and Dunwell JM (1995) Gene dispersal from genetically modified oilseed rape in the field. *Euphytica* 81: 283–289.

Paulmann W and Röbbelen G (1988) Effective transfer of cytoplasmatic male sterility from radish (*Raphanus sativus* L.) to rape (*Brassica napus* L.). *Plant Breed.* 100: 299–309.

Pekrun C, Potter TC, and Lutman PJW (1997) Genotypic variation in the development of secondary dormancy in oilseed rape and its impact on the persistence of volunteer rape. In: British Crop Protection Council (ed.) *The 1997 Brighton Crop Protection Conference—Weeds: Proceedings of an International Conference, Brighton, UK, November 17–20, 1997* (British Crop Protection Council [BCPC]: Farnham, UK), pp. 243–248.

Pekrun C, Hewitt JDJ, and Lutman PJW (1998) Cultural control of volunteer oilseed rape. *J. Agric. Sci.* 130: 155–163.

Pertl M, Hauser TP, Damgaard C, and Jørgensen RB (2002) Male fitness of oilseed rape (*Brassica napus*), weedy *B. rapa*, and their F1 hybrids when pollinating *B. rapa* seeds. *Heredity* 89: 212–218.

Pessel FD and Lecomte J (2000) Towards an understanding of the dynamics of rape populations that have "escaped" from large-scale cultivation in an agricultural region. *OCL* [Oléagineux Corps Gras Lipides] 7: 324–328.

Pessel FD, Lecomte J, Emeriau V, Krouti M, Messéan A, and Gouyon PH (2001) Persistence of oilseed rape (*Brassica napus* L.) outside of cultivated fields. *Theor. Appl. Genet.* 102: 841–846.

Picard-Nizou AL, Pham Delegue MH, Kerguelen V, Doualut P, Marilleau R, Olsen L, Grison R, Toppan A, and Masson C (1995) Foraging behaviour of honeybees

(*Apis mellifera* L.) on transgenic oilseed rape (*Brassica napus* L. var. *oleifera*). *Transgen. Res.* 4: 270–276.

Pinder R, Al-Kaff N, Kreike M, and Dale P (1999) Evaluating the risk of transgene spread from *Brassica napus* to related species. In: Lutman PJW (ed.) *Gene Flow and Agriculture: Relevance for Transgenic Crops; Proceedings of a Symposium Held at the University of Keele, Staffordshire, April 12–14,1999*. BCPC Symposium Proceedings 72 (British Crop Protection Council [BCPC]: Farnham, UK), pp. 275–280.

Pivard S (2008) Characterizing the presence of oilseed rape feral populations on field margins using machine learning. *Ecol. Model.* 212: 147–154.

Plümper B (1995) Somatische und sexuelle Hybridisierung für den Transfer von Krankheitsresistenzen auf *Brassica napus* L. PhD dissertation, Free University of Berlin.

Prakash S and Chopra VL (1988) Introgression of resistance to shattering in *Brassica napus* from *Brassica juncea* through non-homologous recombination. *Plant Breed.* 101: 167–168.

Prakash S and Hinata K (1980) Taxonomy, cytogenetics, and origin of crop *Brassicas*: A review. *Opera. Bot.* 55: 3–57.

Prakash S, Tsunoda S, Raut RN, and Gupta S (1982) Interspecific hybridization involving wild and cultivated genomes in the genus *Brassica*. *Cruciferae Newsl. Eucarpia* 7: 28–29.

Price JS, Hobson RN, Neale MA, and Bruce DM (1996) Seed losses in commercial harvesting of oilseed rape. *J. Agric. Eng. Res.* 65: 183–191.

Quazi MH (1988) Interspecific hybrids between *Brassica napus* L. and *B. oleracea* L. developed by embryo culture. *Theor. Appl. Genet.* 75: 309–318.

Rakow G and Woods DL (1987) Outcrossing in rape and mustard under Saskatchewan prairie conditions. *Can. J. Plant Sci.* 67: 147–151.

Ramsay G, Thompson CE, Neilson S, and Mackay GR (1999) Honeybees as vectors of GM oilseed rape pollen. In: Lutman PJW (ed.) *Gene Flow and Agriculture: Relevance for Transgenic Crops; Proceedings of a Symposium Held at the University of Keele, Staffordshire, April 12–14,1999*. BCPC Symposium Proceedings 72 (British Crop Protection Council [BCPC]: Farnham, UK), pp. 209–214.

Ramsay G, Thompson CE, and Squire G (2003) *Quantifying Landscape-Scale Gene Flow in Oilseed Rape* (Department for Environment, Food, and Rural Affairs [DEFRA]: London, UK). 50 p. www.defra.gov.uk/environment/gm/research/pdf/epg_rg0216.pdf.

Ranito-Lehtimäki A (1995) Aerobiology of pollen and pollen antigens. In: Cox CS and Wathes CM (eds.) *Bioaerosols Handbook* (CRC Press: Boca Raton, FL, USA), pp. 387–406.

Rawsthorne S, Morgan CL, O'Neill CM, Hylton CM, Jones DA, and Frean ML (1998) Cellular expression pattern of the glycine decarboxylase P protein in leaves of an intergeneric hybrid between the C3–C4 intermediate species *Moricandia nitens* and the C3 species *Brassica napus. Theor. Appl. Genet.* 96: 922–927.

Reboud X (2003) Effect of a gap on gene flow between otherwise adjacent transgenic *Brassica napus* crops. *Theor. Appl. Genet.* 106: 1048–1058.

Rhee WY, Cho YH, and Paek KY (1997) Seed formation and phenotypic expression of intra- and inter-specific hybrids of *Brassica* and of intergeneric hybrids obtained by crossing with *Raphanus. J. Korean Soc. Hort. Sci.* 38: 353–360.

Rieger MA, Preston C, and Powles SB (1999) Risks of gene flow from transgenic herbicide-resistant canola (*Brassica napus*) to weedy relatives in southern Australian cropping system. *Aust. J. Agric. Res.* 50: 115–128.

Rieger MA, Potter TD, Preston C, and Powles SB (2001) Hybridization between *Brassica napus* L. and *Raphanus raphanistrum* L. under agronomic field conditions. *Theor. Appl. Genet.* 103: 555–560.

Rieger MA, Lamond M, Preston C, Powles SB, and Roush R (2002) Pollen-mediated movement of herbicide resistance between commercial canola fields. *Science* 296: 2386–2388.

Ringdahl EA, McVetty PBE, and Sernyk JL (1987) Intergeneric hybridization of *Diplotaxis* spp. with *Brassica napus*: A source of new CMS systems? *Can. J. Plant Sci.* 67: 239–243.

Ripley VL and Arnison PG (1990) Hybridization of *Sinapis alba* L. and *Brassica napus* L. via embryo rescue. *Plant Breed.* 104: 26–33.

Roy NN (1980) Species crossability and early-generation plant fertility in interspecific crosses of *Brassica*. SABRAO [Society for the Advancement of Breeding Researches in Asia and Oceania] J. 12: 43–53.

Sacristán MD and Gerdemann M (1986) Different behavior of *Brassica juncea* and *B. carinata* as sources of *Phoma lingam* resistance in experiments of interspecific transfer to *B. napus. Plant Breed.* 97: 304–314.

Saji H, Nakajima N, Aono M, Tamaoki M, Kubo A, Wakiyama S, Hatase Y, and Nagatsu M (2005) Monitoring the escape of transgenic oilseed rape around Japanese ports and roadsides. *Environ. Biosafety Res.* 4: 217–222.

Salisbury PA (1989) Potential utilization of wild crucifer germplasm in oilseed *Brassica* breeding. *Proceedings of the 7th ARAB Workshop, Toowoomba, Queensland, Australia* (Australian Research Assembly on Brassicas [ARAB]: n.p.), pp. 51–53.

———— (2002a) *Gene Flow Between* Brassica napus *and Other Brassicaceae Species.* Report PAS0201 (Institute of Land and Food Resources, University of Melbourne: Melbourne, NSW, Australia). 45 p. http://abe.dynamicweb.dk/images/

files/Gene%20flow%20to%20other%20Brassiceae%20species%20report%20
-%20April%202002.pdf.

——— (2002b) *Genetically Modified Canola in Australia: Agronomic and Environmental Considerations* [report, ed. K Downey] (Australian Oilseeds Federation: Melbourne, NSW, Australia). 107 p. www.jcci.unimelb.edu.au/GMCanola 2007/PS%20GM%20canola%20book.pdf.

Scheffler JA and Dale PJ (1994) Opportunities for gene transfer from transgenic oilseed rape (*Brassica napus*) to related species. *Transgen. Res.* 3: 263–278.

Scheffler JA, Parkinson R, and Dale PJ (1993) Frequency and distance of pollen dispersal from transgenic oilseed rape (*Brassica napus*). *Transgen. Res.* 2: 356–364.

——— (1995) Evaluating the effectiveness of isolation distances for field plots of oilseed rape (*Brassica napus*) using a herbicide-resistance transgene as a selectable marker. *Plant Breed.* 114: 317–321.

SCIMAC [Supply Chain Initiative on Modified Agricultural Crops] (1999) Guidelines for Growing Newly Developed Herbicide Tolerant Crops. 11 p. www.sci mac.org.uk/files/onfarm_guidelines.pdf.

Scott SE and Wilkinson MJ (1998) Transgene risk is low. *Nature* 393: 320.

Siemens J (2002) Interspecific hybridisation between wild relatives and *Brassica napus* to introduce new resistance traits into the oilseed rape gene pool. *Czech J. Genet. Plant Breed.* 38: 155–157.

Simard MJ and Légère A (2004) Synchrony of flowering between canola and wild radish (*Raphanus raphanistrum*). *Weed Sci.* 52: 905–911.

Simard MJ, Légère A, Pageau D, Lajeunesse J, and Warwick S (2002) The frequency and persistence of canola (*Brassica napus*) volunteers in Québec cropping systems. *Weed Technol.* 16: 433–439.

Simpson EC, Norris CE, Law JR, Thomas JE, and Sweet JB (1999) Gene flow in genetically modified herbicide tolerant oilseed rape (*Brassica napus*) in the UK. In: Lutman PJW (ed.) *Gene Flow and Agriculture: Relevance for Transgenic Crops; Proceedings of a Symposium Held at the University of Keele, Staffordshire, April 12–14, 1999.* BCPC Symposium Proceedings 72 (British Crop Protection Council [BCPC]: Farnham, UK), pp. 75–81.

Skarzhinskaya M, Landgren M, and Glimelius K (1996) Production of intertribal somatic hybrids between *Brassica napus* L. and *Lesquerella fendleri* (Gray) Wats. *Theor. Appl. Genet.* 93: 1242–1250.

Snow AA, Andersen B, and Jørgensen RB (1999) Costs of transgenic herbicide resistance introgressed from *Brassica napus* into weedy *B. rapa. Mol. Ecol.* 8: 605–615.

Squire GR, Crawford JW, Ramsay G, Thompson C, and Brown J (1999) Gene flow at the landscape level. In: Lutman PJW (ed.) *Gene Flow and Agriculture: Relevance for Transgenic Crops; Proceedings of a Symposium Held at the Univer-*

sity of Keele, Staffordshire, April 12–14,1999. BCPC Symposium Proceedings 72 (British Crop Protection Council [BCPC]: Farnham, UK), pp. 57–64.

Squire GR, Begg B, Crawford J, Gordon S, Hawes C, Johnstone C, Marshall B, Ramsay G, Thompson C, Wright G, and Young M (2003) Outcrossing among crops and feral descendents—geneflow. In: Scottish Crop Research Institute [SCRI] (ed.) *SCRI Annual Report 2001/2002* (Scottish Crop Research Institute: Dundee, Scotland). pp. 176–180. www.scri.ac.uk/scri/file/individualreports/2002/33OutCr.pdf.

Sridevi O and Sarla N (1996) Reciprocal hybridization between *Sinapis alba* and *Brassica* species. *Cruciferae Newsl. Eucarpia* 18: 16.

Staniland BK, McVetty PBE, Friesen LF, Yarrow S, Freyssinet G, and Freyssinet M (2000) Effectiveness of border areas in confining the spread of transgenic *Brassica napus* pollen. *Can. J. Plant Sci.* 80: 521–526.

Stringam GR and Downey RK (1982) Effectiveness of isolation distance in seed production of rapeseed (*Brassica napus*). *Agron. Abst.* (1982): 136–137.

Sundberg E and Glimelius K (1991) Effects of parental ploidy level and genetic divergence on chromosome elimination and chloroplast segregation in somatic hybrids within Brassicaceae. *Theor. Appl. Genet.* 83: 81–88.

Sutherland JP, Justinova L, and Poppy GM (2006) The responses of crop-wild *Brassica* hybrids to simulated herbivory and interspecific competition: Implications for transgene introgression. *Environ. Biosafety Res.* 5: 15–25.

Sweet JB, Shepperson R, Thomas JE, and Simpson EC (1997) The impact of releases of genetically modified herbicide-tolerant oilseed rape in the UK. In: British Crop Protection Council (ed.) *The 1997 Brighton Crop Protection Conference—Weeds: Proceedings of an International Conference, Brighton, UK, November 17–20, 1997* (British Crop Protection Council [BCPC]: Farnham, UK), pp. 291–302.

Sweet JB, Norris CE, Simpson EC, and Thomas JE (1999) Assessing the impact and consequences of the release and commercialisation of genetically modified crops. In: Lutman PJW (ed.) *Gene Flow and Agriculture: Relevance for Transgenic Crops; Proceedings of a Symposium Held at the University of Keele, Staffordshire, April 12–14,1999.* BCPC Symposium Proceedings 72 (British Crop Protection Council [BCPC]: Farnham, UK), pp. 241–246.

Takahata Y and Hinata K (1983) Studies on cytodemes in subtribe Brassicinae (Cruciferae). *Tohoku J. Agr. Res.* 33: 111–124.

Takahata Y and Takeda T (1990) Intergeneric (intersubtribe) hybridization between *Moricandia arvensis* and *Brassica* **A** and **B** genome species by ovary culture. *Theor. Appl. Genet.* 80: 38–42.

Takahata Y, Takeda T, and Kaizuma N (1993) Wide hybridization between *Moricandia arvensis* and *Brassica* amphidiploid species (*B. napus* and *B. juncea*). *Euphytica* 69: 155–160.

Takeshita M, Masahiro K, and Tokumasu S (1980) Application of ovule culture to the production of intergeneric or interspecific hybrids in *Brassica* and *Raphanus*. *Japan. J. Genet.* 55: 373–387.

Thalmann C, Guadagnuolo R, and Felber F (2001) Search for spontaneous hybridization between oilseed rape (*Brassica napus* L.) and wild radish (*Raphanus raphanistrum* L.) in agricultural zones and evaluation of the genetic diversity of the wild species. *Bot. Helvetica* 111: 107–119.

Thompson CE, Squire G, Mackay GR, Bradshaw JE, Crawford J, and Ramsay G (1999) Regional patterns of gene flow and its consequences for GM oilseed rape. In: Lutman PJW (ed.) *Gene Flow and Agriculture: Relevance for Transgenic Crops; Proceedings of a Symposium Held at the University of Keele, Staffordshire, April 12–14,1999.* BCPC Symposium Proceedings 72 (British Crop Protection Council [BCPC]: Farnham, UK), pp. 95–100.

Timmons AM, O'Brien ET, Charters YM, Dubbels SJ, and Wilkinson MJ (1995) Assessing the risks of wind pollination from fields of *Brassica napus* subsp. *oleifera*. *Euphytica* 85: 417–423.

Timmons AM, Charters YM, Crawford JW, Burn D, Scott SE, Dubbels SJ, Wilson NJ, Robertson A, O'Brien ET, Squire GR, and Wilkinson MJ (1996) Risks from transgenic crops. *Nature* 380: 487.

Tomiuk J, Hauser TP, and Jørgensen RB (2000) A- or C-chromosomes, does it matter for the transfer of transgenes from *Brassica napus*? *Theor. Appl. Genet.* 100: 750–754.

USDA [United States Department of Agriculture] (2008) 7 CFR [Code of Federal Regulations] § 201.76. Federal Seed Act Regulations; Minimum Land, Isolation, Field, and Seed Standards. Source: 59 FR [Federal Register] 64516, Dec. 14, 1994, as amended at 65 FR 1710, Jan. 11, 2000 (Government Printing Office: Washington, DC, USA), pp. 372–378. www.access.gpo.gov/nara/cfr/waisidx_08/7cfr201_08.html.

Vyas P, Prakash S, and Shivanna KR (1995) Production of wide hybrids and backcross progenies between *Diplotaxis erucoides* and crop *Brassicas*. *Theor. Appl. Genet.* 90: 549–553.

Walklate PJ, Hunt JCR, Higson HL, and Sweet JB (2004) A model of pollen-mediated gene flow for oilseed rape. *Proc. Roy. Soc. London, Ser. B, Biol. Sci.* 271: 441–449.

Warwick SI, Francis A, and La Fleche J (2000) *Guide to Wild Germplasm of Brassica and Allied Crops (Tribe Brassiceae, Brassicaceae).* 2nd ed. (Agriculture and Agri-Food Canada: Saskatoon, SK, Canada). www.brassica.info/info/publications/guidewild/brass00.pdf.

Warwick SI, Simard M-J, Légère A, Beckie HJ, Braun L, Zhu B, Mason P, Séguin-Swartz G, and Stewart CN Jr (2003) Hybridization between transgenic *Bras-*

sica napus L. and its wild relatives: *Brassica rapa* L., *Raphanus raphanistrum* L., *Sinapis arvensis* L. and *Erucastrum gallicum* (Willd.) O.E. Schulz. *Theor. Appl. Genet.* 107: 528–539.

Weekes R, Deppe C, Allnutt T, Boffey C, Morgan D, Morgan S, Bilton M, Daniels R, and Henry C (2005) Crop-to-crop gene flow using farm-scale sites of oilseed rape (*Brassica napus* L.) in the U.K. *Transgen. Res.* 14: 749–759.

Wilkinson MJ, Timmons AM, Charters Y, Dubbels S, Robertson A, Wilson N, Scott S, O'Brien E, and Lawson HM (1995) Problems of risk assessment with genetically modified oilseed rape. *Brighton Crop Protection Conference: Weeds 1995; Proceedings of an International Conference Held at the Brighton Centre and the Brighton Metropole Hotel, Brighton, England, November 20–23, 1995*, vol. 3 (British Crop Protection Council [BCPC]: Farnham, UK), pp. 1035–1044.

Wilkinson MJ, Davenport IJ, Charters YM, Jones AE, Allainguillaume J, Butler HT, Mason DC, and Raybould AF (2000) A direct regional-scale estimate of transgene movement from genetically modified oilseed rape to its wild progenitors. *Mol. Ecol.* 9: 983–991.

Wilkinson MJ, Elliott LJ, Allainguillaume J, Shaw MW, Norris C, Welters R, Alexander M, Sweet J, and Mason DC (2003) Hybridization between *Brassica napus* and *B. rapa* on a national scale in the United Kingdom. *Science* 302: 457–459.

Williams IH (1984) The concentration of air-borne pollen over a crop of oilseed rape (*Brassica napus* L.). *J. Agric. Sci.* 103: 353–357.

Yoshimura Y, Beckie HJ, and Matsuo K (2006) Transgenic oilseed rape along transportation routes and port of Vancouver in western Canada. *Environ. Biosafety Res.* 5: 2–9.

Zhu B, Lawrence JR, Warwick SI, Mason P, Braun L, Halfhill MD, and Stewart CN Jr (2004) Stable *Bacillus thuringiensis* (*Bt*) toxin content in interspecific F1 and backcross populations of wild *Brassica rapa* after *Bt* gene transfer. *Mol. Ecol.* 3: 237–241.

Zhu JS, Struss D, and Röbbelen G (1993) Studies on resistance to *Phoma lingam* in *Brassies napus-Brassica nigra* addition lines. *Plant Breed.* 111: 192–197.

SIX

Cassava, Manioc, Yuca
(*Manihot esculenta* Crantz)

Center(s) of Origin and Diversity

Cassava (*M. esculenta* Crantz subsp. *esculenta*) originated in S America, in the S Amazon basin in Brazil (Olsen and Schaal 1999, 2001; Olsen 2004).

Different centers of diversity have been described for the genus *Manihot* Mill. Approximately 80 species are reported from two centers of diversity in Brazil (one in central Brazil, one in NE Brazil). A secondary center of diversity, with 17 species, is found in Mexico and Central America (Rogers and Appan 1973; Jennings 1995; Nassar 2001a). Africa is sometimes considered to be an additional center of diversity for the cultivated species of cassava (Gulick et al. 1983; Pickersgill 1998).

Biological Information

Flowering

Cassava is a monoecious, protogynous species—within the same inflorescence, the (basal) female flowers open first and then the (terminal) male flowers open one to two weeks later (Kawano 1980). Self-pollination can occur because male and female flowers on different branches can open simultaneously (Jennings and Iglesias 2002). Cassava flowers usually begin to open around midday and stay that way for about one day. The stigma remains receptive for about the same period of time (Kawano 1980). Fertilization occurs between 8 and 19 hours after pollination (Chandratna and Nanayakkara 1948).

Pollen Dispersal and Longevity

Cassava is pollinated by insects, mainly by stingless bees (Meliponinae, including *Melipona* Illiger, *Paratrigona* Schwarz, and *Trigona* Jurine) and honeybees (*Apis mellifera*), and to a lesser extent by carpenter bees (*Xylocopa* spp.), bumblebees (*Bombus* spp.), and other native bee and wasp species (Kawano 1980). Nassar and de Carvalho (1990) observed 12 different bee, bumblebee, and wasp species belonging to the genera *Apis* L., *Augochloropsis* Cockerell, *Bombus* Latreille, *Melipona*, *Paratrigona*, *Partamona* Schwarz, *Polybia* Lepeletier, and *Xylocopa* Latreille foraging on wild and cultivated *Manihot* species.

Although cassava pollen has been reported to remain viable from two to six days (Chandratna and Nanayakkara 1948; Nassar and Ortiz 2007), viability declines rapidly after pollen shedding, and plant breeders usually perform pollination within one hour after collection to help ensure successful fertilization (Halsey et al. 2008).

Sexual Reproduction

Cassava ($2n = 36$) is a perennial, predominantly allogamous species (aided by protogyny). Outcrossing rates range from 60% to 100%, but low levels of self-fertilization do occur (Kawano 1980; Meireles da Silva et al. 2003; Hershey, in press). Actual outcrossing rates are influenced by the spatial structure of populations, which in turn depends on how farmers plant clones. In Amazonian Amerindian systems, cassava is often planted in monovarietal patches, with several such patches occurring within the same field. This leads to the development of mating systems that vary greatly in terms of inbreeding (D. McKey, pers. comm. 2008), which sometimes reduces the fertility and seed viability of cultivars (Jennings 1963; Biggs et al. 1986). Under appropriate ecological conditions, however, plenty of seeds are produced.

Some wild species and hybrid cultivars have shown low levels of apomixis, that is, they can produce seed asexually (Nassar 2001b; Nassar and Collevatti 2005).

Vegetative Reproduction

Cassava is, for the most part, vegetatively propagated via stem cuttings. In many traditional cultivation systems, however, sexual propagation is an integral part of the reproductive ecology of domesticated cassava, through the incorporation of volunteer seedlings (Elias and McKey 2000; Halsey et al. 2008). Occasionally, botanical seed has also been used in commercial propagation schemes (Rajendran et al. 2000). A number of rapid propagation systems have been developed, not just for use in research, but also for commercial purposes. These systems can be based on *in vitro* methods, rooted shoots, or modified stem-cutting techniques. All wild species propagate through seed.

Seed Dispersal and Dormancy

Cassava seeds are dispersed mechanically via exploding dehiscent pods (autochory) and are often further disseminated by ants (myrmecochory) that drag the seeds into their underground nests (Elias and McKey 2000). As is the case for many other plant species of the same family (Euphorbiaceae), ants are attracted by a food reward, in the form of an oil body attached to the seed (Ganeshaiah and Uma Shaanker 1988; Lisci and Pacini 1997). Seeds from shattered capsules are also buried in the soil seed bank by rain and by agronomic management practices such as plowing. Seed-eating birds may also act as seed dispersers, since farmers have reported sometimes finding cassava seeds in the droppings of small doves (Elias and McKey 2000). The viability of cassava seeds passing through the digestive tract of these granivorous birds has not yet been studied.

Cassava seeds buried in the soil fall dormant and form persistent seed banks (Pujol et al. 2005a, 2007). Although germination percentages may decline substantially after six months (Rajendran et al. 2000), farmers' reports provide anecdotal evidence that seeds can sometimes remain dormant and viable for over 40 years (Elias and McKey 2000; Elias et al. 2000b).

Volunteers, Ferals, and Their Persistence

Frequently, after recent burnings, numerous volunteers appear from dormant seeds buried in the soil seed bank in slash-and-burn fields (Elias and McKey 2000; Pujol et al. 2005b). Amerindian farmers in S America (and some farmers from Africa) often incorporate cuttings of these volunteers into their stocks of clones and multiply them if they show desirable traits (see the section below on gene flow through farming practices). Feral cassava plants can be observed outside of cultivation, but to our knowledge, their population dynamics and persistence have not been studied so far.

Cassava volunteers do not survive well in abandoned fields or as feral escapes outside of cultivation (Rogers 1965). They do not compete well with plants from subsequent successional stages and usually disappear within a few generations.

Weediness and Invasiveness Potential

Cassava and many of its wild relatives are adapted to open vegetation and moderately disturbed habitats (Jennings 1995). The wild relatives tend to be weedy pioneer plants and may have some invasive potential (Nassar and Ortiz 2007).

Crop Wild Relatives

The neotropical genus *Manihot* contains 98 species, provisionally distributed into 16 sections (Rogers and Appan 1973; Jennings 1995; Allem 2002a). Cassava wild relatives can be found from the SW United States to S Argentina, with major diversity centers in Brazil and Mexico (Nassar 2001a). The chromosome number for all species in the genus studied thus far is $2n = 36$ (Rogers and Appan 1973).

Hybridization

Species relationships in the genus *Manihot* have long been controversial, and they still are, to some extent. In the last decade, an increasing number of biosystematic and genetic studies have contributed greatly to a better understanding both of the phylogenetic relationships within the genus and the evolution of the cultivated crop.

For example, it has long been speculated that cassava is a hybrid (or compilospecies) derived from interbreeding *Manihot* species complexes in Mexico (Rogers 1965; Rogers and Appan 1973) and/or S America (Rogers 1963; Ugent et al. 1986; Sauer 1993). Several wild species from Mexico and Central and S America have been proposed to be cassava's closest wild relatives (reviewed by Renvoize 1972 and Olsen and Schaal 1999; also see Rogers and Appan 1973; Sauer 1993). However, recent genetic studies have shown that cassava domestication probably involves a single wild progenitor, rather than a pool of unidentified hybridizing species, and a single domestication center, rather than many independent sites throughout the neotropics. Studies from a wide range of disciplines—including taxonomy, biosystematics, ethnobotany, and genetics—now mainly concur in their definition of the closest wild relative of cassava and point to southern Brazilian populations of the wild species *M. esculenta* subsp. *flabellifolia* (Pohl) Cif. as the crop's direct ancestor (see, e.g., Fregene et al. 1994; Roa et al. 1997; Second et al. 1997; Allem 1999, 2000a; Olsen and Schaal 1999, 2001; Carvalho and Schaal 2001).

Multiple questions remain open regarding interspecific relationships and the potential for gene exchange among species. Many plant breeders have shown that reproductive isolation barriers in the genus are weak, and that cassava can be crossed with numerous wild congeners (see, e.g., Jennings 1959, 1963; Rogers and Appan 1973; Bryne 1984; Asiedu et al. 1992; Wanyera et al. 1992; CIAT 2006, 2007; Blair et al. 2007). Since all species studied to date share the same genomic constitution ($2n = 36$), no indications of cytogenetic barriers have been found. The extent of cross-compatibility varies widely and depends on species and genotype (see, e.g., Jennings 1976, 1995; Allem 1992, 2002a, b).

Despite these claims, natural hybrids remain primarily undocumented, with existing reports being based on the morphological intermediacy of the putative hybrids (see, e.g., Rogers 1963; Nassar 1989, 2001a; Nassar et al. 1996). Seeds originating from these hybrids have rarely been tested for viability, and the plants have never been characterized on a molecular basis (Allem 1992; Rieseberg and Ellstrand 1993;

Table 6.1 Cassava (*Manihot esculenta* Crantz) and its closest crop wild relatives

Species	Common name	Origin and distribution
Primary gene pool (GP-1)		
M. esculenta Crantz subsp. esculenta	tree species cassava, yuca, manioc, mandioca, tapioca	known only from cultivation
M. esculenta subsp. flabellifolia (Pohl) Cif.		Amazonian rim; putative wild ancestor
M. esculenta subsp. peruviana (Müll. Arg.) Allem		Brazil, Peru; wild
M. pruinosa Pohl		endemic to Brazil; wild
Secondary gene pool (GP-2)		
M. carthagenensis subsp. carthagenensis (Jacq.) Müll. Arg.		Venezuela, Bolivia, Colombia, Argentina, Paraguay, Trinidad and Tobago, Antilles
M. carthagenensis subsp. glaziovii (Müll. Arg.) Allem (= M. glaziovii Müll. Arg.)	Ceará rubber	native to Brazil; cultivated and naturalized elsewhere (Africa, Asia, Pacific Islands)
M. carthagenensis subsp. hahnii Allem		Brazil
M. aesculifolia (Kunth) Pohl	wild yuca	C America (Mexico, Costa Rica, Belize, Panama, El Salvador, Guatemala)
M. anomala Pohl		Brazil, Bolivia, Peru, Argentina, Paraguay
M. brachyloba Müll. Arg.		throughout C and S America (from Nicaragua to Brazil)
M. chlorosticta Standl. & Goldman		Mexico
M. dichotoma Ule	Jequie Manicoba rubber	Brazil
M. epruinosa Pax & K. Hoffm.		Brazil
M. gracilis Pohl		Brazil
M. leptophylla Pax & K. Hoffm.		Brazil, Ecuador, Peru
M. pilosa Pohl		Brazil
M. pohlii Wawra		Brazil
M. tripartita (Spreng.) Müll. Arg.		Brazil, Bolivia, Paraguay
M. triphylla Pohl		Brazil

Source: Govaerts et al. 2000.
Note: The cultivated form is in bold. Gene pool boundaries are according to Allem 1999 and Allem et al. 2001.

Rieseberg 1995). Duputié et al. (2007) demonstrated for the first time, based on molecular evidence, that domesticated cassava has hybridized in nature with a wild relative (considered by those authors as conspecific with *M. esculenta* subsp. *flabellifolia*) at several places in French Guiana and that the hybrids (F1, F2, and possibly backcrosses with the wild relative) were fertile.

Despite limited current knowledge, cassava wild relatives have been grouped into primary, secondary, and tertiary gene pools, based on their genetic compatibility with the crop in crossing experiments. The boundaries of the secondary gene pool, however, are preliminary, because thus far only a few wild species have been tested.

Primary Gene Pool (GP-1)

The primary gene pool of cassava includes the wild species *Manihot esculenta* subsp. *flabellifolia* and subsp. *peruviana* (Müll. Arg.) Allem, and *M. pruinosa* Pohl (table 6.1).

Manihot esculenta subsp. flabellifolia

Cassava is interfertile with its putative ancestor, *Manihot esculenta* subsp. *flabellifolia*, and controlled crosses produce vigorous and fully fertile F1 progeny (Nichols 1947; Bolhuis 1953, 1969; Jennings 1957, 1959, 1963, 1995; Roa et al. 1997). The species is also used in breeding programs, and high root-protein content and resistance to the cassava green mite (*Mononychellus tanajoa* Bondar) have been identified as useful traits for transferral into cassava germplasm (see, e.g., CIAT 2006, 2007; Fregene et al. 2006).

A genetic study undertaken by Olsen and Schaal (1999) showed no evidence of introgression of crop alleles into populations of the wild subspecies after domestication. However, in the wild, the likelihood of intraspecific gene flow and introgression between cultivated cassava and its putative ancestor still warrants further research. Although subsp. *flabellifolia* usually occurs in forested areas (Allem 1994), it can also be found growing in close proximity to the crop in secondary forests, pastures, and along roadsides (D. McKey, pers. comm. 2008). No molecular studies have yet been conducted using wild and domesticated cassava populations from these close-proximity sites.

Manihot esculenta subsp. peruviana

Manihot esculenta subsp. *peruviana* is placed into the primary gene pool due to its close morphological (Allem 1994) and genetic (Colombo et al. 2000) similarities with cassava. To our knowledge, no published information is available about crosses between cassava and this wild relative.

Manihot pruinosa

Allem (1992) included *Manihot pruinosa* in the secondary gene pool at first, but he later placed it into the primary gene pool, due to its close morphological and genetic resemblance to cassava and even more so to its putative wild ancestor, *M. esculenta* subsp. *flabellifolia* (Allem 1999; Allem et al. 2001). The species is thought to be interfertile with both cassava and *M. esculenta* subsp. *flabellifolia* (Olsen and Schaal 2001), but thus far no natural hybrids have been observed in sympatric populations in the field, and no experimental evidence for actual cross-compatibility has been published. Olsen and Schaal (1999) state that, in spite of being adapted to different habitats, *M. pruinosa* and cassava grow in close proximity in the Cerrado-forest transition zone (Brazil), providing an opportunity for interbreeding.

Secondary Gene Pool (GP-2)

The secondary gene pool currently comprises 13 species (table 6.1), but it is likely that more species will be included once the genetic relationships and cross-compatibility of cassava with other wild *Manihot* species are better understood.

Manihot carthagenensis subsp. glaziovii

The NE Brazilian species *Manihot carthagenensis* subsp. *glaziovii* (Müll. Arg.) Allem (= *M. glaziovii* Müll. Arg.), or Ceará rubber, has been cultivated for rubber production in many parts of the world, and it is the most frequently used species in cassava breeding programs (see, e.g., Storey and Nichols 1938; Nichols 1947; Abraham 1957; Jennings 1963, 1995; Magoon et al. 1966; Hahn et al. 1980a, b, 1990; Chavarriaga et al. 2004). It is also widely planted as an ornamental shade tree (D. McKey, pers. comm. 2008). Natural hybrid swarms, based on morphological and isozyme markers, have been identified in Africa, where both species are introduced (Wanyera et al. 1992; Lefèvre and Charrier 1993).

Other GP-2 Species

Crosses of cassava are possible with all other GP-2 species (see, e.g., Nassar et al. 1986, 1996; Bai et al. 1993; Roa et al. 1997), but gene transfer requires human intervention. Several GP-2 species have been used in breeding programs with varying degrees of success, including *Manihot leptophylla* Pax and *M. epruinosa* Pax & K. Hoffm. (IITA 1988; Hahn et al. 1990) and *M. dichotoma* Ule (Nichols 1947; Jennings 1957, 1995). The resulting F_1 hybrids tend to be sterile (Allem et al. 2001).

Tertiary Gene Pool (GP-3)

The remaining *Manihot* species currently belong to GP-3. Several of them have been used in breeding programs: *Manihot brachyandra* Pax & K. Hoffm. (IITA 1988), *M. catingae* Ule (Nichols 1947; Jennings 1957), *M. caerulescens* Pohl (Sheela et al. 2004; Unnikrishnan et al. 2004), *M. walkerae* Croizat (delayed post-harvest physiological deterioration), and *M. crassisepala* Pax & K. Hoffm. (amylose-free or waxy starch) (Tanksley and Nelson 1996; CIAT 2006, 2007; Fregene et al. 2006). The fertility and fitness of F_1 hybrids and later generations has rarely been documented. Moreover, all these reports refer to artificial hybridizations under controlled conditions, so it is difficult, if not impossible, to draw conclusions about the potential for spontaneous hybridization under natural conditions.

There is some anecdotal evidence of putative hybrids between cassava and the following wild species, based on morphological intermediacy: *M. neusana* Nassar (Nassar 1989), *M. reptans* Pax, *M. procumbens* Müll. Arg., and *M. oligantha* Pax & K. Hoffm. (Nassar et al. 1986, 1996).

Some of the above-mentioned wild relatives, or other thus-far-unstudied wild *Manihot* species, may eventually be included in GP-2 once more information about the extent of their cross-compatibility with cassava becomes available.

Outcrossing Rates and Distances

Pollen Flow

Due to their large size, cassava pollen grains are primarily disseminated by insects and, only to a limited extent, by wind (Kawano 1980). Honeybees can play an important role in pollen transfer over long distances, since they are capable of carrying pollen over several kilometers (see, e.g., Eckert 1933; Mesquida et al. 1988; Ramsay et al. 1999). Most foraging, however, occurs between 1 and 5 m, and in field experiments in Colombia, over 90% of the cassava pollen was deposited within a 10-meter radius (Hershey, in press). To our knowledge, no other published information is available about pollen flow distances in cassava, let alone about outcrossing rates at different distances.

Isolation Distances

Based on these reports, a distance of 30 meters is considered to be sufficient to minimize cross-pollination (Kawano et al. 1978; Halsey et al. 2008).

Gene Flow through Farmer Practices

The cassava fields of traditional and indigenous farmers—especially in S America and Africa, but also in parts of Asia—are often polycultures, integrated by clusters of plants belonging to different landraces, and stem cuttings are frequently exchanged within and among neighboring communities (Salick et al. 1997; Elias and McKey 2000; Elias et al. 2000b; Manu-Aduening et al. 2005; Balyejusa Kizito et al. 2007). Although cassava is mainly a vegetatively propagated crop, these polycultural fields are highly heterogeneous and harbor substantial genetic diversity. In the Americas, wild *Manihot* species can occur in close proximity to cultivated cassava fields (Duputié et al. 2007), but this is not common.

Cassava is propagated by stem cuttings, but individual plants are usually fertile and can produce seeds. Cassava is also an allogamous plant, and outcrossing between clusters of different landraces is frequent (Kawano et al. 1978). Natural gene flow within and between cassava fields and, more rarely, with wild relatives can therefore play an important role in maintaining these genotypically diverse and heterogeneous clonal stocks. This is further facilitated by farmers' agronomic management practices, which involve intended or unintended human selection and the exchange of clonal stocks. When the cassava capsules are mature, they shatter and the seeds fall to the ground. If they are not eaten by foraging seed predators, these seeds are often mixed into the soil by plowing, rain, or ants, forming a persistent soil seed bank. Volunteers, emerging from the buried seeds in subsequent cultivation cycles, are often used as new landraces by farmers in S America (Elias et al. 2000a, b, 2001; Pujol et al. 2002, 2007) and Africa (Mkumbira et al. 2003; Balyejusa Kizito et al. 2005).

In one study site in S America, the size of the volunteer plants was correlated with heterozygosity (Pujol et al. 2005; Pujol and McKey 2006). While farmers usually kill small volunteers when weeding, they frequently retain large ones and allow them to mature. When they show desirable traits (e.g., high yield or colored roots), farmers sometimes incorporate cuttings from these volunteers into their stocks of clones, thus enhancing heterozygosity and genetic diversity (Elias and McKey 2000; Elias et al. 2000a, b, 2001; Pujol et al. 2002; Manu-Aduening et al. 2005).

State of Development of GM Technology

A number of different protocols for the genetic transformation of cassava have been developed in the last decade (Raemakers et al. 1997; Fregene and Puonti-Kaerlas 2002; Taylor et al. 2004a; Ihemere et al. 2006). However, the most reliable results were achieved via *Agrobacterium*-mediated transformation, which has therefore become the preferred method. There are about half a dozen GM cassava programs worldwide (in Colombia, Denmark, the Netherlands, Nigeria, Switzerland, and the United States), and proof-of-concept has been provided for several successfully transformed cassava cultivars (Ladino et al. 2001; Taylor et al. 2004a, b). The first field trials, in screenhouses, have been conducted only recently (in Colombia, Kenya, and the United States [Virgin Islands]), and no GM varieties have been released for commercial production thus far. The GM traits researched include insect resistance, virus resistance, herbicide resistance, reduced cyanogenic content, enhanced starch content, waxy starch phenotype, enhanced nutritional quality (increased protein content, increased micronutrient content), extended leaf retention, and reduced post-harvest deterioration.

Agronomic Management Recommendations

There are no specific recommendations.

Crop Production Area

Cassava is grown worldwide, throughout both the wet lowlands and the seasonally dry tropics and subtropics. The total crop production area is estimated at 18.7 million ha, of which about 12 million are in Africa, 2.9 million in Latin America, and 3.8 million in Asia (FAO 2009). About 60% of cassava's world production is concentrated in five countries: Brazil, the Democratic Republic of the Congo, Indonesia, Nigeria, and Thailand. As mentioned above, no commercial production of GM cassava has been reported to date.

Research Gaps and Conclusions

Cassava is of major importance for millions of people living in the world's least developed countries in tropical and subtropical S America, Africa, and Asia. However, because its cultivation in industrialized countries is very limited, research for agronomic improvement has been underfunded in comparison to crops such as maize and wheat (Cock 1985).

This lack of investment has meant that many basic aspects related to the crop's biology and ecology have still not been sufficiently studied. For instance—and relevant for the purpose of this book—due to a paucity of scientific information, it is impossible to thoroughly assess the likelihood of gene flow between cassava and its wild relatives. There is also a lack of knowledge about the extent of outcrossing of different cassava cultivars, and the distances of pollen flow and outcrossing rates need to be investigated further, under both controlled and field conditions.

Farmer practices are of great importance in cassava cultivation, since they can either facilitate or hamper the enhancement of heterozygosity and genetic diversity in the farmers' stocks of clones. In this respect, the potential contribution of volunteer seedlings to crop genetic diversity needs to be quantified by assessing the spatial genetic structure and pollen-mediated gene flow within and between fields. Also, the ecological dynamics of cassava seed dispersal and predation rates need to be better understood, and the fate of shattered seeds should be studied and quantified. If dormancy in a soil seed bank is part of seed biology, then the factors that inhibit and promote the germination of cassava seeds must also be characterized.

Due to the limited amount of information that is available, no satisfactory conclusions can be drawn about the likelihood of gene flow between cassava and all its wild relatives. While there appear to be weak genetic barriers between domesticated cassava and many of its wild relatives, the necessity of synchronous flowering represents a major hurdle for spontaneous hybridization in the wild. However, in areas of sympatry with GP-1 species, gene flow and introgression between cassava and its wild relatives may occur. This could also apply for some of the species in GP-2, depending on the level of F1 hybrid fertility and that of the plants' progeny in later generations. It therefore seems that, in principle, the likelihood of gene introgression between cassava and its wild relatives—assuming physical proximity (less than 30 m) and flowering overlap, and being conditional on the nature of the respective trait(s)—is as follows:

- *moderate* for the GP-1 species *Manihot esculenta* subsp. *flabellifolia*, subsp. *peruviana*, and *M. pruinosa* (all in S America), as well as for the GP-2 species *M. carthagenensis* subsp. *glaziovii*, or Ceará rubber, in Africa and Brazil (map 6.1)

- *low* for some of the GP-2 species, depending on their level of F1 fertility and that of their progeny in later generations

REFERENCES
....

Abraham AS (1957) Breeding of tuber crops in India. *Indian J. Genet. Plant Breed.* 17: 212–217.
Allem AC (1992) *Manihot* germplasm collecting priorities. In: Roca WM and Thro

AM (eds.) *Report of the First Meeting of the International Network for Cassava Genetic Resources* (Centro Internacional de Agricultura Tropical [CIAT]: Cali, Colombia), pp. 87–110.

——— (1994) The origin of *Manihot esculenta* Crantz (Euphorbiaceae). *Genet. Resour. Crop Evol.* 41: 133–150.

——— (1999) The closest wild relatives of cassava (*Manihot esculenta* Crantz). *Euphytica* 107: 123–133.

——— (2000) Ethnobotanical testimony on the ancestors of cassava (*Manihot esculenta* Crantz subsp. *esculenta*). *Plant Genet. Resour. Newsl.* 123: 19–22.

——— (2002a) The origins and taxonomy of cassava (*Manihot esculenta* Crantz subspecies *esculenta*). In: Hillocks RJ, Thresh MJ, and Bellotti AC (eds.) *Cassava: Biology, Production, and Utilization* (CAB International: Oxford, UK), pp. 1–16.

——— (2002b) State of conservation and utilization of wild *Manihot* genetic resources and biodiversity. *Plant Genet. Resour. Newsl.* 131: 16–22.

Allem AC, Mendes RA, Salomão AN, and Burle ML (2001) The primary gene pool of cassava (*Manihot esculenta* Crantz subspecies *esculenta*, Euphorbiaceae). *Euphytica* 120: 127–132.

Asiedu R, Bai KV, Terauchi R, Dixon AGO, and Hahn SK (1992) Status of wide crosses in cassava and yam. In: Thotttappily G (ed.) *Biotechnology: Enhancing Research on Tropical Crops in Africa; Proceedings of an International Conference, Held at the International Institute of Tropical Agriculture (IITA), Ibadan, Nigeria, November 26–30, 1990* (Technical Centre for Agricultural and Rural Cooperation: Wageningen, The Netherlands and International Institute of Tropical Agriculture: Ibadan, Nigeria), pp. 63–68.

Bai KV, Asiedu R, and Dixon AGO (1993) Cytogenetics of *Manihot* species and interspecific hybrids. In: Roca WM and Thro AM (eds.) *1st International Scientific Meeting of the Cassava Biotechnology Network, Cartagena, Colombia, August 25–28, 1992.* CIAT Working Document No. 123 (Centro Internacional de Agricultura Tropical [CIAT]: Cali, Colombia), pp. 51–55.

Balyejusa Kizito E, Bua A, Fregene M, Egwang T, Gullberg U, and Westerbergh A (2005) The effect of cassava mosaic disease on the genetic diversity of cassava in Uganda. *Euphytica* 146: 45–54.

Balyejusa Kizito E, Chiwona-Karltun L, Egwang T, Fregene M, and Westerbergh A (2007) Genetic diversity and variety composition of cassava on small-scale farms in Uganda: An interdisciplinary study using genetic markers and farmer interviews. *Genetica* 130: 301–318.

Biggs BJ, Smith MK, and Scott KJ (1986) The use of embryo culture for the recovery of plants from cassava (*Manihot esculenta* Crantz) seeds. *Plant Cell Tissue Organ Cult.* 6: 229–234.

Blair MW, Fregene MA, Beebe SE, and Ceballos H (2007) Marker-assisted selection in common beans and cassava. In: Guimaraes EP, Ruane J, Scherf BD, Sonnino A, and Dargie JD (eds.) *Marker-Assisted Selection: Current Status and Future Perspectives in Crops, Livestock, Forestry, and Fish* (Food and Agriculture Organization of the United Nations [FAO]: Rome, Italy), pp. 81–115.

Bolhuis GG (1953) A survey of some attempts to breed cassava varieties with a high content of proteins in the roots. *Euphytica* 20: 107–112.

―――― (1969) Intra- and interspecific crosses in the genus *Manihot*. In: Tai EA (ed.) *Proceedings of the [1st] International Symposium on Tropical Root Crops, University of the West Indies. St. Augustine, Trinidad, April 2–8, 1967* (University of the West Indies: St. Augustine, Trinidad), pp. 81–88.

Bryne D (1984) Breeding cassava. In: Jannick J (ed.) *Plant Breeding Reviews*, vol. 2 (AVI Publishing: Westpoint, CT, USA), pp. 73–134.

Carvalho LJCB and Schaal BA (2001) Assessing genetic diversity in the cassava (*Manihot esculenta* Crantz) germplasm collection in Brazil using PCR-based markers. *Euphytica* 120: 133–142.

Chandratna MF and Nanayakkara KDSS (1948) Studies in cassava: II. The production of hybrids. *Trop. Agric.* 104: 59–70.

Chavarriaga P, Prieto S, Herrera CJ, López D, Bellotti A, and Tohme J (2004) Screening transgenics unveils apparent resistance to hornworm (*E. ello*) in the nontransgenic, African cassava clone 60444. In: Alves A and Tohme J (eds.) *Adding Value to a Small-Farmer Crop: 6th International Scientific Meeting of the Cassava Biotechnology Network, CIAT, Cali, Colombia, March 8–14, 2004* (Centro Internacional de Agricultura Tropical [CIAT]: Cali, Colombia), p. 4.

CIAT [Centro Internacional de Agricultura Tropical, or International Center for Tropical Agriculture] (2006) *Annual Report 2005, Project IP3: Improved Cassava for the Developing World* (Centro Internacional de Agricultura Tropical [CIAT]: Cali, Colombia). www.ciat.cgiar.org/yuca/pdf/report_2005/contents.pdf.

―――― (2007) *Annual Report 2006 Project IP3: Improved Cassava for the Developing World* (Centro Internacional de Agricultura Tropical [CIAT]: Cali, Colombia). www.ciat.cgiar.org/yuca/pdf/report_2006/output_9.pdf.

Cock J (1985) *Cassava: New Potential for a Neglected Crop* (Westview Press: Boulder, CO, USA). 240 p.

Colombo C, Second G, and Charrier A (2000) Genetic relatedness between cassava (*Manihot esculenta* Crantz) and *M. flabellifolia* and *M. peruviana* based on both RAPD and AFLP markers. *Genet. Mol. Biol.* 23: 417–423.

Duputié A, David P, Debain C, and McKey D (2007) Natural hybridization between a clonally propagated crop, cassava (*Manihot esculenta* Crantz), and a wild relative in French Guiana. *Mol. Ecol.* 16: 3025–3038.

Eckert JE (1933) The flight range of the honeybee. *J. Apicult. Res.* 47: 257–285.

Elias M and McKey D (2000) The unmanaged reproductive ecology of domesticated plants in traditional agroecosystems: An example involving cassava and a call for data. *Acta Oec.* 21: 223–230.

Elias M, Panaud O, and Robert T (2000a) Assessment of genetic variability in a traditional cassava (*Manihot esculenta* Crantz) farming system, using AFLP markers. *Heredity* 85: 219–230.

Elias M, Rival L, and McKey D (2000b) Perception and management of cassava (*Manihot esculenta* Crantz) diversity among Makushi Amerindians of Guyana (South America). *J. Ethnobiol.* 20: 239–265.

Elias M, Penet L, Vindry P, McKey D, Panaud O, and Robert T (2001) Unmanaged sexual reproduction and the dynamics of genetic diversity of a vegetatively propagated crop plant, cassava (*Manihot esculenta* Crantz), in a traditional farming system. *Mol. Ecol.* 10: 1895–1907.

FAO [Food and Agriculture Organization of the United Nations] (2009) FAOSTAT data. http://faostat.fao.org/site/567/default.aspx.

Fregene MA and Puonti-Kaerlas J (2002) Cassava biotechnology. In: Hillocks RJ, Thresh MJ, and Bellotti AC (eds.) *Cassava: Biology, Production, and Utilization* (CAB International: Oxford, UK), pp. 179–207.

Fregene MA, Vargas J, Ikea J, Angel F, Tohme J, Asiedu RA, Akoroda MO, and Roca WM (1994) Variability of chloroplast DNA and nuclear ribosomal DNA in cassava (*Manihot esculenta* Crantz) and its wild relatives. *Theor. Appl. Genet.* 89: 719–727.

Fregene MA, Morante N, Sánchez T, Marín J, Ospina C, Barrera E, Gutiérrez J, Guerrero J, Bellotti A, Santos L, Alzate A, Moreno S, and Ceballos H (2006) Molecular markers for the introgression of useful traits from wild *Manihot* relatives of cassava: Marker-assisted selection of disease and root quality traits. *J. Root Crops* 32: 1–31.

Ganeshaiah K and Uma Shaanker R (1988) Evolution of a unique seed maturity pattern in *Croton bonplandianum* Baill. strengthens ant-plant mutualism for seed dispersal. *Oecol.* 77: 130–134.

Gulick P, Hershey C, and Esquinas-Alcazar J (1983) Genetic Resources of Cassava and Wild Relatives (International Board for Plant Genetic Resources [IBPGR]: Rome, Italy). 56 p.

Hahn SK, Terry ER, and Leuschner K (1980a) Cassava breeding for resistance to cassava mosaic disease. *Euphytica* 29: 673–683.

Hahn SK, Howland AK, and Terry ER (1980b) Correlated resistance of cassava to mosaic and bacterial blight diseases. *Euphytica* 29: 305–311.

Hahn SK, Bai KV, and Asiedu R (1990) Tetraploids, triploids, and 2*n* pollen from diploid interspecific crosses with cassava. *Theor. Appl. Genet.* 79: 433–439.

Halsey ME, Olsen KM, Taylor NJ, and Chavarriaga-Aguirre P (2008) Reproductive biology of cassava (*Manihot esculenta* Crantz) and isolation of experimental field trials. *Crop Sci.* 48: 49–58.

Hershey CH (in press) *Cassava Genetic Improvement: Theory and Practice* [Food and Agriculture Organization of the United Nations]: Rome, Italy).

Ihemere U, Arias-Garzon D, Lawrence S, and Sayre R (2006) Genetic modification of cassava for enhanced starch production. *Plant Biotech. J.* 4: 453–465.

IITA [International Institute for Tropical Agriculture] (1988) *IITA Annual Report and Research Highlights* (International Institute for Tropical Agriculture [IITA]: Ibadan, Nigeria).

Jennings DL (1957) Further studies in breeding cassava for virus resistance. *East Afr. Agr. J.* 22: 213–219.

—— (1959) *Manihot melanobasis* Muell. Arg.—a useful parent for cassava breeding. *Euphytica* 8: 157–162.

—— (1963) Variation in pollen and ovule fertility in varieties of cassava and the effect of interspecific crossing on fertility. *Euphytica* 12: 69–76.

—— (1976) Cassava. In: Simonds NW (ed.) *Evolution of Crop Plants* (Longman Scientific and Technical: Harlow, UK), pp. 81–84.

—— (1995) Cassava. In: Simmonds S and Smartt J (eds.) *Evolution of Crop Plants*. 2nd ed. (Longman Scientific and Technical: Harlow, UK), pp. 128–132.

Jennings DL and Iglesias CA (2002) Breeding for crop improvement. In: Hillocks RJ, Thresh MJ, and Bellotti AC (eds.) *Cassava: Biology, Production, and Utilization* (CAB International: Oxford, UK), pp. 149–166.

Kawano K (1980) Cassava. In: Fehr WR and Hadley HH (eds.) *Hybridization of Crop Plants* (American Society of Agronomy: Madison, WI, USA), pp. 225–233.

Kawano K, Amaya A, Daza P, and Ríos M (1978) Factors affecting efficiency of hybridization and selection in cassava. *Crop Sci.* 18: 373–380.

Ladino J, Mancilla LI, Chavarriaga P, Tohme J, and Roca WM (2001) Transformation of cassava cv. TMS60444 with *A. tumefaciens* carrying a *cry* 1Ab gene for insect resistance. In: Taylor NJ, Ogbe F, and Fauquet C (eds.) *Cassava, An Ancient Crop for Modern Times: Food, Health, Culture; CBN-5, 5th International Scientific Meeting of the Cassava Biotechnology Network, Donald Danforth Plant Science Center, St. Louis, Missouri, USA, November 4–9, 2001* (Donald Danforth Plant Science Center: St. Louis, MO, USA), CBN-V video archives S7–16. www.danforthcenter.org/media/video/cbnv/session7/S7–16.htm.

Lefèvre F and Charrier A (1993) Isozyme diversity within African *Manihot* germplasm. *Euphytica* 66: 73–80.

Lisci M and Pacini E (1997) Fruit and seed structural characteristics and seed dispersal in *Mercurialis annua* L. (Euphorbiaceae). *Acta Soc. Bot. Pol.* 66: 379–386.

Magoon ML, Jos JS, and Appan SG (1966) Cytomorphology of interspecific hybrid between cassava and Ceará rubber. *Chromosome Infor. Serv.* 7: 8–10.

Manu-Aduening JA, Lamboll RI, Dankyi AA, and Gibson RW (2005) Cassava diversity in Ghanaian farming systems. *Euphytica* 144: 331–340.

Meireles da Silva R, Bandel G, and Sodero Martins P (2003) Mating system in an experimental garden composed of cassava (*Manihot esculenta* Crantz) ethnovarieties. *Euphytica* 134: 127–135.

Mesquida J, Renard M, and Pierre JS (1988) Rapeseed (*Brassica napus* L.) productivity: The effect of honeybees (*Apis mellifera* L.) and different pollination conditions in cage and field tests. *Apidologie* 19: 51–72.

Mkumbira J, Chiwona-Karltun L, Lagercrantz U, Mahungu NM, Saka J, Mhone A, Bokanga M, Brimer L, Gullberg U, and Rosling H (2003) Classification of cassava into "bitter" and "cool" in Malawi: From farmers' perceptions to characterisation by molecular markers. *Euphytica* 132: 7–22.

Nassar NMA (1989) Broadening the genetic base of cassava, *Manihot esculenta* Crantz, by interspecific hybridization. *Can. J. Plant Sci.* 69: 1071–1073.

———— (2001a) Cassava, *Manihot esculenta* Crantz, and wild relatives: Their relationships and evolution. *Genet. Resour. Crop Evol.* 48: 429–436.

———— (2001b) The nature of apomixis in cassava (*Manihot esculenta* Crantz). *Hereditas* 134: 185–187.

Nassar NMA and Collevatti RG (2005) Microsatellite markers confirm high apomixis level in cassava bred clones. *Hereditas* 142: 33–37.

Nassar NMA and de Carvalho CGP (1990) Insetos polinizadores e seus comportamentos nas espécies silvestres da mandioca, *Manihot* spp. *Cienc. Cult.* 42: 703–705.

Nassar NMA and Ortiz R (2007) Cassava improvement: Challenges and impacts. *J. Agric. Sci.* 145: 163–171.

Nassar NMA, da Silva JR, and Vieira C (1986) Hibridação interespecífica entre mandioca e espécies silvestres de *Manihot*. *Cienc. Cult.* 38: 1050–1055.

Nassar NMA, de Carvalho CGP, and Vieira C (1996) Overcoming crossing barriers between cassava, *Manihot esculenta* Crantz, and a wild relative, *M. pohlii* Wawra. *Brazilian J. Genet.* 19: 617–620.

Nichols RFW (1947) Breeding cassava for virus resistance. *East Afr. Agric. J.* 15: 154–160.

Olsen KM (2004) SNPs, SSRs, and inferences on cassava's origin. *Plant Mol. Biol.* 56: 517–526.

Olsen KM and Schaal BA (1999) Evidence on the origin of cassava: Phylogeography of *Manihot esculenta*. *Proc. Nat. Acad. Sci. USA* 96: 5586–5591.

———— (2001) Microsatellite variation in cassava (*Manihot esculenta*, Euphorbiaceae) and its wild relatives: Further evidence for a southern Amazonian origin of domestication. *Am. J. Bot.* 88: 131–142.

Pickersgill B (1998) Crop introductions and the development of secondary areas of diversity. In: Prendergast HD (ed.) *Plants for Food and Medicine* (Royal Botanic Gardens, Kew: Richmond, Surrey, UK), pp. 93–105.

Pujol B and Mckey D (2006) Size asymmetry in intraspecific competition and the density-dependence of inbreeding depression in a natural plant population: A case study in cassava (*Manihot esculenta* Crantz, Euphorbiaceae). *J. Evol. Biol.* 19: 85–96.

Pujol B, Gigot G, Laurent G, Pinheiro-Kluppel M, Elias M, Hossaert-Mckey M, and Mckey D (2002) Germination ecology of cassava (*Manihot esculenta* Crantz, Euphorbiaceae) in traditional agroecosystems: Seed and seedling biology of a vegetatively propagated domesticated plant. *Econ. Bot.* 56: 366–379.

Pujol B, Mühlen G, Garwood N, Horoszowski Y, Douzery EJP, and Mckey D (2005a) Evolution under domestication: Contrasting functional morphology of seedlings in domesticated cassava and its closest wild relatives. *New Phytol.* 166: 305–318.

Pujol B, David P, and Mckey D (2005b) Microevolution in agricultural environments: How a traditional Amerindian farming practice favours heterozygosity in cassava (*Manihot esculenta* Crantz, Euphorbiaceae). *Ecol. Letters* 8: 138–147.

Pujol B, Renoux F, Elias M, Rival L, and Mckey D (2007) The unappreciated ecology of landrace populations: Conservation consequences of soil seed banks in cassava. *Biol. Cons.* 136: 541–551.

Raemakers CJJM, Sofiari E, Jacobsen E, and Visser RGF (1997) Regeneration and transformation in cassava. *Euphytica* 96: 153–161.

Rajendran PG, Ravindran CS, Nair SG, and Nayar TVR (2000) *True Cassava Seeds (TCS) for Rapid Spread of the Crop in Non-Traditional Areas* (Central Tuber Crop Research Institute [CTCRI], Indian Council of Agricultural Research: Kerala, India).

Ramsay G, Thomson CE, Neilson S, and Mackay GR (1999) Honeybees as vectors of GM oilseed rape pollen. In: Lutman PJW (ed.) *Gene Flow and Agriculture: Relevance for Transgenic Crops; Proceedings of a Symposium Held at the University of Keele, Staffordshire, April 12–14, 1999.* BCPC Symposium Proceedings 72 (British Crop Protection Council [BCPC]: Farnham, UK), pp. 209–214.

Renvoize BS (1972) The area of origin of *Manihot esculenta* as a crop plant—a review of the evidence. *Econ. Bot.* 26: 352–360.

Rieseberg LH (1995) The role of hybridization in evolution: Old wine in new skins. *Am. J. Bot.* 82: 944–953.

Rieseberg LH and Ellstrand NC (1993) What can morphological and molecular markers tell us about plant hybridization? *Crit. Rev. Plant Sci.* 12: 213–241.

Roa AC, Maya MM, Duque MC, Tohme J, Allem AC, and Bonierbale MW (1997) AFLP analysis of relationships among cassava and other *Manihot* species. *Theor. Appl. Genet.* 95: 741–750.

Rogers DJ (1963) Studies of *Manihot esculenta* and related species. *Bull. Torrey Bot. Club* 90: 43–54.

——— (1965) Some botanical and ethnological considerations of *Manihot esculenta*. *Econ. Bot.* 19: 369–377.

Rogers DJ and Appan SG (1973) *Manihot, Manihotoides* (Euphorbiaceae). Flora Neotropica Monograph 13 (published for the Organization for Flora Neotropica, New York Botanical Garden, by Hafner Press: New York, NY, USA). 274 p.

Salick J, Cellinese N, and Knapp S (1997) Indigenous diversity of cassava: Generation, maintenance, use, and loss among the Amuesha, Peruvian upper Amazon. *Econ. Bot.* 51: 6–19.

Sauer JD (1993) *Historical Geography of Crop Plants: A Select Roster* (CRC Press: Boca Raton, FL, USA). 309 p.

Second G, Allem AC, Mendes RA, Carvalho LJCB, Emperaire L, Ingram C, and Colombo C (1997) Molecular markers (AFLP)-based *Manihot* and cassava numerical taxonomy and genetic structure analysis in progress: Implications for their dynamic conservation and genetic mapping. *Afr. J. Root Tuber Crops* 2: 140–147.

Sheela MN, Unnikrishnan M, Edison S, and Easwari Amma CS (2004) *Manihot caerulescence*—a new source of resistance to cassava mosaic disease (ICMD). In: Alves A and Tohme J (eds.) *Adding Value to a Small-Farmer Crop: 6th International Scientific Meeting of the Cassava Biotechnology Network, CIAT, Cali, Colombia, March 8–14, 2004* (Centro Internacional de Agricultura Tropical [CIAT]: Cali, Colombia), p. 48. www.ciat.cgiar.org/biotechnology/cbn/sixth_international_meeting/posters.htm.

Storey HH and Nichols RFW (1938) Studies of the mosaic of cassava. *Ann. Appl. Biol.* 25: 790–806.

Tanksley SD and Nelson JC (1996) Advanced backcross QTL analysis: A method for the simultaneous discovery and transfer of valuable QTLs from unadapted germplasm into elite breeding lines. *Theor. Appl. Genet.* 92: 191–203.

Taylor N, Chavarriaga P, Raemakers K, Siritunga D, and Zhang P (2004a) Development and application of transgenic technologies in cassava. *Plant Mol. Biol.* 56: 671–688.

Taylor N, Kent L, and Fauquet C (2004b) Progress and challenges for the deployment of transgenic technologies in cassava. *AgBioForum* 7: 51–56.

Ugent D, Pozorski S, and Pozorski T (1986) Archaeological manioc (*Manihot*) from coastal Peru. *Econ. Bot.* 40: 78–102.

Unnikrishnan M, Easwari Amma CS, Santha V Pillai, Sheela MN, Anantharaman M, and Nair RR (2004) Varietal improvement programme in cassava. *Annual Report 2002–2003* (Central Tuber Crops Research Institute [CTCRI]: Thiruvananthapuram, Kerala, India), pp. 24–26.

Wanyera NMW, Hahn SK, and Aken'ova ME (1992) Introgression of Ceará rubber (*Manihot glaziovii* Muell-Arg.) into cassava (*M. esculenta* Crantz): Morphological and electrophoretic evidence. In: Akoroda MO (ed.) *Root Crops for Food Security in Africa: Proceedings of the 5th Triennial Symposium of the International Society for Tropical Root Crops-Africa Branch (ISTRC-AB), Kampala, Uganda, November 22–28, 1992* (International Society for Tropical Roots Crops, African Branch [ISTRC-AB]: Ibadan, Nigeria), pp. 125–130.

SEVEN

Chickpea
(*Cicer arietinum* L.)

Center(s) of Origin and Diversity

The chickpea plant originated in the Fertile Crescent, specifically in the region between SE Turkey and Syria (Maesen 1987; Ladizinsky 1995).

Vavilov (1951) identified four primary centers of species diversity within the genus *Cicer* L.: the Mediterranean region, central Asia, the Near East, and India. Ethiopia is considered to be a secondary center of *Cicer* diversity.

Biological Information

Flowering

The chickpea inflorescence is a raceme, with one to several flowers per axis. Individual flowers are zygomorphic and typical of the Papilionoideae, with two anterior keel petals that almost completely enclose the male and female organs. Anthers usually dehisce the day before the flowers open, leading to high rates of self-fertilization, although pollen grains are still viable when the flower subsequently opens. Cleistogamy can also occur. The time and duration of flowering are dependent on temperature and photoperiod (Roberts et al. 1980; K. Singh et al. 1997).

Pollen Dispersal and Longevity

Chickpea pollen is disseminated by insects. The flowers are visited by butterflies, honeybees, solitary bees, and bumblebees (Malhotra and Singh 1986; Free 1993; Tayyar et al. 1996).

Chickpea pollen formed at low temperature is usually sterile, and most current cultivars will not set pods if the average daily temperature is below 15°C (Savithri et al. 1980; Srinivasan et al. 1998, 1999; Croser et al. 2003a). To our knowledge, no information is available regarding the longevity of chickpea pollen in terms of hours or days.

Sexual Reproduction

Chickpeas are an annual diploid ($2n = 2x = 16$). They are a self-compatible, highly autogamous crop, with outcrossing rates of less than 1%, depending on environmental conditions and the cropping season (Niknejad and Khosh-Khui 1972; Smithson et al. 1985; Malhotra and Singh 1986; K. Singh 1987; Tayyar et al. 1996).

Vegetative Reproduction

Although cultivated chickpea plants are exclusively propagated by seed, vegetative propagation through stem cuttings is often used for multiplying experimental hybrids (Rupela and Dart 1981; Rupela 1982; Bassiri et al. 1985; Collard et al. 2002; Syed et al. 2002; Dane-hloueipour et al. 2006).

Seed Dispersal and Dormancy

Wild *Cicer* species shed their dehiscent pods, which drop to the ground where they then burst and disperse their seeds (Ladizinsky and Adler 1976a). In cultivars, this feature is suppressed through breeding; pods are retained on the plant and dehiscence is reduced (K. Singh and Bejiga 1991).

Cultivated chickpeas do not show seed dormancy, but some wild *Cicer* species do (K. Singh and Ocampo 1997).

Volunteers, Ferals, and Their Persistence

Chickpeas are not known to occur in the wild, but volunteer plants can appear as weeds in subsequent cropping cycles. Cultivated chickpeas cannot colonize successfully without human intervention.

Weediness and Invasiveness Potential

Cultivated chickpeas do not compete with other plant species in the wild, particularly weeds (Muehlbauer 1993). However, some wild species occur in weedy or disturbed habitats, such as fallows and roadsides (e.g., *C. reticulatum* Ladiz. and *C. bijugum* Rech. f.).

Crop Wild Relatives

The genus *Cicer* contains 43 species and is divided into four sections: *Monocicer* Popov, *Chamaecicer* Popov, *Polycicer* Popov, and *Acanthacicer* Popov (Croser et al. 2003b). Together with eight other annual wild relatives, cultivated chickpeas are placed in subg. *Pseudononis* Popov (Maesen 1987). The remaining 34 wild *Cicer* species are perennial shrubby plants and comprise subg. *Viciastrum* Popov (Maesen 1972, 1987). All nine annual and eight of the 34 perennial *Cicer* species are diploids, with $2n = 2x = 16$ (reviewed by Ahmad et al. 2005).

Hybridization

The wild relatives of chickpeas can be grouped into primary, secondary, and tertiary gene pools, according to cross-compatibility and their level of genetic similarity with cultivated chickpeas (reviewed by Ahmad et al. 2005; table 7.1). The extent of compatibility and the fertility of interspecific hybrids are genotype dependent (Ladizinsky and Adler 1976b; K. Singh et al. 1994; Collard et al. 2003; Yadav et al. 2007).

Primary Gene Pool (GP-1)

Besides cultivated chickpeas (*Cicer arietinum* L.), the primary gene pool includes the chickpea progenitor *C. reticulatum* Ladiz. (Ladizinsky et al. 1988). Both species are fully cross-compatible. Hybrids between them can be easily obtained and are viable and fully fertile (see, e.g., Ladizinsky 1975, 1995; Ladizinsky and Adler 1976b; Pundir and Maesen 1983; K. Singh and Ocampo 1993, 1997; S. Singh et al. 2005).

Gene flow between cultivated chickpeas (*C. arietinum*) and their wild relative *C. reticulatum* may exist under natural conditions; how-

Table 7.1 Cultivated chickpea (*Cicer arietinum* L.) and its wild relatives

Species	Origin and distribution
Primary gene pool (GP-1)	
C. *arietinum* L. (cultivated chickpea)	known only from cultivation
C. *reticulatum* Ladiz.	endemic to SE Turkey; weedy wild progenitor
Secondary gene pool (GP-2)	
C. *echinospermum* P.H. Davis	Turkey, Iraq
Tertiary gene pool (GP-3)	
C. *bijugum* Rech. f.	Iraq, Syria, Turkey; weedy
C. *judaicum* Boiss.	Israel, Lebanon, Palestine, Turkey
C. *pinnatifidum* Jaub. & Spach	Cyprus, Iraq, Turkey, Syria, Lebanon, Armenia
C. *cuneatum* Hochst. *ex* A. Rich.	N and NE Africa (Egypt, Eritrea, Ethiopia, Sudan)
C. *yamashitae* Kitam.	endemic to Afghanistan
C. *chorassanicum* (Bunge) Popov	Afghanistan, Iran
perennial *Cicer* species (34 species)	

Sources: Origin and distribution information is from Robertson et al. 1995; gene pool designations follow Ahmad et al. 2005.

ever, no naturally occurring hybrids have been reported thus far. Iruela et al. (2002) detected the introgression of C. *arietinum* alleles in a C. *reticulatum* population from the International Center for Agricultural Research in the Dry Areas (ICARDA), but this observation may be due to manmade crosses, since accessions from GP-1 are routinely used in breeding programs at ICARDA (SM Udupa, pers. comm. 2008).

Secondary Gene Pool (GP-2)

The secondary gene pool of cultivated chickpeas includes only the wild annual species *Cicer echinospermum* P.H. Davis. A number of studies show that C. *echinospermum* can be crossed more or less readily both with cultivated chickpeas and C. *reticulatum*, generating vegetatively normal, viable F1 progeny (see, e.g., Ladizinsky 1975, 1995; Ladizinsky and Adler 1976a, b; K. Singh and Ocampo 1993, 1997; S. Singh et al. 2005). Some authors therefore consider this wild relative to belong to GP-1 (see, e.g., Ladizinsky et al. 1988; Croser et al. 2003b).

The cross-compatibility of C. *echinospermum* with cultivated chickpeas, however, is somewhat more difficult to achieve than intraspecific crosses of chickpeas (Smartt 1990; K. Singh and Ocampo 1993). Hybridization success depends on several factors, including genotype, flower size, and environment. For instance, in Aleppo, Syria (at ICARDA), outcrossing rates and hybrid fertility are greater than in Hyderabad, India (at the International Crops Research Institute for the Semi-Arid Tropics [ICRISAT]) (SM Udupa, pers. comm. 2008).

Hand pollination results in the production of vegetatively normal F_1 hybrids (Ladizinsky and Adler 1976b; Sheila et al. 1992; Ladizinsky 1995; Pundir and Mengesha 1995). The fertility of F_1 hybrids is reduced and can range from complete sterility up to 54% pollen fertility, with occasional seed set (Ladizinsky and Adler 1976b; Verma et al. 1990; Pundir et al. 1992; Pundir and Mengesha 1995). Some F_2 individuals have been produced with highly variable levels of fertility, including fully fertile as well as completely sterile F_2 plants (Pundir and Mengesha 1995; K. Singh and Ocampo 1997). Gene introgression between cultivated chickpeas and *C. echinospermum* is possible, and the species has been used in chickpea improvement programs (Knights et al. 2002; Yadav et al. 2002; Collard et al. 2003).

Tertiary Gene Pool (GP-3)

The tertiary gene pool of cultivated chickpeas includes the remaining six annual wild species (*Cicer pinnatifidum* Juab. & Spach, *C. judaicum* Boiss., *C. bijugum*, *C. chorassanicum* (Bunge) Popov, *C. yamashitae* Kitam., and *C. cuneatum* Hochst. *ex* A. Rich.) and all of the perennial species.

Cultivated chickpeas are sexually compatible with the first three species, but hybridization is not easy via traditional methods, so artificial techniques such as embryo rescue are used in crossing experiments (Badami et al. 1997; Mallikarjuna 1999; Clarke et al. 2004, 2006). No spontaneous outcrossing has been reported, and the hybrids are completely sterile (Croser et al. 2003b; Ahmad et al. 2005; McNeil et al. 2007). Reports of successful crosses between chickpeas and some of the GP-3 species without the use of sophisticated artificial hybridization techniques (Verma et al. 1990; K. Singh et al. 1994, A. Singh et al. 1999; N. Singh et al. 1999)—which still required growth regulators to stimulate embryo development—have been questioned (Ahmad et al.

2005; McNeil et al. 2007). It seems to be quite unlikely that spontaneous hybrids could occur under natural conditions.

No authentic hybrids, using any conventional or biotechnological procedures, are known between cultivated chickpeas and any of the other annual (*C. cuneatum*, *C. chorassanicum*, *C. yamashitae*) or perennial *Cicer* species (Mallikarjuna 1999; Clarke et al. 2006; Millan et al. 2006; Sharma et al. 2006; McNeil et al. 2007; Toker et al. 2007), due to the presence of strong post-zygotic barriers resulting in embryo abortion at early developmental stages (see, e.g., Mercy and Kakkar 1975; Ahmad et al. 1988; Stamigna et al. 2000; Ahmad and Slinkard 2004; Babb and Muehlbauer 2005).

Outcrossing Rates and Distances

Pollen Flow

To our knowledge, no studies have been published measuring the distances of pollen flow in chickpeas. However, several experiments have been conducted to estimate outcrossing rates between adjacent plants. These studies showed that cross-fertilization in chickpeas is often below 1%. The highest outcrossing rates reported were up to 1.25% by Toker et al. (2006) and up to 1.92% by Gowda (1981).

Isolation Distances

The separation distance established by regulatory authorities for chickpea seed production is 3 m in the United States (CCIA 2009). In OECD countries, chickpea varieties for seed production "shall be isolated from other crops by a definite barrier or a space sufficient to prevent mixture during harvest" (OECD 2008, p. 18).

State of Development of GM Technology

The genetic transformation of chickpeas is currently in process. Several transformation protocols have been developed (Fontana et al. 2004; Polowick et al. 2004; Senthil et al. 2004) and genes conferring a tolerance for insect pests (Kar et al. 1997; Ignacimuthu and Prakash 2006; Indurker et al. 2007) and abiotic stresses (Shukla et al. 2006) have been transferred to chickpeas. Researched GM traits for transforming chickpeas include tolerance for the *Helicoverpa* pod borer, drought, chilly temperatures, and salinity. There are also reports of experimental field trials in Bangladesh and India.

Agronomic Management Recommendations

There are no specific recommendations.

Crop Production Area

Chickpeas are grown in over 50 countries in tropical, subtropical, and temperate regions all over the world, including Africa (Ethiopia, Malawi, and Tanzania), N and S America (Canada, Mexico, the United States, Argentina, Chile, and Peru), Asia (from India and SE Asia to the Middle East and the Mediterranean region), Australia, and S Europe (Ladizinsky 1995; FAO 2009). The major regions of crop production are India, Pakistan, and Turkey—with 64%, 9%, and 6% of the world's production, respectively—and the total crop production area is estimated at 11.7 million ha (FAO 2009). No commercial production of GM chickpeas has been reported.

Research Gaps and Conclusions

Gene flow and introgression from cultivated chickpeas to wild relatives is likely only in the Mediterranean region and the Near East, where wild relatives occur. Although no naturally occurring hybrids have been reported thus far, and chickpeas are selfers, there is evidence that outcrossing rates under natural conditions can exceed the 0.9% threshold of maximum allowable GM trait contamination (Gowda 1981; Toker et al. 2006). Therefore, the likelihood of gene introgression from cultivated chickpeas to their wild relatives—assuming physical proximity (less than 3 m) and flowering overlap, and being conditional on the nature of the respective trait(s)—is as follows:

- *moderate* for the chickpea wild progenitor *Cicer reticulatum* (map 7.1)
- *low* for the wild relative *C. echinospermum* (map 7.1)
- *highly unlikely* for the remaining wild annual and perennial relatives

REFERENCES

Ahmad F and Slinkard AE (2004) The extent of embryo and endosperm growth following interspecific hybridization between *Cicer arietinum* L. and related annual wild species. *Genet. Resour. Crop Evol.* 51: 765–772.

Ahmad F, Slinkard AE, and Scoles GJ (1988) Investigations into the barrier(s) to interspecific hybridization between *Cicer arietinum* L. and eight other annual *Cicer* spp. *Plant Breed.* 100: 193–198.

Ahmad F, Gaur PM, and Croser J (2005) Chickpea (*Cicer arietinum* L.). In: Singh RJ and Jauhar PP (eds.) *Genetic Resources, Chromosome Engineering, and Crop Improvement: Grain Legumes*, vol. 1 (CRC Press: Boca Raton, FL, USA), pp. 187–217.

Babb SL and Muehlbauer FJ (2005) Interspecific cross incompatibility in hybridizations between *Cicer arietinum* L. and *Cicer anatolicum* Alef. In: *Proceedings of the XIII Plant & Animal Genomes Conference, Town & Country Convention Center, San Diego, California, January 15–19, 2005.* www.intl-pag.org/13/ab stracts/PAG13_P472.html.

Badami PS, Mallikarjuna N, and Moss J (1997) Interspecific hybridization between *Cicer arietinum* and *C. pinnatifidum. Plant Breed.* 116: 393–395.

Bassiri A, Slinkard AE, and Ahmad F (1985) Rooting of stem cuttings in *Cicer. Int. Chickpea Newsl.* 13: 10–11.

CCIA [California Crop Improvement Association] (2009) Chickpea Certification Standards. http://ccia.ucdavis.edu/seed_cert/seedcert_index.htm.

Clarke HJ, Kuo I, Kuo J, and Siddique KHM (2004) Abortion and stages for embryo rescue following wide crosses between chickpea (*Cicer arietinum* L.) and *C. bijugum* (K.H. Rech.). In: *5th European Conference on Grain Legumes, 2nd International Conference on Legume Genomics and Genetics: Legumes for the Benefit of Agriculture, Nutrition, and the Environment; Their Genomics, Their Products, and Their Improvement, Dijon, France, June 7–11, 2004* (Association européenne de recherche sur les protéagineux [AEP]: Paris, France), p. 192.

Clarke HJ, Wilson JG, Kuo I, Lülsdorf MM, Mallikarjuna N, Kuo J, and Siddique KHM (2006) Embryo rescue and plant regeneration *in vitro* of selfed chickpea (*Cicer arietinum* L.) and its wild annual relatives. *Plant Cell Tissue Organ Cult.* 85: 197–204.

Collard BCY, Pang ECK, Brouwer JB, and Taylor PWJ (2002) Use of stem cuttings to generate populations for QTL mapping in chickpea. *Int. Chickpea Pigeonpea Newsl.* 9: 30–32.

Collard BCY, Pang ECK, Ades PK, and Taylor PWJ (2003) Preliminary investigation of QTLs associated with seedling resistance to ascochyta blight from *Cicer echinospermum*, a wild relative of chickpea. *Theor. Appl. Genet.* 107: 719–729.

Croser JS, Clarke HJ, Siddique KHM, and Khan TN (2003a) Low temperature stress: Implications for chickpea (*Cicer arietinum* L.) improvement. *Crit. Rev. Plant Sci.* 22: 185–219.

Croser JS, Ahmad F, Clarke HJ, and Siddique KHM (2003b) Utilisation of wild *Cicer* in chickpea improvement—progress, constraints, and prospects. *Aust. J. Agric. Res.* 54: 429–444.

Danehloueipour N, Yan G, Clarke HJ, and Siddique KHM (2006) Successful stem cutting propagation of chickpea, its wild relatives, and their interspecific hybrids. *Aust. J. Exp. Agric.* 46: 1349–1354.

FAO [Food and Agriculture Organization of the United Nations] (2009) FAOSTAT data. http://faostat.fao.org/site/567/default.aspx.

Fontana GS, Santini L, Caretto S, Frugis G, and Mariotti D (2004) Genetic transformation in the grain legume *Cicer arietinum* L. (chickpea). *Plant Cell Rep.* 12: 194–198.

Free JB (1993) *Insect Pollination of Crops.* 2nd ed. (Academic Press: San Diego, CA, USA). 684 p.

Gowda CLL (1981) Natural outcrossing in chickpea. *Int. Chickpea Newsl.* 5: 6.

Ignacimuthu S and Prakash S (2006) *Agrobacterium*-mediated transformation of chickpea with alpha-amylase inhibitor gene for insect resistance. *J. Biosci.* 31: 339–345.

Indurker S, Misra HS, and Eapen S (2007) Genetic transformation of chickpea (*Cicer arietinum* L.) with insecticidal crystal protein gene using particle gun bombardment. *Plant Cell Rep.* 26: 755–763.

Iruela M, Rubio J, Cubero JI, Gil J, and Millán T (2002) Phylogenetic analysis in the genus *Cicer* and cultivated chickpea using RAPD and ISSR markers. *Theor. Appl. Genet.* 104: 643–651.

Kar S, Basu D, Das S, Ramkrishnan NA, Mukherjee P, Nayak P, and Sen SK (1997) Expression of *cry*1A(c) gene of *Bacillus thuringiensis* in transgenic chickpea plants inhibits development of pod-borer (*Heliothis armigera*) larvae. *Transgen. Res.* 6: 177–185.

Knights T, Brinsmead B, Fordyce M, Wood J, Kelly A, and Harden S (2002) Use of the wild relative *Cicer echinospermum* in chickpea improvement. In: McComb JA (ed.) *Plant Breeding for the 11th Millennium: Proceedings of the 12th Australasian Plant Breeding Conference, Perth, Australia, September 15–20, 2002* (Australasian Plant Breeding Association: Perth, WA, Australia), pp. 150–154.

Ladizinsky G (1975) A new *Cicer* from Turkey. *Notes Royal Bot. Gard. Edinburgh* 34: 201–202.

——— (1995) Chickpea. In: Simmonds S and Smartt J (eds.) *Evolution of Crop Plants.* 2nd ed. (Longman Scientific and Technical: Harlow, UK), pp. 258–261.

Ladizinsky G and Adler A (1976a) Genetic relationships among the annual species of *Cicer* L. *Theor. Appl. Genet.* 48: 197–204.

——— (1976b) The origin of chickpea *Cicer arietinum* L. *Euphytica* 25: 211–217.

Ladizinsky G, Pickersgill B, and Yamamoto K (1988) Exploitation of wild relatives of the food legumes. In: Summerfield RJ (ed.) *World Crops: Cool Season Food Legumes* (Kluwer Academic: Dordrecht, The Netherlands), pp. 967–978.

Maesen LJG van der (1972) Cicer *L., A Monograph of the Genus, with Special Reference to the Chickpea (*Cicer arietinum *L.), Its Ecology, and Cultivation.* Communication Agricultural University 72–10 (Mededelingen Landbouwhogeschool: Wageningen, The Netherlands). 133 p.

——— (1987) Origin, history, and taxonomy of chickpea. In: Saxena MC and Singh KB (eds.) *The Chickpea* (CAB International: Wallingford, UK), pp. 11–34.

Malhotra RS and Singh KB (1986) Natural cross pollination in chickpea. *Int. Chickpea Newsl.* 14: 4–5.

Mallikarjuna N (1999) Ovule and embryo culture to obtain hybrids from interspecific incompatible pollinations in chickpea. *Euphytica* 110: 1–6.

McNeil D, Ahmad F, Abbo S, and Bahl PN (2007) Genetics and cytogenetics. In: Yadav SS, Redden R, Chen W, and Sharma B (eds.) *Chickpea Breeding and Management* (CAB International: Wallingford, UK), pp. 321–337.

Mercy ST and Kakkar SK (1975) Barriers to interspecific crossing in *Cicer. Proc. Indian Nat. Sci. Acad., Sect. B* 41: 78–82.

Millan T, Clarke HJ, Siddique KHM, Buhariwalla HK, Gaur PM, Kumar J, Gil J, Kahl G, and Winter P (2006) Chickpea molecular breeding: New tools and concepts. *Euphytica* 147: 81–103.

Muehlbauer FJ (1993) Food and grain legumes. In: Janick J and Simon JE (eds.) *New Crops* (John Wiley and Sons: New York, NY, USA), pp. 256–265.

Niknejad M and Khosh-Khui M (1972) Natural cross-pollination in gram (*Cicer arietinum* L.). *Indian J. Agr. Sci.* 42: 273–274.

OECD [Organization for Economic Cooperation and Development] (2008) OECD Scheme for the Varietal Certification of Grass and Legume Seed Moving in International Trade: Annex VI to the Decision. C(2000)146/FINAL. OECD Seed

Schemes, February 2008. www.oecd.org/document/o/o,3343,en_2649_33905 _1933504_1_1_1_1,00.html.

Polowick P, Baliski D, and Mahon J (2004) *Agrobacterium tumefaciens*-mediated transformation of chickpea (*Cicer arietinum* L.): Gene integration, expression, and inheritance. *Plant Cell Rep.* 23: 485–491.

Pundir RPS and Mengesha MH (1995) Cross compatibility between chickpea and its wild relative, *Cicer echinospermum* Davis. *Euphytica* 83: 241–245.

Pundir RPS and Maesen LJG van der (1983) Interspecific hybridization in *Cicer*. *Int. Chickpea Newsl.* 8: 4–5.

Pundir RPS, Mengesha MH, and Reddy GV (1992) Interspecific hybridization in *Cicer*. *Int. Chickpea Newsl.* 26: 6–8.

Roberts EH, Summerfield RJ, Minchin FR, and Hadley P (1980) Phenology of chickpeas (*Cicer arietinum*) in contrasting aerial environments. *Expl. Agric.* 16: 343–360.

Rupela OP (1982) Rooting chickpea cuttings from field-grown plants. *Int. Chickpea Newsl.* 7: 9–10.

Rupela OP and Dart PJ (1981) Vegetative propagation of chickpea. *Int. Chickpea Newsl.* 4: 12–13.

Savithri KS, Ganapathy PS, and Sinha SK (1980) Sensitivity to low tolerance in pollen germination and fruit set in *Cicer arietinum* L. *J. Exp. Bot.* 31: 475–481.

Senthil G, Williamson B, Dinkins RD, and Ramsay G (2004) An efficient transformation system for chickpea (*Cicer arietinum* L.). *Plant Cell Rep.* 23: 297–303.

Sharma HC, Bhagwat MP, Pampapathy G, Sharma JP, and Ridsdill-Smith TJ (2006) Perennial wild relatives of chickpea as potential sources of resistance to *Helicoverpa armigera*. *Genet. Resour. Crop Evol.* 53: 131–138.

Sheila VK, Moss JP, Gowda CLL, and van Rheenen HA (1992) Interspecific hybridization between *Cicer arietinum* and wild *Cicer* species. *Int. Chickpea Newsl.* 27: 11–13.

Shukla RK, Raha S, Tripathi V, and Chattopadhyay D (2006) Expression of CAP2, an APETALA2–family transcription factor from chickpea, enhances growth and tolerance to dehydration and salt stress in transgenic tobacco. *Plant Physiol.* 142: 113–123.

Singh A, Singh NP, and Asthana AN (1999) Genetic potential of wide crosses in chickpea. *Legume Res.* 22: 19–25.

Singh KB (1987) Chickpea breeding. In: Saxena MC and Singh KB (eds.) *The Chickpea* (CAB International: Oxford, UK), pp. 127–162.

Singh KB and Bejiga G (1991) Evaluation of world collection of chickpea for resistance to pod dehiscence. *J. Genet. Breed.* 45: 93–96.

Singh KB and Ocampo B (1993) Interspecific hybridization in annual *Cicer* species. *J. Genet. Breed.* 47: 199–204.

———— (1997) Exploitation of wild *Cicer* species for yield improvement in chickpea. *Theor. Appl. Genet.* 95: 418–423.

Singh KB, Malhorta RS, Halila H, Knights EJ, and Verma MM (1994) Current status and future strategy in breeding chickpea for resistance to biotic and abiotic stresses. *Euphytica* 73: 137–149.

Singh KB, Pundir RPS, Robertson LD, van Rheenen HA, Singh U, Kelley TJ, Parthasarathy Rao P, Johansen C, and Saxena NP (1997) Chickpea. In: Fuccillo DA, Sears L, and Stapleton P (eds.) *Biodiversity in Trust: Conservation and Use of Plant Genetic Resources in CGIAR [Consultative Group on International Agricultural Research] Centres* (Cambridge University Press: Cambridge, UK), pp. 100–113.

Singh NP, Singh A, Asthana AN, and Singh A (1999) Studies on interspecific cross-ability barriers in chickpea. *Indian J. Pulses Res.* 12: 13–19.

Singh S, Gumber RK, Joshi N, and Singh K (2005) Introgression from wild *Cicer reticulatum* to cultivated chickpea for productivity and disease resistance. *Plant Breed.* 124: 477–480.

Smartt J (1990) *Grain Legumes: Evolution and Genetic Resources* (Cambridge University Press: Cambridge, UK), pp. 229–242.

Smithson JB, Thompson JA, and Summerfield RJ (1985) Chickpea (*Cicer arietinum* L.). In: Summerfield RJ and Roberts EH (eds.) *Grain Legume Crops* (Collins: London, UK), pp. 312–390.

Srinivasan A, Johansen C, and Saxena NP (1998) Cold tolerance during early reproductive growth of chickpea (*Cicer arietinum* L.): Characterisation of stress and genetic variation in pod set. *Field Crops Res.* 57: 181–193.

Srinivasan A, Saxena NP, and Johansen C (1999) Cold tolerance during early reproductive growth of chickpea (*Cicer arietinum* L.): Genetic variation in gamete development and function. *Field Crops Res.* 60: 209–222.

Stamigna C, Crino P, and Saccardo F (2000) Wild relatives of chickpea: Multiple disease resistance and problems to introgression in the cultigen. *J. Genet. Breed.* 54: 213–219.

Syed H, Ahsan-ul-Haq M, and Shah TM (2002) Vegetative propagation of chickpea (*Cicer arietinum* L.) through stem cuttings. *Asian J. Plant Sci.* 1: 218–219.

Tayyar RI, Federici CV, and Waines JG (1996) Natural outcrossing in chickpea (*Cicer arietinum* L.). *Crop Sci.* 36: 203–205.

Toker C, Canci H, and Ceylan F (2006) Estimation of outcrossing rate in chickpea (*Cicer arietinum* L.) sown in autumn. *Euphytica* 151: 201–205.

Toker C, Canci H, and Yildirim T (2007) Evaluation of perennial wild *Cicer* species for drought resistance. *Genet. Resour. Crop Evol.* 8: 1781–1786.

Vavilov NI (1951) The origin, variation, immunity, and breeding of cultivated plants. *Chron. Bot.* 13: 1–366.

Verma MM, Sandhu JS, Brar HS, and Brar JS (1990) Crossability studies in different species of *Cicer* (L.). *Crop Improv.* 17: 179–181.

Yadav SS, Turner NC, and Kumar J (2002) Commercialization and utilization of wild genes for higher productivity in chickpea. In: McComb JA (ed.) *Plant Breeding for the 11th Millennium: Proceedings of the 12th Australasian Plant Breeding Conference, Perth, Australia, September 15–20, 2002* (Australasian Plant Breeding Association: Perth, WA, Australia), pp. 155–160.

Yadav SS, Redden R, Chen W, and Sharma B (eds.) (2007) *Chickpea Breeding and Management* (CAB International: Wallingford, UK). 638 p.

EIGHT

Common Bean
(*Phaseolus vulgaris* L.)

There are five domesticated species in the genus *Phaseolus* (*P. acuti-folius* A. Gray, *P. coccineus* L., *P. lunatus* L., *P. dumosus* Macfad., and *P. vulgaris* L.), each derived from a distinct wild ancestor (Debouck 1991; Smartt 1990). This chapter focuses on the domesticated common bean *P. vulgaris*, since this species comprises more than 85% of the world's bean production and is the one most subject to genetic modification. However, transformation protocols have also been developed for *P. acutifolius* and *P. coccineus* (Dillen et al. 1997; de Clercq et al. 2002; Broughton et al. 2003; see below).

Center(s) of Origin and Diversity

The common bean plant (*Phaseolus vulgaris*) originated in the Americas. There are two major independent centers of origin, which are also the centers of domestication for common beans (Debouck 2000): one in Mesoamerica (probably W Mexico) for smaller seeded, 'S' phaseolin cultivars, and one in the southern Andes of S America for larger seeded, 'T' phaseolin cultivars (Gepts et al. 1986; Gepts and Bliss 1988; Koenig and Gepts 1989a; Sonnante et al. 1994; Chacón et al. 2005). A further minor domestication center may be located in Colombia (Gepts and De-bouck 1991; Chacón et al. 1996; Tohme et al. 1996).

The primary center of diversity for the genus *Phaseolus* lies in Mesoamerica, more concretely in central and S Mexico, bordering Guatemala. The region is home to about 70 to 80 wild *Phaseolus* species and to the ancestors of the 5 domesticated species (Debouck 1986; Freytag and Debouck 2002). The presence of secondary centers of diversification in Spain, Turkey, and Ethiopia (Purseglove 1968; Harlan 1992; Santalla et al. 2002; Ocampo et al. 2005) is controversial (Kami et al. 1995).

Biological Information

Flowering

The flowering behavior of domesticated and wild beans is strongly influenced by photoperiod and temperature. Like most tropical plants, beans are short-day (long-night) plants that flower with 12 hours or less of daylight, with two photoperiod-responsive genes controlling photoperiod sensitivity (Gu et al. 1998; Gepts 1999). During domestication and crop dispersal to other geographical regions, plants were selected for day-length neutrality, and most modern bean cultivars now flower under any daylight period.

Bean flowers—as is typical for the Papilionoideae—have two anterior keel petals that almost enclose the male and female organs (Webster et al. 1977). The stigma becomes receptive a day or two before the flower opens, and pollen is shed from the anthers either the evening before or in the morning when the flower opens. Self-pollination does not require insect visits. However, tripping (vibrating the flower) by visiting insects can cause cross-pollination, through foreign pollen attached to them (Webster et al. 1982; McCormack 2004a). Pollen tubes from foreign pollen can grow more quickly than self pollen, leading to a higher incidence of cross-fertilization (Free 1993). As a rule, fertilization takes place about eight to nine hours after pollination has occurred (Weinstein 1926).

Pollen Dispersal and Longevity

The flowers of common beans are frequently visited by pollinating insects, particularly by different bumblebee and bee species, but also by hummingbirds (Free 1993; Freytag and Debouck 2002). Carpenter bees (*Xylocopa* spp.) are especially attracted by bean flowers, whereas honeybees are less so (Ferreira et al. 2007). Bees, particularly bumblebees, vibrate the flower, causing the pollen to fall onto their body hairs, thus significantly increasing the amount of cross-pollination in normally self-pollinated plants.

To our knowledge, there is no published information about the longevity of common bean pollen.

Sexual Reproduction

The common bean plant is a self-compatible, predominantly autogamous, annual diploid ($2n = 2x = 22$), with varying degrees of outcrossing (Summerfield and Roberts 1985). The outcrossing rate is usually below 5% (reviewed by Ibarra-Pérez et al. 1997), but rates of more than 50%, and even up to 80%, have been reported in some cases (Brunner and Beaver 1988; Wells et al. 1988; Gepts et al. 2000).

Self- and cross-pollinating wild *Phaseolus* species can be found, but self-pollination is more frequent. Cleistogamy is observed in a few species (Freytag and Debouck 2002).

Vegetative Reproduction

There are no reports of vegetative propagation in common beans. However, some species with tuberous roots (e.g., *Phaseolus coccineus*) can resprout from the roots after a frost (P. Gepts, pers. comm. 2007).

Seed Dispersal and Dormancy

Seed dispersal in wild beans occurs through exploding dehiscent pods. During the domestication process, this characteristic has been lost for bean cultivars and landraces. The pods of cultivated beans contain considerably fewer or no fibers and therefore do not open at maturity. The dispersal of domesticated beans is exclusively by humans.

Similarly, seed dormancy is common in wild bean species but close to absent in the domesticated forms (Smartt 1990). For example, soil seed banks of wild lima beans (*P. lunatus*) have been shown to favor sympatric growth with domesticated populations, thus increasing the possibility of genetic exchange between the two gene pools (Degreef et al. 2002; Martínez-Castillo et al. 2006). Based on observations from the Central Valley of Costa Rica, Baudoin et al. (2004) estimate that all *P. lunatus* seeds would germinate within three years of their dispersal. In Mexico, however, farmers have reported *P. lunatus* seeds remaining viable in the soil for up to eight years (Martínez-Castillo et al. 2006).

Volunteers, Ferals, and Their Persistence

Domesticated common beans usually do not produce significant numbers of volunteers or ferals, due to a loss of seed dormancy. However, at times ferals can be observed outside of cultivation. These plants probably arise from spilt seed.

During the domestication process, beans lost many characters important for survival in the wild. Among the most important are pod dehiscence (i.e., seed dispersal) and seed dormancy (Smartt 1978; Gepts 2002). The absence of many of these characters makes it difficult for domesticated beans to persist outside of cultivation for more than a few years. However, some domesticated beans have been observed to survive for several years after fields have been abandoned. For instance, domesticated *P. coccineus* runner beans are perennials—although they can be cultivated as annuals—and can persist in the field for several years, due to their perennial, thickened roots (Freytag and Debouck 2002). Also, domesticated year beans (*P. dumosus*) can occasionally escape and become weedy outside of cultivation, particularly in the Andes (Debouck 1991).

Weediness and Invasiveness Potential

In general, escaped domesticated beans rarely survive in the wild, and thus are not considered to be important as weeds (Smartt 1978). Their wild relatives, and hybrids between wild and domesticated beans, do appear as weeds in and near fields, but they can be easily controlled by traditional hand weeding or by the use of herbicides; as a result, they are not considered to be significant weeds, either. Fieldwork in Costa Rica showed that weedy (i.e., hybrid) forms of common beans can survive for only one to six generations in fields; they disappear or revert to wild forms as soon as human intervention ceases (Lentini et al. 2006).

Crop Wild Relatives

The genus *Phaseolus* currently comprises more than 70 species, distributed in the Americas (Asian *Phaseolus* species have been reclassified as *Vigna* Savi), although the number of accepted species and subdivisions within the genus has not been resolved conclusively (see, e.g., Maréchal et al. 1978; Lackey 1983; Delgado-Salinas 1985; Delgado-Salinas et al. 1999). In the most recent and comprehensive botanical review of the genus, Freytag and Debouck (2002) recognize 75 species, classified into 15 sections, that are distributed throughout N, Central, and S America, from the SW United States to N Argentina (table 8.1).

All species of the genus are diploid, and most have 22 chromosomes ($2n = 2x = 22$); a few species have only 20 (Gepts 2002). Recently, a molecular systematic study combining *matK-trnK* cpDNA and nuclear ITS data confirmed the monophyly of the genus *Phaseolus* and identified two major lineages (A and B) within the genus. All domesticated species were contained in one of the major lineages (Delgado-Salinas et al. 2006).

Hybridization

In general, gene transfer is rather difficult with beans, which has been a limitation for bean breeding thus far (Debouck 1991). Within the genus *Phaseolus*, each cultivated form (i.e., each of the five domesticated bean species), along with its wild ancestral type, forms a primary gene pool where full or almost full genetic compatibility exists (Smartt 1990). The wider gene pool of domesticated common beans (*P. vulgaris*) expands upon this, and includes a larger number of domesticated and wild relatives within the genus, which can be divided into secondary (GP-2), tertiary (GP-3) and quaternary (GP-4) gene pools (table 8.1).

Primary Gene Pool (GP-1)

The primary gene pool of common beans comprises both *P. vulgaris* cultivars and wild populations. Wild *P. vulgaris* forms are the immediate ancestors of common bean cultivars (Berglund-Brücher and Brücher

Table 8.1 Domesticated common bean (*Phaseolus vulgaris* L.) and its crop wild relatives

Species	Common name	Origin and distribution
Primary gene pool (GP-1)		
sect. *Phaseolus*		
P. vulgaris L. var. vulgaris	common bean, navy bean, French bean, snap bean	widely cultivated throughout N and S America, Europe, Middle East, W and E Asia, and Africa
P. vulgaris L. (wild varieties)*	wild common bean	wild; distributed from Mexico to Costa Rica and from Andean S America down to Argentina
Secondary gene pool (GP-2)		
P. dumosus Macfad.	year bean	weedy and domesticated from C Mexico to S Peru; wild forms are rare in Guatemala and Mexico
P. albescens McVaugh ex R. Ramírez & A. Delgado		W Mexico
P. costaricensis Freytag & Debouck		Costa Rica and NW Panama
sect. *Coccinei* Freytag		
P. coccineus L. subsp. coccineus (includes 1 domesticated and 11 wild forms)	scarlet runner, runner bean	wild from N Mexico to Panama; cultivated in Colombia, Europe, and some African highland areas
P. coccineus subsp. *striatus* (Brandegee) Freytag		Mexico (Puebla)
Tertiary gene pool (GP-3)		
sect. *Acutifolii* Freytag		
P. acutifolius var. latifolius G.F. Freeman	tepary bean	cultivated from SW USA down to Panama; some putative wild types are found in NW Mexico, SW USA, and Texas
P. acutifolius A. Gray var. *acutifolius*		wild in SW USA and NW Mexico
P. acutifolius var. *tenuifolius* A. Gray		wild in SW USA and NW Mexico
P. parvifolius Freytag		wild from W Mexico to C Guatemala

Quaternary gene pool (GP-4)

sect. *Rugosi* Freytag

  P. angustissimus A. Gray — common in SW USA and N Mexico

  P. carteri Freytag & Debouck — Mexico

  P. filiformis Benth. — common in SW USA and NW Mexico

sect. *Paniculati* Freytag

  P. lunatus L. — lima bean, sieva bean, butter or Madagascar bean — wild from N Mexico to NW Argentina; domesticated

Source: Freytag and Debouck 2002.

Note: Domesticated forms are in bold. Gene pool designations are in relation to common beans (*P. vulgaris* var. *vulgaris*) and follow Singh 2001a and Freytag and Debouck 2002.

*Freytag and Debouck 2002 list two wild varieties for *P. vulgaris*—var. *aborigineus* (Burkart) Baudet and var. *mexicanus* Freytag (*nom. illeg.*)—as well as a potential third wild variety, var. *ecuatorianus, ined.*

1976; Brücher 1988), and they are found from N Mexico to NW Argentina (Toro et al. 1990; Singh 2001a).

Within the domesticated and wild populations, two major gene pools can be distinguished: an Andean and a Mesoamerican gene pool (Gepts and Bliss 1985; Koenig and Gepts 1989a; Khairallah et al. 1990). Not all domesticated and wild *P. vulgaris* forms are sexually compatible, and F1 hybrid weakness is found, for the most part, in crosses between some Mesoamerican and Andean parents (both domesticated and wild forms). These hybridization barriers are due to the presence of incompatibility genes in some Mesoamerican (Dl-1) and Andean (Dl-2) domesticated and wild populations (van Rheenen 1979; Shii et al. 1980; Singh and Gutiérrez 1984; Vieira et al. 1989). These genes lead to unviable F1 hybrids (Shii et al. 1980; Gepts and Bliss 1985) and hybrid weakness and phenotypic abnormalities in later generations (Sprecher and Khairallah 1989; Koinange and Gepts 1992; Debouck et al. 1993; Freyre et al. 1996; Singh and Molina 1996). Gene exchange and introgression between certain wild or domesticated populations belonging to these different gene pools is therefore limited (Nienhuis and Singh 1986; Kornegay et al. 1992; Mumba and Galwey 1998, 1999).

Apart from this genetic barrier in some populations, domesticated and wild Andean and Mesoamerican forms are fully cross-compatible (Singh 2001b), and intraspecific hybridization results in fully fertile progeny, both in F1 as well as in subsequent generations (Motto et al. 1978; Koenig and Gepts 1989b; Singh et al. 1995). Wild-weedy-domesticated hybrid swarms, and individual spontaneous intermediates and weedy types between compatible forms, are frequently observed at the margins of bean fields in Central and S America, where they occur in sympatry (see, e.g., Delgado-Salinas et al. 1988; Debouck et al. 1989; Acosta et al. 1994; Beebe et al. 1997; Araya Villalobos et al. 2001; Lentini et al. 2006). Hybrid swarms can persist in cultivated areas for several years (Freyre et al. 1996), but they disappear once bean cropping is interrupted or abandoned, since they are not able to compete outside of cultivation (Lentini et al. 2006).

The direction and extent of short- and long-term gene flow between domesticated and wild common beans has been debated for some time. In Mexico, Vanderborght (1983) reported introgression rates of up to 50% in wild populations. Lentini et al. (2006) noted that the predominant direction of gene flow in Costa Rica is mainly from the wild into the domesticated type, due to the longer flowering period of wild beans and the floral characteristics that render them less prone to outcrossing. This is supported by studies of chloroplast gene flow by means of allozyme and microsatellite markers (Singh et al. 1991; González-Torres 2004). In contrast, other molecular studies found that (nuclear) gene introgression from domesticated to wild types was three times more prevalent than vice versa (Papa and Gepts 2003; Chacón et al. 2005; González et al. 2005; Payró de la Cruz et al. 2005; Zizumbo-Villareal et al. 2005). This asymmetric gene flow can be due to differences in population size (domesticated > wild), dominance of the wild phenotype, and/or human selection against hybrids in domesticated populations (Papa and Gepts 2003, 2004; Papa et al. 2005).

In lima beans (*P. lunatus*), the magnitude of both recent (Martínez-Castillo et al. 2007) and long-term (Hardy et al. 1997; Maquet et al. 2001; Ouédraogo and Baudoin 2002) intraspecific gene flow between domesticated and wild populations is low in Mexico and Costa Rica. Albeit also low, gene flow in Mexico was considerably more important at the local level—that is, between closely situated or intermingled domesticated and wild *P. lunatus* populations—than at interregional and intraregional levels, and it was almost three times higher from the domesticated to the wild gene pool than vice versa (Martínez-Castillo et al. 2007).

Secondary Gene Pool (GP-2)

The secondary gene pool of common beans comprises *P. coccineus*, *P. dumosus* (= *P. polyanthus* Greenm.), and *P. costaricensis* Freytag & Debouck. *Phaseolus albescens* McVaugh *ex* R. Ramírez & A. Delgado, which is placed in the same section (sect. *Phaseolus*) as *P. vulgaris*, *P. dumosus*, and *P. costaricensis*, may belong to GP-2, but to our knowledge no information has been published thus far about crossability and genetic compatibility with common beans (or any of the other species in GP-2).

These three species—*P. coccineus*, *P. dumosus*, and *P. costaricensis*—cross easily with each other and with domesticated common beans under both natural and controlled conditions, mainly when the common bean plant is the female parent (see, e.g., Baggett 1956; Camarena and Baudoin 1987; Singh et al. 1997). When common bean plants are the pollen donor, it is difficult to obtain a cross, and their progeny tend to revert to the female-parent genotype (Smartt 1970; Ockendon et al. 1982; Hucl and Scoles 1985; Debouck 1999). In all cases, hybrid viability and fertility are dependent on the combination of parental genotypes and, therefore, are variable among crosses. Some crosses result in partially sterile progeny (Ibrahim and Coyne 1975; Manshardt and Bassett 1984).

Phaseolus coccineus

Phaseolus coccineus is interfertile with common beans (see, e.g., Al-Yasiri and Coyne 1966; Le Marchand 1971; Smartt 1979, 1990; Manshardt and Bassett 1984; Hoover et al. 1985; Debouck and Smartt 1995). Introgression is possible in both directions (Smartt 1970; Ibrahim and Coyne 1975; Ockendon et al. 1982). Haka et al. (1995) found that in their *P. coccineus* × *P. vulgaris* crosses, meiosis and seed fertility were almost normal in the F1, F2, and most of the BC1 progenies. However, the level of cross-compatibility between these two species varies, and seems to be influenced by the parental genotypes (Bemis and Kedar 1961; Smartt and Haq 1972; Zagorcheva and Poriazov 1983). H. Thomas (1964) suggested that two complementary dominant genes were involved. Hybridization in the wild therefore seems probable. However, thus far no evidence for spontaneous hybridization has been found—at least not in the area of central Jalisco (Mexico), where these species are sympatric. This may indicate the presence of a genetic isolation mechanism between the two species (Miranda Colín and Evans 1973; Freytag and Debouck 2002).

Natural hybrids between *P. coccineus* and *P. dumosus* are abundant in areas of sympatry in Colombia and Ecuador, but not in Guatemala (Freytag and Debouck 2002). Experimental evidence shows that the hybrid progeny have high fertility (Maréchal et al. 1978; Delgado-Salinas 1988).

Phaseolus dumosus (= P. polyanthus)

Phaseolus dumosus is cross-compatible with both common beans and *P. coccineus*, to the point where some authors have concluded that *P. dumosus* could be the result of natural hybridization between the two (Hernández et al. 1959; Piñero and Eguiarte 1988). However, *P. dumosus* is now recognized as a species in its own right. Although it is closely related to both common beans and *P. coccineus*, it seems to be closer to the latter (Schmit et al. 1992; Llaca et al. 1994).

Hybridization of *P. dumosus* with common beans can be achieved easily (see, e.g., Le Marchand 1971; Maréchal et al. 1978; Baudoin et al. 1985, 1986; Camarena and Baudoin 1987), although F1 hybrids sometimes have reduced fertility (Maréchal et al. 1978) or may even be completely sterile (Lecomte et al. 1998; Geerts et al. 2002). Natural *P. vulgaris* × *P. dumosus* hybrids have been found in Colombia (Debouck and Smartt 1995).

As mentioned above, natural *P. dumosus* × *P. coccineus* hybrids occur frequently, and experimental tests demonstrate almost total fertility for their progeny. However, introgression seems limited, since the domesticated forms of both species are frequently grown together by farmers in Central America and still maintain their genetic integrity (Debouck 1991).

Phaseolus costaricensis

Hybrids between *Phaseolus costaricensis* and common beans can be obtained by hand pollination (Singh et al. 1997). In addition, natural hybrids between *P. costaricensis* and *P. dumosus* have been reported from field observations in Costa Rica (Debouck in Freytag and Debouck 2002).

Tertiary Gene Pool (GP-3)

The tertiary gene pool includes *Phaseolus acutifolius* and *P. parvifolius* Greene (Debouck 1999). Crosses between common beans and these two species are possible through embryo rescue. Viable interspecific F1 progeny are usually sterile (see, e.g., Mok et al. 1978; Hwang 1979; Rabakoarihanta et al. 1979; Alvarez et al. 1981; Pratt 1983; C. Thomas et al. 1983; Andrade-Aguilar and Jackson 1988; Haghighi and Ascher 1988; Waines et al. 1988; Mejía-Jiménez et al. 1994; Singh et al. 1998). Hybrid fertility can be restored via chromosome doubling or by one or more generations of backcrossing to the recurrent common bean parent (Prendota et al. 1982; Pratt et al. 1985; C. Thomas and Waines 1984). In exceptional cases, fertile hybrids have been obtained (Honma 1956).

The level of cross-compatibility between *P. vulgaris* and *P. acutifolius* is strongly influenced by genotype, due to the action of two complimentary dominant genes that are present in some populations (Parker and Michaels 1986a, b). Choosing compatible common bean genotypes for the female parent facilitates interspecific hybridization and increases hybrid fitness (Federici and Waines 1988). Smartt (1970) even obtained a few viable but completely sterile F1 plants from reciprocal crosses by hand pollination, without the use of embryo culture. Singh (1991) suggested that the F1 hybrid incompatibility observed between *P. vulgaris* × *P. acutifolius*, *P. vulgaris* × *P. coccineus*, and other interspecific crosses may perhaps be controlled by the same loci (*Dl-1* and *Dl-2*).

Quaternary Gene Pool (GP-4)

Common beans have a so-called quaternary gene pool that includes *Phaseolus lunatus*, *P. carteri* Freytag & Debouck, *P. filiformis* Benth., and *P. angustissimus* A. Gray. Crosses of common beans have been attempted with *P. filiformis* (Maréchal and Baudoin 1978; Weilenmann de Tau et al. 1986; Petzold and Dickson 1987; Federici and Waines 1988; Baudoin et al. 2001), *P. angustissimus* (Belivanis and Doré 1986; Petzold and Dickson 1987; Baudoin et al. 2001), and *P. lunatus* (Al-Yasiri and Coyne 1966; Mok et al. 1978; Rabakoarihanta et al. 1979; Leonard et al. 1987; Federici and Waines 1988; Cabral and Crocomo 1989; Kuboyama et al. 1991), but none of these reports documented that viable and fertile hybrid progeny could be obtained. Thus, for purposes of gene introgression, these species should be considered in the quaternary gene pool of common beans.

Bridge Species

Phaseolus coccineus has been mentioned as a bridge species for gene introgression from *P. acutifolius*—a GP-3 species—to common beans (Alvarez et al. 1981; Debouck 1991). However, it is very unlikely that bridging would occur under natural conditions in the field, because there is only a very restricted area where all three species are either cultivated simultaneously or are sympatric in the wild (N and NW Mexico).

Outcrossing Rates and Distances

Pollen Flow

Common beans are a self-compatible, mainly self-pollinating crop, and outcrossing rates are usually below 5% in adjacent rows (reviewed in Ibarra-Pérez et al. 1997). Pollen flow and outcrossing rates for common beans, like those for lima beans, are strongly influenced by genotype, local climate and growth conditions, wind direction and speed, the presence and abundance of pollinating insects, and proximity to other pollinator-attracting plants (Wells et al. 1988; Zoro Bi et al. 2005).

Furthermore, outcrossing rates within the same variety can be highly heterogeneous, and under certain conditions they can be greater for individual plants than normal (see, e.g., Kristofferson 1921 [up to 47%]; Mackie and Smith 1935 [9%–35%]; Bliss 1980 [15%–20%]; Brunner and Beaver 1988 [16%–39%]; Wells et al. 1988 [up to 85%]; Ibarra-Pérez et al. 1997 [up to 78%]). Unfortunately, the pollinating species present during those trials, and their population size, were reported in only a few of the outcrossing studies. More research is needed to determine pollen flow distances and the extent of outcrossing at different locations and times.

Moreover, the distances that common bean pollen can travel is a topic that has not received much attention, and only a few studies have been conducted investigating the levels of outcrossing in relation to distance. The only study we found (Ferreira et al. 2007) detected pollen flow at up to 10 m, albeit at very low frequencies (0.05%). In the Central Valley of Costa Rica, pollen flow between wild lima bean populations did not exceed distances of 6 m (Baudoin et al. 1998).

Isolation Distances

The separation distances established by regulatory authorities for the seed production of common beans range from 0 to 20 m in N and S America, and up to 50 m for organic seed production (table 8.2). In OECD countries, bean varieties for seed production "shall be isolated from other crops by a definite barrier or a space sufficient to prevent mixture during harvest" (OECD 2008, p. 18).

Table 8.2 Isolation distances for common bean (*Phaseolus vulgaris* L.)

Threshold*	Minimum isolation distance (m)	Country	Reference	Comment
0.01% / 0.0%	3	USA	CCIA 2009	foundation (basic) seed requirements
0.05% / 0.05%	3	USA	CCIA 2009	registered seed requirements
0.05 %/ 0.1%	3	USA	CCIA 2009	certified seed requirements
0.05% / 0.05%	0	USA	USDA 2008	foundation (basic) seed requirements
0.1% / 0.1%	0	USA	USDA 2008	registered seed requirements
0.25% / 0.2%	0	USA	USDA 2008	certified seed requirements
n.a.	50	USA	McCormack 2004a, b	for organic seed production
0.01% / n.a.	20	Canada	CSGA 2009	foundation (basic) seed requirements
0.01% / n.a.	3	Canada	CSGA 2009	registered seed requirements
0.05% / n.a.	3	Canada	CSGA 2009	certified seed requirements
n.a.	5	Brazil	Vieira et al. 1999	certified seed requirements

*Maximum percentage of off-type plants permitted in the field / seed impurities of other varieties or off-types permitted. n.a. = not available.

State of Development of GM Technology

The genetic transformation of common beans is currently under development, and protocols are still being optimized and fine-tuned. *Agrobacterium*-mediated transformation seems to be the method of choice, since stable transformation via particle bombardment has only been achieved at low frequencies (Broughton et al. 2003). Emphasis is being placed on the transformation of tepary beans (*P. acutifolius*), due to their higher transformation efficiency (de Clercq et al. 2002; Zambre et al. 2003); the objective here is the subsequent introgression to common beans (*P. vulgaris*) through artificial hybridization using embryo rescue (Mejía-Jiménez et al. 1994; Dillen et al. 1997). Transformation of *P. coccineus* is being studied for the same purpose (Broughton et al.

2003). The GM traits researched thus far are virus tolerance, herbicide resistance, and enhanced nutritional quality (increased seed protein content). Laboratory trials have taken or are taking place in Belgium, Brazil, Colombia, Italy, Malaysia, Mexico, the Netherlands, and the United States, and field trials in Brazil.

Agronomic Management Recommendations

There are no specific recommendations.

Crop Production Area

Common beans are the most important legume species worldwide for direct human consumption, and they are cultivated throughout N and S America, Europe, the Middle East, W and E Asia, and Africa. The total crop production area is estimated at 27.8 million ha, 26.9 in dry beans plus 0.9 in green beans (FAO 2009). The main dry-bean-producing countries are Brazil, India, China, Myanmar, Mexico, and the United States, whereas green beans are mainly grown in China, Indonesia, and around the Mediterranean (FAO 2009).

There is no commercial production of GM beans at this point.

Research Gaps and Conclusions

From a biological standpoint, the reproductive dynamics in common beans and their wild relatives need to be better understood, including the role played by the floral characteristics of wild bean species; estimates of the outcrossing rates in domesticated and wild populations under different conditions associated with the presence, abundance, and species composition of pollinators; and determinations of effective isolation distances. Studies are particularly necessary to assess the year-to-year and location-to-location dynamics of outcrossing between domesticated and wild bean populations, as well as the rates of introgression into wild beans.

Gene flow from the domesticated common bean *Phaseolus vulgaris* to its wild and weedy relatives through outcrossing will most probably occur only with the species mentioned above. Pollen-mediated gene flow to these species is relatively low, generally below 5% at distances of less than 10 m.

Also, domesticated and wild beans have been sympatric for at least 2500 years (Kaplan and Lynch 1999), and, overall, they seem to maintain their genetic integrity. Common beans are therefore often considered to be a low-risk species with respect to introgression from GM varieties.

However, hybridization does occur relatively frequently under field conditions, producing partly or fully fertile progeny. In such instances, although the extent of gene flow can be low, at least localized individual outcrossing rates (i.e., between adjacent plants) can be very high (up to 80%). Also, the predominant direction of gene flow in the wild is likely to be from domesticated beans to their wild relatives, since most wild bean populations are small and suffer from habitat destruction (Debouck et al. 1993; Freyre et al. 1996; Gepts and Papa 2003). Furthermore, several studies have demonstrated that the genetic diversity of domesticated types is reduced compared to that of wild forms (see, e.g., Gepts et al. 1986; Sonnante et al. 1994). Asymmetrical domesticated-to-wild gene flow, albeit low, can thus create a drastic reduction in the genetic diversity of these wild populations and even lead to local extinctions (Martínez-Castillo et al. 2007). Gene introgression to wild relatives in the primary and secondary gene pools—even at low rates—can therefore pose a serious risk to local genetic diversity in areas of sympatry (Papa et al. 2005).

It could be argued that many of the domesticated characteristics of common beans would be negatively selected for in the wild and therefore would not persist (e.g., lack of seed dormancy, dwarfing, dependence on nutrient-rich soils). On the other hand, as can be observed in the case of the immediate wild ancestor of common beans (wild *P. vulgaris*), hybridization with wild relatives may lead to increased weediness because of the inheritance of the dominant wild traits, thus providing hybrids with an adaptive advantage (Koinange et al. 1996; Papa and Gepts 2004; Papa et al. 2005). In particular, GM traits conferring fitness advantages such as insect or disease resistance may spread at increased rates to weedy and wild bean populations (Ellstrand 2003; Payró de la Cruz et al. 2005).

As Papa and Gepts (2004) and Papa et al. (2005) have pointed out, hybridization between domesticated and wild beans would probably lead to differential selection pressures in cultivated and wild environments. On the one hand, the entry of wild alleles into domesticated populations is likely to be restricted. This is to be expected, due to the strong selection pressure exercised by farmers against hybrids between domesticated and wild species—for example, against certain seed colors, shapes, and sizes—because of their lack of consumer appeal (Zizumbo-Villareal et al. 2005). Moreover, the fact that domesticated and wild beans have maintained their genetic integrity over centuries is attributed to recurrent human selection, rather than to innate genetic barriers to introgression (Papa et al. 2005). In wild populations, on the other hand, F1 progeny would probably have a higher fitness, due to the predominantly wild phenotype of the hybrids, thus facilitating the introgression of domesticated alleles to F2 and later generations in wild populations.

Although the extent of gene flow through hybridization can be small and localized, it is likely to spread more widely, facilitated by the particularity of traditional farming practices in Central and S America and the dynamics of a vital, informal seed-exchange system among farmers, similar to the one known for maize (Zizumbo-Villareal et al. 2005). Therefore, gene flow and possibly introgression of bean alleles to the sexually compatible wild relatives mentioned above is quite likely. In countries where no bean wild relatives occur naturally, Lentini et al. (2006) believe minimum isolation distances of 50 m are adequate to minimize or even prevent pollen flow and intraspecific gene exchange between GM and conventional common bean cultivars.

Based on the above considerations, the likelihood of gene flow and introgression to wild relatives of common beans exists throughout the natural area of distribution of the sexually compatible wild species (i.e., in the Americas). In spite of common beans being predominantly selfing, the probability of gene introgression from domesticated common beans to their domesticated and wild relatives—assuming physical proximity (less than 10 m) and flowering overlap, and being conditional on the nature of the respective trait(s)—is as follows (map 8.1):

- *high* for the wild and weedy forms of common beans (*Phaseolus vulgaris*), for the domesticated and wild forms of *P. dumosus* (year bean) and *P. coccineus* (runner bean), and for the wild relatives *P. costaricensis* and, most probably, *P. albescens* (note that although the focus of this chapter is on *P. vulgaris*, the likelihood of introgression is also *high* between wild and domesticated *P. lunatus*)

- *low* for the domesticated and wild forms of *P. acutifolius* (tepary bean) and for its wild relative *P. parvifolius* (in map 8.1, part of this area is hidden by red areas representing a high likelihood of gene flow and introgression)

REFERENCES

Acosta J, Gepts P, and Debouck D (1994) Observations on wild and weedy accessions of common beans in Oaxaca, Mexico. *Annu. Rep. Bean Improv. Coop.* 37: 137–138.

Álvarez MN, Ascher PD, and Davis DW (1981) Interspecific hybridization in *Euphaseolus* through embryo rescue. *Hort. Sci.* 16: 541–543.

Al-Yasiri SA and Coyne DP (1966) Interspecific hybridization in the genus *Phaseolus. Crop Sci.* 6: 59–60.

Andrade-Aguilar JA and Jackson MT (1988) Attempts at interspecific hybridization between *Phaseolus vulgaris* L. and *P. acutifolius* A. Gray using embryo rescue. *Plant Breed.* 101: 173–175.

Araya Villalobos R, González Ugalde WG, Camacho Chacón F, Sánchez Trejos P, and Debouck DG (2001) Observations on the geographic distribution, ecology, and conservation status of several *Phaseolus* bean species in Costa Rica. *Genet. Resour. Crop Evol.* 48: 221–232.

Baggett JR (1956) The inheritance of resistance to strains of bean yellow mosaic virus in the interspecific cross *Phaseolus vulgaris* × *P. coccineus. Plant Dis. Rptr.* 40: 702–707.

Baudoin JP, Maréchal R, Otoul E, and Camarena F (1985) Interspecific hybridizations within *Phaseolus vulgaris* L.-*Phaseolus coccineus* L. complex. *Annu. Rep. Bean Improv. Coop.* 28: 64–65.

Baudoin JP, Camarena F, and Maréchal R (1986) Interspecific crosses between *Phaseolus coccineus* subsp. *polyanthus* M[arechal], M[ascherpa] & S[tainer], as seed parent, and *Phaseolus vulgaris* L. *Annu. Rep. Bean Improv. Coop.* 29: 64.

Baudoin JP, Degreef J, Hardy O, Janart F, and Zoro Bi I (1998) Development of an *in situ* conservation strategy for wild lima bean (*Phaseolus lunatus* L.) populations in the Central Valley of Costa Rica. In: Owens SJ and Rudall PJ (eds.) *Reproductive Biology in Systematics, Conservation, and Economic Botany* (Royal Botanic Gardens, Kew: Richmond, Surrey, UK), pp. 417–426.

Baudoin JP, Camarena F, Lobo M, and Mergeai G (2001) Breeding *Phaseolus* for intercrop combinations in Andean highlands. In: Cooper HD, Spillane C, and Hodgkin T (eds.) *Broadening the Genetic Base of Crop Production* (CAB International: Wallingford, UK), pp. 373–384.

Baudoin JP, Rocha O, Degreef J, Maquet A, and Guarino L (2004) *Ecogeography,*

Demography, Diversity, and Conservation of Phaseolus lunatus L. *in the Central Valley of Costa Rica.* Systematic and Ecogeographic Studies on Crop Genepools 12 (International Plant Genetic Resources Institute [IPGRI]: Rome, Italy). 94 p.

Beebe S, Toro O, González AV, Chacon MI, and Debouck DG (1997) Wild-weed-crop complexes of common bean (*Phaseolus vulgaris* L., Fabaceae) in the Andes of Peru and Colombia, and their implications for conservation and breeding. *Genet. Resour. Crop Evol.* 44: 73–91.

Belivanis T and Doré C (1986) Interspecific hybridization of *Phaseolus vulgaris* L. and *P. angustissimus* A. Gray using *in vitro* embryo culture. *Plant Cell Rep.* 5: 329–331.

Bemis WP and Kedar N (1961) Inheritance of morphological abnormalities in seedlings of two species of *Phaseolus. J. Hered.* 52: 171–178.

Berglund-Brücher O and Brücher H (1976) The South American wild bean (*Phaseolus aborigineus* Burk.) as ancestor of the common bean. *Econ. Bot.* 30: 257–272.

Bliss FA (1980) Common bean. In: Fehr W and Hadley HH (eds.) *Hybridization of Crop Plants* (American Society of Agronomy: Madison, WI, USA), pp. 273–284.

Broughton WJ, Hernández G, Blair M, Beebe S, Gepts P, and Vanderleyden J (2003) Beans (*Phaseolus* spp.)—model food legumes. *Plant & Soil* 252: 55–128.

Brücher H (1988) The wild ancestor of *Phaseolus vulgaris* in South America. In: Gepts P (ed.) *Genetic Resources of* Phaseolus *Beans: Their Maintenance, Domestication, Evolution, and Utilization* (Kluwer Academic Publishers: Dordrecht, The Netherlands), pp. 185–214.

Brunner BR and Beaver JS (1988) Estimation of outcrossing of dry beans in Puerto Rico. *Annu. Rep. Bean Improv. Coop.* 31: 42–43.

Cabral JB and Crocomo OJ (1989) Interspecific hybridization of *Phaseolus vulgaris*, *P. acutifolius*, and *P. lunatus* using *in vitro* technique. *Turrialba* 39: 243–246.

Camarena F and Baudoin JP (1987) Obtention des premiers hybrids interspécifiques entre *Phaseolus vulgaris* et *Phaseolus polyanthus* avec le cytoplasme de cette dernière forme. *Bull. Rech. Agron. Gembloux* 22: 43–55.

CCIA [California Crop Improvement Association] (2009) Bean Certification Standards. http://ccia.ucdavis.edu/seed_cert/seedcert_index.htm.

Chacón MI, González AV, Gutiérrez JP, Beebe S, and Debouck DG (1996) Increased evidence for common bean (*Phaseolus vulgaris*) domestication in Colombia. *Annu. Rep. Bean Improv. Coop.* 39: 201–202.

Chacón MI, Pickersgill B, and Debouck DG (2005) Domestication patterns in common bean (*Phaseolus vulgaris* L.) and the origin of the Mesoamerican and Andean cultivated landraces. *Theor. Appl. Genet.* 110: 432–444.

CSGA [Canadian Seed Growers Association] (2009) Canadian Regulations and Procedures for Pedigreed Seed Crop Production. Circular 6–2005/Rev. 01.4–2009

(Canadian Seed Growers Association [CSGA]: Ottawa, ON, Canada). www .seedgrowers.ca/cropcertification/circular.asp.

Debouck DG (1986) Primary diversification of *Phaseolus* in the Americas: Three centres? *Plant Genet. Resour. Newsl.* 67: 2–8.

────── (1991) Systematics and morphology. In: van Schoonhoven A and Voysest O (eds.) *Common Beans: Research for Crop Improvement* (CAB International: Wallingford, UK), pp. 55–118.

────── (1999) Diversity in *Phaseolus* species in relation to the common bean. In Singh SP (ed.) *Common Bean Improvement in the Twenty-First Century* (Kluwer Academic Publishers: Dordrecht, The Netherlands), pp. 25–52.

────── (2000) Biodiversity, ecology, and genetic resources of common *Phaseolus* beans—seven answered and unanswered questions. In: Oono K (ed.) *Wild Legumes: Proceedings of the 7th Ministry of Agriculture, Forestry, and Fisheries [MAFF], Japan International Workshop on Genetic Resources, October 13–15, 1999* (National Institute of Agro-biological Resources [NIAR]: Tsukuba, Japan), pp. 95–123.

Debouck DG and Smartt J (1995) Beans. In: Smartt J and Simmonds NW (eds.) *Evolution of Crop Plants.* 2nd ed. (Longman Scientific and Technical: Harlow, UK), pp. 287–294.

Debouck DG, Gamarra Flores M, Ortiz Arriola V, and Tohme J (1989) Presence of a wild-weed-crop complex in *Phaseolus vulgaris* L. in Peru? *Annu. Rep. Bean Improv. Coop.* 32: 64–65.

Debouck DG, Toro O, Paredes OM, Jonson WC, and Gepts P (1993) Genetic diversity and ecological distribution of *Phaseolus vulgaris* in northwestern South America. *Econ. Bot.* 47: 408–423.

de Clercq J, Zambre M, van Montagu M, Dillen W, and Angenon G (2002) An optimized *Agrobacterium*-mediated transformation procedure for *Phaseolus acutifolius* A. Gray. *Plant Cell Rep.* 21: 333–340.

Degreef J, Rocha OJ, Vanderborght T, and Baudoin JP (2002) Soil seed bank and seed dormancy in wild populations of lima bean (Fabaceae): Considerations for *in situ* and *ex situ* conservation. *Am. J. Bot.* 89: 1644–1650.

Delgado-Salinas A (1985) Systematics of the genus *Phaseolus* (Leguminosae) in North and Central America. PhD dissertation, University of Texas at Austin. 363 p.

────── (1988) Variation, taxonomy, domestication, and germplasm potentialities in *Phaseolus coccineus.* In: Gepts P (ed.) *Genetic Resources of* Phaseolus *Beans: Their Maintenance, Domestication, Evolution, and Utilization* (Kluwer Academic Publishers: Dordrecht, The Netherlands), pp. 441–463.

Delgado-Salinas A, Bonet A, and Gepts P (1988) The wild relative of *Phaseolus vulgaris* in Middle America. In: Gepts P (ed.) *Genetic Resources of* Phaseolus

Beans: Their Maintenance, Domestication, Evolution, and Utilization (Kluwer Academic Publishers: Dordrecht, The Netherlands), pp. 163–184.

Delgado-Salinas A, Turley T, Richman A, and Lavin M (1999) Phylogenetic analysis of the cultivated and wild species of *Phaseolus* (Fabaceae). *Syst. Bot.* 24: 438–460.

Delgado-Salinas A, Bibler R, and Lavin M (2006) Phylogeny of the genus *Phaseolus* (Leguminosae): A recent diversification in an ancient landscape. *Syst. Bot.* 31: 779–791.

Dillen W, de Clercq J, Goosens A, van Montagu M, and Angenon G (1997) *Agrobacterium*-mediated transformation of *Phaseolus acutifolius* A. Gray. *Theor. Appl. Genet.* 94: 151–158.

Ellstrand NC (2003) *Dangerous Liaisons? When Cultivated Plants Mate with Their Wild Relatives* (Johns Hopkins University Press: Baltimore, MD, USA). 244 p.

FAO [Food and Agriculture Organization of the United Nations] (2009) FAOSTAT data. http://faostat.fao.org/site/567/default.aspx.

Federici CT and Waines JG (1988) Interspecific hybrid compatibility of selected *Phaseolus vulgaris* L. lines with *P. acutifolius* A. Gray, *P. lunatus* L., and *P. filiformis* Bentham. *Annu. Rep. Bean Improv. Coop.* 31: 201–202.

Ferreira J, de Souza Carneiro JE, Lara Teixeira A, Fortunata de Lanes F, Cecon PR, and Borém A (2007) Gene flow in common bean (*Phaseolus vulgaris* L.). *Euphytica* 153: 165–170.

Free JB (1993) *Insect Pollination of Crops.* 2nd ed. (Academic Press: San Diego, CA, USA). 684 p.

Freyre R, Ríos R, Guzmán L, Debouck DG, and Gepts P (1996) Ecogeographic distribution of *Phaseolus* spp. (Fabaceae) in Bolivia. *Econ. Bot.* 50: 195–215.

Freytag GF and Debouck DG (2002) *Taxonomy, Distribution, and Ecology of the Genus* Phaseolus *(Leguminosae-Papilionoideae) in North America, Mexico, and Central America.* Sida, Botanical Miscellany 23 (Botanical Research Institute of Texas [BRIT]: Forth Worth, TX, USA). 298 p.

Geerts P, Toussaint A, Mergeai G, and Baudoin JP (2002) Study of the early abortion in reciprocal crosses between *Phaseolus vulgaris* and *Phaseolus polyanthus* Greenm. *Biotech. Agron. Soc. Environ.* 6: 109–119.

Gepts P (1999) What can molecular markers tell us about the process of domestication in common bean? In: Damania A, Valkoun J, Willcox G, and Qualset CO (eds.) *The Origins of Agriculture and the Domestication of Crop Plants: The Harlan Symposium, Aleppo, Syria, May 10–14, 1997.* Genetic Resources Conservation Program, Division of Agriculture and Natural Resources, University of California, Report No. 21 (International Center for Agricultural Research in the Dry Areas [ICARDA], International Plant Genetic Resources Institute [IPGRI], Food and Agriculture Organization of the United Nations [FAO], and

Genetic Resources Conservation Program, Division of Agriculture and Natural Resources, University of California [UC/GRCP]: Aleppo, Syria), pp. 198–209. www2.bioversityinternational.org/publications/Web_version/47/.

——— (2002) *Phaseolus vulgaris* (beans). In: Brenner S and Miller JH (ed.) *Encyclopedia of Genetics* (Academic Press: San Diego, CA, USA), pp. 1444–1445.

Gepts P and Bliss FA (1985) F1 hybrid weakness in the common bean: Differential geographic origin suggests two gene pools in cultivated bean germplasm. *J. Hered.* 76: 447–450.

——— (1988) Dissemination pathways of common bean (*Phaseolus vulgaris*, Fabaceae) deduced from phaseolin electrophoretic variability: II. Europe and Africa. *Econ. Bot.* 42: 86–104.

Gepts P and Debouck DG (1991) Origin, domestication, and evolution of the common bean (*Phaseolus vulgaris* L.). In: van Schoonhoven A and Voysest O (eds.) *Common Beans: Research for Crop Improvement* (CAB International: Wallingford, UK), pp. 7–53.

Gepts P and Papa R (2003) Possible effects of (trans) gene flow from crops on the genetic diversity from landraces and wild relatives. *Environ. Biosafety Res.* 2: 89–103.

Gepts P, Osborn TC, Rashka K, and Bliss FA (1986) Phaseolin-protein variability in wild forms and landraces of the common bean (*Phaseolus vulgaris* L.): Evidence for multiple centres of domestication. *Econ. Bot.* 40: 451–468.

Gepts P, González AV, Papa R, Acosta J, Wong A, and Delgado-Salinas A (2000) Outcrossing in Mexican wild and domesticated populations of common bean. *Annu. Rep. Bean Improv. Coop.* 43: 25–26.

González-Mejía A, Wong A, Delgado-Salinas A, Papa R, and Gepts P (2005) Assessment of inter simple sequence repeat markers to differentiate sympatric wild and domesticated populations of common bean (*Phaseolus vulgaris* L.). *Crop Sci.* 45: 606–615.

González-Torres RI (2004) Estimación de flujo de genes en *Phaseolus vulgaris* L. mediante marcadores moleculares: Microsatélites y polimorfismo de ADN de cloroplasto. Master's thesis, Universidad Nacional de Colombia.

Gu W, Zhu J, Wallace D, Singh S, and Weeden N (1998) Analysis of genes controlling photoperiod sensitivity in common bean using DNA markers. *Euphytica* 102: 125–132.

Haghighi KR and Ascher PD (1988) Fertile intermediate hybrids between *Phaseolus vulgaris* and *P. acutifolius* from congruity back-crossing. *Sex. Plant Reprod.* 1: 51–58.

Haka MN, Lane GR, and Smartt J (1995) The cytogenetics of *Phaseolus vulgaris* L., *Phaseolus coccineus* L., their interspecific hybrids, and derived amphidiploid and backcross progeny in relation to their potential exploitation in breeding. *Cytology* 45: 791–798.

Hardy O, Dubois S, Zoro Bi I, and Baudoin JP (1997) Gene dispersal and its consequences on the genetic structure of wild populations of lima bean (*Phaseolus lunatus*) in Costa Rica. *Plant Genet. Resour. Newsl.* 109: 1–6.

Harlan JR (1992) *Crops and Man.* 2nd ed. (American Society of Agronomy: Madison, WI, USA). 284 p.

Hernández XE, Miranda Colín S, and Prywer C (1959) El origin de *Phaseolus coccineus* L. *darwinianus* Hdz. X. & Miranda C., subspecies nova. *Rev. Soc. Mex. Hist. Nat.* 20: 99–121.

Honma S (1956) A bean interspecific hybrid. *J. Hered.* 47: 217–220.

Hoover EE, Brenner ML, and Ascher PD (1985) Comparison of development of two bean crosses. *Hort. Sci.* 20: 884–886.

Hucl P and Scoles GJ (1985) Interspecific hybridization in the common bean: A review. *Hort. Sci.* 20: 352–357.

Hwang JK (1979) Interspecific hybridization between *Phaseolus vulgaris* L. and *Phaseolus acutifolius* A. Gray. PhD dissertation, Kansas State University, 1978. Diss. Abst. Int., DAI-B 39/10, 4672. 33 p.

Ibarra-Pérez FJ, Ehdaie B, and Waines JG (1997) Estimation of outcrossing rate in common bean. *Crop Sci.* 37: 60–65.

Ibrahim AM and Coyne DP (1975) Genetics of stigma shape, cotyledon position, and flower color in reciprocal crosses between *Phaseolus vulgaris* L. and *Phaseolus coccineus* (Lam.) and implications in breeding. *J. Am. Soc. Hort. Sci.* 100: 622–626.

Kami J, Becerra Velásquez B, Debouck DG, and Gepts P (1995) Identification of presumed ancestral DNA sequences of phaseolin in *Phaseolus vulgaris*. *Proc. Natl. Acad. Sci. USA* 92: 1101–1104.

Kaplan L and Lynch T (1999) *Phaseolus* (Fabaceae) in archaeology: AMS radiocarbon dates and their significance for pre-Columbian agriculture. *Econ. Bot.* 53: 261–272.

Khairallah MM, Adams MW, and Sears BB (1990) Mitochondrial DNA polymorphisms of Malawian bean lines: Further evidence for two major gene pools. *Theor. Appl. Genet.* 80: 753–761.

Koenig R and Gepts P (1989a) Allozyme diversity in wild *Phaseolus vulgaris*: Further evidence for two major centers of genetic diversity. *Theor. Appl. Genet.* 78: 809–817.

——— (1989b) Segregation and linkage of genes for seed proteins, isozymes, and morphological traits in common bean (*Phaseolus vulgaris*). *J. Hered.* 80: 455–459.

Koinange EMK and Gepts P (1992) Hybrid weakness in wild *Phaseolus vulgaris* L. *J. Hered.* 83: 135–139.

Koinange EMK, Singh SP, and Gepts P (1996) Genetic control of the domestication syndrome in common-bean. *Crop Sci.* 36: 1037–1045.

Kornegay J, White JW, and Ortíz de la Cruz O (1992) Growth habit and gene pool effects on inheritance of yield in common bean. *Euphytica* 62: 171–180.

Kristofferson KB (1921) Spontaneous crossing in the garden bean, *Phaseolus vulgaris*. *Hereditas* 2: 395–400.

Kuboyama T, Shintaku Y, and Takeda G (1991) Hybrid plant of *Phaseolus vulgaris* L. and *P. lunatus* L. obtained by means of embryo rescue and confirmed by restriction endonuclease analysis of rDNA. *Euphytica* 54: 177–182.

Lackey JA (1983) A review of generic concepts in American Phaseolinae (Fabaceae, Fabaoideae). *Iselya* 2: 21–64.

Lecomte B, Longly B, Crabbe J, and Baudoin JP (1998) Étude comparative du développement de l'ovule chez deux espèces de *Phaseolus*, *P. polyanthus* et *P. vulgaris*: Développement de l'ovule dans le genre *Phaseolus*. *Biotech. Agron. Soc. Environ.* 2: 77–84.

Le Marchand G (1971) Observations sur quelques hybrids dans le genre *Phaseolus*: I. Le problème des incompatibilitiés interspécifiques. *Bull. Rech. Agron. Gembloux* 6: 441–452.

Lentini Z, Debouck D, Espinoza AM, and Araya R (2006) Gene flow analysis into wild/weedy relatives from crops with center of origin/diversity in tropical America. In: *Proceedings of the 9th International Symposium on Biosafety of Genetically Modified Organisms (ISBGMO), Jeju Island, South Korea, September 24–29, 2006* (International Society for Biosafety Research [ISBR], n.p.), pp. 152–156. www.isbr.info/isbgmo/docs/9th_isbgmo_program.pdf.

Leonard MF, Stephens LC, and Summers WL (1987) Effect of maternal genotype on development of *Phaseolus vulgaris* L. × *P. lunatus* L. interspecific hybrid embryos. *Euphytica* 36: 327–332.

Llaca V, Delgado-Salinas A, and Gepts P (1994) Chloroplast DNA as an evolutionary marker in the *Phaseolus vulgaris* complex. *Theor. Appl. Genet.* 88: 646–652.

Mackie WW and Smith FL (1935) Evidence of field hybridization in beans. *J. Am. Soc. Agron.* 27: 903–909.

Manshardt RM and Bassett MJ (1984) Inheritance of stigma position in *Phaseolus vulgaris* × *P. coccineus* hybrid populations. *J. Hered.* 75: 45–50.

Maquet A, Masumbuko B, Ouedraogo M, Zoro Bi I, and Baudoin JP (2001) Estimation of gene flow among wild populations of *Phaseolus lunatus* L. using isozyme markers. *Annu. Rep. Bean Improv. Coop.* 44: 27–28.

Maréchal R and Baudoin JP (1978) Observations sur quelques hybrides dans le genre *Phaseolus*: IV. L'hybride *Phaseolus vulgaris* × *P. filiformis*. *Bull. Rech. Agron. Gembloux* 13: 233–240.

Maréchal R, Mascherpa JM, and Stainier F (1978) Étude taxonomique d'un groupe complexe d'espèces des genres *Phaseolus* et *Vigna* (Papilionaceae) sur la base de données morphologiques et polliniques, traitées par l'analyse informatique. *Boissiera* 28: 1–273.

Martínez-Castillo J, Zizumbo-Villarreal D, Gepts P, Delgado-Valerio P, and Colunga-GarcíaMarín P (2006) Structure and genetic diversity of wild populations of lima bean (*Phaseolus lunatus* L.) from the Yucatan Peninsula, Mexico. *Crop Sci.* 46: 1071–1080.

Martínez-Castillo J, Zizumbo-Villarreal D, Gepts P, and Colunga-GarcíaMarín P (2007) Gene flow and genetic structure in the wild-weedy-domesticated complex of *Phaseolus lunatus* L. in its Mesoamerican center of domestication and diversity. *Crop Sci.* 47: 58–66.

McCormack JH (2004a) *Bean Seed Production: An Organic Seed Production Manual for Seed Growers in the Mid-Atlantic and Southern U.S.* 14p. www .savingourseed.org/pdf/BeanSeedProductionVer_1pt4.pdf.

——— (2004b) *Isolation Distances: Principles and Practices of Isolation Distances for Seed Crops: An Organic Seed Production Manual for Seed Growers in the Mid-Atlantic and Southern U.S.* 21p. www.savingourseed.org/pdf/Isolation DistancesVer_1pt5.pdf.

Mejía-Jiménez A, Muñoz C, Jacobsen HJ, Roca WM, and Singh SP (1994) Interspecific hybridization between common and tepary beans: Increased hybrid embryo growth, fertility, and efficiency of hybridization through recurrent and congruity backcrossing. *Theor. Appl. Genet.* 88: 324–331.

Miranda Colín S and Evans AM (1973) Exploring the genetic isolating mechanisms between *Phaseolus vulgaris* L. and *P. coccineus* Lam. *Annu. Rep. Bean Improv. Coop.* 16: 39–41.

Mok DWS, Mok MC, and Rabakoarihanta A (1978) Interspecific hybridization of *Phaseolus vulgaris* with *P. lunatus* and *P. acutifolius. Theor. Appl. Genet.* 52: 209–215.

Motto M, Sorresi GP, and Salamini F (1978) Seed size inheritance in a cross between wild and cultivated common beans (*Phaseolus vulgaris* L.). *Genetica* 49: 31–36.

Mumba LE and Galwey NW (1998) Compatibility of crosses between gene pools and evolutionary classes in the common bean (*Phaseolus vulgaris* L.). *Genet. Resour. Crop Evol.* 45: 69–80.

——— (1999) Compatibility between wild and cultivated common bean (*Phaseolus vulgaris* L.) genotypes of the Mesoamerican and Andean gene pools: Evidence from the inheritance of quantitative characters. *Euphytica* 108: 105–119.

Nienhuis J and Singh SP (1986) Combining ability analyses and relationships among yield, yield components, and architectural traits in dry bean. *Crop Sci.* 26: 21–27.

Ocampo CH, Martín JP, Sánchez-Yélamo MD, Ortiz JM, and Toro O (2005) Tracing the origin of Spanish common bean cultivars using biochemical and molecular markers. *Genet. Resour. Crop Evol.* 52: 33–40.

Ockendon DJ, Currah L, and Taylor JD (1982) Transfer of resistance to halo-blight (*Pseudomonas phaseolicola*) from *Phaseolus vulgaris* to *P. coccineus*. *Annu. Rep. Bean Improv. Coop.* 25: 84–85.

OECD [Organization for Economic Cooperation and Development] (2008) OECD Scheme for the Varietal Certification of Grass and Legume Seed Moving in International Trade: Annex VI to the Decision. C(2000)146/FINAL. OECD Seed Schemes, February 2008. www.oecd.org/document/0/0,3343,en_2649_33905 _1933504_1_1_1_1,00.html.

Ouédraogo M and Baudoin JP (2002) Comparative analysis of genetic structure and diversity in wild lima bean populations from the Central Valley of Costa Rica, using microsatellite and isozyme markers. *Annu. Rep. Bean Improv. Coop.* 45: 240–241.

Papa R and Gepts P (2003) Asymmetry of gene flow and differential geographical structure of molecular diversity in wild and domesticated common bean (*Phaseolus vulgaris* L.) from Mesoamerica. *Theor. Appl. Genet.* 106: 239–250.

——— (2004) Asymmetric gene flow and introgression between wild and domesticated populations. In: den Nijs HCM, Bartsch D, and Sweet J (eds.) *Introgression from Genetically Modified Plants into Wild Relatives* (CAB International: Wallingford, UK), pp. 125–138.

Papa R, Acosta J, Delgado-Salinas A, and Gepts P (2005) A genome-wide analysis of differentiation between wild and domesticated *Phaseolus vulgaris* from Mesoamerica. *Theor. Appl. Genet.* 111: 1147–1158.

Parker JP and Michaels T (1986a) Genetic control over hybrid plant development in interspecific crosses between *P. vulgaris* L. and *P. acutifolius* A. Gray. *Annu. Rep. Bean Improv. Coop.* 29: 22–23.

——— (1986b) Simple genetic control of hybrid plant development in interspecific crosses between *Phaseolus vulgaris* L. and *P. acutifolius* A. Gray. *Plant Breed.* 97: 315–323.

Payró de la Cruz E, Gepts P, Colunga-GarcíaMarín P, and Zizumbo-Villarreal D (2005) Spatial distribution of genetic diversity in wild populations of *Phaseolus vulgaris* L. from Guanajuato and Michoacán, Mexico. *Genet. Resour. Crop Evol.* 52: 589–599.

Petzold R and Dickson MH (1987) Interspecific hybridization of *Phaseolus vulgaris* with *P. angustissimus*, *P. filiformis*, and *P. ritensis*. *Annu. Rep. Bean Improv. Coop.* 30: 94–95.

Piñero D and Eguiarte L (1988) The origin and biosystematic status of *Phaseolus coccineus* subsp. *polyanthus*: Electrophoretic evidence. *Euphytica* 37: 199–203.

Pratt RC (1983) Gene transfer between tepary and common beans (*Phaseolus acutifolius*, *Phaseolus vulgaris*). *Desert Plants* 5: 57–63.

Pratt RC, Bressan RA, and Hasegawa PM (1985) Genotypic diversity enhances recovery of hybrids and fertile backcrosses of *Phaseolus vulgaris* L. × *P. acutifolius* A. Gray. *Euphytica* 34: 329–344.

Prendota K, Baudoin JP, and Maréchal R (1982) Fertile allopolyploids from the cross *Phaseolus acutifolius* × *Phaseolus vulgaris. Bull. Rech. Agron. Gembloux* 17: 177–189.

Purseglove JW (1968) *Tropical Crops: Dicotyledons*, vol. 1 (Longman Scientific and Technical: Harlow, UK). 719 p.

Rabakoarihanta A, Mok DWS, and Mok MC (1979) Fertilization and early embryo development in reciprocal interspecific crosses of *Phaseolus. Theor. Appl. Genet.* 54: 55–59.

Santalla M, Rodiño AP, and de Ron AM (2002) Allozyme evidence supporting southwestern Europe as a secondary center of genetic diversity for the common bean. *Theor. Appl. Genet.* 104: 934–944.

Schmit V, Baudoin JP, and Wathelet B (1992) Contribution à l'étude des relations phylétiques au sein du complexe *Phaseolus vulgaris* L.-*Phaseolus coccineus* L.-*Phaseolus polyanthus* Greenman. *Bull. Rech. Agron. Gembloux* 27: 199–207.

Shii CT, Mok MC, Temple SR, and Mok DWS (1980) Expression of developmental abnormalities in hybrids of *Phaseolus vulgaris* L.: Interaction between temperature and allelic dosage. *J. Hered.* 71: 218–222.

Singh SP (1991) Bean genetics. In: van Schoonhoven A and Voysest O (eds.) *Common Beans: Research for Crop Improvement* (CAB International: Wallingford, UK), pp. 199–286.

――― (2001a) Broadening the genetic base of common bean cultivars: A review. *Crop Sci.* 41: 1659–1675.

――― (2001b) Use of germplasm in breeding. In: de la Cuadra C, de Ron AM, and Schachl R (eds.) *Handbook on Evaluation of* Phaseolus *Germplasm*. PHASE-LIEU-FAIR-PL97–3463 (Misión Biológica de Galicia, Spanish Council for Scientific Research [Consejo Superior de Investigaciones Científicas, or CSIC]: Pontevedra, Spain), pp. 65–77.

Singh SP and Gutiérrez JA (1984) Geographical distribution of *Dl-1* and *Dl-2* genes causing hybrid dwarfism in *Phaseolus vulgaris* L., their association with seed size, and their significance to breeding. *Euphytica* 33: 337–345.

Singh SP and Molina A (1996) Inheritance of crippled trifoliolate leaves occurring in interracial crosses of common bean and its relationship with hybrid dwarfism. *J. Hered.* 87: 464–469.

Singh SP, Nodari R, and Gepts P (1991) Genetic diversity in cultivated common bean: I. Allozymes. *Crop Sci.* 31: 19–23.

Singh SP, Molina A, and Gepts P (1995) Potential of wild common bean for seed yield improvement of cultivars in the tropics. *Can. J. Plant Sci.* 75: 807–813.

Singh SP, Debouck DG, and Roca WM (1997) Successful interspecific hybridization between *Phaseolus vulgaris* L. and *P. costaricensis* Freytag & Debouck. *Annu. Rep. Bean Improv. Coop.* 40: 40–41.

—— (1998) Interspecific hybridization between *Phaseolus vulgaris* L. and *P. parvifolius* Freytag. *Annu. Rep. Bean Improv. Coop.* 41: 7–8.

Smartt J (1970) Interspecific hybridization between cultivated American species of the genus *Phaseolus. Euphytica* 19: 480–489.

—— (1978) The evolution of pulse crops. *Econ. Bot.* 32: 185–198.

—— (1979) Interspecific hybridization in the grain legumes: A review. *Econ. Bot.* 33: 329–337.

—— (1990) The New World pulses: *Phaseolus* species. In: *Grain Legumes: Evolution and Genetic Resources* (Cambridge University Press: Cambridge, UK), pp. 85–139.

Smartt J and Haq N (1972) Fertility and segregation of the amphidiploid *Phaseolus vulgaris* L. × *P. coccineus* L. and its behaviour in backcrosses. *Euphytica* 21: 496–591.

Sonnante G, Stockton T, Nodari RO, Becerra Velásquez VL, and Gepts P (1994) Evolution of genetic diversity during the domestication of common bean (*Phaseolus vulgaris* L.). *Theor. Appl. Genet.* 89: 629–635.

Sprecher S and Khairallah M (1989) Association of male sterility with gene pool recombinants in bean. *Annu. Rep. Bean Improv. Coop.* 32: 56–67.

Summerfield RJ and Roberts EH (1985) *Phaseolus vulgaris.* In: Halevy AH (ed.) *CRC Handbook of Flowering*, vol. 1 (CRC Press: Boca Raton, FL, USA), pp. 139–148.

Thomas CV and Waines JG (1984) Fertile backcross and allotetraploid plants from crosses between tepary beans and common beans (*Phaseolus acutifolius, Phaseolus vulgaris*). *J. Hered.* 75: 93–98.

Thomas CV, Manshardt RM, and Waines JG (1983) Teparies as a source of useful traits for improving common beans (*Phaseolus acutifolius*, hybridization with *Phaseolus vulgaris*). *Desert Plants* 5: 43–48.

Thomas H (1964) Investigations into the inter-relationships of *Phaseolus vulgaris* L. and *P. coccineus* Lam. *Genetica* 35: 59–74.

Tohme J, González DO, Beebe S, and Duque MC (1996) AFLP analysis of gene pools of a wild bean core collection. *Crop Sci.* 36: 1375–1384.

Toro O, Tohme J, and Debouck DG (1990) *Wild Bean (*Phaseolus vulgaris *L.) Description and Distribution* (Centro Internacional de Agricultura Tropical [CIAT]: Cali, Colombia). 106 p.

USDA [United States Department of Agriculture] (2008) 7 CFR [Code of Federal Regulations] § 201.76. Federal Seed Act Regulations; Minimum Land, Isolation, Field, and Seed Standards. Source: 59 FR [Federal Register] 64516, Dec. 14, 1994, as amended at 65 FR 1710, Jan. 11, 2000 (Government Printing Office:

Washington, DC, USA), pp. 372–378. www.access.gpo.gov/nara/cfr/waisidx_08/7cfr201_08.html.

Vanderborght T (1983) Evaluation of *Phaseolus vulgaris* wild and weedy forms. *Plant Genet. Resour. Newsl.* 54: 18–25.

van Rheenen HA (1979) A sub-lethal combination of two dominant factors in *Phaseolus vulgaris* L. *Annu. Rep. Bean Improv. Coop.* 22: 67–69.

Vieira AL, Patto Ramalho MA, and dos Santos JB (1989) Crossing incompatibility in some bean cultivars utilized in Brazil. *Rev. Brasil. Genet.* 12: 169–171.

Vieira C, Borém A, and Ramalho MAP (1999) Melhoramento do feijão. In: Borém A (ed.) *Melhoramento de Espécies Cultivadas.* 1st ed. (Editora UFV: Viçosa, Brazil), pp. 273–349.

Waines JG, Manshardt RM, and Wells WC (1988) Interspecific hybridization between *Phaseolus vulgaris* and *P. acutifolius.* In: Gepts P (ed.) *Genetic Resources of* Phaseolus *Beans: Their Maintenance, Domestication, Evolution, and Utilization* (Kluwer Academic Publishers: Dordrecht, The Netherlands), pp. 485–502.

Webster BD, Tuker CL, and Lynch P (1977) A morphological study of the development of reproductive structures of *Phaseolus vulgaris* L. *J. Am. Soc. Hort. Sci.* 102: 640–643.

Webster BD, Ross RM, and Evans T (1982) Nectar and the nectary of *Phaseolus vulgaris* L. *J. Am. Soc. Hort. Sci.* 107: 497–503.

Weilenmann de Tau E, Baudoin JP, and Maréchal R (1986) Obtention d'allopolyploïdes fertiles chez le croisement entre *Phaseolus vulgaris* et *Phaseolus filiformis. Bull. Rech. Agron. Gembloux* 21: 35–46.

Weinsten AI (1926) Cytological studies on *Phaseolus vulgaris. Am. J. Bot.* 13: 248–263.

Wells WC, Isom WH, and Waines JG (1988) Outcrossing rates of six common bean lines. *Crop Sci.* 28: 177–178.

Zagorcheva L and Poriazov I (1983) Hybrid development and fertility in *Phaseolus vulgaris* × *Phaseolus coccineus. Annu. Rep. Bean Improv. Coop.* 26: 92–93.

Zambre M, Terryn N, De Clercq J, De Buck S, Dillen W, Van Montagu M, Van Der Straeten D, and Angenon G (2003) Light strongly promotes gene transfer from *Agrobacterium tumefaciens* to plant cells. *Planta* 216: 580–586.

Zizumbo-Villarreal D, Colunga-GarcíaMarín P, Payró de la Cruz E, Delgado-Valerio P, and Gepts P (2005) Population structure and evolutionary dynamics of wild-weedy-domesticated complexes of common bean in a Mesoamerican region. *Crop Sci.* 45: 1073–1083.

Zoro Bi I, Maquet A, and Baudoin JP (2005) Mating system of wild *Phaseolus lunatus* L. and its relationships with population size. *Heredity* 94: 153–158.

NINE

Cotton
(*Gossypium hirsutum* L. and *Gossypium barbadense* L.)

Four cotton (*Gossypium* L.) species are cultivated: upland cotton (*Gossypium hirsutum* L.) and three other species that were domesticated independently as sources of textile fiber (Wendel 1995; Brubaker et al. 1999a)—the New World tetraploid *G. barbadense* L. (Egyptian cotton, **AADD** genome), and the two African-Asian diploid tree cottons *G. arboreum* L. and *G. herbaceum* L. (both **AA** genome). This chapter focuses on the two tetraploid cottons, *G. hirsutum* and *G. barbadense* (both **AADD** genome), since they are the most widely cultivated cotton species, and also the ones most frequently used for genetic transformation.

Center(s) of Origin and Diversity

Gossypium hirsutum seems to have originated from germplasm accessions imported from S Mexico and Guatemala. Its center of domestication is considered to be the Yucatan peninsula in Mexico (Wendel et al. 1992; Brubaker and Wendel 1994; Alvarez et al. 2005). The center of origin of *G. barbadense* lies in western S America, and its center of domestication is in NW Peru and SW Ecuador (Piperno and Pearsall 1998; Westengen et al. 2005).

Both *G. hirsutum* and *G. barbadense*, tetraploid domesticated New World cottons, originated through hybridization of ancestral diploid species that presently have allopatric ranges in Africa-Asia (A-genome species) and the American tropics and subtropics (D-genome species) (Wendel et al. 1995).

The origins of *G. arboreum* and *G. herbaceum* might be in Africa and the Indus Valley, respectively (Brubaker et al. 1999a), although the evidence is not conclusive.

The genus *Gossypium* has three primary centers of diversity: one in W-central and S Mexico (18 species; **D** genome), one in NE Africa and Arabia (14 species; **A**, **B**, **E**, and **F** genomes), and the third in Australia (17 species; **C**, **G**, and **K** genome) (Small and Wendel 2000; Wendel and Cronn 2003; OGTR 2008). Two centers of diversity have been recognized for domesticated upland cotton (*G. hirsutum*) and its progenitors: one in S Mexico-Guatemala and a second one in the Caribbean (Wendel et al. 1992).

Biological information

Flowering

Upland cotton plants flower from the bottom to the top, over a period of several weeks. Each flower is receptive for only a single day, from sunrise to midafternoon (McGregor 1976).

Most wild cotton species have showy, attractive flowers adapted for insect-mediated cross-pollination. Some wild relatives (G-genome species) produce both chasmogamous (cross-pollinated) and cleistogamous (100% self-fertilized) flowers.

Pollen Dispersal and Longevity

The pollen of upland cotton is relatively large and sticky and forms bulks that are too heavy to be dispersed by wind (McGregor 1976; Fryxell 1979; Jenkins 1994). The primary vectors of pollen dispersal are therefore native insects, often bees—such as bumblebees (*Bombus* spp.) and *Melissodes* Latreille, *Tetralonia* Spinola, and *Andrena* Fabr. species (Moffett et al. 1975; Mamood et al. 1990; Sen et al. 2004)—but also beetles and other insects (Llewellyn and Fitt 1996; Llewellyn et al. 2007), depending on the native insect fauna of each location.

Honeybees (*Apis mellifera*) are often considered to be only secondary pollinators (Waller et al. 1985; Rao et al. 1996). Although they visit the flowers of upland cotton frequently in quest of nectar, their preference is for the pollen from other crops (e.g., maize, sorghum, and safflower). When honeybees collect cotton pollen, the pollen loads they transport are significantly smaller than those of pollen from other crops (Vaissière et al. 1984; Rhodes 2002). This is attributable to the morphological characteristics of cotton pollen grains, which are loose and thus gather in smaller loads, resulting in little or no pollen transfer (Loper and DeGrandi-Hoffman 1994; Vaissière and Vinson 1994).

Moths may pollinate some wild cotton species, such as *G. tomentosum* Nutt. *ex* Seem. (Fryxell 1979). Under normal field conditions, the viability of upland cotton pollen declines to zero over a period of 12 hours after release (Govila and Rao 1969; Richards et al. 2005).

Sexual Reproduction

Gossypium hirsutum and *G. barbadense* are both allotetraploids ($2n = 4x = 52$, **AADD** genome). Upland cotton is grown as a perennial crop in tropical regions, but it dies from frost damage during the winter in temperate environments such as Japan and the United States. Both species are fully self-compatible and mainly autogamous, but they act as opportunistic outcrossers when insect pollinators are present. In upland cotton, outcrossing rates are often below 10% (Meredith and Bridge 1973; Gridley 1974; Smith 1976), but they can reach up to 47% (and even 80%), depending on insect prevalence and environmental conditions (Richmond 1951; Simpson and Duncan 1956; Oosterhuis and Jernstedt 1999).

Vegetative Reproduction

There are no reports of vegetative reproduction in tetraploid domesticated cotton (Llewellyn and Fitt 1996).

Seed Dispersal and Dormancy

Cotton seeds (unprocessed seed cotton, fuzzy seed, or delinted black seed) are spread by the wind or by humans as a byproduct during production and post-production stages, for example, through spillage at sowing, during transportation, or during feeding (with cotton livestock feed, also known as fuzzy seeds). Occasionally, seeds are dispersed by animals (birds and mammals, Stephens and Rick 1966) and water (e.g., in *G. darwinii*, Porter 1983, 1984a, b; Stephens 1958).

Information about the persistence of domesticated cotton seeds is scarce. The formation of significant soil seed banks is unlikely, since seeds that do not germinate right away quickly become weathered, leading to significant decreases in their viability (Halloin 1975; Woodstock et al. 1985).

Volunteers, Ferals, and Their Persistence

Gossypium hirsutum volunteers and feral plants occur in all upland cotton growing areas, and they are quite common where cotton seeds are used as livestock feed.

Feral plants of upland cotton often reach reproductive maturity and produce open bolls. In Australia, the relative dryness of uncultivated habitats limits the germination and growth of subsequent generations (OGTR 2008). Also, plants are destroyed by roadside management practices and/ or are grazed by livestock, thereby restricting their potential to reproduce and become weedy. Nonetheless, with an adequate supply of fresh water and protection from fire and grazing animals, individual feral plants and/ or small naturalized feral populations of upland and other domesticated cottons can persist for decades outside of cultivation. This has occurred in several tropical regions, including N Australia, Asia (Vietnam), Mexico, S United States, and Hawaii (Hawkins et al. 2005; OGTR 2008).

Weediness and Invasiveness Potential

Although upland cotton is recognized as weedy or invasive in some regions, it is generally not considered to be a significant agricultural or environmental weed (Keeler et al. 1996).

Crop Wild Relatives

The genus *Gossypium* includes 50 recognized species distributed throughout the tropical and subtropical regions of the world (Fryxell 1979; 1992). Five species, including the two most important domesticated cottons (*G. hirsutum* and *G. barbadense*), are allotetraploids ($2n$

= $4x$ = 52). The remaining species, all but two of which are wild, are diploids ($2n$ = $2x$ = 26) (Brubaker et al. 1999a).

Gossypium species are currently grouped into four subgenera and seven sections (Brubaker and Brown 2001). Furthermore, eight diploid genomic groups (**A-G** and **K**), and one tetraploid genomic group (**AD**) can be distinguished, based on chromosome pairing relationships in interspecific hybrids (table 9.1).

Hybridization

Historically, hybridization has occurred time and time again among Gossypium lineages. It is thought that about one-quarter of modern Gossypium species have experienced historical interspecific cytoplasmic (and perhaps nuclear) introgression (Cronn and Wendel 2003). For example, the allotetraploid lineage (**AD**-genome species) evolved from a hybridization of a New World **D**-genome diploid with an **A**-genome diploid from Africa (Beasley 1940; Harland 1940; Percy and Wendel 1990; Brubaker and Wendel 1994). This is corroborated by fluorescent in situ hybridization, which shows that **A**-genome-specific dispersed repetitive sequences are found in the **D** genome of natural Gossypium polyploids (Hanson et al. 1998; Zhao et al. 1998).

Today, the descendents of these ancient lineages are strongly isolated by geographic and intrinsic genetic barriers (Adams and Wendel 2004). Hybridization can still occur between Gossypium species sharing the same genome (intragenomic), but only rarely between species belonging to different genomic groups (intergenomic). If the latter occurs, naturally or artificially, intergenomic F1 hybrids are always sterile (Endrizzi et al. 1985; Cronn and Wendel 2003).

According to their cross-compatibility with tetraploid domesticated cotton, Gossypium species can be grouped into primary, secondary, and tertiary gene pools (Brubaker and Brown 2001).

Primary Gene Pool (GP-1)

The primary gene pool of tetraploid domesticated cottons comprises species sharing the same genome (**AD**), that is, wild forms and landraces of Gossypium hirsutum and G. barbadense, and the three wild tetra-

Table 9.1 Domesticated cotton (*Gossypium hirsutum* L.) and its crop wild relatives

Taxonomic level	Species	Common name	Genome	Origin and distribution
Primary gene pool (GP-1) subg. *Karpas* Raf.			allotetraploid ($2n = 4x = 52$)	New World tropics and subtropics (including Hawaii)
	G. hirsutum L. (including 7 races, some of them wild)	upland cotton, wild cotton	AADD	wild and cultivated; native to C America, Caribbean, USA (S Florida), West Indies, Samoa, and the Pacific
	G. barbadense L.	pima cotton, Egyptian cotton, creole cotton, Sea Island cotton, S American cotton	AADD	native to S America; introduced/ naturalized in USA (Hawaii, Puerto Rico, Virgin Islands); wild populations currently only in Ecuador and Peru; cultivated in USA, Australia, Peru, Israel, and Egypt
	G. darwinii G. Watt		AADD	endemic to Galápagos Islands (Ecuador)
	G. mustelinum Miers *ex* G. Watt		AADD	endemic to NE Brazil
	G. tomentosum Nutt. *ex* Seem.	Hawaiian cotton	AADD	endemic to Hawaii
Secondary gene pool (GP-2) subg. *Houzingenia* Fryxell			diploid ($2n = 2x = 26$)	New World species
sect. *Houzingenia*				Mexico and Galápagos Islands (Ecuador)
	G. thurberi Tod.	Thurber's cotton, Arizona wild cotton	D	N Mexico and S USA (Arizona)

Taxon	Common name	Genome	Distribution
G. trilobum (DC.) Skovst.		D	C Mexico
G. klotzschianum Andersson		D	endemic to Galapagos Islands (Ecuador)
G. davidsonii Kellogg, G. armourianum Kearney, and G. harknessii Brandegee		D	Baja California (Mexico)
G. turneri Fryxell		D	Mexico
sect. Erioxylum (Rose & Standl.) Prokh.			Mexico and Peru
G. aridum (Rose & Standl.) Skovst.		D	tree; NW to SW Mexico
G. laxum L. Ll. Phillips, G. lobatum Gentry, and G. schwendimanii Fryxell & S.D. Koch		D	trees; endemic to Guerrero and Michoacán (Mexico)
G. raimondii Ulbr.		D	endemic to Peru
G. gossypioides (Ulbr.) Standl.	Mexican cotton	D	endemic to Oaxaca (Mexico)
subg. Gossypium		diploid (2n = 2x = 26)	African-Asian-Arabian species
sect. Gossypium			Asian-African
G. herbaceum L. subsp. herbaceum	tree cotton, Levant cotton	A	cultivated; native to S Africa; introduced in SE Europe, Near East, and Asia; small-scale cultivation in Africa and Asia, from Ethiopia to W India

(continued)

Table 9.1 (continued)

Taxonomic level	Species	Common name	Genome	Origin and distribution
	G. herbaceum subsp. africanum (G. Watt) J.B. Hutch. ex S.C. Harland		A	wild; endemic to S Africa
	G. arboreum L.	tree cotton	A	known only from cultivation; from China, India, and Korea to N Africa Africa, Cape Verde
	G. anomalum Wawra, G. capitis-viridis Mauer, and G. triphyllum (Harv.) Hochr.		B	
	G. longicalyx J.B. Hutch. & B.J.S. Lee		F	E Africa
Tertiary gene pool (GP-3)	G. areysianum Deflers and G. stocksii Mast.		E	Arabian Peninsula, NE Africa, SW Asia
	G. benadirense Mattei, G. bricchettii (Ulbr.) Vollesen, G. somalense (Gürke) J.B. Hutch., and G. vollesenii Fryxell		E	Africa
	G. trifurcatum Vollesen		unknown diploid (2n = 2x = 26)	endemic to Africa (E Somalia)
sect. Serratta Fryxell subg. Sturtia (R. Br.) Tod. sect. Sturtia (R. Br.) Tod.	G. robinsonii F. Muell. and G. sturtianum J.H. Willis		C	Australia

sect. *Hibiscoidea* Tod.	*G. australe* F. Muell., *G. bickii* Prokh., and *G. nelsonii* Fryxell	**G**
sect. *Grandicalyx* Fryxell (Fryxell)	*G. anapoides* J.M. Stewart et al., *nom. inval.*, *G. costulatum* Tod., *G. cunninghamii* Tod., *G. enthyle* Fryxell et al., *G. exiguum* Fryxell et al., *G. londonderriense* Fryxell et al., *G. marchantii* Fryxell et al., *G. nobile* Fryxell et al., *G. pilosum* Fryxell, *G. populifolium* (Benth.) F. Muell. *ex* Tod., *G. pulchellum* (C.A. Gardner) Fryxell, and *G. rotundifolium* Fryxell et al.	**K**

Sources: Fryxell 1979; Stewart et al. 1997; Percival et al. 1999.
Note: Domesticated forms are in bold. Genome designations follow Endrizzi et al. 1985 and Stewart 1995. Gene pool delimitations are according to Brubaker and Brown 2001.

ploid species, G. *darwinii* G. Watt, G. *mustelinum* Miers *ex* G. Watt, and G. *tomentosum*. Hybridization among these species is feasible, and it can occur spontaneously in the wild. The frequency of genetic recombination is high, and this has been used to introduce agronomically important traits—such as blight resistance, boll weevil resistance, boll worm resistance, cleistogamy, and *Fusarium* and *Verticillium* resistance—from wild tetraploid cottons into cultivars of domesticated cotton (see, e.g., Endrizzi et al. 1985; Meredith 1991; Stewart 1995).

Hybridization between upland cotton (G. *hirsutum*) and the other GP-1 species is discussed in detail below.

Gossypium barbadense

Gossypium hirsutum and domesticated G. *barbadense* hybridize readily, both under controlled and natural conditions, resulting in vigorous and fully fertile hybrid progeny (Kerr 1960; Phillips 1961; Shepherd 1974; Stephens 1974, 1976). Hybrid F2 progeny and subsequent generations often contain either depauperate types or plants that closely resemble one of the parents, perhaps due to a cryptic structural differentiation in the chromosomes of the two species (Stephens 1950; Phillips 1961; Waghmare et al. 2005). Several authors have suggested that isolating mechanisms exist between the species, including differences in the timing of pollen shed (Stephens and Phillips 1972), selective fertilization (Kearney and Harrison 1932), partial ecological isolation (Mauer 1930), and selective elimination of donor-parent genes (Stephens 1949; Lewis and McFarland 1952).

On the other hand, ample molecular evidence proves that gene introgression between G. *barbadense* and cultivated or feral G. *hirsutum* has occurred in both directions in areas of sympatry, for example, in Central America and the Caribbean (Percy and Wendel 1990; Wendel et al. 1992; Brubaker et al. 1993; Brubaker and Wendel 1994; Reinisch et al. 1994; Wang et al. 1995; Brubaker et al. 1999a; Abdalla et al. 2001; Alvarez et al. 2005). Based on genome-wide analysis (Jiang et al. 2000), levels and patterns of persistence of G. *barbadense* chromatin, following recurrent backcrossing to G. *hirsutum*, vary widely, with some regions of G. *hirsutum* resistant to G. *barbadense* introgression and others actually favoring it. A worldwide sampling of G. *barbadense* cultivars consistently shows introgression of the genome from G. *hirsutum* in a total of five specific chromosomal regions, suggesting that some selective advantage is conferred by introgression into these locations, although the precise basis for it remains unknown (Wang et al. 1995).

In the United States, restrictions have been imposed on the sale and distribution of GM *Bt* cotton, prohibiting its use in southern Florida, where feral populations of *G. hirsutum* and *G. barbadense*, and hybrid swarms of the two, persist (US EPA 2000). Similar regulations have been considered for Puerto Rico and the Virgin Islands, where naturalized *Gossypium* populations of both species are scattered throughout the islands.

Gossypium tomentosum

The closely related AD-genome species *Gossypium tomentosum* is completely cross-compatible with *G. hirsutum*, producing fully fertile F1 plants (Stephens 1964; DeJoode and Wendel 1992; Hawkins et al. 2005). This wild relative, a Hawaiian endemic, has at times been used in breeding programs (Meyer and Meredith 1978; Waghmare et al. 2005; Saha et al. 2006), but there are no published reports of naturally occurring hybrids to date.

Moreover, thus far molecular evidence neither proves nor rejects gene introgression from domesticated cotton to *G. tomentosum* (Hawkins et al. 2005). Although *G. tomentosum* flowers remain open at night and seem to be pollinated by lepidopterans—presumably moths (Fryxell 1979)—during the day, both visits by nonnative honeybees and carpenter bees and flowering overlap with upland cotton have been observed (Hawkins et al. 2005). Given that populations are in reasonable proximity, gene flow between these two species is possible.

As with *G. barbadense*, genome-wide analysis (Waghmare et al. 2005) detected some regions of *G. hirsutum* that were resistant to *G. tomentosum* introgression, and others that were actually favorable to it. In most cases, these incompatibilities between *G. tomentosum* and *G. hirsutum*, and between *G. barbadense* and *G. hirsutum*, are in disparate genomic regions.

Still, morphological evidence led Fryxell (1979) to suggest that *G. tomentosum* is at risk of extinction as a result of its hybridization with *G. hirsutum*. Also, because of the close genetic relationship between both species, the likelihood of gene flow and introgression cannot be excluded. Therefore, in the United States, the cultivation of commercial

Bt-cotton has been prohibited in Hawaii as a precaution, since both species occur in sympatry there (US EPA 2000).

Natural hybridization of *G. tomentosum* with *G. barbadense*—which, in turn, is fully interfertile with *G. hirsutum*—has also been reported (Stephens 1964; Münster and Wieczorek 2007). Since *G. tomentosum* is more closely related to *G. hirsutum* than *G. barbadense* (DeJoode and Wendel 1992; Lacape et al. 2007), interspecific hybrids between *G. tomentosum* and *G. barbadense* could act as bridges for gene introgression from *G. hirsutum*. In such cases, both domesticated tetraploids would need to be cultivated in sympatry with *G. tomentosum* in Hawaii.

Gossypium darwinii

The Galápagos endemic *Gossypium darwinii* is closely related to *G. barbadense* and the two taxa had been considered to be conspecific for a while (Percy and Wendel 1990; Wendel and Percy 1990).

On the basis of morphology, Fryxell (1979) suggested that *G. darwinii* is at risk of extinction as a result of hybridization with domesticated tetraploid cotton. Although no natural hybrids of *G. darwinii* with *G. hirsutum* have been reported thus far, allozyme studies have shown significant introgression of *G. hirsutum* alleles into *G. darwinii*, apparently derived from an indirect gene transfer through introduced *G. hirsutum*-introgressed *G. barbadense* (Wendel and Percy 1990). Thus, as is the case for *G. tomentosum*, interspecific *G. hirsutum* × *G. barbadense* hybrids may act as genetic bridges for gene transfer from domesticated *G. hirsutum* cotton to *G. darwinii*.

Gossypium mustelinum

The Brazilian endemic *Gossypium mustelinum* is genetically the most divergent species within the tetraploid cotton clade (DeJoode and Wendel 1992; Small et al. 1998). There is little evidence of interspecific introgression of alleles from domesticated cottons into *G. mustelinum*, despite several centuries of sympatric cultivation with *G. barbadense* and *G. hirsutum* (Wendel et al. 1994). However, the possibility of gene transfer from either *G. hirsutum* or *G. barbadense* to *G. mustelinum* cannot be eliminated. There is still insufficient information about outcrossing and hybridization behavior among these species, but a genome-wide analysis of levels and patterns of persistence of *G. mustelinum* chromatin in *G. hirsutum* is underway (A. Paterson, pers. comm. 2008).

Secondary Gene Pool (GP-2)

The secondary gene pool of tetraploid domesticated cottons is comprised of diploid *Gossypium* species that possess **A** or **D** progenitor genomes, as well as **B** or **F** genomes, whose chromosomes are structurally similar to those of the **A** genome (Phillips 1966; Phillips and Strickland 1966). Hybridization with tetraploid cottons can occur naturally, although sexual compatibility varies, depending on the species and the genotype. Pre- or post-zygotic hybridization barriers are frequently encountered, and they lead to embryo abortion in early developmental stages (see, e.g., Weaver 1957).

Assuming that there are no unreduced gametes, any F_1 hybrid formed between tetraploid domesticated cotton and a wild diploid GP-2 species will be a triploid and, thus, usually sterile. Some female and even male fertility can nonetheless be observed (see, e.g., Skovsted 1935; Amin 1940; M. Brown 1951; Meyer 1974; Lee 1986). Fertility can also be easily restored by chromosome doubling (Stewart 1995). Once a fertile hybrid is obtained, the frequency of homoeologous recombination is high enough to permit gene introgression. Successful cases of introgression from GP-2 species into domesticated cotton include genes for fiber strength, disease resistance, and male sterility (Endrizzi et al. 1985; Meredith 1991; Stewart 1995).

Tertiary Gene Pool (GP-3)

The tertiary gene pool of tetraploid domesticated cottons includes the Australian **C**, **G**, and **K** genomes and the African-Arabian **E** genome. In general, these species show a high degree of incompatibility with **AD**-genome species; to obtain viable seeds from hybridizations of GP-3 species with domesticated cotton, significant human intervention is required. This includes the application of plant hormones (gibberellic acid) to retain fruits that would otherwise be aborted, as well as other artificial techniques, such as ovary culture and embryo rescue (Stewart and Hsu 1978; Stewart 1981, 1995; A. Brown et al. 1997). Triploid F_1 plants can show low vigor

and hence are difficult to maintain, and most of their surviving F₁ progeny have no functional pollen fertility (Maréchal 1974, 1983).

Nonetheless, some GP-3 species have been used in breeding programs
to introgress genes into domesticated cotton (see, e.g., Altman et al. 1987;
Zhu and Li 1993; He and Sun 1994; Mergeai et al. 1995; Vroh Bi et al.
1998; Brubaker et al. 2002). Intergenomic hybrids can be obtained if the
plant with the lowest ploidy is used as the pollen source (Beasley 1941). C-
genome species are somewhat easier to cross artificially with domesticated
cotton than the K-, G-, and E-genome species. For example, in one instance, natural hybrids between *G. hirsutum* and *G. sturtianum* J.H. Willis (C genome) were reported in Australia, but the F₁ plants were function-
ally sterile and aborted their flowers before the fruit set (OGTR 2008).
C- and G-genome species are of particular interest, since they are the only
source of genes delaying the synthesis of terpenoid aldehydes, which are
undesirable because they render cotton seeds toxic to ruminants (Brubaker
et al. 1996). Knowledge about the hybridization of domesticated cotton
and the Asian-African E-genome species is largely lacking.

Nor is much information available about the hybridization of domesticated cotton with K-genome species (J. Zhang and Stewart 1997;
Brubaker et al. 1999b). Partially fertile (artificial) hybrids, obtained via
human intervention and artificial hybridization techniques, have been
reported between *G. hirsutum* and three K-genome species (*G. exiguum*
Fryxell et al., *G. nobile* Fryxell et al., and *G. populifolium* (Benth.) F.
Muell. *ex* Tod.). Some triploid hybrids between *G. hirsutum* and Australian K-genome species have exhibited measurable levels of pollen that
can stain (up to 12% for some combinations), although there is no confirmation of the ability of these pollen grains to effect fertilization.
Brubaker et al. (1999b) and J. Zhang and Stewart (1997) have observed
restricted levels of female fertility. In both cases, it is assumed that this
reflects the occasional production of unreduced gametes.

Outcrossing Rates and Distances

Pollen Flow

An indirect estimation of pollen flow in cotton, through the assessment
of pollinator visits, can overestimate cross-pollination rates, because

many insects visit cotton flowers to collect sugar from the nectaries outside of the flowers without transporting pollen (see above).

All in all, outcrossing rates in cotton are strongly influenced by the abundance of insect pollinators, which vary with location and time (Fryxell 1956; Elfawal et al. 1976; Moffett et al. 1976; Moresco et al. 1999). For example, outcrossing rates measured in the United States are consistently higher than those in Australia, which might be a reflection of differences in pollinator species, particularly because of the absence of bumblebees (*Bombus* spp.) in Australia (Llewellyn and Fitt 1996). Since pollinator species and their population densities vary both geographically and seasonally, the calculation of outcrossing rates—and hence pollen dispersal, subsequent hybridization, and gene introgression—needs to be assessed on a regional basis.

In cotton, outcrossing through pollen dispersal can be described as a leptocurtic curve, with the highest frequencies localized around the pollen source (about 1 m), exponentially decreasing with distance (Kareiva et al. 1994). The majority of field-based measures result in outcrossing rates of 10% or less within 1 m from the pollen source (table 9.2), although higher estimates (16.5% to 25%) have been reported in a few cases (Smith 1976; Moresco et al. 1999). Under certain conditions (e.g., a large abundance of pollinators), outcrossing rates can reach 47% (Simpson and Duncan 1956), and even up to 80% (Richmond 1951; Oosterhuis and Jernstedt 1999). This decrease in the outcrossing rate with increasing distance is slightly less pronounced in the presence of pollinators, but the tendency is the same, and outcrossing rates beyond 10 m are generally less than 1% (table 9.2; Kareiva et al. 1994). In rare instances, pollen has been detected at distances as far as 1625 m, in which case the frequency of outcrossing was extremely low (< 0.05%; van Deynze et al. 2005).

Isolation Distances

Isolation standards for commercial cotton-seed production are quite similar in most cotton-producing countries, with a minimum separation distance of 200 m for certified seed and 400 m for registered and foundation seed (table 9.3). Evidence from experimental field trials (fig. 9.1) and recent field-scale studies in the United States (Berkey et al. 2003; van Deynze et al. 2005; Llewellyn et al. 2007) indicates that these distances are ade-

Table 9.2 Some representative studies on pollen dispersal distances and outcrossing rates of domesticated cotton (*Gossypium hirsutum* L. and *G. barbadense* L.)

Pollen flow from	Pollen dispersal distance (m)	% outcrossing (max. values)	Country	Reference
G. hirsutum	< 3, 3, 4–8	5, < 0.01, 0	Australia	Thomson 1966
G. hirsutum	1, 3, 5, 10, 15, 20	0.15, 0.08, 0, 0, 0, 0	Australia	Llewellyn and Fitt 1996
G. hirsutum	1, 3, 5, 10, 15, 20	0.4, 0.13, 0.1, 0.06, 0, 0	Australia	Llewellyn and Fitt 1996
G. hirsutum	1, > 1, 32, 53	1.7, < 1, 0, 0.3	Australia	Mungomery and Glassop 1969*
G. hirsutum	50, 55	1.38, 0.19	Emerald, QLD, Australia	Llewellyn et al. 2007
G. hirsutum	1, 10, 25, 50	22.7, 1.59, 0.90, 0.76	Kununurra, WA, Australia	Llewellyn et al. 2007
G. hirsutum	1, 7, 22	5.7, < 1, 0.7	USA	Umbeck et al. 1991*
G. hirsutum	35–350	up to 5.9	USA	Meredith and Bridge 1973
G. hirsutum (experimental plot, without bees)	0.3, 1, 3, 9, 30	4.86, 0.3, 0.03, 0.03, 0.03	Shafter, CA, USA	van Deynze et al. 2005
G. hirsutum (experimental plot, with bees)	0.3, 1, 3, 9, 30	7.65, 3.1, 1.6, 0.67, 0.32	Kearney, CA, USA	van Deynze et al. 2005
G. hirsutum (averaged from various commercial fields, with bees)	30, 200, 400, 800, 1625	1.0, 0.07, 0.3, 0.2, 0.04	CA, USA	van Deynze et al. 2005
G. hirsutum	1, 2, 10	1.67–2.67, 1.42, < 0.1 (outcrossing rates estimated using gland status as genetic marker)	Greece	Xanthopoulos and Kechagia 2000

Species				
G. hirsutum	1, 2, 10	3.85, 2.79, 0.31 (outcrossing rates estimated using red-leaf trait as genetic marker)	Greece	Xanthopoulos and Kechagia 2000
G. hirsutum (with bees)	0.7, 1.3, 2.6, 3.9, 5.2, 6.5, 7.8, 8.5	4.87, 2.98, 1.23, 0.87, 0.24, 0.26, 0.15, 0.03	Turkey	Sen et al. 2004
G. hirsutum	1, 35	< 10, < 0.5	Israel	Yasour et al. 2002
G. hirsutum	0, 1.22, 2.44, 3.66, 4.88, 6.1	6.29, 1.63, 0.67, 0.27, 0.08, 0.04	S Africa	Theron and van Staden 1975
G. hirsutum	1, 5, 10, 20, 50, 100, 150	11.2, 0.61, 0.16, 0.09, 0.03, 0, 0	China	C. Zhang et al. 1997
G. hirsutum	0–6 (up to 36)	1.85–11.05	China	Shen et al. 2001
G. hirsutum	5, 10, 25	6, 4.7, 0.6	n.a.[†]	Green and Jones 1953
G. hirsutum	1.1, 9.6, 10.7	19.5, 2.6, 1	n.a.	Green and Jones 1953*
G. barbadense	1.1, 35.2	7.8, 0.16	Egypt	Galal et al. 1972
G. barbadense (without bees)	1, 1.8, 2.7, 3.6, 4.5, 6.3, 8, 9	0.9, 0.15, 0, 0, 0, 0, 0.2, 0.35	Egypt	Elfawal et al. 1976
G. barbadense (with bees)	1, 1.8, 2.7, 3.6, 4.5, 6.3, 8, 9	2.2, 1, 0.2, 0.4, 0.2, 0, 1.2, 1	Egypt	Elfawal et al. 1976
G. barbadense	0–6 (up to 72)	1.42–8.67	China	Shen et al. 2001

*Using buffer rows.
[†]n.a. = not available.

Table 9.3 Isolation distances for domesticated cotton (*Gossypium hirsutum* L.)

Threshold*	Minimum isolation distance (m)	Country	Reference	Comment
1 plant in 30 m² / 0.2%	400†	OECD	OECD 2008	foundation (basic) seed requirements (non-hybrids)
1 plant in 10 m² / 0.2%	200†	OECD	OECD 2008	certified seed requirements (non-hybrids)
0.2% / 0.5%	600†	OECD	OECD 2008	foundation (basic) seed requirements (hybrids)
0.5% / 0.5%	200†	OECD	OECD 2008	certified seed requirements (hybrids)
0.01% / 0.0%	400	USA	CCIA 2009	foundation (basic) seed requirements
0.02% / 0.0%	400	USA	CCIA 2009	registered seed requirements
0.1% / 1 seed per pound	200	USA	CCIA 2009	certified seed requirements
0.01% / 0.03%	400‡	USA	USDA 2008	foundation (basic) seed requirements
0.02% / 0.05%	400‡	USA	USDA 2008	registered seed requirements
0.1% / 0.1%	200‡	USA	USDA 2008	certified seed requirements
1 plant in 30 m² / 0.3%	400	EU	EC 2002	foundation (basic) seed requirements
1 plant in 10 m² / 0.3%	200	EU	EC 2002	certified seed requirements

*Maximum percentage of off-type plants permitted in the field / seed impurities of other varieties or off-types permitted.
†*G. barbadense:* 800 m for foundation (basic) non-hybrid seed, 600 m for certified hybrid seed.
‡Other cotton species: 30 m.

quate for maintaining seed purity standards, as well as remaining under the maximum allowable amount (0.9%) of GM trait contamination.

Yet the U.S. Biopesticides Registration Action Document states that "upon approval by EPA, test plots and/or breeding nurseries in Hawaii, the U.S. Virgin Islands, and Puerto Rico may be established without restrictions if alternative measures, such as insecticide applications, are shown to effectively mitigate gene flow" (US EPA 2000, p. 61). In Aus-

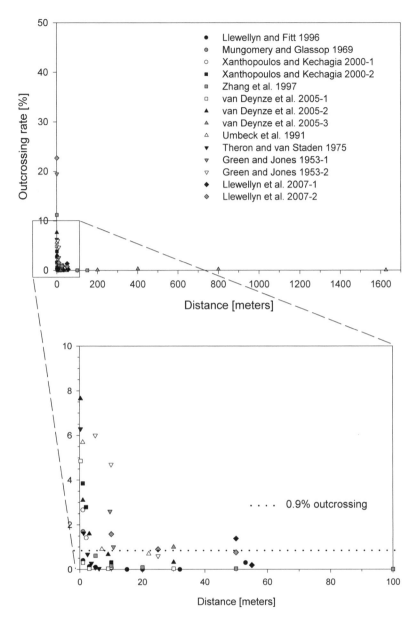

Fig. 9.1 Pollen dispersal distances and outcrossing rates of domesticated upland cotton (*Gossypium hirsutum* L.)

Xanthopoulus and Kechagia 2000–1: using glandless traits as a genetic marker; Xanthopoulus and Kechagia 2000–2: using red-leaf traits as a genetic marker; van Deynze et al. 2005–1: without bees, experimental plot in Shafter, CA, USA; van Deynze et al. 2005–2: with bees, experimental plot in Kearney, CA, USA; van Deynze et al. 2005–3: with bees, averaged from various commercial fields in California, USA; Green and Jones 1953–1: with cotton buffer rows; Green and Jones 1953–2: without buffer rows; Llewellyn et al. 2007–1: Emerald, QLD, Australia; Llewellyn et al. 2007–2: Kununurra, WA, Australia.

tralia, a 20-meter buffer zone is considered adequate for GM trait containment, and recent field studies have confirmed this (Llewellyn et al. 2007). All in all, these measures might increase the likelihood of gene flow from cultivated to feral (*G. hirsutum* and *G. barbadense*) and wild (*G. tomentosum*, *G. thurberi*) cotton in those locations in which they coexist, and even more so in the United States, where there seems to be a higher pollinator abundance than in Australia.

State of Development of GM Technology

Genetic modification of cotton has long been established. Reports about researched GM traits mention insect resistance (*Bt*), enhanced yield, improved fiber quality (increased fiber length and strength), and fiber color. Field trials are or have been conducted in Burkina Faso, France, Greece, Guatemala, Japan, Kenya, the Philippines, Spain, Uganda, and Vietnam. The commercial production of GM cotton covers about 15 million ha in Argentina, Australia, Brazil, Burkina Faso, China, Colombia, Costa Rica, Egypt, India, Indonesia, Mexico, Pakistan, Paraguay, South Africa, and the United States (James 2008).

Agronomic Management Recommendations

There are no specific recommendations.

Crop Production Area

Gossypium hirsutum is the most widely cultivated of the four domesticated cotton species. It accounts for more than 90% of the world's cotton production and is grown in over 40 countries in both tropical and temperate latitudes, ranging from 47° N in the Ukraine and 37° N in the United States to 32° S in S America and Australia. Four countries account for 70% of the world's upland cotton production: China (32%), the United States (17%), India (13%), and Pakistan (9%) (FAO 2009).

　　Gossypium barbadense accounts for less than 10% of the world's cotton production and is grown primarily in China, Egypt, India, Sudan, and the United States. Tree cotton (*G. arboreum* and *G. herbaceum*) is cultivated only to a very limited extent, mostly in small-scale farming systems in some regions of Asia (India and Pakistan) and Africa.

The total crop production area (of seed cotton) is estimated at 33.8 million ha (FAO 2009), out of which 46% (15.5 million ha) is in GM cotton (James 2008).

Research Gaps and Conclusions

For cotton, there are three main ways gene introgression to other domesticated or wild cotton plants occurs:

1 through pollen flow from cotton fields

2 through volunteers in fields previously planted with cotton

3 through seed dispersal during cropping, harvesting, handling, storage, and transport of the crop and the subsequent establishment of ferals outside of cultivated areas

If flowering overlap is probable with any of the cotton species identified above as being susceptible to gene introgression, no plants should be in the area at distances shorter than those specified by legal regulations. The isolation distances established by regulators have been shown to be effective for the containment of pollen-mediated gene flow in cotton and should be taken into consideration. In particular, a minimum isolation distance of 200 m has proven to be adequate both for effective containment of pollen dispersal and for staying within the limits established by regulators for the maximum allowable percentages of seed impurities. Twenty-meter-wide buffer zones of cotton varieties that flower at the same time seem to be effective alternatives in limiting pollen dispersal.

The likelihood of gene flow and introgression from *Gossypium hirsutum* to other cotton species—assuming physical proximity and flowering overlap, and being conditional on the nature of the respective trait(s)—is as follows:

- *high* for cultivated and wild *Gossypium barbadense* (pima or Egyptian cotton) and for the wild cotton species *G. tomentosum* (Hawaiian cotton), *G. darwinii*, and *G. mustelinum* (in the Americas) (map 9.1)

- *moderate* for the diploid wild GP-2 cotton species, that is, species bearing the **A** (2 Asian-African species), **B** (3 species endemic to Cape Verde), **D** (13 New World species; in map 9.1, most of this area is hidden by the high-likelihood areas), or **F** genome (1 species endemic to E Africa) (map 9.1)

- *highly unlikely* for the diploid wild GP-3 cotton species, that is, species bearing the **C, G, K** (17 Australian species) or **E** genome (5–6 African-Arabian species; in map 9.1, part of this area is hidden by the moderate-likelihood areas)

REFERENCES

Abdalla AM, Reddy OUK, El-Zik KM, and Pepper AE (2001) Genetic diversity and relationships of diploid and tetraploid cottons revealed using AFLP. *Theor. Appl. Genet.* 102: 222–229.

Adams KL and Wendel JF (2004) Exploring the genomic mysteries of polyploidy in cotton. *Biol. J. Linn. Soc.* 82: 573–581.

Altman DW, Stelly DM, and Kohel RJ (1987) Introgression of the glanded-plant and glandless-seed trait from *Gossypium sturtianum* Willis into cultivated upland cotton using ovule culture. *Crop Sci.* 27: 880–884.

Alvarez I, Cronn R, and Wendel JF (2005) Phylogeny of the New World diploid cottons (*Gossypium* L., Malvaceae) based on sequences of three low-copy nuclear genes. *Plant Syst. Evol.* 252: 199–214.

Amin KC (1940) Interspecific hybridization between Asiatic and New World cottons. *Indian J. Agric. Sci.* 10: 404–413.

Beasley JO (1940) The production of polyploids in *Gossypium*. *J. Hered.* 31: 39–48.

—— (1941) Hybridization, cytology, and polyploidy of *Gossypium*. *Chron. Bot.* 6: 394–395.

Berkey DA, Savoy BR, Jeanes VR, and Lehman JD (2003) Pollen dispersal from transgenic cotton fields. In: *Proceedings of the Beltwide Cotton Conference, Nashville, Tennessee, USA, January 6–10, 2003* (National Cotton Council of America: Memphis, TN, USA), CD-ROM.

Brown AHD, Brubaker CL, and Kilby MJ (1997) Assessing the risk of cotton transgene escape into wild Australian *Gossypium* species. In: McLean GD, Waterhouse PM, Evans G, and Gibbs MJ (eds.) *Commercialisation of Transgenic Crops: Risk, Benefit, and Trade Considerations* (Bureau of Resource Sciences: Canberra, ACT, Australia), pp. 83–93.

Brown MS (1951) The spontaneous occurrence of amphiploidy in species hybrids of *Gossypium. Evolution* 5: 25–41.

Brubaker CL and Brown AHD (2001) An Evaluation of the Potential for Gene Flow between Commercial Cotton Cultivars and Wild Australian Cotton Species. (Commonwealth Scientific and Industrial Research Organisation [CSIRO], Centre for Plant Biodiversity Research, CSIRO Plant Industry: Canberra, ACT, Australia).

Brubaker CL and Wendel JF (1994) Reevaluating the origin of domesticated cotton (*Gossypium hirsutum*, Malvaceae) using nuclear restriction fragment length polymorphisms (RFLP). *Am. J. Bot.* 81: 1309–1326.

Brubaker CL, Koontz JA, and Wendel JF (1993) Bidirectional cytoplasmic and nuclear introgression in the New World cottons, *Gossypium barbadense* and *G. hirsutum* (Malvaceae). *Am. J. Bot.* 80: 1203–1208.

Brubaker CL, Benson CG, Miller C, and Leach DN (1996) Occurrence of terpenoid aldehydes and lysigenous cavities in the glandless seeds of Australian *Gossypium* species. *Aust. J. Bot.* 44: 601–612.

Brubaker CL, Brown AHD, Grace JP, and Kilby MJ (1996) Using wild Australian *Gossypium* germplasm in cotton breeding. In: *Cotton on to the Future: 8th Australian Cotton Conference, Broadbeach, Queensland, August 14–16, 1996* (Australian Cotton Growers Research Association: Narrabri, NSW, Australia), pp. 619–624.

Brubaker CL, Bourland FM, and Wendel JF (1999) The origin and domestication of cotton. In: Smith CW and Cothren JT (eds.) *Cotton: Origin, History, Technology, and Production* (John Wiley and Sons: New York, NY, USA), pp. 3–31.

Brubaker CL, Brown AHD, Stewart JM, Kilby MJ, and Grace JP (1999) Production of fertile hybrid germplasm with diploid Australian *Gossypium* species for cotton improvement. *Euphytica* 108: 199–213.

CCIA [California Crop Improvement Association] (2009) Cotton Certification Standards. http://ccia.ucdavis.edu/seed_cert/seedcert_index.htm.

Cronn R and Wendel JF (2003) Cryptic trysts, genomic mergers, and plant speciation. *New Phytol.* 161: 133–142.

DeJoode DR and Wendel FF (1992) Genetic diversity and origin of the Hawaiian Island cotton, *Gossypium tomentosum. Am. J. Bot.* 79: 1311–1319.

EC [European Commission] (2002) Council Directive 2002/57/EC of 13 June 2002 on the Marketing of Seed of Oil and Fibre Plants. OJ No. L 193, 20.7.2002. www.legaltext.ee/text/en/U70003.htm.

Elfawal MA, Bishr MA, and Hassoub EK (1976) Natural cross-pollination in Egyptian cotton (*Gossypium barbadense* L.). *J. Agric. Sci.* 86: 205–209.

Endrizzi JE, Turcotte EL, and Kohel RJ (1985) Genetics, cytology, and evolution of *Gossypium*. *Adv. Genet.* 23: 271–375.

FAO [Food and Agriculture Organization of the United Nations] (2009) FAOSTAT data. http://faostat.fao.org/site/567/default.aspx.

Fryxell PA (1956) Effect of varietal mass on percentage of outcrossing in *Gossypium hirsutum* in New Mexico. *J. Hered.* 48: 299–301.

—— (1979) *The Natural History of the Cotton Tribe (Malvaceae, Tribe Gossypieae)* (Texas A&M University Press: College Station, TX, USA). 245 p.

—— (1992) A revised taxonomic interpretation of the genus *Gossypium*. *Rheedea* 2: 108–165.

Galal HE, Abou-el-fittouh HA, and Morshed G (1972) Effect of direction and distance on cross-pollination in Egyptian cotton (*Gossypium barbadense* L.). *Exp. Agric.* 8: 67–71.

Govila OP and Rao CH (1969) Studies on the *in vitro* germination and storage of cotton pollen. *J. Palynol.* 5: 37–41.

Green JM and Jones MD (1953) Isolation of cotton for seed increase. *Agron. J.* 45: 366–368.

Gridley HE (1974) Natural and artificial crossing in upland cotton at Namulonge, Uganda. *Cotton Growing Rev.* 51: 149–152.

Halloin JM (1975) Solute loss from deteriorated cotton seed: Relationships between deterioration, seed moisture, and solute loss. *Crop Sci.* 15: 11–15.

Hanson RE, Zhao X-P, Islam-Faridi MN, Paterson AH, Zwick MS, Crane CF, McKnight TD, Stelly DM, and Price HJ (1998) Evolution of interspersed repetitive elements in *Gossypium* (Malvaceae). *Am. J. Bot.* 85: 1364–1368.

Harland SC (1940) New allopolyploids in cotton by the use of colchicine. *Trop. Agric.* 17: 53–55.

Hawkins JS, Pleasants J, and Wendel JF (2005) Identification of AFLP markers that discriminate between cultivated cotton and the Hawaiian Island endemic, *Gossypium tomentosum* Nuttall *ex* Seeman. *Genet. Resour. Crop Evol.* 52: 1069–1078.

He J and Sun C (1994) A scheme for introgression of delayed gland morphogenesis gene from wild *Gossypium* into cultivated cotton (*G. hirsutum*). *Act. Genet. Sin.* 21: 52–58.

James C (2008) *Global Status of Commercialized Biotech/GM Crops 2008*. ISAAA Brief 39 (The International Service for the Acquisition of Agri-biotech Applications [ISAAA]: Ithaca, NY, USA). 243 p.

Jenkins JN (1994) Cotton. In: Organization for Economic Cooperation and Development [OECD] (ed.) *Traditional Crop Breeding Practices: An Historical Review to Serve as a Baseline for Assessing the Role of Modern Biotechnology* (Organization for Economic Cooperation and Development [OECD]: Paris, France), pp. 61–70.

Jiang CX, Chee PW, Draye X, Morrell PL, Smith CW, and Paterson AH (2000) Multi-locus interactions restrict gene introgression in interspecific populations of polyploid *Gossypium* (cotton). *Evolution* 54: 798–814.

Kareiva P, Morris W, and Jacobi CM (1994) Studying and managing the risk of cross-fertilization between transgenic crops and wild relatives. *Mol. Ecol.* 3: 15–21.

Kearney TH and Harrison GJ (1932) Pollen antagonism in cotton. *J. Agric. Res.* 44: 191–226.

Keeler KH, Turner CE, and Bolick MR (1996) Movement of crop transgenes into wild plants. In: Duke SO (ed.) *Herbicide Resistant Crops: Agricultural, Environmental, Economic, Regulatory, and Technical Aspects* (CRC Press: Boca Raton, FL, USA), pp. 303–330.

Kerr T (1960) The potentials of *barbadense* cottons. *Proceedings of the 12th Annual Cotton Improvement Conference* (National Cotton Council of America: Memphis, TN, USA), pp. 57–60.

Lacape JM, Dessauw D, Rajab M, Noyer JL, and Hau B (2007) Microsatellite diversity in tetraploid *Gossypium* germplasm: Assembling a highly informative genotyping set of cotton SSRs. *Mol. Breed.* 19: 45–58.

Lee JA (1986) An early example of a viable hybrid from a cross of *Gossypium barbadense* L. and *G. davidsonii* Kell. *J. Hered.* 77: 56–57.

Lewis CF and McFarland EF (1952) The transmission of marker genes in intraspecific backcrosses of *Gossypium hirsutum* L. *Genetics* 37: 353–358.

Llewellyn D and Fitt G (1996) Pollen dispersal from two field trials of transgenic cotton in the Namoi Valley, Australia. *Mol. Breed.* 2: 157–166.

Llewellyn D, Tyson C, Constable G, Duggan B, Beale S, and Steel P (2007) Containment of regulated genetically modified cotton in the field. *Agric. Ecosys. Environ.* 121: 419–429.

Loper GM and DeGrandi-Hoffman G (1994) Does in-hive pollen transfer by honey bees contribute to cross-pollination and seed set in hybrid cotton? *Apidology* 25: 94–102.

Mamood AN, Waller GD, and Hagler JR (1990) Dispersal of upland and pima cotton pollen by honey bees (Hymenoptera: Apideae) visiting upland male-sterile flowers. *Environ. Entomol.* 19: 1034–1036.

Maréchal R (1974) Analyses de la conjugaison méiotique chez les hybrides triploïdes entre *Gossypium hirsutum* L. et des espèces sauvages australiennes. *Bull. Rech. Agron. Gembloux* 9: 193–204.

——— (1983) Une collection d'hybrides interspécifiques du genre *Gossypium*. *Coton et Fibres Tropicales* 38: 240–246.

Mauer FM (1930) The cottons of Mexico, Guatemala, and Colombia. *Bull. Appl. Bot. Genet. Plant Breed.* 47 (Suppl.): 543–553.

McGregor SE (1976) *Insect Pollination of Cultivated Crop Plants*. U.S. Department of Agriculture, Agricultural Research Service, Agriculture Handbook 496 (Government Printing Office: Washington DC, USA). 411 p.

Meredith WR Jr (1991) Contributions of introductions to cotton improvement. In: Shands HL and Wiesner LS (eds.) *Use of Plant Introductions in Cultivar Development* (Crop Science Society of America: Madison, WI, USA), pp. 127–146.

Meredith WR Jr and Bridge RR (1973) Natural crossing in cotton (*Gossypium hirsutum* L.). *Crop Sci.* 13: 551–552.

Mergeai G, Vroh Bi I, du Jardin P, and Baudoin JP (1995) Introgression of glanded-plant and glandless-seed trait from *G. sturtianum* Willis into tetraploid cotton plants. In: *Proceedings of the Beltwide Cotton Conference, San Antonio, Texas, USA, January 4–7, 1995* (National Cotton Council of America: Memphis, TN, USA), pp. 513–514.

Meyer VG (1974) Interspecific cotton breeding. *Econ. Bot.* 28: 56–60.

Meyer VG and Meredith WR Jr (1978) New germplasm for crossing upland cotton (*Gossypium hirsutum* L.) with *G. tomentosum*. *J. Hered.* 69: 183–187.

Moffett JO, Stith LS, Burkhart CC, and Shipman CW (1975) Honey bee visits to cotton flowers. *Environ. Entomol.* 4: 203–206.

—— (1976) Fluctuation of wild bee and wasp visits to cotton flowers. *Arizona Acad. Sci.* 11: 64–68.

Moresco ER, Vello NA, Aguiar PH, Griddi-Papp IL, Freire EC, Farias FJCF, Marques MF, and Souza MC de (1999) Determinação da taxa de alogamia no algodoeiro herbáceo no cerrado de Mato Grosso. In: *Anais: Congresso Brasileiro de Algodão, Ribeirão Preto, SP, 5–10 Setembro 1999; O Algodão no Século XX, Perspectivas para o Século XXI* (Empresa Brasileira de Pesquisa Agropecuária [EMBRAPA]: Campina Grande, PB, Brazil), p. 98.

Mungomery VE and Glassop AJ (1969) Natural cross-pollination of cotton in central Queensland. *Queensland J. Agric. Animal Sci.* 26: 69–74.

Münster P and Wieczorek AM (2007) Potential gene flow from agricultural crops to native plant relatives in the Hawaiian Islands. *Agric. Ecosys. Environ.* 119: 1–10.

OECD [Organization for Economic Cooperation and Development] (2008) OECD Scheme for the Varietal Certification of Crucifer Seed and Other Oil or Fibre Species Seed Moving in International Trade: Annex VII to the Decision. C(2000)146/FINAL. OECD Seed Schemes, February 2008. www.oecd.org/document/o/o,3343,en_2649_33905_1933504_1_1_1_1,00.html.

OGTR [Office of the Gene Technology Regulator] (2008) *The Biology and Ecology of Cotton (Gossypium hirsutum) in Australia* (Office of the Gene Technology Regulator [OGTR]: Canberra, ACT, Australia). 91 p. www.ogtr.gov.au/internet/ogtr/publishing.nsf/Content/riskassessments-1.

Oosterhuis DM and Jernstedt J (1999) Morphology and anatomy of the cotton plant.

In: Smith W and Cothren JS (eds.) *Cotton: Origin, History, Technology, and Production* (John Wiley and Sons: New York, NY, USA), pp. 175–206.

Percival AE, Wendel JF, and Stewart JM (1999) Taxonomy and germplasm resources. In: Smith CW and Cothren JT (eds.) *Cotton: Origin, History, Technology, and Production* (John Wiley and Sons: New York, NY, USA), pp. 33–63.

Percy RG and Wendel JF (1990) Allozyme evidence for the origin and diversification of *Gossypium barbadense* L. *Theor. Appl. Genet.* 79: 529–542.

Phillips LL (1961) The cytogenetics of speciation in Asiatic cotton. *Genetics* 46: 77–83.

——— (1966) The cytology and phylogenetics of the diploid species of *Gossypium*. *Am. J. Bot.* 53: 328–335.

Phillips LL and Strickland MA (1966) The cytology of a hybrid between *Gossypium hirsutum* and *G. longicalyx. Can. J. Genet. Cytol.* 8: 91–95.

Piperno DR and Pearsall DM (1998) *The Origins of Agriculture in the Lowland Neotropics* (Academic Press: San Diego, CA, USA). 400 p.

Porter DM (1983) Vascular plants of the Galápagos: Origins and dispersal. In: Bowman RI, Berson M, and Leviton AE (eds.) *Patterns of Evolution in Galápagos Organisms* (American Association for the Advancement of Science [AAAS]: San Francisco, CA, USA), pp. 33–96.

——— (1984a) Endemism and evolution in terrestrial plants. In: Perry R (ed.) *Key Environments—Galápagos* (Pergamon Press: Oxford, UK), pp. 85–100.

——— (1984b) Relationships of the Galápagos flora. *Biol. J. Linn. Soc.* 21: 243–251.

Rao GM, Nadre KR, and Suryanarayana MC (1996) Studies on the utility of honey bees on production of foundation seed of cotton cv. NCMHH-20. *Indian Bee J.* 58: 13–15.

Reinisch JA, Dong J, Brubaker CL, Stelly DM, Wendell JF, and Paterson AH (1994) A detailed RFLP map of cotton, *Gossypium hirsutum* × *Gossypium barbadense*: Chromosome organization and evolution in disomic polyploid genome. *Genetics* 138: 829–847.

Rhodes R (2002) Cotton pollination by honeybees. *Aust. J. Agric. Res.* 42: 513–518.

Richards JS, Stanley JN, and Gregg PC (2005) Viability of cotton and canola pollen on the proboscis of *Helicoverpa armigera*: Implications for spread of transgenes and pollination ecology. *Ecol. Entomol.* 30: 327–333.

Richmond TR (1951) Procedures and methods of cotton breeding with special reference to American cultivated species. *Adv. Genet.* 4: 213–245.

Saha S, Raska DA, and Stelly DM (2006) Upland cotton (*Gossypium hirsutum* L.) × Hawaiian cotton (*G. tomentosum* Nutt. *ex* Seem.) F1 hybrid hypoaneuploid chromosome substitution series. *J. Cotton Sci.* 10: 263–272.

Sen I, Oglakci M, Bolek Y, Cicek B, Kiskurek N, and Aydin S (2004) Assessing the out-crossing ratio, isolation distance, and pollinator insects in cotton. *Asian J. Plant Sci.* 3: 724–727.

Shen FF, Yu YJ, Zhang XK, Bi JJ, and Yin CY (2001) *Bt* gene flow from transgenic cotton. *Act. Genet. Sin.* 28: 562–567 [English abstract].

Shepherd RL (1974) Transgressive segregation for root-knot nematode resistance in cotton. *Crop Sci.* 14: 872–875.

Simpson DM and Duncan EN (1956) Varietal response to natural crossing in cotton. *Agron. J.* 48: 74–75.

Skovsted A (1935) Some new interspecific hybrids in the genus *Gossypium* L. *J. Genet.* 30: 447–463.

Small RL and Wendel JF (2000) Phylogeny, duplication, and intraspecific variation of *Adh* sequences in New World diploid cottons (*Gossypium* L., Malvaceae). *Mol. Phylogen. Evol.* 16: 73–84.

Small RL, Ryburn JA, Cronn RC, Seelanan T, and Wendel JF (1998) The tortoise and the hare: Choosing between noncoding plastome and nuclear ADH sequences for phylogeny reconstruction in a recently diverged plant group. *Am. J. Bot.* 85: 1301–1315.

Smith WC (1976) Natural cross-pollination of cotton. *Arkansas Farm Res.* 25: 6.

Stephens SG (1949) The cytogenetics of speciation in *Gossypium*: I. Selective elimination of the donor parent genotype in interspecific backcrosses. *Genetics* 34: 627–637.

—— (1950) The internal mechanism of speciation in *Gossypium*. *Bot. Rev.* 16: 115–149.

—— (1958) Salt water tolerance of seeds of *Gossypium* species as a possible factor in seed dispersal. *Am. Nat.* 92: 83–92.

—— (1964) Native Hawaiian cotton (*Gossypium tomentosum* Nutt.). *Pacific Sci.* 18: 385–398.

—— (1974) The use of two polymorphic systems, nectary fringe hairs and corky alleles, as indicators of phylogenetic relationships in New World cottons. *Biogeotropica* 6: 194–201.

—— (1976) The origin of Sea Island cotton. *Agric. Hist.* 50: 391–399.

Stephens SG and Phillips LL (1972) The history and geographical distribution of a polymorphic system in New World cottons. *Biotropica* 4: 49–60.

Stephens SG and Rick CM (1966) Problems on the origin, dispersal, and establishment of the Galápagos cottons. In: Bowman RI (ed.) *The Galápagos* (University of California Press: Berkeley, CA, USA), pp. 201–208.

Stewart JMcD (1981) *In vitro* fertilization and embryo rescue. *Environ. Exp. Bot.* 21: 301–315.

—— (1995) Potential for crop improvement with exotic germplasm and genetic engineering. In: Constable GA and Forrester NW (eds.) *Challenging the Future: Proceedings of the World Cotton Research Conference, Brisbane, Australia,*

February 14–17, 1994 (Commonwealth Scientific and Industrial Research Organisation [CSIRO]: Melbourne, VIC, Australia), pp. 313–327.

Stewart JMcD and Hsu CL (1978) Hybridization of diploid and tetraploid cottons through *in-ovulo* embryo culture. *J. Hered.* 69: 404–408.

Stewart JMcD, Craven LA, and Wendel JF (1997) A new Australian species of *Gossypium*. In: *Proceedings of the Beltwide Cotton Conference, New Orleans, Louisiana, USA, January 6–10, 1997* (National Cotton Council of America: Memphis, TN, USA), p. 448.

Theron CC and van Staden WH (1975) Natural cross pollination of cotton at Upington (Natuurlike kruisbestuiwing van katoen te Upington). *Agroplantae* 7: 91–92.

Thomson NJ (1966) Cotton variety trials in the Ord Valley, north-western Australia: 4. Natural crossing of cotton. *Empire Cotton Growing Rev.* 43: 18–21.

Umbeck PF, Barton KA, Nordheim EV, McCarty JC, Parrott WL, and Jenkins JN (1991) Degree of pollen dispersal by insects from a field test of genetically engineered cotton. *J. Econ. Entomol.* 84: 1943–1991.

USDA [United States Department of Agriculture] (2008) 7 CFR [Code of Federal Regulations] § 201.76. Federal Seed Act Regulations: Minimum Land, Isolation, Field, and Seed Standards. Source: 59 FR [Federal Register] 64516, Dec. 14, 1994, as amended at 65 FR 1710, Jan. 11, 2000 (Government Printing Office: Washington, DC, USA), pp. 372–378. www.access.gpo.gov/nara/cfr/waisidx_08/7cfr201_08.html.

US EPA [United States Environmental Protection Agency] (2000) Biopesticides Registration Action Document: *Bacillus thuringiensis* Cry2Ab2 Protein and Its Genetic Material Necessary for Its Production in Cotton (Chemical PC Code 006487), Amended (EPA, Office of Pesticide Programs, Biopesticides and Pollution Prevention Division, U.S. Environmental Protection Agency [EPA]: Washington D.C., Washington, USA). 75 p. www.epa.gov/oppbppd1/biopesticides/ingredients/tech_docs/brad_006487.pdf.

Vaissière BE and Vinson SB (1994) Pollen morphology and its effect on pollen collection by honey bees, *Apis mellifera* L. (Hymenoptera: Apideae), with special reference to upland cotton, *Gossypium hirsutum* L. (Malvaceae). *Grana* 33: 128–138.

Vaissière BE, Moffet JO, and Loper GM (1984) Honey bees as pollinators for hybrid cotton seed production in the Texas High Plains. *Agron. J.* 76: 1005–1010.

van Deynze AE, Sundstrom FJ, and Bradford KJ (2005) Pollen-mediated gene flow in California cotton depends upon pollinator activity. *Crop Sci.* 45: 1565–1570.

Vroh Bi I, Baudoin JP and Mergeai G (1998) Cytogenetics of the "glandless-seed and glanded-plant" trait from *Gossypium sturtianum* Willis introgressed into upland cotton (*Gossypium hirsutum* L.). *Plant Breed.* 117: 235–241.

Waghmare VN, Rong J, Rogers CJ, Pierce GJ, Wendel JF, and Paterson AH (2005) Genetic mapping of a cross between *Gossypium hirsutum* (cotton) and the Hawaiian endemic, *Gossypium tomentosum. Theor. Appl. Genet.* 111: 665–676.

Waller GD, Moffet JO, Loper GM, and Martin JH (1985) Evaluation of honey bees' foraging activity and pollination efficacy for male-sterile cotton. *Crop Sci.* 215: 211–214.

Wang GL, Dong JM, and Paterson AH (1995) The distribution of *Gossypium hirsutum* chromatin in *G. barbadense* germplasm: Molecular analysis of introgressive plant breeding. *Theor. Appl. Genet.* 91: 1153–1161.

Weaver JB Jr (1957) Embryological studies following interspecific crosses in *Gossypium*: I. *G. hirsutum* × *G. arboreum. Am. J. Bot.* 44: 209–214.

Wendel JF (1995) Cotton. In: Simmonds S and Smartt J (eds.) *Evolution of Crop Plants.* 2nd ed. (Longman Scientific and Technical: Harlow, UK), pp. 358–366.

Wendel JF and Cronn RC (2003) Polyploidy and the evolutionary history of cotton. *Adv. Agron.* 78: 139–186.

Wendel JF and Percy RG (1990) Allozyme diversity and introgression in the Galápagos-Islands endemic *Gossypium darwinii* and its relationship to continental *Gossypium barbadense. Biochem. Syst. Ecol.* 18: 517–528.

Wendel JF, Brubaker CL, and Percival AE (1992) Genetic diversity in *Gossypium hirsutum* and the origin of upland cotton. *Am. J. Bot.* 79: 1291–1310.

Wendel JF, Rowley R, and Stewart JMcD (1994) Genetic diversity in and phylogenetic relationships of the Brazilian endemic cotton *Gossypium mustelinum. Plant Syst. Evol.* 192: 49–59.

Wendel JF, Schnabel A, and Seelanan T (1995) An unusual ribosomal DNA sequence from *Gossypium gossypioides* reveals ancient, cryptic, intergenomic introgression. *Mol. Phylogenet. Evol.* 4: 298–313.

Westengen OT, Huamán Z, and Heun M (2005) Genetic diversity and geographic pattern in early South American cotton domestication. *Theor. Appl. Genet.* 110: 392–402.

Woodstock LW, Furman K, and Leffler HR (1985) Relationship between weathering deterioration and germination, respiratory metabolism, and mineral leaching from cotton seeds. *Crop Sci.* 25: 459–466.

Xanthopoulos FP and Kechagia UE (2000) Natural crossing in cotton (*Gossypium hirsutum* L.). *Aust. J. Agric. Res.* 51: 979–983.

Yasour H, Tsuk G, and Rubin B (2002) Potential benefits and risks associated with the introduction of herbicide-resistant transgenic cotton in Israel. In: Spafford JH, Dodd J, and Moore JH (eds.) *Papers and Proceedings, 13th Australian Weeds Conference, Sheraton Perth Hotel, Perth, Western Australia, September 8–13, 2002* (Plant Protection Society of Western Australia: Victoria Park, WA, Australia), pp. 634–637.

Zhang CQ, Lu QY, Wang ZX, and Jia SR (1997) Frequency of 2,4–D resistant gene flow of transgenic cotton. *Scientia Agric. Sin.* 30: 92–93 [English abstract].

Zhang J and Stewart JMcD (1997) Hybridization of new Australian *Gossypium* species (section *Grandicalyx*) with cultivated tetraploid cotton. In: *Proceedings of the Beltwide Cotton Conference, New Orleans, Louisiana, USA, January 6–10, 1997* (National Cotton Council of America: Memphis, TN, USA), pp. 487–490.

Zhao XP, Si Y, Hanson RE, Crane CF, Price HJ, Stelly DM, Wendel JF, and Paterson AH (1998) Dispersed repetitive DNA has spread to new genomes since polyploid formation in cotton. *Genome Res.* 8: 479–492.

Zhu S and Li B (1993) Studies of introgression of the "glandless seeds-glanded plant" trait from *Gossypium bickii* into cultivated upland cotton (*G. hirsutum*). *Coton et Fibres Tropicales* 48: 195–199.

TEN

Cowpea
(*Vigna unguiculata* (L.) Walp.)

Center(s) of Origin and Diversity

The domesticated cowpea plant, *Vigna unguiculata* (L.) Walp. subsp. *unguiculata* var. *unguiculata*, originated in Africa, somewhere in the broad sub-Saharan belt (Steele 1976; Smartt 1990). Several hypotheses have been brought forward regarding the exact location of the center of origin of cowpeas, including W Africa (Faris 1965; Rawal 1975; Vaillancourt and Weeden 1992; Ba et al. 2004), Ethiopia (Vavilov 1926; Pasquet 2000), S Africa (Sauer 1952; Baudoin and Maréchal 1985) or a diffuse domestication in the savanna (Chevalier 1944; Steele 1976; Garba and Pasquet 1998).

The primary center of diversity for the genus *Vigna* Savi is in sub-Saharan Africa. The highest concentration of *Vigna* species is found in the Democratic Republic of the Congo, where 80% of all African *Vigna* species are found (White 1983). Asia, and particularly India, is considered a secondary center of diversity for cultivated cowpeas (Steele 1976).

Biological Information

Domesticated cowpeas are a grain legume grown mainly in the savanna regions of the tropics and subtropics in Africa, Asia, and S America. The species includes five cultivar groups (table 10.1): cv. *unguiculata*, cv. *biflora*, and cv. *melanophthalmus* are grown for grain and fodder; cv. *sesquipedialis* for its immature pods; and cv. *textilis* for fiber extraction from its peduncles (Westphal 1974; Pasquet 1998).

Table 10.1   Domesticated cowpea (*Vigna unguiculata* (L.) Walp.) and its closest wild relatives

Species	Common name	Lifespan and life form	Sexual reproduction	Origin and distribution
Primary gene pool (GP-1) *V. unguiculata* var. *unguiculata* (L.) Walp.		annual, cultivated	mainly autogamous	wild populations native to Africa, but naturalized in Asia (Thailand and Japan); cultivated in Africa and Asia
cv. *unguiculata*	cowpea, China pea, African bean, marble pea, southern pea			most widespread and economically most important group of species; cultivated as a pulse in the tropics and subtropics
cv. *biflora*	Catjang cowpea, Hindu cowpea			cultivated as a pulse or green vegetable, forage, and cover crop in tropical Asia (India, Sri Lanka) and Africa
cv. *sesquipedalis*	asparagus bean, yard-long bean, bodi bean, snake bean			cultivated as a vegetable, forage, and green manure in tropical Asia, as well as in W Africa and C America (Caribbean)
cv. *melanophthalmus** cv. *textilis**	black-eyed bean/pea			cultivated as a pulse, mainly in W Africa cultivated for fiber of its floral peduncles as well as for fodder or seed
V. unguiculata var. *spontanea* (Schweinf.) Pasquet*	weedy cowpea	annual, wild/weedy	mainly autogamous	ANG, BOT, BUR, CHA, CMN, CON, ERI, KEN, MLW, MOZ, NAM, NGA, NGR, SEN, SUD, ZAF, ZAI, ZAM, ZIM
V. unguiculata subsp. *alba* (G. Don) Pasquet*		perennial, wild	mixed	ANG, CON, GAB, GGI, TAN, ZAI, ZIM
V. unguiculata subsp. *dekindtiana* (Harms) Verdc.		perennial, wild	mixed	ANG, BEN, BKN, BOT, BUR, CAF, CHA, CMN, CON, ERI, ETH, GAB, GAM, GGI, GHA, GNB, GUI, IVO, KEN, MLI, MLW, MOZ, NAM, NGA, NGR, SEN, SIE, SOM, SUD, SWZ, TAN, UGA, ZAF, ZAI, ZAM, ZIM

V. unguiculata subsp. pubescens (R. Wilczek) Pasquet*	perennial, wild	mixed	BUR, GHA, KEN, MLW, MOZ, NGA, SUD, TAN, UGA, ZAF, ZAI, ZIM
V. unguiculata subsp. stenophylla (Harv.) Maréchal et al.	perennial, wild	mixed	BOT, BUR, MOZ, NAM, SWZ, ZAF, ZAM, ZIM
V. unguiculata subsp. tenuis (E. Mayer) Maréchal et al.	perennial, wild	mixed	KEN, MLW, MOZ, ZAF, ZAM, ZIM
Secondary gene pool (GP-2)			
V. unguiculata subsp. aduensis Pasquet*	perennial, wild	allogamous	endemic to ETH
V. unguiculata subsp. baoulensis (A. Chev.) Pasquet*	perennial, wild	allogamous	CMN, GHA, IVO, LBR, NGA, SIE, TOG, ZAM
V. unguiculata subsp. burundiensis Pasquet*	perennial, wild	allogamous	BUR, KEN, RWA, UGA, ZAI
V. unguiculata subsp. letouzeyi Pasquet*	perennial, wild	allogamous	CAF, CMN, GAB, ZAI
V. unguiculata subsp. pawekiae Pasquet*	perennial, wild	allogamous	ANG, BOT, BUR, CMN, ERI, ETH, GGI, KEN, MLW, MOZA, NGA, SIE, TAN, UGA, ZAI, ZAM, ZIM

Sources: The taxonomy used in this study is based on Maréchal et al. 1978, but adopts the classification used by Maxted et al. 2004 in the most recent and comprehensive study of the genus. Maxted et al. emended Maréchal's classification by embracing taxa that were described later and Pasquet's 2001 concept of the cultivated forms of *V. unguiculata*. For a detailed revision of the cultivated forms, see Westphal 1974 and Pasquet 1998.

Note: Gene pool boundaries are according to Pasquet and Baudoin 1997 and Kouadio et al. 2007. Abbreviations of African countries follow Brummitt 2001: ANG, Angola; BEN, Benin; BKN, Burkina; BOT, Botswana; BUR, Burundi; CAF, Central African Republic; CHA, Chad; CMN, Cameroon; CON, Democratic Republic of Congo; ERI, Eritrea; ETH, Ethiopia; GAB, Gabon; GAM, Gambia; GGI, Gulf of Guinea Islands; GHA, Ghana; GNB, Guinea-Bissau; GUI, Guinea; IVO, Cote d'Ivoire; KEN, Kenya; LBR, Liberia; MLI, Mali; MDG, Madagascar; MLW, Malawi; MOZ, Mozambique; NAM, Namibia; NGA, Nigeria, NGR, Niger; RWA, Rwanda; SEN, Senegal; SIE, Sierra Leone; SOM, Somalia; SUD, Sudan; SWZ, Swaziland; TAN, Tanzania; TOG, Togo; UGA, Uganda; ZAF, South Africa; ZAI, Zaire; ZAM, Zambia; ZIM, Zimbabwe.

*Non Maréchal et al. 1978.

Flowering

In cultivated cowpeas, the flowers open at the end of the night and close late in the morning. The stamens are enclosed in a tight keel, with only a small hole at the top near the tip of the stamen. Thus only large insects can trip the flowers or gain access to the pollen. The anthers dehisce several hours before the flowers open, although the stigma remains receptive for two days (Ladeinde and Bliss 1977). Most pollen is therefore shed within the flower; consequently, self-fertilization is predominant in cultivated cowpeas (Rachie et al. 1975).

In contrast, wild cowpeas flower around dawn. This means that pollinators that visit wild cowpeas cannot transfer pollen to cultivated cowpeas, since the cultivated cowpea flowers have already closed (L. Murdock, pers. comm. 2008).

Pollen Dispersal and Longevity

Cowpea flowers are visited by a number of different bees, flies, and ants (Duke 1981; Vaz et al. 1998). However, only the larger species, such as *Xylocopa* carpenter bees and megachilids, are sizable enough to trip the cowpea flowers or gain access to the pollen (R.S. Pasquet, pers. comm. 2008).

Cowpea pollen can remain viable for 12 to 15 hours after anthesis for controlled crosses under experimental conditions (Ebong 1972; Myers 1996).

Sexual Reproduction

The cowpea plant is a self-compatible, predominantly autogamous, annual diploid ($2n = 2x = 22$) (Rachie et al. 1975). According to Purseglove (1968), outcrossing is uncommon, but it can occur, as evidenced by the presence of hybrid seed types in certified seed lots when good isolation practices have not been followed in seed production. Duke (1981) reported outcrossing rates of up to 15%.

Among wild subspecies, mating systems vary, and mixed mechanisms, including some predominantly allogamous forms, are found (Lush 1979; Pasquet 1996).

Vegetative Reproduction

As is the case for most legume crops, cowpeas are mainly propagated by seed, but vegetative propagation through stem cuttings is rather easy, and this is often used for experimental crosses and multiplications of cowpeas (Myers 1996).

Seed Dispersal and Dormancy

As is the case for wild relatives of chickpeas and common beans, wild cowpea forms have vigorously dehiscent pods that explosively shed their seeds. In contrast, dehiscence is reduced or completely suppressed in domesticated forms (Lush and Evans 1980). Ants, beetles, locusts, birds (*Francolinus coqui* A. Smith and *Streptopelia capicola* Sundevall), rodents (gerbils), and large mammals are all cowpea seed predators, and some of them (e.g., ants and rodents) are likely to be involved in seed dispersal (R.S. Pasquet, pers. comm. 2008).

Initial feeding tests showed that domesticated cowpea seeds do not survive passage through the digestive tract of birds. The seeds of domesticated cowpeas eaten by large mammals—such as cows, goats, horses, and pigs—also seem to be entirely destroyed by digestion, yet 10% to 30% of ingested wild cowpea seeds can survive passage through the digestive tract of large mammals. Based on these findings, large mammals may be the primary long-distance seed dispersers for up to 10% of the ingested seeds (R.S. Pasquet, pers. comm. 2008).

As with most other wild pulses, wild cowpea forms exhibit seed dormancy, whereas domesticated varieties do not share this attribute (Lush and Evans 1981).

Volunteers, Ferals, and Their Persistence

Although not reported in the literature, cowpea volunteers have been observed in subsequent cropping cycles (N. Cissé, pers. comm. 2008). Feral cowpea escapes show weedy traits and can establish themselves outside of cultivation. Feral populations are often seen in Africa (Coulibaly et al. 2002; Feleke et al. 2006), and they have also been documented in Thailand and Japan (Bervillé et al. 2005).

To our knowledge, no information has been published regarding the persistence of volunteers or ferals.

Weediness and Invasiveness Potential

Weedy cowpea forms are found along roadsides and in disturbed areas in Africa (Rawal 1975; Ba et al. 2004; Bervillé et al. 2005), but they are not considered to be problematic weeds in ruderal habitats or cultivated fields (Bervillé et al. 2005). In W Africa, weedy cowpeas are even harvested for fodder (Feleke et al. 2006).

Crop Wild Relatives

The genus *Vigna*—as defined by Maréchal et al. (1978) and emended by Pasquet (2001)—contains around 80 species that are widely distributed throughout the tropics. The domesticated cowpea plant belongs to subg. *Vigna* sect. *Catiang* (DC.) Verdc. (Ng 1995) and, together with its closest relatives (several cultivated, wild, and weedy forms), it is classified under a single botanical species, *V. unguiculata* (table 10.1); however, the subspecies that is the direct ancestor of domesticated cowpeas has not yet been determined. Several hypotheses exist that point toward *V. unguiculata* subsp. *unguiculata* var. *spontanea* (Schweinf.) Pasquet (= subsp. *dekindtiana sensu* Verdc.; Ng 1995; Padulosi and Ng 1997; Pasquet 1999), *V. vexillata* (L.) A. Rich., or *V. reticulata* Hook. f. (Vaillancourt and Weeden 1996; Jaaska 1999).

Hybridization

Within the large *Vigna* genus, the *V. unguiculata* complex belongs to an ancient phylogenetic lineage whose species, and its other relatives in the genus, evolved in near-complete genetic isolation (Smartt 1990). *Vigna unguiculata* is considered to be incompatible with other members of the genus outside of sect. *Catiang*; to date, studies of 35 such species have shown complete or partial incompatibility with *V. unguiculata* (see, e.g., Singh et al. 1964; Rawal 1975; Evans 1976; Ng 1995). Whereas fertilization does occur with some of these *Vigna* species, embryos are aborted at very early stages of development, making embryo rescue difficult (Ng 1990; Barone et al. 1992).

The intraspecific relationships among the subspecies and varieties of *V. unguiculata* remain unclear. For decades, many investigators regarded the species' members as interfertile, with negligible cross-compatibility variations, depending on the genotype (see, e.g., Rawal et al. 1976; Steele 1976; Smartt 1979, 1981; Baudoin and Maréchal 1985; Mithen 1987; Ng 1990, 1995; Sakupwanya et al. 1990). Others have shown that there are significant pre- and post-zygotic barriers to hybridization between certain subspecies (Rawal 1975; Fatokun 1991; Ng 1995; Fatokun et al. 1997; Pasquet and Baudoin 1997; Kouadio et al. 2006a, b, 2007). The wild relatives of domesticated cowpeas are grouped into primary, secondary and tertiary gene pools, according to their cross-compatibility with the crop (table 10.1).

Primary Gene Pool (GP-1)

The primary gene pool of cowpeas includes the cultivated forms of *V. unguiculata* subsp. *unguiculata* var. *unguiculata*, the weedy annual *V. unguiculata* subsp. *unguiculata* var. *spontanea*, and the five allo-autogamous perennial wild subspecies (table 10.1). Intraspecific cross-fertilization between cowpeas and these conspecifics is easy using hand pollination, which results in viable and fertile F1 hybrids. Moderate pre-and post-zygotic hybridization barriers exist, resulting in varying levels of cross-compatibility and hybrid fertility among and within subspecies, depending on the subspecies and the genotype.

Vigna unguiculata subsp. *unguiculata* var. *spontanea*

Cowpeas are morphologically and genetically closest to their only weedy relative and supposed ancestor, *Vigna unguiculata* subsp. *unguiculata* var. *spontanea* (= subsp. *dekindtiana sensu* Verdc.), with which they are highly cross-compatible (Panella and Gepts 1992; Vaillancourt et al. 1993; Pasquet 1999, 2000). The *spontanea* variety may have evolved through hybridization between subsp. *unguiculata* and subsp. *mensensis* (Schweinf.) Maréchal et al. (= subsp. *burundiensis* Pasquet), as evidenced by its intermediate morphological characteristics and its weedy trait (Ng and Maréchal 1985; Smartt 1990). It is an annual, primarily inbred weed found in fields or disturbed habitats almost everywhere in Africa.

Outcrossing rates of up to 79% were obtained by hand pollination in crossing studies between domesticated cowpeas and var. *spontanea* (Kouadio et al. 2007). The fitness of F1 wild-domesticated hybrids is much higher when the domesticated cowpea plant is the maternal parent (Feleke et al. 2006). Molecular marker studies provide evidence for a widely distributed cowpea crop-weed complex throughout the African continent, where spontaneous hybridization exists between these two conspecifics (Pasquet 1999; Coulibaly et al. 2002; Ba et al. 2004; Feleke et al. 2006).

Weedy-domesticated hybrid swarms are more frequent in W Africa than elsewhere on the continent, because farmers in W Africa often tolerate weedy cowpea plants in their fields (Feleke et al. 2006), since cowpeas are also cultivated for fodder in this region, and fodder from weedy cowpeas is as good as fodder from domesticated cowpeas. In addition, farmers occasionally harvest wild/weedy-domesticated F1 seeds and use them for their next sowing.

Perennial, Allo-Autogamous Subspecies

Partial incompatibility exists between domesticated cowpeas and *Vigna unguiculata* subsp. *pubescens* (R. Wilczek) Pasquet. When domesticated cowpeas are the pollen donor, crosses fail and flowers drop within 24 hours after cross-pollination. In contrast, crosses are successful when cowpeas are the maternal parent, although—depending on the genotype—embryo rescue may be necessary in certain cross-combinations (Fatokun and Singh 1987).

Crossing studies between cowpeas and subsp. *mensensis* (= subsp. *burundiensis*) and subsp. *dekindtiana* (Rawal 1975; Lush 1979) also provide evidence of some kind of genetic crossing barrier between these forms.

Outcrossing rates of up to 58% were obtained in hand-pollination studies between cowpeas and subsp. *alba* (G. Don) Pasquet, whose F1 plants are vigorous but whose fertility is considerably reduced (Kouadio et al. 2007).

Hand pollination between cowpeas and subsp. *stenophylla* (Harv.) Maréchal et al. results in lower (up to 40%) rates of successful hybridization (Kouadio et al. 2007), although F1 hybrids between these two conspecifics are generally very vigorous and completely fertile (Sakupwanya et al. 1990; Kouadio et al. 2007).

Secondary Gene Pool (GP-2)

The secondary gene pool includes the five perennial obligate outcrossing subspecies of *Vigna unguiculata*: subsp. *aduensis* Pasquet, subsp. *baoulensis* (A. Chev.) Pasquet, subsp. *burundiensis*, subsp. *letouzeyi* Pasquet, and subsp. *pawekiae* Pasquet (table 10.1). Cross-fertilization between domesticated cowpeas and these subspecies is possible, but the outcrossing rates are much lower when compared with the subspecies in the primary gene pool, due to considerable pre- and post-zygotic hybridization barriers (see, e.g., Kouadio et al. 2006a, b). For instance, hand pollination between *Vigna unguiculata* subsp. *unguiculata* cv. *biflora* and subsp. *baoulensis* resulted in hybrid formation (outcrossing rate of 20%), but no seeds were set (Kouadio et al. 2007).

Tertiary Gene Pool (GP-3)

The tertiary gene pool includes the remaining species of the genus, all of which are genetically remote from domesticated cowpeas. The nature of the hybridization barriers acting between cowpeas and these species is still undetermined (Smartt 1990). Several crossing experiments have shown that pollen-tube germination and subsequent fertilization occur normally in various situations, which means that embryo rescue could be successful. However, spontaneous hybridization in the wild seems very unlikely, because of strong pre- and post-zygotic hybridization barriers that lead to embryo abortion at very early developmental stages, or to the pods aborting a few days after their emergence. More detailed information for the most prominent of these wild relatives is given below.

Vigna schlechteri

Vigna schlechteri Harms (= *V. nervosa* Markötter) might be the closest relative of *V. unguiculata*, and some investigators placed it in GP-2, despite of a lack of experimental support (Mithen 1987; Baudoin and Maréchal 1991). In fact, hybridization between the two species has not succeeded because of strong post-zygotic barriers, causing embryo abortion at early developmental stages (Mithen 1987).

Vigna rhomboidea

Similarly to *Vigna schlechteri*, the wild relative *V. rhomboidea* Burtt Davy has been proposed for GP-2 (Ng and Singh 1997) and even for GP-1 (Aliyu 2005). Hand crosses between domesticated cowpeas and this wild relative resulted in partially fertile F1 hybrids when cowpeas were used as the female parent (Ng and Singh 1997). Furthermore, Aliyu (2005, 2007) reported that *V. rhomboidea* was reciprocally crossable with domesticated cowpeas. The outcrossing rates obtained in these studies were 5.7% and 22.6%, respectively, depending on whether cowpeas or *V. rhomboidea* were used as the pollen source (Aliyu 2005).

Vigna vexillata

Several hundred crosses between domesticated cowpeas and *Vigna vexillata* resulted in pods that withered and aborted within a few days and in unviable hybrid seed. This suggests that there is strong cross-incompatibility between the two species (Barone and Ng 1990; Fatokun 1991, 2002; Barone et al. 1992). However, because pollen-tube germination and subsequent fertilization occur normally (Fatokun 1991), an embryo-rescue technique was developed for the interspecific hybrids between these two species (Murdock 1992). Also, Gomathinayagam et al. (1998) reported partial success in growing immature embryos (10–12 days old); validation of their results is pending.

Vigna subterranea var. *subterranea*

In a hand-pollination crossing experiment, successful hybridization between *Vigna subterranea* (L.) Verduc. var. *subterranea* (Bambara groundnut) and domesticated cowpeas was reported, where cowpeas were used as the female parent (Begemann 1997). These results have not yet been replicated.

Outcrossing Rates and Distances

Pollen Flow

To our knowledge, there are no published studies measuring cowpea pollen flow distances, and information on outcrossing rates is scarce.

Cowpeas are mainly self-fertilizing plants, but outcrossing rates of up to 15% have been reported (Duke 1981).

Isolation Distances

The separation distances established by regulatory authorities for cowpea seed production range from 0 to 3 m in the United States (CCIA 2009; USDA 2008). In OECD countries, cowpea varieties for seed production "shall be isolated from other crops by a definite barrier or a space sufficient to prevent mixture during harvest" (OECD 2008, p. 18). In spite of these regulations, seed producers in California have been plagued by outcrossing problems, and a revision of the isolation standards may be necessary (J. Ehlers, pers. comm. 2008).

State of Development of GM Technology

Methods to genetically transform cowpeas have been in process for several years in Australia, India, the Netherlands, Nigeria, and the United States (García et al. 1986a, b; Penza et al. 1991; Muthukumar et al. 1996; Sahoo et al. 2001; Somers et al. 2003). Efficient transformation protocols are now available (Ikea et al. 2003; Popelka et al. 2006; Chaudhury et al. 2007). Insect resistance (*Bt*) is the only trait on which GM research has focused thus far, and the first experimental field trials are currently being conducted in Puerto Rico (T. Higgins, pers. comm. 2008).

Agronomic Management Recommendations

There are no specific recommendations.

Crop Production Area

Domesticated cowpeas are widely cultivated throughout the tropical and subtropical regions of the world (Ehlers and Hall 1997). They are an important crop in most African countries—mainly in the Sahel belt—as well as in SE Asia, Australia, NE Brazil, S China, Cuba, India, the southern United States, and Venezuela (Rachie 1985; Langyintuo et al. 2003). The crop is also cultivated in some parts of E Europe and the

Mediterranean. The total crop production area is about 11.3 million ha (FAO 2009). About 60% of the 5.4 million tons of the world's dry grain cowpea production comes from Nigeria alone, followed by Niger (19%) and Burkina Faso (8%) (FAO 2009). *Vigna unguiculata* subsp. *unguiculata* cv. *sesquipedialis* may be grown on more than 300,000 ha in Asia (J. Ehlers, pers. comm. 2008).

To date, there is no commercial cultivation of GM cowpeas.

Research Gaps and Conclusions

The boundaries of the domesticated cowpea gene pool are not adequately delineated, and the degree of knowledge about the interrelationships and cross-compatibility of cowpeas and their wild relatives is far from satisfactory. Even for the closest wild relatives (i.e., those that are conspecific with domesticated cowpeas), data on outcrossing rates, hybrid vigor, and fertility are available for only a few of the approximately 10 subspecies included in GP-1 and GP-2. The situation is even more disturbing when considering *Vigna* species other than those in the *V. unguiculata* complex. Hybridization experiments under natural and controlled conditions are needed to gain a better understanding of which species are cross-compatible with cowpeas—and, if they are, to what extent—and to determine the fate of introgressed alleles in subsequent generations. Apart from morphological evaluations, molecular studies should be conducted to determine the predominant direction of gene flow and to quantify introgression (Maxted et al. 2004).

Gene flow and introgression from domesticated cowpeas to their sexually compatible wild relatives is likely in sub-Saharan Africa, where wild relatives are sympatric with the crop. Naturally occurring hybrid swarms have been observed throughout the continent, and although cowpeas are a predominantly selfing crop, outcrossing rates through pollen flow under natural conditions will very likely exceed the threshold of maximum allowable GM trait contamination (0.9%). Another situation where gene flow and introgression from cowpeas to their wild relatives could occur is seed dispersal by animals (and humans). While ants, rodents, and large mammals play an important role in controlling feral and wild cowpea populations, they also contribute to short- and long-distance seed dispersal.

Based on the knowledge available to date about these relationships, the likelihood of gene introgression from domesticated cowpeas to their wild and weedy relatives—assuming physical proximity (less than 3 m) and flowering overlap, and being conditional on the nature of the respective trait(s)—is as follows (also see map 10.1):

- *high* for its weedy relative *Vigna unguiculata* subsp. *unguiculata* var. *spontanea* and the perennial wild conspecifics subsp. *alba*, subsp. *dekindtiana*, subsp. *pubescens*, subsp. *stenophylla*, and subsp. *tenuis*

- *moderate* for the perennial wild conspecifics subsp. *aduensis*, subsp. *baoulensis*, subsp. *burundiensis*, subsp. *letouzeyi*, and subsp. *pawekiae* (in map 10.1, this area is hidden by the areas in red, corresponding to high likelihood). There may also be a *low* to *moderate* likelihood of gene flow and introgression to some other *Vigna* species that have not yet been adequately studied (e.g., *V. rhomboidea*)

REFERENCES

Aliyu B (2005) Crossability of *V. rhomboidea* Burtt Davy with cowpea (*Vigna unguiculata* (L.) Walp.). *Genet. Resour. Crop Evol.* 52: 447–451.

——— (2007) Heritability and gene effects for incorporating pubescence into cowpea (*Vigna unguiculata* (L) Walp.) from *V. rhomboidea* Burtt Davy. *Euphytica* 155: 295–303.

Ba FS, Pasquet RS, and Gepts P (2004) Genetic diversity in cowpea [*Vigna unguiculata* (L.) Walp.] as revealed by RAPD markers. *Genet. Resour. Crop Evol.* 51: 539–550.

Barone A and Ng NQ (1990) Embryological study of crosses between *Vigna unguiculata* and *V. vexillata*. In: Ng NQ and Monti LM (eds.) *Cowpea Genetic Resources* (International Institute of Tropical Agriculture [IITA]: Ibadan, Nigeria), pp. 151–161.

Barone A, Delgiudice A, and Ng NQ (1992) Barriers to interspecific hybridization between *Vigna unguiculata* and *Vigna vexillata*. *Sex. Plant Reprod.* 5: 195–200.

Baudoin JP and Maréchal R (1985) Genetic diversity in *Vigna*. In: Singh SR and Rachie KO (eds.) *Cowpea Research, Production, and Utilization* (John Wiley and Sons: Chichester, UK), pp. 3–11.

———— (1991) Taxonomy and wide crosses of pulse crops, with special reference to *Phaseolus* and *Vigna*. In: Ng NQ, Perrino P, Attere F, and Zedan H (eds.) *Crop Genetic Resources of Africa*, vol. 2 (International Board for Plant Genetic Resources [IBPGR]: Rome, Italy), pp. 287–299.

Begemann F (1997) An experiment to cross Bambara groundnut and cowpea. In: Heller J, Begemann F, and Mushonga J (eds.) *Bambara Groundnut (Vigna subterranea (L.) Verdc.): Proceedings of the Workshop on Conservation and Improvement of Bambara Groundnut (Vigna subterranea (L.) Verdc.), Harare, Zimbabwe, November 14–16, 1995.* Promoting the Conservation and Use of Underutilized and Neglected Crops 9 (Department of Research and Specialist Services: Causeway, Harare, Zimbabwe; International Plant Genetic Resources Institute [IPGRI]: Rome, Italy), pp. 135–138.

Bervillé A, Breton C, Cunliffe K, Darmency H, Good AG, Gressel J, Hall LM, McPherson MA, Médail F, Pinatel C, Vaughan DA, and Warwick SI (2005) Issues of ferality or potential for ferality in oats, olives, the pigeon-pea group, ryegrass species, safflower, and sugarcane. In: Gressel J (ed.) *Crop Ferality and Volunteerism: A Threat to Food Security in the Transgenic Era?* (CRC Press: Boca Raton, FL, USA), pp. 231–255.

Brummitt RK (2001) *World Geographical Scheme for Recording Plant Distributions.* 2nd ed. Plant Taxonomic Database Standards No. 2 (published for the International Working Group on Taxonomic Databases for Plant Sciences [TDWG] by the Hunt Institute for Botanical Documentation: Carnegie Mellon University, Pittsburgh, PA, USA). 137 p. www.tdwg.org/TDWG_geo2.pdf.

CCIA [California Crop Improvement Association] (2009) Cowpea Certification Standards. http://ccia.ucdavis.edu/seed_cert/seedcert_index.htm.

Chaudhury D, Madnpotra S, Jaiwal R, Saini R, Ananda Kumar P, and Jaiwal PK (2007) *Agrobacterium tumefaciens*-mediated high frequency genetic transformation of an Indian cowpea (*Vigna unguiculata* (L.) Walp.) cultivar and transmission of transgenes into progeny. *Plant Sci.* 172: 692–700.

Chevalier A (1944) La dolique de Chine en Afrique. *Rev. Bot. Appl. Agric. Trop.* 24: 128–152.

Coulibaly S, Pasquet RS, Papa R, and Gepts P (2002) AFLP analysis of the phenetic organization and genetic diversity of *Vigna unguiculata* (L.) Walp. reveals extensive gene flow between wild and domesticated types. *Theor. Appl. Genet.* 104: 358–366.

Duke JA (1981) *Handbook of Legumes of World Economic Importance* (Plenum Press: New York, NY, USA). 345 p.

Ebong UU (1972) Optimum time for artificial pollination in cowpea, *Vigna sinensis* Endl. *Samaru Agric. Newsl.* 14: 31–35.

Ehlers JD and Hall AE (1997) Cowpea (*Vigna unguiculata* (L.) Walp.). *Field Crops Res.* 53: 187–204.

Evans AM (1976) Species hybridization in the genus *Vigna*. In: Luse RA and Rachie KO (eds.) *Proceedings of IITA Collaborators' Meeting on Grain Legume Improvement Held at IITA on June 9–13, 1975* (International Institute of Tropical Agriculture [IITA]: Ibadan, Nigeria), pp. 337–347.

FAO [Food and Agriculture Organization of the United Nations] (2009) FAOSTAT data. http://faostat.fao.org/site/567/default.aspx.

Faris DG (1965) The origin and evolution of the cultivated forms of *Vigna sinensis*. *Can. J. Genet. Cytol.* 7: 433–452.

Fatokun CA (1991) Wide crossing in cowpea: Problems and prospects. *Euphytica* 54: 137–140.

——— (2002) Breeding cowpea for resistance to insect pests: Attempted crosses between cowpea and *V. vexillata*. In: Fatokun CA, Tarawali SA, Singh BB, Kormawa PM, and Tamo M (eds.) *Challenges and Opportunities for Enhancing Sustainable Cowpea Production: Proceedings of the World Cowpea Conference III, International Institute of Tropical Agriculture (IITA), Ibadan, Nigeria, September 4–8, 2000* (International Institute of Tropical Agriculture [IITA]: Ibadan, Nigeria), pp. 52–61.

Fatokun CA and Singh BB (1987) Interspecific hybridization between *Vigna pubescens* and *V. unguiculata* (L.) Walp. through embryo rescue. *Plant Cell Tissue Organ Cult.* 9: 229–233.

Fatokun CA, Perrino P, and Ng NQ (1997) Wide crossing in African *Vigna* species. In: Singh BB, Mohan Raj DR, Dashiel KE, and Jackai LEN (eds.) *Advances in Cowpea Research* (International Institute of Tropical Agriculture [IITA]: Ibadan, Nigeria; Japan International Research Centre for Agricultural Sciences [JIRCAS]: Tsukuba, Ibaraki, Japan), pp. 50–58.

Feleke Y, Pasquet RS, and Gepts P (2006) Development of PCR-based chloroplast DNA markers that characterize domesticated cowpea (*Vigna unguiculata* subsp. *unguiculata* var. *unguiculata*) and highlight its crop-weed complex. *Plant Syst. Evol.* 262: 75–87.

Garba M and Pasquet RS (1998) Isozyme polymorphism within section *Reticulatae* of genus *Vigna* (Tribe Phaseoleae: Fabaceae). *Biochem. Syst. Ecol.* 26: 297–308.

García JA, Hillie J, and Goldbach R (1986a) Transformation of cowpea *Vigna unguiculata* cells with an antibiotic resistance gene using Ti-plasmid derived vectors. *Plant Sci.* 44: 37–46.

——— (1986b) Transformation of cowpea *Vigna unguiculata* cells with a full-length DNA copy of cowpea mosaic virus mRNA. *Plant Sci.* 44: 89–98.

Gomathinayagam P, Ganeshram S, Rathnaswany R, and Ramaswamy NW (1998) Interspecific hybridisation between *V. unguiculata* (L.) Walp. and *V. vexillata* (L.) A. Rich. through *in vitro* embryo culture. *Euphytica* 102: 203–209.

Ikea J, Ingelbrecht I, Uwaifo A, and Thottappilly G (2003) Stable gene transformation in cowpea (*Vigna unguiculata* (L.) Walp.) using particle gun method. *Afr. J. Biotech.* 2: 211–218.

Jaaska V (1999) Isoenzyme diversity and phylogenetic affinities among the African beans of the genus *Vigna* Savi (Fabaceae). *Biochem. Syst. Ecol.* 27: 569–589.

Kouadio D, Toussaint A, Pasquet RS, and Baudoin JP (2006a) Barrières d'incompatibilité pré-zygotiques chez les hybrides entre formes sauvages du niébé, *Vigna unguiculata* (L.) Walp. *Biotech. Agron. Soc. Environ.* 10: 1–9.

Kouadio D, Echikh N, Toussaint A, Pasquet RS, and Baudoin JP (2006b) Barrières pré-zygotiques chez les hybrides entre formes sauvages du niébé, *Vigna unguiculata* (L.) Walp. *Biotech. Agron. Soc. Environ.* 10: 33–41.

―――― (2007) Organisation du pool génique de *Vigna unguiculata* (L.) Walp.: Croisements entre les formes sauvages et cultivées du niébé. *Biotech. Agron. Soc. Environ.* 11: 47–57.

Ladeinde TOA and Bliss FA (1977) Identification of the bud stage for pollinating without emasculation in cowpea (*Vigna unguiculata* (L.) Walp.). *Nigerian J. Sci.* 11: 183–195.

Langyintuo AS, Lowenberg-DeBoer J, Faye M, Lambert D, Ibro G, Moussa B, Kergna A, Kushwaha S, Musa S, and Ntoukam G (2003) Cowpea supply and demand in West Africa. *Field Crops Res.* 82: 215–231.

Lush WM (1979) Floral morphology of wild and cultivated cowpeas. *Econ. Bot.* 33: 442–447.

Lush WM and Evans LT (1980) The seed coats of cowpeas and other grain legumes: Structure in relation to function. *Field Crops Res.* 3: 267–286.

―――― (1981) The domestication and improvement of cowpeas (*Vigna unguiculata*) (L.) Walp). *Euphytica* 30: 579–578.

Maréchal R, Mascherpa J, and Stainier F (1978) Étude taxonomique d'un groupe complexe d'espèces des genres *Phaseolus* et *Vigna* (Papilionaceae) sur la base de données morphologiques et polliniques, traitées par l'analyse informatique. *Boissiera* 28: 1–273.

Maxted N, Mabuza-Diamini P, Moss H, Padulosi S, Jarvis A, and Guarino L (2004) *An Ecogeographic Study of African* Vigna. Systematic and Ecogeographic Studies on Crop Genepools 11 (International Plant Genetic Resources Institute [IPGRI]: Rome, Italy). 454 p.

Mithen R (1987) The African genepool of *Vigna*: I. *V. nervosa* and *V. unguiculata* from Zimbabwe. *Plant Genet. Resour. Newsl.* 70: 13–19.

Murdock LL (1992) Improving insect resistance in cowpea through biotechnology:

Initiatives at Purdue University, USA. In: Thottappilly G, Monti LM, Mohan DR, and Moore AW (eds.) *Biotechnology: Enhancing Research on Tropical Crops in Africa* (International Institute of Tropical Agriculture [IITA]: Ibadan, Nigeria), pp. 313–320.

Muthukumar B, Mariamma M, Veluthambi K, and Gnanam A (1996) Genetic transformation of cotyledon explants of cowpea (*Vigna unguiculata* (L.) Walp.) using *Agrobacterium tumefaciens*. *Plant Cell Rep.* 15: 980–985.

Myers GO (1996) *Hand Crossing of Cowpeas*. IITA Research Guide 42 (International Institute of Tropical Agriculture [IITA]: Ibadan, Nigeria). 18 p. www.iita.org/cms/details/trn_mat/irg42/irg42.html.

Ng NQ (1990) Recent developments in cowpea germplasm collection, conservation, evaluation, and research at the Genetic Resources Unit, IITA. In: Ng NQ and Monti LM (eds.) *Cowpea Genetic Resources* (International Institute of Tropical Agriculture [IITA]: Ibadan, Nigeria), pp. 13–28.

――― (1995) Cowpea. In: Smartt J and Simmonds NW (eds.) *Evolution of Crop Plants*. 2nd ed. (Longman Scientific and Technical: Harlow, UK), pp. 326–331.

Ng NQ and Maréchal R (1985) Cowpea taxonomy, origin, and germplasm. In: Singh SR and Rachie KO (eds.) *Cowpea Research, Production, and Utilization* (John Wiley and Sons: Chichester, UK), pp. 11–21.

Ng NQ and Singh BB (1997) Cowpea. In: Fuccillo DA, Sears L, and Stapleton P (eds.) *Biodiversity in Trust: Conservation and Use of Plant Genetic Resources in CGIAR [Consultative Group on International Agricultural Research] Centres* (Cambridge University Press: Cambridge, UK), pp. 82–99.

OECD [Organization for Economic Cooperation and Development] (2008) OECD Scheme for the Varietal Certification of Grass and Legume Seed Moving in International Trade: Annex VI to the Decision. C(2000)146/FINAL. OECD Seed Schemes, February 2008. www.oecd.org/document/0/0,3343,en_2649_33905 _1933504_1_1_1_1,00.html.

Padulosi S and Ng NQ (1997) Origin, taxonomy, and morphology of *Vigna unguiculata* (L.) Walp. In: Singh BB, Mohan Raj DR, Dashiell KE, and Jackai LEN (eds.) *Advances in Cowpea Research* (International Institute of Tropical Agriculture [IITA]: Ibadan, Nigeria; Japan International Research Centre for Agricultural Sciences [JIRCAS]: Tsukuba, Ibaraki, Japan), pp. 1–12.

Panella L and Gepts P (1992) Genetic relationship within *Vigna unguiculata* (L.) Walp. based on isozyme analyses. *Genet. Resour. Crop Evol.* 39: 71–88.

Pasquet RS (1996) Wild cowpea (*Vigna unguiculata*) evolution. In: Pickersgill B and Lock JM (eds.) *Legumes of Economic Importance*. Advances in Legume Systematics 8 (Royal Botanic Gardens, Kew: Richmond, Surrey, UK), pp. 95–100.

――― (1998) Morphological study of cultivated cowpea *Vigna unguiculata* (L.)

Walp.: Importance of ovule number and definition of cv. gr *melanophthalmus*. *Agronomie* 18: 61–70.

—— (1999) Genetic relationships among subspecies of *Vigna unguiculata* (L.) Walp. based on allozyme variation. *Theor. Appl. Genet.* 98: 1104–1119.

—— (2000) Allozyme diversity of cultivated cowpea *Vigna unguiculata* (L.) Walp. *Theor. Appl. Genet.* 101: 211–219.

—— (2001) *Vigna* Savi. In: Mackinder B, Pasquet R, Polhill R, and Verdcourt B (eds.) *Flora Zambesiaca: Phaseoleae*, vol. 3, part 5 (Royal Botanic Gardens, Kew: Richmond, Surrey, UK), pp. 121–156.

Pasquet RS and Baudoin JP (1997) Le niébé, *Vigna unguiculata*. In: Charrier A, Jacquot M, Hammon S, and Nicolas D (eds.) *L'Amélioration des Plantes Tropicales*. Comptes rendus de l'Académie d'agriculture de France 84, no. 7 (Cooperation internationale en recherche agronomique pour le développement [CIRAD]-Institut français de recherche scientifique pour le développement en cooperation [ORSTOM]: Montpellier, France), pp. 483–505.

Penza R, Lurquin PF, and Fillippone E (1991) Gene transfer by co-cultivation of mature embryos with *Agrobacterium tumefaciens*: Application to cowpea. *J. Plant Physiol.* 138: 39–43.

Popelka JC, Gollasch S, Moore A, Molvig L, and Higgins TJV (2006) Genetic transformation of cowpea (*Vigna unguiculata* L.) and stable transmission of the transgenes to progeny. *Plant Cell Rep.* 25: 304–312.

Purseglove JW (1968) *Tropical Crops: Dicotyledons*, vol. 1 (Longman Scientific and Technical: London, UK). 719 p.

Rachie KO (1985) Introduction. In: Singh SR and Rachie KO (eds.) *Cowpea Research, Production, and Utilization* (John Wiley and Sons: Chichester, UK), pp. xxi–xxvii.

Rachie KO, Rawal K, Franckowiak JD, and Akinpelu MA (1975) Two outcrossing mechanisms in cowpea, *Vigna unguiculata* (L.) Walp. *Euphytica* 24: 159–163.

Rawal KM (1975) Natural hybridization among wild, weedy, and cultivated *Vigna unguiculata* (L.) Walp. *Euphytica* 24: 699–705.

Rawal KM, Rachie KO, and Francowiak JO (1976) Reduction in seed size in crosses between wild and cultivated cowpeas. *J. Hered.* 67: 253–254.

Sahoo L, Sushma T, Sugla T, Singh ND, and Jaiwal PK (2001) *In vitro* plant regeneration and recovery of cowpea (*Vigna unguiculata*) transformants via *Agrobacterium*-mediated transformation. *Plant Cell Biotech. Mol. Biol.* 1: 47–51.

Sakupwanya S, Mithen R, and Mhlanga M (1990) Studies on the African *Vigna* genepool: II. Hybridization studies with *Vigna unguiculata* var. *tenuis* and var. *stenophylla*. *Plant Genet. Resour. Newsl.* 78/79: 5–10.

Sauer CO (1952) *Agricultural Origins and Dispersal* (American Geographical Society: New York, NY, USA). 110 p.

Singh B, Khanna AN, and Vaidyn SM (1964) Crossability studies in the genus *Phaseolus. J. Postgrad. School* (India) 2: 47–50.

Smartt J (1979) Interspecific hybridization in the grain legumes: A review. *Econ. Bot.* 33: 329–337.

——— (1981) Genepools in *Phaseolus* and *Vigna* cultigens. *Euphytica* 30: 445–449.

——— (1990) The Old World pulses: *Vigna* species. In: Smartt J (ed.) *Grain Legumes: Evolution and Genetic Resources* (Cambridge University Press: Cambridge, UK), pp. 140–175.

Somers DA, Samac DA, and Olhoft PM (2003) Recent advances in legume transformation. *Plant Physiol.* 131: 892–899.

Steele WM (1976) Cowpeas. In: Simmonds NW (ed.) *Evolution of Crop Plants* (Longman Scientific and Technical: London, UK), pp. 183–185.

USDA [United States Department of Agriculture] (2008) 7 CFR [Code of Federal Regulations] § 201.76. Federal Seed Act Regulations: Minimum Land, Isolation, Field, and Seed Standards. Source: 59 FR [Federal Register] 64516, Dec. 14, 1994, as amended at 65 FR 1710, Jan. 11, 2000 (Government Printing Office: Washington, DC, USA), pp. 372–378. www.access.gpo.gov/nara/cfr/waisidx_08/7cfr201_08.html.

Vaillancourt RE and Weeden NF (1992) Chloroplast DNA polymorphism suggests Nigerian center of domestication for the cowpea, *Vigna unguiculata*, Leguminosae. *Am. J. Bot.* 79: 1194–1199.

——— (1996) *Vigna unguiculata* and its position within the genus *Vigna*. In: Pickersgill B and Lock JM (eds.) *Legumes of Economic Importance*. Advances in Legume Systematics 8 (Royal Botanic Gardens, Kew: Richmond, Surrey, UK), pp. 89–93.

Vaillancourt RE, Weeden NF, and Barnard J (1993) Isozyme diversity in the cowpea species complex. *Crop Sci.* 33: 606–613.

Vavilov NI (1926) Studies on the origin of cultivated plants. *Bull. Appl. Bot. Genet. Plant Breed.* 16: 1–245.

Vaz GV, de Olivera D, and Ohashi OS (1998) Pollinator contribution to the production of cowpea on the Amazon. *Hort. Sci.* 33: 1157–1159.

Westphal E (1974) *Pulses in Ethiopia: Their Taxonomy and Agricultural Significance*. Agricultural Research Report 815 (Centre for Agricultural Publishing and Documentation: Wageningen, The Netherlands).

White F (1983) *The Vegetation of Africa: A Descriptive Memoir to Accompany the UNESCO/AETFAT/UNSO Vegetation Map of Africa* (United Nations Educational, Scientific, and Cultural Organization [UNESCO]: Paris, France).

ELEVEN

Finger Millet
(*Eleusine coracana* (L.) Gaertn.)

Center(s) of Origin and Diversity

The cereal finger millet, *Eleusine coracana* (L.) Gaertn. subsp. *coracana*, is native to E Africa; most likely it originated in Ethiopia, Uganda, or neighboring areas around Lake Victoria (Rachie and Peters 1977; de Wet et al. 1984).

The primary center of diversity for the genus *Eleusine* Gaertn. is E Africa, where 8 of its approximately 10 species occur. India is considered to be a secondary center of diversity (Hilu and de Wet 1976b).

Biological Information

Flowering

The digitate inflorescence of domesticated finger millet is monoecious, that is, the individual flowers are either male or female, but both sexes can be found on the same plant. By anthesis, the floral organs of each spikelet are completely enveloped by the glumes (Liu et al. 2007), thus facilitating self-fertilization.

Pollen Dispersal and Longevity

Finger millet pollen is wind dispersed (Phillips and Renvoize 1974; de Wet et al. 1984).

To our knowledge, no information has yet been published regarding the longevity of finger millet pollen.

Sexual Reproduction

Finger millet is an almost entirely self-pollinating allo-tetraploid ($2n = 4x = 36$, **AABB** genome).

Seed Dispersal and Dormancy

While seeds of wild finger millet relatives shatter easily when mature, cultivated forms lost this trait during domestication and some—but not all—cultivars completely lack the ability to disperse seed naturally through shedding (de Wet et al. 1984).

By and large, the seeds of domesticated finger millet are nondormant and germinate within a few weeks, although several landraces do show some seed dormancy (Kulkarni and Basavaraju 1976). Likewise, some wild species have dormant seeds that remain viable in soil seed banks for two years or longer, such as *E. indica* (L.) Gaertn. (Chua et al. 2004).

Volunteers, Ferals, and Their Persistence

Finger millet volunteers are common in fields planted in earlier seasons with the crop (C. Oduori and M. Dida, pers. comm. 2008). Feral finger millet plants are rather uncommon, as the crop is a poor competitor and is quickly smothered by weeds unless there is human intervention (i.e., weeding) (C. Oduori, pers. comm. 2008).

Weediness and Invasiveness Potential

Domesticated finger millet suffers from weed competition and requires hand weeding or chemical control to survive. Neither domesticated finger millet nor any of its wild relatives are considered to be noxious weeds (GISD 2007), but the wild ancestor of finger millet, *E. coracana* subsp. *africana* (Kenn.-O'Byrne) Hilu & de Wet, does have some weediness potential and is frequently associated with finger millet fields (Hiremath and Salimath 1991). The natural hybrids between domesticated finger millet and its wild ancestor are weedy, aggressive colonizers (de Wet et al. 1984; de Wet 1995). Another less closely related wild relative, *E. in-*

dica, is a ubiquitous weed in tropical and subtropical regions (Holm et al. 1977).

Crop Wild Relatives

The grass genus *Eleusine* comprises approximately 10 annual and perennial species that are native to E Africa, except for the vicariant *E. tristachya* (Lam.) Lam. that is widely distributed in S and Central America (Phillips 1972; Hilu and de Wet 1976a). The genus includes diploids and tetraploids with chromosome numbers of 8, 9, and 10 (Hiremath and Chennaveeraiah 1982; Hiremath and Salimath 1991; table 11.1).

Cultivated finger millet is a tetraploid (**AABB**) and was domesticated from its conspecific ancestor, wild finger millet, or *Eleusine coracana* subsp. *africana* (Chennaveeraiah and Hiremath 1974; Hilu and de Wet 1976a, b; Hilu et al. 1978; Hilu 1988). *Eleusine coracana* may have arisen from hybridization of two diploid species; wild goosegrass (*E. indica*) is the putative donor of the **A** genome (Hilu 1988; Hiremath and Salimath 1992; Hilu and Johnson 1992; Werth et al. 1994), while the donor of the **B** genome remains unknown. At some point, *E. floccifolia* (Forrsk.) Spreng. was suggested as a candidate donor of the **B** genome (Bisht and Mukai 2001), but this was later refuted (Neves et al. 2005).

Hybridization

Neither the taxonomic classification nor the phylogenetic and biosystematic relationships among species in the genus *Eleusine* are conclusively resolved thus far. Also, to our knowledge, there is no information available about the boundaries of the primary, secondary, and tertiary gene pools for finger millet.

Eleusine coracana

Based on taxonomic, phylogenetic and biosystematic studies, the primary gene pool would surely include the domesticated and wild forms of finger millet, *Eleusine coracana* subsp. *coracana* (domesticated) and subsp. *africana* (wild). The latter is the crop's putative progenitor and is the only close relative known to cross naturally with cultivated finger

Table 11.1 Cultivated finger millet (*Eleusine coracana* (L.) Gaertn.) and its wild relatives

Species	Common name	Ploidy	Genome	Life form	Origin and distribution
E. coracana (L.) Gaertn. subsp. *coracana*	finger millet, birdsfoot millet, African millet, ragi	tetraploid ($2n = 4x = 36$)	**AABB**	annual	widely cultivated in tropical, sub-tropical, and warm-temperate regions of Africa and Asia
E. coracana subsp. *africana* (Kenn.-O'Byrne) Hilu & de Wet	wild finger millet	tetraploid ($2n = 4x = 36$)	**AABB**	annual	wild progenitor, native to Africa and Yemen; weedy, also in crop fields
E. kigeziensis S.M. Phillips		tetraploid ($2n = 4x = 36, 38$)	**AABB**	perennial	E Africa (Uganda, Rwanda, Burundi, and adjacent parts of Congo)
E. floccifolia (Forssk.) Spreng.		diploid ($2n = 2x = 18$)	**AfAf**	perennial	wild; native to Eritrea, Ethiopia, Somalia, and Yemen; naturalized in Egypt and Kenya
E. indica (L.) Gaertn.	yard grass, wire grass, goose grass	diploid ($2n = 2x = 18$)	**AiAi**	annual	widely distributed throughout the tropics and subtropics; ubiquitous crop weed
E. tristachya (Lam.) Lam.	threespike grass	diploid ($2n = 2x = 18$)	**AtAt**	annual	native to S America (Brazil, Bolivia, Argentina, Chile, Paraguay, Uruguay); rarely found as an adventive in some locations in NE Africa
E. intermedia S.M. Phillips		diploid/tetraploid ($2n = 2x = 18$, $2n = 4x = 36$)	**BB**	perennial	E Africa
E. multiflora Hochst. ex A. Rich.		diploid ($2n = 2x = 16$)	**CC**	annual	wild; native to Eritrea, Ethiopia, Kenya, Tanzania, and Yemen
E. jaegeri Pilg.	Manyatta grass	diploid ($2n = 2x = 20$)	**CC**	perennial	wild; native to Ethiopia, Kenya, Tanzania, and Uganda
E. semisterilis S.M. Phillips		n.a.	n.a.	perennial	SE Kenya; perhaps now extinct

Source: Phillips 1972.
Note: Ploidy levels follow Hiremath and Chennaveeraiah 1982 and Hiremath and Salimath 1992. Genome designations are according to Hiremath and Chennaveeraiah 1982 and Hiremath and Salimath 1992. n.a. = not available.

millet (Mehra 1962). Both of these subspecies are allotetraploid ($2n = 4x = 36$) and are completely cross-compatible; hybridization results in fully fertile F1, F2, and F3 hybrids (Mehra 1962; Chennaveeraiah and Hiremath 1974; Hiremath 1974; Hiremath and Salimath 1992; de Wet 1995). F1 hybrids show morphologies ranging from the cultivated crop to wild finger millet, and they are aggressive colonizers and noxious weeds (de Wet et al. 1984; de Wet 1995). They are often found in and near finger millet fields and along roadsides in areas of sympatry in Africa (Mehra 1962; Phillips 1972). Results of microsatellite marker studies provide evidence that confirms introgression between the two subspecies (Dida and Devos 2006).

Eleusine kigeziensis

Eleusine kigeziensis S.M. Phillips is another tetraploid species closely related to finger millet (Phillips 1972; Neves et al. 2005). This species has a similar chromosome number ($2n = 4x = 36$, some forms 38) to the two subspecies of *E. coracana*—the crop and its wild ancestor—and is cross-compatible with both (M. Dida, pers. comm. 2008). Initial crossing studies resulted in the production of male-sterile F1 hybrids. Yet if they were crossed with a male-fertile *E. coracana* plant, they would be likely to set seed. The results of these crossing experiments, together with unpublished molecular studies, indicate that *E. kigeziensis* may carry the **AABB** genome (M. Dida, pers. comm. 2008).

Other *Eleusine* Species

All other wild relatives are genetically isolated from cultivated finger millet, due to differences in their ploidy numbers. In consequence, no natural hybrids between finger millet and any of these other wild relatives have been reported.

Triploid—and thus sterile—hybrids with most of finger millet's diploid wild relatives can be obtained quite easily by hand pollination. In a crossing experiment by Hiremath and Salimath (1992), hybridization with the wild diploids *E. indica*, *E. floccifolia*, *E. multiflora* Hochst. *ex* A. Rich., *E. tristachya*, and *E. intermedia* (Chiov.) S.M. Phillips was successful when the wild relatives were used as male parents. Hybridization rates ranged from 0.5% to 5.1%, and a few F1 plants reached maturity for all crosses except those involving *E. intermedia*. The resulting F1 hybrids were triploids and were either completely seed sterile (*E. indica*, *E. floccifolia*, *E. multiflora*) or did not flower at all (*E. tristachya*) (Chennaveeraiah and Hiremath 1974; Hilu and de Wet 1976a; Hiremath and Salimath 1992).

We found no information concerning hybridization studies between finger millet and the diploid species *E. jaegeri* Pilg. (*n* = 10) and *E. semisterilis* S.M. Phillips. Neither the ploidy level, the chromosome number, nor the genome type of *E. semisterilis* have been described.

Outcrossing Rates and Distances

Pollen Flow

No reports of studies measuring the distances for finger millet pollen flow have been found, and no information is available about outcrossing rates at different distances. Although finger millet is mainly self-fertilizing, spontaneous outcrossing with its progenitor *E. coracana* subsp. *africana* is often observed in the wild. Experiments under natural and controlled conditions are necessary to determine the extent of outcrossing between cultivated finger millet and its wild relatives.

Isolation Distances

In the United States, no separation distances are required for the seed production of inbreeding millet species such as finger millet (USDA 2008). In OECD countries, isolation from other cereal crops "by a definite barrier or a space sufficient to prevent mixture during harvest" is recommended (OECD 2008, p. 18).

State of Development of GM Technology

Methods to genetically transform finger millet are in process; therefore, efficient and stable transformation protocols are not yet available (Latha et al. 2005). Traits of interest for the genetic modification of finger millet are those conferring disease resistance, in particular to leaf and finger blast.

Agronomic Management Recommendations

There are no specific recommendations.

Crop Production Area

Finger millet and pearl millet together account for most of the world's millet production (FAO 2009). Finger millet is grown in 4.0 million ha in the highlands and savannas of several countries in E and S Africa and Asia. The major producers of finger millet are Kenya, Tanzania, Uganda, and Zimbabwe in Africa and India and Nepal in Asia (K. Devos, pers. comm. 2008).

Research Gaps and Conclusions

Finger millet and its wild relatives are among the least scientifically analyzed crops. Because of this, *Eleusine* would greatly benefit from the development of more genetic and genomic tools that would certainly contribute to increased knowledge and information regarding the crop and its relationships with its wild relatives.

Pollen-mediated gene flow and introgression from cultivated finger millet to its wild relatives is likely in Africa, where the plants occur in sympatry. Also, in spite of the fact that finger millet is self-fertilizing for the most part, spontaneous weedy hybrids are found in and near finger millet fields. Unfortunately, there is no information available about the role of pollinator vectors (wind and insects), the distances that finger millet pollen can travel, and expected outcrossing rates at varying distances from the pollen source. Moreover, further studies are needed to investigate the rates of natural outcrossing between finger millet and its wild relatives, especially those concerning subsp. *africana*, where no information is available on hybrid vigor and fertility, the extent and main direction of introgression, and the fate of introgressed alleles in later generations.

Seed exchange constitutes another form of gene flow and introgression between finger millet and its wild relatives that needs to be considered. In many tropical and subtropical regions, such as in India, farmers often exchange seed both within a village and with farmers in neighboring villages (Nagarajan and Smale 2007). These informal seed systems facilitate gene flow not only within the crop (landraces and modern cultivars), but also between domesticated and wild finger millet.

Based on the very limited knowledge available about the relationships of finger millets, the likelihood of gene flow and introgression from domesticated finger millet to its wild relatives—assuming physical proximity and flowering overlap, and being conditional on the nature of the respective trait(s)—is as follows (also see map 11.1):

- *high* for its wild ancestor *Eleusine coracana* subsp. *africana* and probably eventually for its tetraploid wild relative *E. kigeziensis* (map 11.1)

- *low* for its diploid wild relatives *E. floccifolia*, *E. indica*, *E. intermedia*, *E. jaegeri*, *E. multiflora*, and *E. tristachya* (in map 11.1, part of this area is hidden by the red areas representing a high likelihood of gene flow and introgression). Although this has not yet been sufficiently studied, there may also be some likelihood of gene flow to its wild relative *E. semisterilis*

REFERENCES

Bisht MS and Mukai Y (2001) Genomic *in situ* hybridization identifies genome donor of finger millet (*Eleusine coracana*). *Theor. Appl. Genet.* 102: 825–832.

Chennaveeraiah MS and Hiremath SC (1974) Genome analysis of *Eleusine coracana* (L.) Gaertn. *Euphytica* 23: 489–495.

Chuah TS, Salmijah S, Teng YT, and Ismail BS (2004) Changes in seed bank size and dormancy characteristics of the glyphosate-resistant biotype of goosegrass (*Eleusine indica* (L.) Gaertn.). *Weed Biol. Managem.* 4: 114–121.

de Wet JMJ (1995) Finger millet. In: Simmonds S and Smartt J (eds.) *Evolution of Crop Plants.* 2nd ed. (Longman Scientific and Technical: Harlow, UK), pp. 137–140.

de Wet JMJ, Rao KEP, Brink DE, and Mengesha MH (1984) Systematics and evolution of *Eleusine coracana* (Gramineae). *Am. J. Bot.* 71: 550–557.

Dida MM and Devos KM (2006) Finger millet. In: Kole C (ed.) *Cereals and Millets.*

Vol. 1 in *Genome Mapping and Molecular Breeding in Plants* (Springer: Berlin and Heidelberg, Germany), pp. 333–343.

FAO [Food and Agriculture Organization of the United Nations] (2009) FAOSTAT data. http://faostat.fao.org/site/567/default.aspx.

GISD [Global Invasive Species Database] (2007) *Pennisetum.* www.issg.org/data base/welcome.

Hilu KW (1988) Identification of the A genome of finger millet using chloroplast DNA. *Genetics* 118: 163–167.

Hilu KW and de Wet JMJ (1976a) Domestication of *Eleusine coracana. Econ. Bot.* 30: 199–208.

Hilu KW and de Wet JMJ (1976b) Racial evolution in *Eleusine coracana* subsp. *coracana* (finger millet). *Am. J. Bot.* 63: 1311–1318.

Hilu KW and Johnson JL (1992) Ribosomal DNA variation in finger millet and wild species of *Eleusine* (Poaceae). *Theor. Appl. Genet.* 83: 895–902.

Hilu KW, de Wet JMJ, and Seigler D (1978) Flavonoid patterns and systematics in *Eleusine. Biochem. Syst. Ecol.* 6: 247–249.

Hiremath SC (1974) Inheritance of pigmentation in an interspecific cross between *Eleusine coracana* and *E. africana. Curr. Sci.* 43: 557–558.

Hiremath SC and Chennaveeraiah MS (1982) Cytogenetical studies in wild and cultivated species of *Eleusine* (Gramineae). *Caryologia* 35: 57–69.

Hiremath SC and Salimath SS (1991) On the origin of *Eleusine africana* K. O'Byme (Poaceae). *Bothalia* 21: 61–62.

———— (1992) The A genome donor of *Eleusine coracana* (L.) Gaertn. (Gramineae). *Theor. Appl. Genet.* 84: 747–754.

Holm LG, Plucknett DL, Pancho JV, and Herberger JP (1977) *The World's Worst Weeds: Distribution and Biology* (University Press of Hawaii: Honolulu, HI, USA). 609 p.

Kulkarni GN and Basavaraju V (1976) Firm seeds in the finger millet (*Eleusine coracana* G[aertn].). *Curr. Sci.* 45: 425–426.

Latha AM, Rao KV, and Reddy VD (2005) Production of transgenic plants resistant to leaf blast disease in finger millet (*Eleusine coracana* (L.) Gaertn.). *Plant Sci.* 169: 657–667.

Liu Q, Peterson PM, Columbus JT, Zhao N, Hao G, and Zhang D (2007) Inflorescence diversification in the "finger millet clade" (Chloridoideae, Poaceae): A comparison of molecular phylogeny and developmental morphology. *Am. J. Bot.* 94: 1230–1247.

Mehra KL (1962) Natural hybridization between *Eleusine coracana* and *E. africana* in Uganda. *J. Indian Bot. Soc.* 41: 531–539.

Nagarajan L and Smale M (2007) Village seed systems and the biological diversity of millet crops in marginal environments of India. *Euphytica* 155: 167–182.

Neves SS, Swire-Clark G, Hilu KW, and Baird WV (2005) Phylogeny of *Eleusine* (Poaceae: Chloridoideae) based on nuclear ITS and plastid *trnT-trnF* sequences. *Mol. Phylogenet. Evol.* 35: 395–419.

OECD [Organization for Economic Cooperation and Development] (2008) OECD Scheme for the Varietal Certification of Cereal Seed Moving in International Trade: Annex VIII to the Decision. C(2000)146/FINAL. OECD Seed Schemes, February 2008. www.oecd.org/document/0/0,3343,en_2649_33905_1933504 _1_1_1_1,00.html.

Phillips SM (1972) A survey of the genus *Eleusine* Gaertn. (Gramineae) in Africa. *Kew Bull.* 27: 251–270.

Phillips SM and Renvoize SA (1974) Gramineae. In: Milne-Redhead E and Polhill RM (eds.) *Flora of Tropical East Africa, Part 2* (A.A. Balkema on behalf of the East African Governments: Rotterdam, The Netherlands). 449 p.

Rachie KO and Peters LV (1977) *The* Eleusines: *A Review of the World Literature* (International Crops Research Institute for the Semiarid Tropics [ICRISAT]: Hyderabad, India). 179 p.

USDA [United States Department of Agriculture] (2008) 7 CFR [Code of Federal Regulations] § 201.76. Federal Seed Act Regulations: Minimum Land, Isolation, Field, and Seed Standards. Source: 59 FR [Federal Register] 64516, Dec. 14, 1994, as amended at 65 FR 1710, Jan. 11, 2000 (Government Printing Office: Washington, DC, USA), pp. 372–378. www.access.gpo.gov/nara/cfr/waisidx_08 /7cfr201_08.html.

Werth CR, Hilu KW, and Langner CA (1994) Isozymes of *Eleusine* (Gramineae) and the origin of finger millet. *Am. J. Bot.* 81: 1186–1197.

TWELVE

Maize, Corn
(*Zea mays* L.)

Center(s) of Origin and Diversity

The origin and domestication of maize is controversial (see review in OECD 2003). The most accepted hypothesis at the moment is that domesticated maize (*Zea mays* subsp. *mays* L.) arose from a single domestication event in Mexico—in the central Balsas River Valley in Michoacán and Guerrero states—as a result of human selection of the annual teosinte *Z. mays* subsp. *parviglumis* H.H. Iltis & Doebley (Doebley et al. 1987, 2006; Matsuoka et al. 2002; Doebley 2004; Matsuoka 2005).

Several primary centers of diversity for domesticated maize exist in Latin America. Mexico and Central America are a center of diversity for the most commercially significant dent types, the highlands of S Mexico and Guatemala for the long-eared flint types, central Mexico for the pyramidal or Conico types, and NW Mexico and SW United States for the eight-rowed complex. The greatest source of maize diversity for kernel, cob, and plant colors and kernel sizes is found at mid- and high elevations in the central Andean region. The Caribbean Islands and northern S America are important centers of diversity for most tropical flint, semi-flint, and dent races (Goodman 1995; Sánchez et al. 2000a, b, 2006, 2007).

Biological Information

Flowering

Maize is monoecious, that is, the male (tassel) and female (silk) flowers are borne on separate plant organs. Maize pollen is released in very large quantities, between 4.5 and 25 million pollen grains per plant (Goss 1968; Paterniani and Stort 1974; Jarosz et al. 2005), over a period of 5 to 20 days, depending on climate zone (Ogden et al. 1969; Westgate et al. 2003). In many traditional maize varieties and landraces, the time periods for flower development, pollen shedding, and fruiting (silking and tasseling) are separated and thus ensure cross-pollination with neighboring plants. In most improved maize lines and varieties, however, these events overlap notably or are even synchronized (S. Taba, pers. comm. 2008).

In Mexico, most teosinte populations from the Valley of Mexico, the Nabogame Valley in northern Mexico, and the Central Plateau (*Zea mays* subsp. *mexicana* H.H. Iltis) occur as weeds in or near maize fields. In these areas, the flowering of teosintes, maize landraces, and commercial hybrids is synchronized. In contrast, teosinte from the Balsas River basin (*Zea mays* subsp. *parviglumis*) flowers two to three weeks later than maize landraces and commercial varieties. All in all, the flowering time of teosintes is quite long and commonly overlaps with that of cultivated maize (Wilkes 1977; Berthaud and Gepts 2004; Rodríguez et al. 2006).

Pollen Dispersal and Longevity

Maize pollen is mainly wind-dispersed and is one of the largest and heaviest (90–125 μm long, 250×10^{-9} g weight) wind-borne pollens (Kiesselbach 1999). Pollination by bees and, occasionally, by beetles, flies, and leafhoppers does occur (Ortega 1987), but only to a very limited extent, since the female flowers do not produce nectar and thus are not attractive to insect pollinators.

The longevity of maize pollen depends mainly on temperature and humidity (see, e.g., Barnabas 1984; Schoper et al. 1987; Traore et al. 2000). Under warm and dry conditions, maize pollen remains viable for a period from 10 to 30 minutes to up to two hours (Dumas and Mogensen 1993; Luna et al. 2001). In more temperate, humid conditions, maize pollen can stay viable for several days (Burris 2001).

Rodríguez et al. (2006) studied about 200 accessions and found that teosinte pollen is 15%–30% smaller than maize pollen and more susceptible to desiccation after being exposed to atmospheric conditions, although these characteristics differed among regions and within races. For example, *Zea mays* subsp. *mexicana* had larger pollen grains, while subsp. *parviglumis* was more likely to dry out after being exposed to the elements.

Sexual Reproduction

Domesticated maize is an annual diploid ($2n = 2x = 20$). It is an allogamous crop, typically with 95% cross-pollination (Sleper and Poehlman 2006), despite the plants being completely self-compatible.

Vegetative Reproduction

Domesticated maize does not reproduce vegetatively, but the two perennial teosinte wild relatives (*Zea perennis* (Hitchc.) Reeves & Mangelsd. and *Z. diploperennis* H.H. Iltis et al.) can do so via rhizomes.

Seed Dispersal and Dormancy

During its domestication from teosinte, maize lost its ability to disperse seed and thus survive in the wild. Its seeds (kernels) remain on the cob after ripening, and they do not shatter or otherwise scatter naturally (Doebley et al. 1990). However, small amounts of maize seed, and even whole cobs, can be distributed by animals (birds, rodents) and humans during mechanical planting, harvesting, grain handling, transportation, and storage operations.

Maize seeds dispersed during the harvesting process can only survive for up to one year in the soil, due to their poor dormancy. Also, domesticated maize requires warm conditions in order to grow and does not tolerate prolonged cold and frost. Maize seeds therefore have a poor chance of survival in many (continental) maize-growing regions.

Teosinte seeds fed to cattle as fodder have been shown to survive passage through the intestinal tract (Wilkes 1977, 2004).

Volunteers, Ferals, and Their Persistence

Domesticated maize volunteers are commonly found in fields the year following maize cultivation, due to spilt grain. They can also appear as feral plants in abandoned fields and along roadsides.

Table 12.1 Domesticated maize (*Zea mays* L.) and its crop wild relatives

Species	Common name	Genome	Origin and distribution
sect. *Zea*		diploid	
Z. mays L. subsp. mays	domesticated maize, corn	$2n = 2x = 20$	known only from cultivation
Z. mays subsp. *mexicana* (Schrad.) H.H. Iltis (races Chalco, Central Plateau, Nobogame, and Durango*)	Mexican annual teosinte	$2n = 2x = 20$	ancestor; C and N Mexico
Z. mays subsp. *parviglumis* H.H. Iltis & Doebley (races Balsas and Jalisco*)	Mexican annual teosinte, Balsas teosinte	$2n = 2x = 20$	ancestor; SW Mexico; crop weed
Z. mays subsp. *huehuetenangensis* (H.H. Iltis & Doebley) Doebley (race Huehuetenango)	annual teosinte	n.a.	endemic to NW Guatemala
sect. *Luxuriantes* Doebley & H.H. Iltis		diploid or tetraploid	
Z. diploperennis H.H. Iltis et al.	diploperennial teosinte	$2n = 2x = 20$	endemic to Mexico (Sierra de Manantlán)
Z. luxurians (Durieu & Asch.) R.M. Bird	Guatemala or Florida teosinte (race Guatemala)	$2n = 2x = 20$	native to SE Guatemala, Honduras, Nicaragua
Z. nicaraguensis H.H. Itis & B.F. Benz	Nicaraguan teosinte	$2n = 2x = 20$	native to Nicaragua
Z. perennis (Hitch.) Reeves & Mangelsd.	perennial teosinte	$2n = 4x = 40$	endemic to Mexico (Jalisco)
Other genera			
Tripsacum spp.	gamagrass		16 species distributed from Paraguay to USA

Sources: Doebley and Iltis 1980; Iltis and Doebley 1980.
Note: The cultivated form is in bold. n.a. = not available.
*These races are not yet widely accepted.

Domesticated maize is incapable of sustained reproduction outside of cultivation, and thus is noninvasive in natural habitats (Eastham and Sweet 2002). This is due to the plant's inability to shed its seed naturally, its limited dormancy period, and its susceptibility to low temperatures. Any volunteers or ferals are usually killed by common mechanical, pre-planting soil preparation practices or by frost the following winter. Even in regions where winter temperatures are not low enough to kill off volunteers (e.g., Mediterranean countries), maize volunteers and feral plants do not form persistent populations.

Weediness and Invasiveness Potential

Domesticated maize is a weak competitor outside of cultivation and, consequently, is not an invasive plant. For this reason, crop plants are very rarely found outside of cultivation and cannot be considered to be noxious weeds (GISD 2007).

Teosintes, in contrast, do shed their seeds naturally, in considerable quantities, and can be weeds in some parts of Mexico. In the Valley of Mexico (Chalco, Toluca, and Puebla) and Central Plateau areas, *Zea mays* subsp. *mexicana* has been reported to invade newly planted maize fields, since teosinte seeds fed to cattle as fodder survive in the intestinal tract, and their manure is used to fertilize new maize fields (Wilkes 1977, 2004; Berthaud and Gepts 2004).

Crop Wild Relatives

The closest wild relatives of domesticated maize belong to a group of four annual and perennial diploid ($2n = 2x = 20$) and tetraploid species ($2n = 4x = 40$) within the genus *Zea*, commonly called teosintes (table 12.1; see also Doebley 1990a; Sánchez et al. 1998; Wilkes 1967, 2004). The name teosinte is derived from the Aztec name *teocintle* and has been translated as "grain of the gods." Teosintes are native to a region from N Mexico to W Nicaragua. They are found in isolated populations of varying sizes, from less than 1 ha to several km². Some races (Nobogame in N Mexico) have a narrow geographic distribution and consist of only a few local populations (Sánchez and Ordaz 1987; Sánchez and Ruiz 1997; Sánchez et al. 1998; Wilkes 2004).

Teosintes comprise seven taxa that are now divided into two sections and five species (table 12.1). Teosintes in sect. *Luxuriantes* Doebley & H.H. Iltis are genetically and taxonomically distinct from those in sect. *Zea* (which also includes domesticated maize). Two of them are perennials that propagate vegetatively through rhizomes.

Another group of wild relatives of domesticated maize are species in the perennial grass genus *Tripsacum* L.—the most closely related genus to *Zea*—found from Massachusetts (United States) to Paraguay (de Wet et al. 1982, 1983). The basic chromosome number for *Tripsacum* is $x = 18$, and it is represented by diploid, triploid, tetraploid, and higher ploidy levels (Doebley 1983).

Tripsacum species boundaries are often unclear, due to intermediate forms that are the result of common natural hybridization. Currently, 16 species are recognized in the genus, divided into two sections: sect. *Fasciculata* Hitchc. (five species) and sect. *Tripsacum* (eleven species). Twelve of these species are native to Mexico and Guatemala, with *T. dactyloides* (L.) L. extending throughout the eastern half of the United States, including the corn belt in the Midwest (Gould 1968), and four species native to S America (Berthaud et al. 1997; Zuloaga et al. 2003; Wilkes 2004).

Hybridization

Maize is sexually compatible and can form fertile intra- and interspecific hybrids with most of its wild teosinte relatives (Eastham and Sweet 2002). Hand pollination readily yields vigorous and highly fertile hybrids and F2 progeny, particularly if one of the teosintes is the pollen donor (Kermicle 1997).

In Mexico, teosintes often grow near or in maize fields and spontaneously hybridize with maize (Wilkes 1977; Kermicle and Allen 1990). Although natural hybrids are known to occur in the field, only a few experimental field studies and genetic analyses have been conducted thus far, and actual rates of outcrossing, as well as the main direction and extent of gene flow, are virtually unknown. Also, later generations of hybrid progeny have rarely been observed, and there is no information on how long maize genes can persist in teosinte populations.

The fact that maize and teosintes have coexisted in Mexico and

Guatemala since ancient times raises questions about the degree and direction of gene flow and introgression. The hypothesis is that gene flow is mainly unidirectional from teosintes to maize, with either insignificant introgression from maize to teosintes or none at all in either direction (see, e.g., Ting 1963; Doebley 1984; Kato 1997; Engels et al. 2006). This is supported by studies detecting unilateral pre-zygotic genetic barriers in several teosinte strains (similar to those known from popcorn and other maize varieties), leading to reduced or no seed set, which prevents high rates of introgression (Kermicle and Allen 1990; Evans and Kermicle 2001; Baltazar et al. 2005; Kermicle 2006; Kermicle et al. 2006b). Other genetic studies provide evidence that gene flow and introgression between maize and several of its teosinte relatives may occur more frequently than was previously estimated (Blancas et al. 2002; Ellstrand et al. 2007). However, the situation is even more complex. Fukunaga et al. (2005) showed that the genetic composition of teosintes varies from population to population, even within subspecies. Similarly, Kermicle et al. (2006b) found that the compatibility of domesticated maize with teosintes is influenced by the maize genotype and varies among landrace cultivars, and even among populations. Thus the results of any study investigating genetic compatibility between maize and its teosinte wild relatives need to be analyzed in light of these findings. The hybridization and introgression potential of maize and its wild teosinte relatives are discussed in more detail below.

Zea mays subsp. *mexicana*

Zea mays subsp. *mexicana* is the teosinte most commonly associated with maize fields. Hybrid swarms have been reported to occur naturally in these fields, with as many as up to 10% of the teosinte plants being annual maize-teosinte hybrids that occur year after year in the same location (Wilkes 1977, 1985).

Plants from subsp. *mexicana* (races Chalco and Central Plateau), when hand pollinated with maize, set seed poorly (< 1%) or not at all, whereas a reciprocal cross with pure pollen generally results in successful hybridization (Ting 1963; Kermicle and Allen 1990; Kermicle 1997; Baltazar et al. 2005). When maize is hand pollinated with a mixture of teosinte and maize pollen, few or no hybrids are formed, indicating poor teosinte pollen competition (Castro-Gil 1970). Hence genetic studies using isozymes, allozymes, and SSR markers have detected very low levels of sporadic gene introgression between maize and subsp. *mexicana* (Smith et al. 1985; Doebley et al. 1987; Doebley 1990b; Fukunaga et al. 2005; Ellstrand et al. 2007).

The level of genetic incompatibility between maize and subsp. *mexicana* varies among teosinte populations, and pre-zygotic genetic barriers have been detected in some subsp. *mexicana* stocks, particularly from the Central Plateau and Chalco races (Kermicle and Allen 1990; Kermicle 2006; Kermicle et al. 2006a, b). This genetic incompatibility is asymmetric, being very strong when maize is the pollen parent and weaker when subsp. *mexicana* is the pollen parent (Kermicle and Evans 2005). Incompatibility genes in local subsp. *mexicana* populations may thus be responsible for the low rates of hybridization and introgression, thereby preventing a more complete homogenization of maize and subsp. *mexicana* gene pools where they grow in sympatry (Evans and Kermicle 2001).

Although these incompatibility genes can prevent maize from fertilizing local populations of subsp. *mexicana*, this barrier is not universal (Mangelsdorf and Reeves 1931; Evans and Kermicle 2001). Kermicle and colleagues (2006b) noted that it is not clear whether such incompatibility functions bilaterally *in situ*; unless this incompatibility is complete, genetic isolation is not effective. A recent allozyme study of five populations, representing the geographical range of subsp. *mexicana* (Valley of Mexico and Central Plateau), detected levels of hybridization and introgression that were higher than previously estimated (Blancas et al. 2002). Furthermore, under field conditions, F1 plants have higher maternal fitness than their wild parents and are able to backcross with either parent (Guadagnuolo et al. 2006). While F2 and BC plants have only rarely been observed in the field, hybridization may be sufficiently recurrent to make the introgression of neutral or beneficial maize alleles probable (Ellstrand et al. 2007).

Zea mays subsp. *parviglumis*

Similar to *Zea mays* subsp. *mexicana*, Balsas teosinte (*Z. mays* subsp. *parviglumis*) is quite widely distributed throughout Mexico and is often found in and near maize fields. This subspecies is the teosinte most closely related to maize genetically, and available evidence indicates that it is its direct ancestor (Doebley 1990a, 2004; Matsuoka et al. 2002; Doebley et al. 2006).

Hand pollination generally yields very high frequencies of maize × subsp. *parviglumis* hybrids (Kermicle 1997; Molina and García 1997), and plants with intermediate phenotypes between maize and subsp. *parviglumis* often occur spontaneously in and near Mexican maize fields (Wilkes 1977; Sánchez and Ordaz 1987).

Still, the extent of hybridization and—more importantly—introgression between maize and this teosinte type is still unknown. In a genetic study using SSR markers, Fukunaga et al. (2005) identified 56 out of 117 teosinte plants containing 20% or more of maize germplasm, thus being admixed with maize. In their opinion, these results could reflect the recent origin of maize from subsp. *parviglumis*, and thus not necessarily indicate the existence of fully differentiated gene pools, rather than an admixture through introgression.

Also, it is unclear whether subsp. *parviglumis* presents genetic barriers to hybridization with maize, as does subsp. *mexicana*. Kermicle and Allen (1990) found that no such genetic barrier was present in the subsp. *parviglumis* populations they studied. Yet Doebley and Stec (1993), in a study of an F2 cross of maize and subsp. *parviglumis*, found that the teosinte progenitor had the *Gametophyte-1* (*Ga1*) incompatibility allele. Another study (Kermicle 2006) shows that some subsp. *parviglumis* populations were unreceptive to maize pollen because they carry the allele *Tcb1-s* (*Teosinte crossing barrier-1*).

There is no definitive data assessing the extent of outcrossing and introgression between maize and subsp. *parviglumis*. However, in view of the facts that hybridization occurs spontaneously in the field and that hand pollination yields high rates of fertile hybrids, maintaining a precautionary standpoint is both wise and prudent.

Zea mays subsp. *huehuetenangensis*

There seem to be no reports of spontaneously occurring hybrids between cultivated maize and *Zea mays* subsp. *huehuetenangensis* in the field, although Wilkes (1977) mentioned that this teosinte rarely hybridizes with maize, and that any hybrids with maize that do occur are restricted to plants that invade maize fields. Using SSR, Fukunaga et al. (2005) found no evidence of gene flow or admixtures between subsp. *huehuetenangensis* and maize. Kermicle et al. (2006b) reported that subsp. *huehuetenangensis* was polymorphic for the *Gametophyte-1* locus, that is, there are plants both with and without barriers to hybridization.

This teosinte type is very closely related to Balsas teosinte (*Z. mays* subsp. *parviglumis*), and it was originally classified under the same subspecies. In consequence, it may be assumed that—similar to subsp. *parviglumis*—the possibility of gene flow and introgression between this teosinte and maize is high.

Zea luxurians

No published reports of naturally occurring hybrids between maize and the Guatemalan perennial teosinte *Zea luxurians* (Durieu & Asch.) R.M. Bird were found, except a short note by Bird and Beckett (1980) documenting the presence of putative maize × *Z. luxurians* hybrid plants in Guatemala, based on observations of intermediate morphological characteristics. Wilkes (1977) stated that *Z. luxurians* might hybridize with maize, albeit rarely.

An older study (Beadle 1932) reported that the Florida form of teosinte—now known as *Z. luxurians*—(1) had chromosomes that were cytologically distinct from those of maize, (2) that maize × *Z. luxurians* hybrids exhibited two or more unpaired chromosomes during metaphase, and (3) that these hybrids were partially sterile. Indirect evidence from genetic studies of isozymes indicates that maize-specific alleles appear in perennial *Z. luxurians* teosinte populations (Doebley et al. 1984; Doebley 1990b). However, allele frequencies are too low to be considered positive evidence of introgression from the crop to this perennial teosinte. These results are further confirmed by chromosome and molecular studies, where no evidence of gene flow or admixture between maize and *Z. luxurians* was found (Fukunaga et al. 2005).

To conclusively evaluate the likelihood and extent of outcrossing and introgression between *Z. luxurians* and maize, more experimental field studies and genetic analyses should be conducted.

Zea diploperennis

The perennial teosinte *Zea diploperennis*, a Mexican endemic, is closely related to the Guatemalan teosinte *Z. luxurians*. As opposed to the latter, several publications have reported observing *Z. diploperennis* growing near maize fields and forming fertile hybrids with cultivated maize (Iltis et al. 1979; Lorente-Adame and Sánchez-Velasquez 1996; Benz et al. 1990). However, we have found no studies directly investigating the likelihood and frequency of outcrossing between these two species, nor have there been any indications of the existence of a genetic barrier to hybridization between maize and this perennial teosinte.

Field observations reporting the occurrence of fertile hybrids are confirmed by genetic markers (isozyme, SSR, rDNA) that provide evidence for the admixture of *Z. diploperennis* populations with maize (Doebley et al. 1984; Doebley 1990a; Buckler and Holtsford 1996; Fukunaga et al. 2005). Yet even with these studies, the extent of actual introgression cannot easily be deduced, since the results only allow indistinct interpretations, based on selection, migration, and ancestry (Kato 1997). Kato and Sánchez (2002) examined the chromosome constitutions of *Z. diploperennis* and sympatric maize collections obtained within the limits of the Sierra de Manantlán and found no evidence of natural introgression between maize and *Z. diploperennis*. Moreover, there are no clear indications of the existence of a genetic barrier to hybridization between maize and this perennial teosinte. More detailed studies should be conducted to determine actual outcrossing rates and the fitness of hybrids. Subsequent backcross generations should be examined to better evaluate the extent of introgression between maize and this perennial teosinte species.

Zea perennis

The genetic relationship between cultivated maize and *Zea perennis*, a perennial Mexican endemic, has been documented and studied extensively (see, e.g., Wilkes 1967; Mangelsdorf 1974; Molina and García 1999b; Poggio et al. 1999), and whole-chromosome introgression from *Z. perennis* to maize has been achieved by producing artificial F2 hybrids and BC (e.g., Tang et al. 2005). However, there are strong reproductive barriers between maize and the tetraploid teosinte Z. *perennis*, and there is no evidence that natural hybridization occurs between these taxa (Doebley 1990b). Genetic incompatibility is due to the different ploidy levels of Z. *perennis* ($4x$) and maize ($2x$), leading to interspecific hybrids ($2n = 30$) with abnormal meiosis behavior and high sterility ($< 5\%$ fertile).

Zea nicaraguensis

No published reports of natural or artificial hybridization between maize and the Nicaraguan perennial teosinte *Zea nicaraguensis* H.H. Iltis & B.F. Benz are known.

Tripsacum

Tripsacum species are widely distributed across Central and S America. They can cross freely with each other, as long as the parents have the same ploidy level (de Wet et al. 1972). Hybridization of maize with *Tripsacum* species is not known to occur in the wild, except for one *Tripsacum* species—the sterile *T. andersonii* J.R. Gray—which arose from a unique spontaneous hybridization event between two *Tripsacum* species and *Zea luxurians* (Talbert et al. 1990; Barré et al. 1994).

Some *Tripsacum* species can cross successfully with domesticated maize, with the most compatible being *T. dactyloides* (L.) L. However, crossing experiments between these two species have yielded variable results. In several cases, hybridization failed in both directions and no viable seeds were obtained (see, e.g., Randolph 1959; Berthaud and Savidan 1999; Molina and García 1999a). Other reports indicate viable and fertile F1 progeny were produced after applying maize pollen using artificial techniques (Galinat and Rao 1969; de Wet et al. 1973; Jatimliansky et al. 2004; Molina et al. 2006), or hand or open pollination (Mangelsdorf and Reeves 1931; Farquharson 1957). The resulting F1 hybrids are mainly sexually sterile, but low percentages of viable seeds can be obtained by apomixis (Leblanc et al. 1995b; Sokolov et al. 2000).

Eubanks (1995) claimed to have transferred genes from *Tripsacum* into maize using tripsacorn, a cross between *T. dactyloides* and the perennial teosinte *Z. diploperennis*, to generate maize-tripsacorn hybrids. His claim included the transmission of several traits—such as apomixis, totipotency, perennialism, and adaptation to both adverse soil conditions and a carbon-dioxide-enriched atmosphere—to maize via maize × (*T. dactyloides* × *Z. diploperennis*) and vice versa (MacNeish and Eubanks 2000; Eubanks 2001). These results, however, have been challenged, due to the lack of molecular evidence (Bennetzen et al. 2001) and the fact that thus far, other scientists have not been able to replicate them.

These mixed findings, however, do not exclude the likelihood of spontaneous hybridization between domesticated maize and *T. dactyloides* in the field, although this appears to be very low.

Hybridization of domesticated maize with other *Tripsacum* species is possible, but only artificially and with difficulty, due to their differing chromosome numbers (de Wet et al. 1973; Bernard and Jewell 1985; Furini and Jewell 1995). In general, tetraploid *Tripsacum* species and varieties cross much more readily than diploids (de Wet et al. 1984). The progeny from crosses between *Z. mays* and *Tripsacum* show varying levels of sterility (Mangelsdorf 1974; Russell and Hallauer 1980; Galinat 1988). However, after backcrossing, small portions of the *Tripsacum* genome can be incorporated into that of maize (Goodman 1995), and several scientists have investigated the possibility of introgressing useful characteristics from *Tripsacum* to maize (Bergquist 1981; Ray et al. 1999). These include the transfer of apomixis genes (see, e.g., Burson et al. 1990; Leblanc et al. 1995a; Grimanelli et al. 1998), which have resulted in several patents for apomictic maize in recent years (Eubanks 1994; Kindiger and Sokolov 1998; Savidan et al. 2007).

Outcrossing Rates and Distances

Pollen Flow

Devos et al. (2005) produced a comprehensive review of pollen flow and outcrossing in maize, focusing on Europe, and we recommend it for further reading. The following is a short summary of their findings.

Maize pollen is one of the largest and heaviest pollens transported by wind; therefore, it is not easily dispersed. The impaction rate (set-

Table 12.2 A sample of studies of pollen dispersal distances and outcrossing rates of domesticated maize (*Zea mays* L.)

Pollen dispersal distance (m)	% outcrossing (max. values)	Country	Reference
2, 10, 25, 45, 65, 95	50, 1.2, 0.87, 0.54, 0.49, 0.21	Italy	Della Porta et al. 2008
46, 82, 183	0.23, 0.26, 0.05	USA	Byrne and Fromherz 2003
2, 10, 20, 35, 200	2.0, 2.0, 1.8, 1.7, 1.1	USA	Burris 2001
1.8, 9.4, 20.6, 35.8, 200	2.5, 2.4, 1.9, 1.6, 1.3	USA	Ireland et al. 2006
30, 35, 40, 100, 110, 350	1.6, 0.9, 1.1, 0.7, 1.4, 0	USA	Jemison and Vayda 2001
0, 25, 75, 125, 200, 300, 400, 500	29, 14, 5.8, 2.3, 1.2, 0.5 0.2, 0.2	USA	Jones and Brooks 1950
24, 60, 125, 254, 500, 743	0.30, 0.10, 0.03, 0.01, 0.0004, 0.0004	USA	Halsey et al. 2005
1, 5, 10, 14, 19, 24, 28, 33, 36	62, 13, 16, 4.5, 2.7, 3.1, 3.2, 2.1, 0.6	Canada	Ma et al. 2004
1.8, 6, 8.5, 13, 19, 22, 23, > 23.8	57, 42, 3.5, 6, 2.7, 0.3, 1.6, 0	UK	Bateman 1947
50, 150, 200, 650	> 0.3, > 0.3, 0.42, 0.14	UK	Henry et al. 2003
1, 2, 5, 10, 30, 60	4.3, 1.5, 0.5, 0.2, 0.1, 0.1	Spain	Messeguer et al. 2003
0, 2, 5, 10, 20, 40, 80	12, 6, 2.3, 2.4, 0.6, 0.4, 0.2	Spain	Pla et al. 2006
52, 85, 150, 287, 370, 400, 458, 4440	0.009, 0.015, 0.016, 0.005, 0.008, 0.005, 0.0002, 0.0005	Switzerland	Bannert and Stamp 2007
105, 125, 150, 200, 4125	0.003, 0.01, 0.007, 0.009, 0.006	Switzerland	Bannert and Stamp 2007
10, 50, 100, 150, 200, 400, 500, 600, 700, 800	3.3, 0.3, 0.4, 0.3, 0.5, 0.02, 0.08, 0.79, 0.18, 0.21	Chechenia	Salamov 1940
1, 5, 10, 20, 30, 34	0.4, 0.1, 0.01, 0.01, 0.005, 0.003	Brazil	Paterniani and Stort 1974
> 184	0.01	Mexico	García et al. 1998
1, 50, 100, 200, 400	22.6–56.8, 1.2, 0.23, 0.06, 0.04	Japan	Matsuo et al. 2004
100, 200, 300, 400, 500, 600	2.9, 0.5, 0.15, 0.06, 0.01, 0.001	India	Narayanaswarmy et al. 1997
125, 200, 427	8.9, < 3, 1	n.a.*	Jones and Newell 1948

*n.a. = not available.

tling velocity) of maize pollen varies between 20 and 40 cm/s (Sears and Stanley-Horn 2000; Aylor 2002; Jarosz et al. 2003), and under normal climatic conditions, about 90% of maize pollen is deposited within 5 m of its source (Hutchroft 1958). Approximately 96%–99% of maize pollen remains within a 25–50 m radius, and 100% of the pollen is deposited within 100 m (see, e.g., Raynor et al. 1972; Sears and Stanley-Horn 2000; Bannert et al. 2008). Actual pollen dispersal distances, however, are quite variable and depend on the prevailing wind direction, as well as wind speed, humidity, and temperature (Aylor 2004; Ma et al. 2004; Halsey et al. 2005). Low concentrations of viable pollen, however, have been measured as far as 1 km downwind of the pollen source (Jarosz et al. 2005) and more than 1.8 km in the air, above the atmospheric edge (Brunet et al. 2003).

Outcrossing has been detected up to 800 m from the pollen source (table 12.2), albeit at low frequencies (< 0.5%). In a recent study, very low levels of outcrossing (< 0.01%) through long-distance pollen dispersal were found more than 4 km from the pollen source (Bannert and Stamp 2007). Although this is the maximum distance researched thus far, there seems to be no clear cutoff distance beyond which outcrossing levels reach zero. However, several authors have developed maize-specific models to understand pollen dispersal and predict cross-pollination rates (see, e.g., Aylor et al. 2003; Klein et al. 2003; Loos et al. 2003; Lipsius et al. 2006).

A few viable pollen grains spread over considerable distances may not achieve cross-pollination, due to strong pollen competition by pollen shedding in the receptor plants/field (Bannert and Stamp 2007). However, in seed production fields with female plants with their tassels removed or with cytoplasmic male sterility, cross-pollination through long-distance dispersal may be higher than in fields with fully fertile maize plants (Feil et al. 2003). As has been observed for other crops, large maize fields result in larger pollen dispersal distances than small fields, due to the higher concentration of pollen in the atmosphere at a given time.

Isolation Distances

The current minimum separation distance for all seed production categories in most countries is 200 m (table 12.3), and this is deemed to be

Table 12.3 Isolation distances for domesticated maize (*Zea mays* L.)

Threshold*	Minimum isolation distance (m)	Country	Reference	Comment
0.1% / n.a.	200	EU	EC 2001	foundation (basic) seed requirements (inbred lines, hybrids)
0.2% / n.a.	200	EU	EC 2001	certified seed requirements (inbred lines, hybrids)
0.5% / n.a.	200	EU	EC 2001	foundation (basic) seed requirements (open pollinated)
1.0% / n.a.	200	EU	EC 2001	certified seed requirements (open pollinated)
0.5% / n.a.	300	UK	Ingram 2000	foundation (basic) seed requirements (non-hybrids)
1.0% / n.a.	200	UK	Ingram 2000	certified seed requirements (non-hybrids) and organic crops
0.5% / 0.5%	200	OECD	OECD 2008	foundation (basic) seed requirements (non-hybrids)
1.0% / 1.0%	200	OECD	OECD 2008	certified seed requirements (non-hybrids)
0.5% / 0.1%	200	OECD	OECD 2008	foundation (basic) seed requirements (hybrids)
1.0% / 0.2%	200	OECD	OECD 2008	certified seed requirements (hybrids)
0.1% / 0.1%	200	USA	USDA 2008	foundation (basic) seed requirements (inbred lines)
0.5% / 0.5%	200	USA	USDA 2008	certified seed requirements (open pollinated)
0.1% / 0.5%	200	USA	USDA 2008	certified seed requirements (hybrid)
0.05% / n.a.	200	Canada	CSGA 2009	certified seed requirements (open pollinated)
0.1% / n.a.	200†	Canada	CSGA 2009	certified seed requirements (hybrids)

*Maximum percentage of off-type plants permitted in the field / seed impurities of other varieties or off-types permitted. n.a. = not available.
†Isolation distances can be reduced if adequate numbers of male-plant border rows are provided.

sufficient to obtain maize seeds with purity levels between 99% and 99.9% (Eastham and Sweet 2002; Tolstrup et al. 2003).

In the United Kingdom, the Advisory Committee on Releases to the Environment (ACRE 2001) concluded that separation distances of 80, 130, and 290 m were adequate to ensure tolerance thresholds for the adventitious presence of GM material of 1%, 0.5% and 0.1%, respectively. These isolation distances comply with those established by the Supply Chain Initiative on Modified Agricultural Crops (SCIMAC 1999), and they were employed for and confirmed by a Farm Scale Evaluation (FSE) trial in the United Kingdom (Henry et al. 2003).

In a number of studies at the field-scale level, no samples collected beyond 30 m had levels of outcrossing above the maximum GM threshold of 0.9% (Henry et al. 2003; Brookes et al. 2004; Messeguer et al. 2006; Weber et al. 2007). Perhaps a separation distance of 25 m, or—alternatively—the planting of 20-meter-wide pollen barriers of conventional maize are adequate measures to effectively contain pollen flow in maize below the 0.9% threshold under real-life, large-scale agronomical conditions. Separation distances of 80 m may be great enough to ensure that levels of admixture are kept below the 0.3% threshold.

Devos et al. (2005) and Sanvido et al. (2008) reviewed maize pollen flow and outcrossing in Europe, and both sets of authors propose that isolation distances between 10 and 50 m may be sufficient to keep GM seed impurities "well below" the 0.9% threshold defined by the European Union.

When comparing published studies of maize pollen flow, outcrossing rates have been shown to rapidly decrease within the first 50 m (fig. 12.1). Nonetheless, outcrossing rates of up to 8.9% can often be detected as far as 150 m away from the pollen source (see, e.g., Jones and Brooks 1950, 1952; Narayanaswarmy et al. 1997; Jemison and Vayda 2001). Also, outcrossing rates at greater distances are, at times, higher than those closer to the pollen source (see, e.g., Bateman 1947; Jemison and Vayda 2001; Luna et al. 2001). These observations show that pollen dispersal in maize is strongly influenced by spatially and temporally varying—and often uncontrollable—local factors such as humidity, the effects of landscape and air currents, and the size of the donor and receptor fields. To minimize

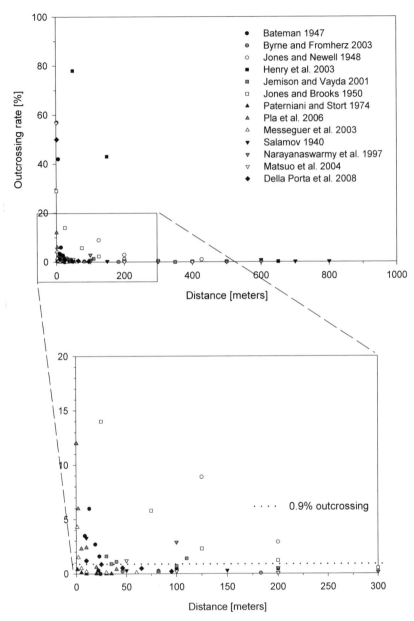

Fig. 12.1 Pollen dispersal distances and outcrossing rates of domesticated maize (*Zea mays* L.)

maize pollen flow and outcrossing, isolation distances should not be less than the 200 m suggested by most countries.

Obviously, these recommendations are primarily valid for commercial-scale maize production in industrialized countries. In other nations, maize is grown either at a small scale or for subsistence farming, and the size of the fields is often less than 1 ha, with short or no separation distances between neighboring fields. In cases such as these, where isolation distances are not viable, other containment measures, such as pollen barriers, should be considered.

Gene Flow through Seed Exchange

Most research about maize gene flow and introgression has been oriented to the study of gene barriers and the estimation of distances that pollen can travel while remaining viable. On the other hand, gene flow via seed dispersal is another important perspective, one where farmers play a crucial role. In Latin America, for example, farmers facilitate gene flow either by the deliberate or inadvertent mixing of seeds (see, e.g., Bellon and Brush 1994), which often happens because many small adjacent fields are planted with very diverse maize types. In Mesoamerica—particularly in Mexico—this situation deserves careful consideration, since the region is also the center of crop origin and diversity, and there is a substantial risk of gene flow and introgression into wild relatives of maize.

In small-scale farming systems in Mesoamerica, maize is grown largely for subsistence. Small farmers, peasants, and indigenous ethnic communities cultivate a broad range of maize landraces and modern varieties that are adapted to particular ecological niches and suited to special types of consumption (Louette et al. 1997; Perales et al. 2003b). Most landraces have specific qualities for food and other uses, which are often appreciated more than simple yield.

The dynamic process of informal seed systems among farmers is an important aspect of gene flow. Mexican farmers frequently trade seeds, sow mixtures of seeds from different sources (including landraces and commercial varieties), and often allow (and even intend) outcrossing between different types (Louette 1997; Perales et al. 2003a; Turrent and Serratos 2004). These autochthonous farming practices stimulate extensive gene flow, and further gene flow into teosinte wild relatives is only a small step away. Most teosinte types are closely associated with maize fields, and outcrossing with the crop is frequently observed (see above). Several studies have stated that the spread of GM traits into Mexican maize landraces would be inevitable once they enter the country—and thus also enter the informal seed system (see, e.g., Serratos et al. 2004; Cleveland et al. 2005; Raven 2005; Soleri et al. 2006; Mercer and Wainwright 2008).

Farmers' practices in other Latin American countries, as well as in Africa, are very similar to those in Mexico (Soleri et al. 2005). Thus seed exchange and cross pollination in and between traditional landraces and commercial varieties in farmers' fields can lead to high rates of gene flow and introgression between different maize types (Smale and de Groote 2003). There are, however, no wild relatives of maize in Africa.

State of Development of GM Technology

Field trials of GM maize have been and are being conducted in many countries in both the North and South: Argentina, Austria, Belgium, Brazil, Bulgaria, Canada, China, Colombia, the Czech Republic, Denmark, Egypt, France, Germany, Greece, Honduras, Hungary, India, Japan, Mexico, the Netherlands, New Zealand, the Philippines, Poland, Portugal, South Africa, Spain, Uruguay, and the United States (including Hawaii), amongst others.

The GM traits that have heretofore been either researched or incorporated into cultivated maize are insect resistance (*Bt*), herbicide tolerance, *Bt*-herbicide tolerance, virus resistance, better grain quality and composition, more efficient nitrogen fixation, drought tolerance, lignin modification, early flowering, and medicinal uses (gastric lipase in seeds). A teosinte-crossing-barrier trait has been patented in the United States for use in inbred, hybrid, haploid, apomictic, and/or GM maize plants (Kermicle et al. 2006a).

Agronomic Management Recommendations

There are no specific recommendations.

Crop Production Area

The total area of maize production is estimated at 158.9 million ha (FAO 2009), of which about 23% (37.3 million ha) is planted with GM maize (James 2008).

Maize is cultivated in the tropical, subtropical, and temperate regions of the world, in almost equal proportions between the tropical-subtropical areas and the temperate areas. Five mega-environments are recognized for maize production: lowland tropics, subtropics, mid-altitude tropical zones, tropical highlands, and temperate zones. Nearly 70% of the world's maize production comes from only three countries: Brazil, China, and the United States. Commercial production of GM maize is found in Argentina, Brazil, Bulgaria, Canada, Chile, Colombia, the Czech Republic, Egypt, France, Germany, Honduras, the Philippines, Poland, Portugal, Romania, Slovakia, South Africa, Spain, Uruguay, and the United States (including Hawaii) (James 2008).

Conservation Status of Maize Wild Relatives and Protected Areas

Teosintes (wild relatives of maize) occur exclusively in Mesoamerica, from N Mexico to W Nicaragua. Most of them have very restricted distribution ranges and often consist of only a few local populations (Sánchez and Ordaz 1987; Sánchez and Ruiz 1997). Some teosintes are considered to be weeds in places; most of them are rare and endangered. It is difficult, however, to precisely estimate the danger of extinction for teosintes, except in the Balsas Basin, where several populations have practically disappeared. A recent survey of S and N Guatemala teosinte populations found that they were nearing extinction (Wilkes 2007). Recent monitoring by the International Maize and Wheat Improvement Center (CIMMYT) in Mexico showed that the Huehuetenango teosinte race is almost extinct, due to intensive coffee cultivation. The Guatemala teosinte race may also be endangered, and populations of the Chalco race have been considerably reduced (S. Taba, pers. comm. 2008). Because of these threats, permanent monitoring programs and *in situ* conservation projects (with participation by farmer communities) are crucial. In the short term, there is an urgent need to collect specimens for *ex situ* conservation in Guatemala, Nicaragua, and several sites in Mexico (J.J. Sánchez-González, pers. comm. 2008). *In situ* conservation activities for teosintes are reported from Costa Rica, Guatemala, and Nicaragua. In Mexico, a reserve in Sierra de Manantlán (Jalisco) was established to protect the endemic perennial teosinte *Zea diploperennis* (Engels et al. 2006). Farmers' activities in Mexico, such as slash-and-burn practices and the creation of rotational or alternating pasture areas, have stimulated the establishment of some teosinte populations that thrive under recurrent (anthropogenic) disturbances, such as the annual Balsas-type teosinte (*Z. mays* subsp. *parviglumis*) and the endemic perennial *Z. diploperennis* (Serratos et al. 1997; Sánchez-Velasquez et al. 2001, 2002). In spite of conservation activities in Mesoamerica, work regarding and resources spent on these wild maize relatives are frequently uncoordinated and dispersed, so invaluable information and field materials have been lost (Turrent and Serratos 2004; Vibrans and Mondragón-Pichardo 2005).

In contrast, *ex situ* conservation activities have yielded important results during the past 25 years (Sánchez and Ordaz 1987; Taba 1997; Sánchez et al. 1998). Together, the Instituto Nacional de Investigaciones Forestales Agrícolas y Pecuarias (INIFAP, Mexico), the International Maize and Wheat Improvement Center (CIMMYT, Mexico), the North Central Regional Plant Introduction Station in the U.S. Department of Agriculture's Agricultural Research Service (USDA-ARS, United States), and the Universidad de Guadalajara (Mexico) conserve about 800 teosinte accessions in genebanks. Unfortunately, only partial data from these accessions are available for users, and only CIMMYT and USDA-ARS have long-term storage facilities. Also, most populations are represented by small quantities of seeds, mostly for genetic and morphological studies. For long-term conservation, larger sample sizes are necessary for many of these rare and endangered populations.

Research Gaps and Conclusions

A comprehensive assessment of the likelihood and ecological impact of gene flow and introgression between maize and its wild relatives reveals several research gaps.

Biologically, a better understanding of the population dynamics of sexually compatible teosintes (wild maize relatives) is necessary. To date, only partial genetic analyses have been carried out to evaluate and monitor temporal changes in the genetic composition and demographic history of teosinte populations. To reduce the risk of extinction, it is crucial to detect severe reductions in size (bottlenecks) for these populations.

Also, questions remain regarding the process of hybridization between maize and its teosinte wild relatives, particularly with respect to outcrossing frequencies, genetic barriers, hybrid vigor and fertility, and rates of introgression. Direct (cross-pollination frequency) and indirect (FST parameter) measurements of gene flow between maize and teosinte in different environments in Mesoamerica could provide answers. Special attention should be given to assessing the frequencies of known and potential incompatibility genes in different teosinte and landrace populations. Furthermore, the fate of outcrossing events in subsequent generations of hybrids needs to be studied, in order to determine the extent of gene introgression into teosintes.

In addition to these genetic factors, biological, environmental, and physical ones influence gene flow between maize and its wild relatives. For example, atmospheric water affects pollen longevity and thus has a significant important impact on the pollination process, for which notable differences exist between and within species and races of *Zea*. Flowering synchrony *in situ* and *ex situ*, as well as thermal requirements, are also aspects that require further study.

Based on existing information, it is difficult to predict the consequences of introducing GM maize varieties with respect to permanent introgression into the crop's teosinte relatives. The possibility of hybridization between maize and most of its wild relatives ranges from moderate to high. Furthermore, the potential for gene flow via seed, grain, and pollen between cultivated maize types is also high, due to the characteristics of traditional farming practices and the dynamics of the growers' informal seed system.

Therefore, it appears that gene flow and introgression of maize alleles to wild teosintes through outcrossing and hybridization is very likely. In countries where no maize wild relatives occur naturally, isolation distances of at least 200 m, or adequate pollen barriers, should minimize pollen flow between maize fields.

The likelihood of gene flow and introgression from domesticated maize to its wild relatives exists only in Mesoamerica (N Mexico to W Nicaragua), where wild teosintes occur. The likelihood of gene introgression from domesticated maize to its wild relatives—assuming physical proximity (less than 200 m) and flowering overlap, and being conditional on the nature of the respective trait(s)—is as follows (see map 12.1):

- *high* for the annual teosintes *Zea mays* subsp. *huehuetenangensis* (race Huehuetenango), *Z. mays* subsp. *mexicana* (races Nobogame and Durango), and *Z. mays* subsp. *parviglumis* (races Balsas and Jalisco), and for the perennial teosinte *Z. diploperennis* (diploperennial teosinte)

- *moderate* for the annual teosinte *Z. mays* subsp. *mexicana* (races Central Plateau and Chalco; in map 12.1 this area is hidden by the sections in red, corresponding to high likelihood)

- *low* for the perennial teosinte *Z. perennis* (perennial teosinte; in map 12.1 this area is hidden by the sections in red, corresponding to high likelihood)

- there may also be a *moderate* to *high* likelihood of gene flow to the perennial teosintes *Z. luxurians* (Guatemala or Florida teosinte) and *Z. nicaraguensis* (Nicaraguan teosinte), but further studies are needed to either confirm or refute this (in map 12.1, both species are included in the high-likelihood category until more information is available).

REFERENCES

ACRE [Advisory Committee on Releases to the Environment] (2001) *Cross-Pollination in Relation to Farm-Scale Evaluations of Genetically Modified Maize in Wales.* www.defra.gov.uk/environment/acre/advice/advice12.htm.

Aylor DE (2002) Settling speed of corn (*Zea mays*) pollen. *J. Aerosol. Sci.* 33: 1601–1607.

——— (2004) Survival of maize (*Zea mays*) pollen exposed in the atmosphere. *Agric. Forest Meteorol.* 123: 125–133.

Aylor DE, Schultes NP, and Shields EJ (2003) An aerobiological framework for assessing cross-pollination in maize. *Agric. Forest Meteorol.* 119: 111–129.

Baltazar BM, Sánchez JJ, de la Cruz L, and Schoper JB (2005) Pollination between maize and teosinte: An important determinant of gene flow in Mexico. *Theor. Appl. Genet.* 110: 519–526.

Bannert M and Stamp P (2007) Cross-pollination of maize at long distance. *Europ. J. Agron.* 27: 44–51

Bannert M, Vogler A, and Stamp P (2008) Short-distance cross-pollination of maize in a small-field landscape as monitored by grain color markers. *Europ. J. Agron.* 29: 29–32.

Barnabas B (1984) Effect of water loss on germination ability of maize (*Zea mays* L.) pollen. *Ann. Bot.* (Oxford) 55: 201–204.

Barré M, Berthaud J, González de León D, and Savidan YH (1994) Evidence for the tri-hybrid origin of *Tripsacum andersonii* Gray. *MNL* [Maize Newsletter] 68: 58–59.

Bateman AJ (1947) Contamination of seed crops: II. Wind pollination. *Heredity* 1: 235–246.

Beadle GW (1932) Studies of *Euchlaena* and its hybrids with *Zea*: I. Chromosome behavior in *Euchlaena mexicana* and its hybrids with *Zea mays*. *Z. Abst. Vererb.* 62: 291–304.

Bellon M and Brush SB (1994) Keepers of maize in Chiapas, Mexico. *Econ. Bot.* 48: 196–209.

Bennetzen J, Buckler E, Chandler V, Doebley J, Dorweiler J, Gaut B, Freeling M, Hake S, Kellogg E, Poethig RS, Walbot V, and Wessler S (2001) Genetic evidence and the origin of maize. *Latin Am. Antiquity* 12: 84–86.

Benz BF, Sánchez-Velásquez LR, and Santana-Michel F (1990) Ecology and ethnobotany of *Zea diploperennis*: Preliminary investigations. *Maydica* 35: 1–14.

Bergquist RR (1981) Transfer from *Tripsacum dactyloides* to corn of a major gene locus conditioning resistance to *Puccinia sorghi*. *Phytopathol.* 71: 518–520.

Bernard S and Jewell DC (1985) Crossing maize with sorghum, *Tripsacum*, and millet: The products and their level of development following pollination. *Theor. Appl. Genet.* 70: 474–483.

Berthaud J and Gepts P (2004) Assessment of effects on genetic diversity. In: CEC [Commission for Environmental Cooperation] (ed.) *Maize and Biodiversity: The Effects of Transgenic Maize in Mexico*. CEC Secretariat Article 13 Report (Communications Department of the CEC Secretariat: Québec, PQ, Canada), chapter 3. www.cec.org/maize/resources/chapters.cfm?varlan=english/.

Berthaud J and Savidan Y (1999) Genetic resources of *Tripsacum* and gene transfer to maize. In: Mujeeb-Kazi A and Sitch LA (eds.) *Review of Advances in Plant Biotechnology, 1985–1988: Second International Symposium on Genetic Manipulation in Crops, El Batan, Mexico, August 29–31, 1988* (International Maize and Wheat Improvement Center [CIMMYT]: México, DF, Mexico; International Rice Research Institute [IRRI]: Manila, Philippines), pp. 121–131.

Berthaud J, Savidan Y, Barré M, and Leblanc O (1997) *Tripsacum*. In: Fuccillo DA, Sears L, and Stapleton P (eds.) *Biodiversity in Trust: Conservation and Use of Plant Genetic Resources in CGIAR [Consultative Group on International Agricultural Research] Centres* (Cambridge University Press: Cambridge, UK), pp. 227–233.

Bird RMcK and Beckett JB (1980) Notes on *Zea luxurians* (Durieu) Bird and some requests. *MNL* [Maize Newsletter] 54: 62–63.

Blancas L, Arias D, and Ellstrand NC (2002) Patterns of genetic diversity in sympatric and allopatric populations of maize and its wild relative teosinte in Mexico: Evidence for hybridization. In: Snow AA (ed.) *Scientific Methods Workshop: Ecological and Agronomic Consequences of Gene Flow from Transgenic Crops to Wild Relatives; Meeting Proceedings, University Plaza Hotel and Conference Center, Ohio State University, Columbus, Ohio, March 5–6, 2002* (Ohio State University: Columbus, OH, USA), pp. 31–38. www.agbios.com /docroot/articles/02-280-003.pdf.

Brookes G, Barfoot P, Melé E, Messeguer J, Bénétrix F, Bloc D, Foueillassar X, Fabié A, and Poeydomenge C (2004) *Genetically Modified Maize: Pollen Movement and Crop Coexistence* (PG Economics: Dorchester, UK). 20 p. www.pgeco nomics.co.uk/pdf/Maizepollennov2004final.pdf.

Brunet Y, Foueillassar X, Audran A, Garrigou D, Dayau S, and Tardieu L (2003) Evidence for long-range transport of viable maize pollen grains. In: Boelt B (ed.) *1st European Conference on the Coexistence of Genetically Modified Crops with Conventional and Organic Crops, Helsingor, Denmark, November 13–14, 2003* (Danish Institute of Agricultural Sciences [DIAS]: Slagelse, Denmark), pp. 74–76.

Buckler ES IV and Holtsford TP (1996) *Zea* systematics: Ribosomal ITS evidence. *Mol. Biol. Evol.* 13: 612–622.

Burris JS (2001) *Adventitious Pollen Intrusion into Hybrid Maize Seed Production Fields*. White Paper for the Association of Official Seed Certifying Agencies [AOSCA], funded by USDA/FAS/ICD/RSED [U.S. Department of Agriculture, Foreign Agricultural Service, International Cooperation and Development, Research and Scientific Exchanges Division] and ASTA [American Seed Trade Association]. www.amseed.com/govt_statementsDetail.asp?id=69.

Burson BL, Voight PW, Sherman RA, and Dewald CL (1990) Apomixis and sexuality in eastern gamagrass. *Crop Sci.* 30: 86–89.

Byrne PF and Fromherz S (2003) Can GM and non-GM crops coexist? Setting a precedent in Boulder County, Colorado, USA. *Food Agric. Environ.* 1: 258–261.

Castro-Gil M (1970) Frequencies of maize by teosinte crosses in a simulation of a natural association. *Maize Genet. Coop. Newsl.* 44: 21–24.

Cleveland DA, Soleri D, Aragón Cuevas F, Crossa J, and Gepts P (2005) Detecting (trans)gene flow to landraces in centers of crop origin: Lessons from the case of maize in Mexico. *Environ. Biosafety Res.* 4: 197–208.

CSGA [Canadian Seed Growers Association] (2009) Canadian Regulations and Procedures for Pedigreed Seed Crop Production. Circular 6–2005/Rev.01.4–2009 (Canadian Seed Growers Association [CSGA]: Ottawa, Canada). www.seed growers.ca/cropcertification/circular.asp.

Della Porta G, Ederle D, Bucchini L, Prandi M, Verderio A, and Pozzi C (2008) Maize pollen mediated gene flow in the Po Valley (Italy): Source-recipient distance and effect of flowering time. *Europ. J. Agron.* 28: 255–265.

Devos Y, Reheul D, and de Schrijver A (2005) The coexistence between transgenic and non-transgenic maize in the European Union: A focus on pollen flow and cross-fertilization. *Environ. Biosafety Res.* 4: 71–87.

de Wet JMJ, Engle LM, Grant CA, and Tanaka ST (1972) Cytology of maize-*Tripsacum* introgression. *Am. J. Bot.* 59: 1026–1029.

de Wet JMJ, Harlan JR, Engle LM, and Grant CA (1973) Breeding behaviour of maize-*Tripsacum* hybrids. *Crop Sci.* 13: 254–256.

de Wet JMJ, Harlan JR, and Brink DE (1982) Systematics of *Tripsacum dactyloides* (Gramineae). *Am. J. Bot.* 69: 1251–1257.

de Wet JMJ, Brink DE, and Cohen CE (1983) Systematics of *Tripsacum* section *Fasciculata* (Gramineae). *Am. J. Bot.* 70: 1139–1146.

de Wet JMJ, Newell CA, and Brink DE (1984) Counterfeit hybrids between *Tripsacum* and *Zea* (Gramineae). *Am. J. Bot.* 71: 245–251.

Doebley JF (1983) The taxonomy and evolution of *Tripsacum* and teosinte, the closest relatives of maize. In: Gordon DT, Knoke JK, and Nault LR (eds.) *Proceedings of the International Maize Virus Disease Colloquium and Workshop, Wooster, Ohio, August 2–6, 1982* (Ohio Agricultural Research and Development Center, Ohio State University: Wooster, OH, USA), pp. 15–28.

——— (1984) Maize introgression into teosinte—a reappraisal. *Ann. Missouri Bot. Gard.* 71: 1100–1113.

——— (1990a) Molecular evidence and the evolution of maize. *Econ. Bot.* 44: 6–27.

——— (1990b) Molecular evidence for gene flow among *Zea* species—genes transformed into maize through genetic engineering could be transferred to its wild relatives, the teosintes. *BioScience* 40: 443–448.

——— (2004) The genetics of maize evolution. *Annu. Rev. Genet.* 38: 37–59.

Doebley JF and Iltis HH (1980) Taxonomy of *Zea* (Gramineae): I. A subgeneric classification with key to taxa. *Am. J. Bot.* 67: 982–993.

Doebley JF and Stec A (1993) Inheritance of the morphological differences between maize and teosinte: Comparison of results for two F2 populations. *Genetics* 134: 559–570.

Doebley JF, Goodman MM, and Stuber CW (1984) Isoenzymatic variation in *Zea* (Gramineae). *Syst. Bot.* 9: 203–218.

——— (1987) Patterns of isozyme variation between maize and Mexican annual teosinte. *Econ. Bot.* 41: 234–246.

Doebley JF, Stec A, Wendel J, and Edwards M (1990) Genetic and morphological analysis of a maize-teosinte F2 population: Implications for the origin of maize. *Proc. Natl. Acad. Sci. USA* 87: 9888–9892.

Doebley JF, Gaut BS, and Smith BD (2006) The molecular genetics of crop domestication. *Cell* 127: 1309–1321.

Dumas C and Mogensen HL (1993) Gametes and fertilization: Maize as a model system for experimental embryogenesis in flowering plants. *Plant Cell* 5: 1337–1348.

Eastham K and Sweet J (2002) *Genetically Modified Organisms (GMOs): The Significance of Gene Flow through Pollen Transfer.* Environmental Issue Report No. 28 (European Environment Agency [EEA]: Copenhagen, Denmark). 75 p. http://reports.eea.europa.eu/environmental_issue_report_2002_28/en/GMOs%20for%20www.pdf.

EC [European Commission] (2001) European Communities (Cereal Seed) Regulations, 2001: S.I. [Statutory Instrument] No. 640/2001, Amending the Council Directive No. 66/402/EEC of 14 June 1966 on the Marketing of Cereal Seed. OJ No. L125 2309–66, 11 July 1966. www.irishstatutebook.ie/2001/en/si/0640.html.

Ellstrand NC, Garner LC, Hegde S, Guadagnuolo R, and Blancas L (2007) Spontaneous hybridization between maize and teosinte. *J. Hered.* 98: 183–187.

Engels JMM, Ebert AW, Thormann I, and de Vicente MC (2006) Centres of crop diversity and/or origin, genetically modified crops, and implications for plant genetic resources conservation. *Genet. Resour. Crop Evol.* 53: 1675–1688.

Eubanks MW (1994) Method and Materials for Conferring *Tripsacum* Genes in Maize. US Patent 5330547, filed September 14, 1992, and issued July 19, 1994. www.patentstorm.us/patents/5330547-description.html.

——— (1995) A cross between two maize relatives: *Tripsacum dactyloides* and *Zea diploperennis* (Poaceae). *Econ. Bot.* 49: 172–182.

——— (2001) An interdisciplinary perspective on the origin of maize. *Latin Am. Antiquity* 12: 91–98.

Evans MMS and Kermicle JL (2001) Teosinte crossing barrier1, a locus governing hybridization of teosinte with maize. *Theor. Appl. Genet.* 103: 259–265.

FAO [Food and Agriculture Organization of the United Nations] (2009) FAOSTAT data. http://faostat.fao.org/site/567/default.aspx.

Farquharson LI (1957) Hybridization of *Tripsacum* and *Zea*. *J. Hered.* 48: 295–299.

Feil B, Weingartner U, and Stamp P (2003) Controlling the release of pollen from genetically modified maize and increasing its grain yield by growing mixtures of male-sterile and male-fertile plants. *Euphytica* 130: 163–165.

Fukunaga K, Hill J, Vigouroux Y, Matsuoka Y, Sánchez JJ, Liu K, Buckler ES, and Doebley JF (2005) Genetic diversity and population structure of teosinte. *Genetics* 169: 2241–2254.

Furini A and Jewell C (1995) Somatic embryogenesis and plant regeneration of maize/*Tripsacum* hybrids. *Maydica* 40: 205–210.

Galinat WC (1988) The origin of corn. In: Sprague GF and Dudley JW (eds.) *Corn and Corn Improvement*. 3rd ed. (American Society of Agronomy, Crop Science Society of America, and Soil Science Society of America: Madison, WI, USA), pp. 1–31.

Galinat WC and Rao BGS (1969) Gene exchange between corn and *Tripsacum*. *MNL* [Maize Newsletter] 43: 94.

García M, Figueroa J, Gómez R, Townsend R, and Schoper J (1998) Pollen control during transgenic hybrid maize development in Mexico. *Crop Sci.* 38: 1597–1602.

GISD [Global Invasive Species Database] (2007) Online database. www.issg.org/database.

Goodman MM (1995) Maize *Zea mays* (Gramineae-Maydeae). In: Smartt J and Simmonds NW (eds.) *Evolution of Crop Plants*. 2nd ed. (Longman Scientific and Technical: Harlow, UK), pp. 192–202.

Goss JA (1968) Development, physiology, and biochemistry of corn and wheat pollen. *Bot. Rev.* 34: 333–358.

Gould FW (1968) *Grass Systematics* (McGraw-Hill: New York, NY, USA). 396 p.

Grimanelli D, Leblanc O, Espinosa E, Perotti E, González de León D, and Savidan Y (1998) Non-Mendelian transmission of apomixis in maize-*Tripsacum* hybrids caused by a transmission ratio distortion. *Heredity* 80: 40–47.

Guadagnuolo R, Clegg J, and Ellstrand NC (2006) Relative fitness of transgenic vs. non-transgenic maize × teosinte hybrids, a field evaluation. *Ecol. Appl.* 16: 1967–1974.

Halsey ME, Remund KM, Davis CA, Qualls M, Eppard PJ, and Berberich SA (2005) Isolation of maize from pollen-mediated gene flow by time and distance. *Crop Sci.* 45: 2172–2185.

Henry C, Morgan D, Weekes R, Daniels R, and Boffey C (2003) *Farm-Scale Evaluations of GM Crops: Monitoring Gene Flow from GM Crops to Non-GM Equivalent Crops in the Vicinity (contract reference EPG 1/5/138); Part I, Forage Maize* (Department for Environment, Food, and Rural Affairs [DEFRA]: London, UK). 25 p. www.defra.gov.uk/Environment/gm/research/pdf/epg_1-5-138.pdf.

Hutchcroft CD (1958) Contamination in seed fields of corn resulting from incomplete detasseling. *Agron. J.* 50: 267–271.

Iltis HH and Doebley JF (1980) Taxonomy of *Zea* (Gramineae): II. Subspecific categories in the *Zea mays* complex and a generic synopsis. *Am. J. Bot.* 67: 994–1004.

Iltis HH, Doebley JF, Guzmán R, and Pazy B (1979) *Zea diploperennis* (Gramineae): A new teosinte from Mexico. *Science* 303: 186–188.

Ingram J (2000) *Report on the Separation Distances Required to Ensure Cross-Pollination Is Below Specified Limits in Non-Seed Crops of Sugar Beet, Maize, and Oilseed Rape.* Report prepared for the Ministry of Agriculture, Fisheries, and Food [MAFF], UK, Project No. RG0123 (Ministry of Agriculture, Fisheries and Food [MAFF]: London, UK). 14 p. www.defra.gov.uk/science/project_data/DocumentLibrary/RG0123/RG0123_2916_FRP.doc.

Ireland DS, Wilson DO, Westgate ME Jr, Burris JS, and Lauer MJ (2006) Managing reproductive isolation in hybrid seed corn production. *Crop Sci.* 46: 1445–1455.

James C (2008) *Global Status of Commercialized Biotech/GM Crops 2008.* ISAAA Brief 39 (International Service for the Acquisition of Agri-biotech Applications [ISAAA]: Ithaca, NY, USA). 243 p.

Jarosz N, Loubet B, Durand B, McCartney HA, Foueillassar X, and Huber L (2003) Field measurements of airborne concentration and deposition rate of maize pollen (*Zea mays* L.) downwind of an experimental field plot. *Agric. Forest Meteor.* 119: 37–51.

Jarosz N, Loubet B, Durand B, Foueillassar X, and Huber L (2005) Variations in maize pollen emission and deposition in relation to microclimate. *Environ. Sci. Technol.* 39: 4377–4384.

Jatimliansky JR, García MD, and Molina MC (2004) Response to chilling of *Zea mays, Tripsacum dactyloides,* and their hybrid. *Biol. Plant.* 48: 561–567.

Jemison JM Jr and Vayda ME (2001) Cross pollination from genetically engineered corn: Wind transport and seed source. *AgBioForum* 4: 87–92.

Jones MD and Brooks JS (1950) *Effectiveness of Distance and Border Rows in Preventing Outcrossing in Corn.* Oklahoma Agricultural Experiment Station Technical Bulletin No. T-38 (Oklahoma Agricultural Experiment Station: Stillwater, OK, USA). 18 p.

—— (1952) *Effect of Tree Barriers on Outcrossing in Corn.* Oklahoma Agricultural Experiment Station Technical Bulletin No. T-45 (Oklahoma Agricultural Experiment Station: Stillwater, OK, USA). 11 p.

Jones MD and Newell LC (1948) Longevity of pollen and stigmas of grasses: Buffalograss, *Buchloe dactyloides* (Nutt.) Engelm., and corn, *Zea mays* L. *J. Am. Soc. Agron.* 40: 195–204.

Kato TA (1997) Review of introgression between maize and teosinte. In: Serratos JA, Willcox MC, and Castillo F (eds.) *Proceedings of a Forum: Gene Flow among Maize Landraces, Improved Maize Varieties, and Teosinte; Implications for Transgenic Maize* (International Maize and Wheat Improvement Center [CIMMYT]: México, DF, Mexico), pp. 44–53.

Kato TA and Sánchez JJ (2002) Introgression of chromosome knobs from *Zea diplo-perennis* into maize. *Maydica* 47: 33–50.

Kermicle JL (1997) Cross-compatibility within the genus *Zea*. In: Serratos JA, Will-cox MC, and Castillo F (eds.) *Proceedings of a Forum: Gene Flow among Maize Landraces, Improved Maize Varieties, and Teosinte; Implications for Transgenic Maize* (International Maize and Wheat Improvement Center [CIMMYT]: México, DF, Mexico), pp. 40–43.

——— (2006) A selfish gene governing pollen-pistil compatibility confers reproductive isolation between maize relatives. *Genetics* 172: 499–506.

Kermicle JL and Allen JO (1990) Cross-incompatibility between maize and teosinte. *Maydica* 35: 399–408.

Kermicle JL and Evans MMS (2005) Pollen-pistil barriers to crossing in maize and teosinte result from incongruity rather than active rejection. *Sex. Plant Reprod.* 18: 187–194.

Kermicle JL, Evans MMS, and Gerrish SR (2006a) Cross-Incompatibility Traits from Teosinte and Their Use in Corn. US Patent 7074984, filed March 20, 2001, and issued July 11, 2006. www.patentstorm.us/patents/7074984–fulltext.html.

Kermicle JL, Taba S, and Evans MMS (2006b) The Gametophyte-1 locus and reproductive isolation among *Zea mays* subspecies. *Maydica* 51: 219–226.

Kiesselbach TA (1999) *The Structure and Reproduction of Corn* (Cold Spring Harbor Laboratory Press: Cold Spring Harbor, NY, USA). 102 p.

Kindiger BK and Sokolov V (1998) Apomictic Maize. US Patent 5710367, filed September 22, 1995, and issued January 20, 1998. www.patentstorm.us/patents/5710367.html.

Klein EK, Lavigne C, Foueillassar X, Gouyon PH, and Laredo C (2003) Corn pollen dispersal: Quasi-mechanistic models and field experiments. *Ecol. Monogr.* 73: 131–150.

Leblanc O, Grimanelli D, González de León D, and Savidan Y (1995a) Detection of the apomictic mode of reproduction in maize-*Tripsacum* hybrids using maize RFLP markers. *Theor. Appl. Genet.* 90: 1198–1203.

Leblanc O, Peel MD, Carman JG, and Savidan Y (1995b) Megasporogenesis and megagametogenesis in several *Tripsacum* species (Poaceae). *Am. J. Bot.* 82: 57–63.

Lipsius K, Wilhelm R, Richter O, Schmalstieg KJ, and Schiemann J (2006) Meteorological input data requirements to predict cross-pollination of GMO maize with Lagrangian approaches. *Environ. Biosafety Res.* 5: 151–168.

Loos C, Seppelt R, Meier-Bethke S, Schiemann J, and Richter O (2003) Spatially explicit modelling of transgenic maize pollen dispersal and cross-pollination. *J. Theor. Biol.* 225: 241–255.

Lorente-Adame RG and Sánchez-Velasquez LR (1996) Dinámica estacional del banco de frutos del teocintle *Zea diploperennis* (Gramineae). *Biotropica* 28: 267–272.

Louette D (1997) Seed exchange among farmers and gene flow among maize varieties in traditional agricultural systems. In: Serratos JA, Willcox MC, and Castillo F (eds.) *Proceedings of a Forum: Gene Flow among Maize Landraces, Improved Maize Varieties, and Teosinte; Implications for Transgenic Maize* (International Maize and Wheat Improvement Center [CIMMYT]: México, DF, Mexico), pp. 56–66.

Louette D, Charrier A, and Berthaud J (1997) *In situ* conservation of maize in Mexico: Genetic diversity and maize seed management in a traditional community. *Econ. Bot.* 51: 20–38.

Luna S, Figueroa VJ, Baltazar MB, Gomez MR, Townsend LR, and Schoper JB (2001) Maize pollen longevity and distance isolation requirements for effective pollen control. *Crop Sci.* 41: 1551–1557.

Ma B, Subedi K, and Reid L (2004) Extent of cross-fertilization in maize by pollen from neighboring transgenic hybrids. *Crop Sci.* 44: 1273–1282.

MacNeish RS and Eubanks MW (2000) Comparative analysis of the Río Balsas and Tehuacán models for the origin of maize. *Latin Am. Antiquity* 11: 3–20.

Mangelsdorf PC (1974) *Corn: Its Origin, Evolution, and Improvement* (Harvard University Press: Cambridge, MA, USA). 288 p.

Mangelsdorf PC and Reeves RG (1931) Hybridization of maize, *Tripsacum*, and *Euchlaena*. *J. Hered.* 22: 321–343.

Matsuo K, Amano K, Shibaike H, Yoshimura Y, Kawashima S, Uesugi S, Misawa T, Miura Y, Ban Y, and Oka M (2004) Pollen dispersal and outcrossing in *Zea mays* populations: A simple identification of hybrids detected by xenia using conventional corn in simulation of transgene dispersion of GM corn. In: International Society for Biosafety Research (ed.) *Proceedings of the 8th International Symposium on Biosafety of Genetically Modified Organisms (ISB-GMO), Montpellier, France, September 26–30, 2004* (International Society for Biosafety Research [ISBR]: n.p.), p. 281. www.isbr.info/symposia/.

Matsuoka Y (2005) Origin matters: Lessons from the search for the wild ancestor of maize. *Breed. Sci.* 55: 383–390.

Matsuoka Y, Vigouroux Y, Goodman MM, Sánchez JJ, Buckler E, and Doebley J (2002) A single domestication for maize shown by multilocus microsatellite genotyping. *Proc. Natl. Acad. Sci. USA* 99: 6080–6084.

Mercer KL and Wainwright JD (2008) Gene flow from transgenic maize to landraces in Mexico: An analysis. *Agric. Ecosys. Environ.* 123: 109–115.

Messeguer J, Ballester J, Peñas G, Olivar J, Alcalde E, and Melé E (2003) Evaluation of gene flow in a commercial field of maize. In: Boelt B (ed.) *1st European Con-*

ference on the Coexistence of Genetically Modified Crops with Conventional and Organic Crops, Helsingor, Denmark, November 13–14, 2003 (Danish Institute of Agricultural Sciences [DIAS]: Slagelse, Denmark), p. 220.

Messeguer J, Peñas G, Ballester J, Bas M, Serra J, Salvia J, Palaudelmàs M, and Melé E (2006) Pollen-mediated gene flow in maize in real situations of coexistence. Plant Biotech. J. 4: 633–645.

Molina MC and García MD (1997) Cytogenetic study of diploid and triploid F1 hybrids of Zea mays subsp. mays × Zea parviglumis. MNL [Maize Newsletter] 71: 54–55.

—— (1999a) Cytogenetic studies in hybrids between Zea mays and Tripsacum dactyloides. MNL [Maize Newsletter] 73: 61–62.

—— (1999b) Influence of ploidy levels on phenotypic and cytogenetic traits in maize and Zea perennis hybrids. Cytologia 64: 101–109.

Molina MC, García MD, and Chorzempa SE (2006) Meiotic study of Zea mays subsp. mays (2n = 40) × Tripsacum dactyloides (2n = 72) hybrid and its progeny. Electron. J. Biotech. 9(3), doi:10.2225/vol9–issue3–fulltext-25. www.ejbiotechnology.info/content/vol9/issue3/full/25/index.html.

Narayanaswarmy S, Jagadish GV, and Ujinnaiah US (1997) Determination of isolation distance for hybrid maize seed production. Curr. Res. 26: 193–195.

OECD [Organization for Economic Cooperation and Development] (2003) Consensus Document on the Biology of Zea mays subsp. mays (maize). Consensus Documents for the Work on Harmonisation of Regulatory Oversight in Biotechnology No. 27, ENV/JM/MONO(2003)11 (Organization for Economic Cooperation and Development [OECD]: Paris, France).

OECD [Organization for Economic Cooperation and Development] (2008) OECD Scheme for the Varietal Certification of Maize and Sorghum Seed Moving in International Trade: Annex XI to the Decision. C(2000)146/FINAL. OECD Seed Schemes, February 2008. www.oecd.org/document/0/0,3343,en_2649_3 3905_1933504_1_1_1_1,00.html.

Ogden EC, Hayes JV, and Raynor GS (1969) Diurnal patterns of pollen emission in Ambrosia, Phleum, Zea, and Ricinus. Am. J. Bot. 56: 16–21.

Ortega A (1987) Insect Pests of Maize: A Guide for Field Identification (International Maize and Wheat Improvement Center [CIMMYT]: México, DF, Mexico). 106 p.

Paterniani E and Stort AC (1974) Effective maize pollen dispersal in the field. Euphytica 23: 129–134.

Perales H, Brush SB, and Qualset CO (2003a) Dynamic management of maize landraces in central Mexico. Econ. Bot. 57: 21–34.

—— (2003b) Maize landraces of central Mexico: An altitudinal transect. Econ. Bot. 57: 7–20.

Pla M, La Paz JL, Peñas G, García N, Palaudelmàs M, Esteve T, Messeguer J, and Melé E (2006) Assessment of real-time PCR-based methods for quantification of pollen-mediated gene flow from GM to conventional maize in a field study. *Transgen. Res.* 15: 219–228.

Poggio L, Confalonieri V, Comas C, González G, and Naranjo CA (1999) Genomic affinities of *Zea luxurians*, *Z. diploperennis*, and *Z. perennis*: Meiotic behavior of their F1 hybrids and genomic *in situ* hybridization (GISH). *Genome* 42: 993–1000.

Randolph LF (1959) The origin of maize. *Indian J. Genet. Plant Breed.* 19: 1–12.

Raven PH (2005) Transgenes in Mexican maize: Desirability or inevitability? *Proc. Natl. Acad. Sci. USA* 102: 13003–13004.

Ray JD, Kindiger B, and Sinclair TR (1999) Introgressing root aerenchyma into maize. *Maydica* 44: 113–117.

Raynor GS, Ogden EC, and Hayes JV (1972) Dispersion and deposition of corn pollen from experimental sources. *Agron. J.* 64: 420–427.

Rodríguez JG, Sánchez JJ, Baltazar B, de la Cruz L, Santacruz-Ruvalcaba F, Ron J, and Schoper JB (2006) Characterization of floral morphology and synchrony among *Zea* species in Mexico. *Maydica* 51: 383–398.

Russell WA and Hallauer AR (1980) Corn. In: Fehr WR and Hadley HH (eds.) *Hybridization of Crop Plants* (American Society of Agronomy: Madison, WI, USA), pp. 299–312.

Salamov AB (1940) About isolation in corn. *Sel. i. Sem.* 3: 25–27 [English translation by AGROSCOPE, Switzerland].

Sánchez JJ and Ordaz L (1987) *El Teocintle en Mexico: Distribución y Situación Actual de las Poblaciones.* Systematic and Ecogeographic Studies on Crop Genepools 2 (International Board for Plant Genetic Resources [IBPGR]: Rome, Italy). 50 p. www.bioversityinternational.org/publications/Web_version/266/begin.htm#Contents.

Sánchez JJ and Ruiz JA (1997) Teosinte distribution in Mexico. In: Serratos JA, Willcox MC, and Castillo F (eds.) *Proceedings of a Forum: Gene Flow among Maize Landraces, Improved Maize Varieties, and Teosinte; Implications for Transgenic Maize* (International Maize and Wheat Improvement Center [CIMMYT]: México, DF, Mexico), pp. 18–35.

Sánchez JJ, Kato TA, Aguilar-Sanmiguel M, Hernández JM, López A, and Ruiz JA (1998) *Distribución y Caracterización del Teocintle.* Libro Técnico No. 2 (Instituto Nacional de Investigaciones Forestales Agrícolas y Pecuarias [INIFAP]: Guadalajara, Jalisco, Mexico). 165 p.

Sánchez JJ, Stuber CW, and Goodman MM (2000a) Isozymatic diversity of the races of maize of the Americas. *Maydica* 45: 185–203.

Sánchez JJ, Goodman MM, and Stuber CW (2000b) Isozymatic and morphological diversity in the races of maize of Mexico. *Econ. Bot.* 54: 43–59.

Sánchez JJ, Goodman MM, Bird RMcK, and Stuber CW (2006) Isozyme and morphological variation in maize of five Andean countries. *Maydica* 51: 25–42.

Sánchez JJ, Goodman MM, and Stuber CW (2007) Racial diversity of maize in Brazil and adjacent areas. *Maydica* 52: 13–30.

Sánchez-Velasquez LR, Jimenez G, Genoveva R, and Benz BF (2001) Population structure and ecology of a tropical rare rhizomatous species of teosinte *Zea diploperennis* (Gramineae). *Rev. Biol. Trop.* 49: 249–258.

Sánchez-Velasquez LR, Ezcurra E, Martinez-Ramos M, Alvarez-Buylla E, and Lorente R (2002) Population dynamics of *Zea diploperennis*, an endangered perennial herb: Effect of slash and burn practice. *J. Ecol.* 90: 684–692.

Sanvido O, Widmer F, Winzeler M, Streit B, Szerencsits E, and Bigler F (2008) Definition and feasibility of isolation distances for transgenic maize cultivation. *Transgen. Res.* 17: 317–335.

Savidan Y, Grimanelli D, Perotti E, and Leblanc O (2007) Means for Identifying Nucleotide Sequences Involved in Apomixis. US Patent Application 20080155712, filed September 7, 2007. www.patentstorm.us/applications/20080155712/fulltext.html.

Schoper JB, Lambert RJ, and Vasilas BL (1987) Pollen viability, pollen shedding, and combining ability for tassel heat tolerance in maize. *Crop Sci.* 27: 27–31.

SCIMAC [Supply Chain Initiative on Modified Agricultural Crops] (1999) Guidelines for Growing Newly Developed Herbicide Tolerant Crops. www.scimac.org.uk/files/onfarm_guidelines.pdf.

Sears MK and Stanley-Horn D (2000) Impact of *Bt* corn pollen on monarch butterfly populations. In: Fairbairn C, Scoles G, and McHughen A (eds.) *Proceedings of the 6th International Symposium on the Biosafety of Genetically Modified Organisms (ISBGMO), Saskatoon, Saskatchewan, Canada, July 8–13, 2000* (University Extension Press, University of Saskatchewan: Saskatoon, SK, Canada), pp. 120–130. www.isbr.info/symposia/.

Serratos JA, Willcox MC, and Castillo-González F (eds.) (1997) *Proceedings of a Forum: Gene Flow among Maize Landraces, Improved Maize Varieties, and Teosinte; Implications for Transgenic Maize* (International Maize and Wheat Improvement Center [CIMMYT]: México, DF, Mexico). www.cimmyt.org/abc/geneflow/geneflow_pdf_engl/contents.htm.

Serratos JA, Islas F, Buendía E, and Berthaud J (2004) Gene flow scenarios with transgenic maize in Mexico. *Environ. Biosafety Res.* 3: 149–157.

Sleper D and Poehlman JM (2006) *Breeding Field Crops*. 5th ed. (Blackwell: Oxford, UK). 448 p.

Smale M and de Groote H (2003) Diagnostic research to enable adoption of transgenic crop varieties by smallholder farmers in sub-Saharan Africa. *African J. Biotech.* 2: 586–595.

Smith JSC, Goodman MM, and Struber CW (1985) Relationships between maize and teosinte of Mexico and Guatemala: Numerical data analysis of allozyme data. *Econ. Bot.* 39: 12–24.

Sokolov VA, Dewald CL, and Khatypova IV (2000) The genetic programs of nonreduction and parthenogenesis in corn-gamagrass hybrids are inherited and expressed in an independent manner. *MNL* [Maize Newsletter] 74: 55–57.

Soleri D, Cleveland DA, Aragón F, Fuentes MR, Ríos H, and Sweeney S (2005) Understanding the potential impact of transgenic crops in traditional agriculture: Maize farmers' perspectives in Cuba, Guatemala, and Mexico. *Environ. Biosafety Res.* 4: 141–166.

Soleri D, Cleveland DA, and Aragón F (2006) Transgenic crops and crop varietal diversity: The case of maize in Mexico. *BioScience* 56: 503–514.

Taba S (1997) Teosinte. In: Fuccillo DA, Sears L, and Stapleton P (eds.) *Biodiversity in Trust: Conservation and Use of Plant Genetic Resources in CGIAR [Consultative Group on International Agricultural Research] Centres* (Cambridge University Press: Cambridge, UK), pp. 234–242.

Talbert LE, Doebley JF, Larson S, and Chandler VL (1990) *Tripsacum andersonii* is a natural hybrid involving *Zea* and *Tripsacum*: Molecular evidence. *Am. J. Bot.* 77: 722–726.

Tang Q, Rong T, Song Y, Yang J, Pan G, Li W, Huang Y, and Cao M (2005) Introgression of perennial teosinte genome into maize and identification of genomic *in situ* hybridization and microsatellite markers. *Crop Sci.* 45: 717–721.

Ting YC (1963) A preliminary report on the 4th chromosome male gametophyte factor in teosintes. *Maize Genet. Coop. Newsl.* 37: 6–7.

Tolstrup K, Andersen SB, Boelt B, Buus M, Gylling M, Holm PB, Kjellsson G, Pedersen S, Ostergard H, and Mikkelsen SA (2003) *Report from the Danish Working Group on the Coexistence of Genetically Modified Crops with Conventional and Organic Crops.* Danish Institute for Agricultural Sciences [DIAS] Report, Plant Production, No. 94 (Danish Research Centre for Organic Farming [DARCOF]: Tjele, Denmark). 275 p.

Traore SB, Carlson RE, Pilcher CD, and Rice M (2000) *Bt* and non-*Bt* maize growth and development as affected by temperature and drought. *Agron. J.* 92: 1027–1035.

Turrent A and Serratos JA (2004) Context and background on maize and its wild relatives in Mexico. In: CEC [Commission for Environmental Cooperation] (ed.) *Maize and Biodiversity: The Effects of Transgenic Maize in Mexico.* CEC Secretariat Article 13 Report (Communications Department of the CEC Secretariat: Québec, PQ, Canada), chapter 1. www.cec.org/maize/resources/chapters.cfm?varlan=english/.

USDA [United States Department of Agriculture] (2008) 7 CFR [Code of Federal

Regulations] § 201.76. Federal Seed Act Regulations: Minimum Land, Isolation, Field, and Seed Standards. Source: 59 FR [Federal Register] 64516, Dec. 14, 1994, as amended at 65 FR 1710, Jan. 11, 2000 (Government Printing Office: Washington, DC, USA), pp. 372–378. www.access.gpo.gov/nara/cfr/waisidx_08/7cfr201_08.html.

Vibrans H and Mondragón-Pichardo J (2005) Ethnobotany of the balsas teosinte (*Zea mays* subsp. *parviglumis*). *Maydica* 50: 123–128.

Weber WE, Bringezu T, Broer I, Eder J, and Holz F (2007) Coexistence between GM and non-GM maize crops—tested in 2004 at the field-scale level (Erprobungsanbau 2004). *J. Agron. Crop Sci.* 193: 79–92.

Westgate ME, Lizaso J, and Batchelor W (2003) Quantitative relationship between pollen-shed density and grain yield in maize. *Crop Sci.* 43: 934–942.

Wilkes HG (1967) Teosinte: The closest relative of maize. PhD dissertation, Harvard University. 159 p.

——— (1977) Hybridization of maize and teosinte in Mexico and Guatemala and the improvement of maize. *Econ. Bot.* 31: 254–293.

——— (1985) Teosinte: The closest relative of maize revisited. *Maydica* 30: 209–223.

——— (2004) Corn, strange and marvelous: But is a definite origin known? In: Smith CW (ed.) *Corn: Origin, History, Technology, and Production* (John Wiley and Sons: New York, NY, USA), pp. 3–63.

——— (2007) Urgent notice to all maize researchers: Disappearance and extinction of the last wild teosinte population is more than half completed; A modest proposal for teosinte evolution and conservation *in situ*; the Balsas, Guerrero, Mexico. *Maydica* 52: 49–58.

Zuloaga FO, Morrone O, Davidse G, Filgueiras TS, Peterson PM, Soreng RJ, and Judziewicz E (2003) *Catalogue of New World Grasses (Poaceae): III. Subfamilies Panicoideae, Aristidoideae, Arundinoideae, and Danthonioideae* (Smithsonian Institution: Washington, DC, USA). 662 p.

THIRTEEN

Oat
(*Avena sativa* L.)

In addition to the hexaploid common oat (*A. sativa* L.), there are other domesticated species within the genus *Avena* L., including hexaploid (*A. byzantina* K. Koch), tetraploid (*A. abyssinica* Hochst.), and diploid (*A. strigosa* Schreb.) forms. Most of them are seldom cultivated, but others are becoming popular, such as *A. strigosa* for organic agriculture in Europe (I. Loskutov, pers. comm. 2008). However, this chapter will focus on the most widely cultivated oat, hexaploid *A. sativa*.

Center(s) of Origin and Diversity

Like wheat and barley, oats originated in the region of the Fertile Crescent in the Near East (SW Asia) (Coffman 1946; Vavilov 1951, 1992). Oats were probably scattered as weed components, primarily of emmer wheat, and they eventually became established as a crop (Thomas 1995; Zhou et al. 1999). A secondary center of origin can be found in Mongolia and NW China for naked (hull-less) forms of hexaploid oat (*A. sativa* subsp. *nudisativa* (Husn.) Rod. & Sold.) (Loskutov 2007).

The region of greatest diversity for the genus *Avena* is the Mediterranean basin, reaching from Europe to the Near East and N Africa, with Morocco being a particular hot spot for diversity (Zeven and de Wet 1982; Baum and Fedak 1985).

Biological Information

Flowering

The inflorescence of the domesticated oat plant is a loose open panicle where spikelets are differentiated from top to bottom (Bonnett 1966). In covered oats, each spikelet usually contains two fertile florets, whereas naked oats have three to five fertile florets. The bisexual florets are enclosed by two bracts, which are also known as the hulls on harvested oat grain (Brown 1980). Stigmas are usually receptive one day before anthesis, and they remain that way for as long as three to five days. Anthers usually shed their pollen just before or at the time the florets open (Brown 1980). The extent to which the anthers are retained inside of or exerted from the spikelets at anthesis is strongly influenced by climatic conditions (see, e.g., Coffman 1937; Brown and Shands 1956). Also, self-fertilization rates are higher in closed-flowering cultivars than in open-flowering types.

Pollen Dispersal and Longevity

Oat pollen, like that of most grasses, is dispersed by the wind. Its longevity is about 10 to 15 minutes and is strongly influenced by climatic conditions (I. Loskutov, pers. comm. 2008).

Sexual Reproduction

Domesticated oats are an allohexaploid ($2n = 6x = 42$, **AACCDD**) annual grass species with diploid-like chromosome pairing. They are self-pollinating, with a limited degree of outcrossing (0.5%–1%, maximum of 10%; Jensen 1961). As a rule, wild oats (*Avena fatua* L.) have outcrossing rates that are usually less than 1% but can reach up to 12% in pure stands (Imam and Allard 1965; Murray et al. 2002).

Vegetative Reproduction

Domesticated and wild oats are not capable of vegetative regeneration, except for their only perennial wild relative, *A. macrostachya* Balansa *ex* Coss. & Durieu.

Seed Dispersal and Dormancy

The seeds of most wild oats shatter at maturity, whereas cultivated forms retain their grains in the plant (Thomas 1995). Humans are the main dispersal agent for the grains of domesticated oats.

In general, wild oat seeds are dormant when they shatter, but not those of the cultivated forms (Thomas 1995). Dormancy is broken by warm, dry conditions after the seeds ripen. The seeds germinate if moisture is available; otherwise, they become dormant again. Although 85%–95% of wild oat seeds germinate within two years, some of them show deep-seated dormancy and remain viable in sufficient numbers at the end of five years to seriously reinfest the area (Banting 1965).

Volunteers, Ferals, and Their Persistence

Volunteers of both domesticated oat and feral oat plants are often found within and near crop fields (Johnson 1935).

Domesticated oats rarely persist as feral populations, because their seeds are not dormant and spontaneous germination is a dominant trait (Johnson 1935). Furthermore, ferals of domesticated oats cannot compete with wild oat populations, which frequently occur in the same habitats.

Weediness and Invasiveness Potential

Domesticated oats lack weedy characteristics, such as dormancy or seed shattering, and they do not persist for more than a few generations outside of cultivation. In contrast, their two closest wild relatives—*Avena fatua* and *A. sterilis* L.—are very noxious weeds.

Wild oats (*Avena fatua*) are a common grass species, and they are one of the world's worst weeds in cereal crops throughout the globe (Holm et al. 1977; M. Sharma and van den Born 1978). The species is highly competitive, its seeds remain viable in the soil bank for up to nine years, and it has a high propensity to propagate vegetatively. In barley fields, more than 1000 wild oat seeds/m^2 can enter the seed bank, and in wheat fields this number is estimated to be twice as high (Scursoni et al. 1999).

Table 13.1 Domesticated oat (*Avena sativa* L.) and its closest wild relatives

Species	Common name	Genome	Origin and distribution
Primary gene pool (GP-1)		hexaploid ($2n = 6x = 42$)	
A. *sativa* L.*	cultivated oat	AACCDD	known only from cultivation, cultivated worldwide
A. *fatua* L.	wild oat, oat grass, flax grass	AACCDD	crop weed; native to N Africa, Asia, the Russian Federation, and Europe; naturalized worldwide in temperate regions, including S Africa, N America, Oceania, and Australia
A. *occidentalis* Durieu	western oat	AACCDD	wild; native to Mediterranean region (Canary Islands, S Europe, Saudi Arabia, N Africa)
A. *sterilis* L.	wild oat, sterile oat, animated oat	AACCDD	wild; noxious weed in crops; native to Mediterranean region; naturalized worldwide in temperate regions, including Europe, Africa, Australia, SW Asia, and N and S America
A. *ludoviciana* Durieu		AACCDD	wild; native to Mediterranean region and Caucasia
Secondary gene pool (GP-2)		tetraploid ($2n = 4x = 28$)	
A. *maroccana* Gand.	Moroccan oat	AACC	endemic to Morocco
A. *murphyi* Ladiz.		AACC	endemic to S Spain
A. *insularis* Ladiz.		AACC	native to Sicily, Tunisia
A. *macrostachya* Balansa ex Coss. & Durieu	Algerian oat	CCCC	wild; native to Algeria

Source: Loskutov 2008.

Note: The cultivated form is in bold. Genome designations follow Rajhathy and Thomas 1974. Gene pool designations follow Rajhathy and Thomas 1974 and Loskutov, pers. comm. 2008.

*Including the cultivated subspecies red oat (A. *sativa* subsp. *byzantina* (K. Koch) Romero Zarco) and naked oat (A. *sativa* subsp. *nudisativa* (Husnot.) Rod. & Sold.).

Avena sterilis, an annual grass, formerly was a valuable pasture plant. Today, however, it is a significant weed for cultivated cereals. Its grains are highly viable and can persist in the soil for several years. They are also dispersed by humans, by grazing animals, and as impurities in harvested cereals.

Crop Wild Relatives

The genus *Avena* is distributed throughout the temperate regions of the world. Domesticated oats are allohexaploids with diploid-like chromosome pairing (21 bivalents), whereas wild *Avena* species include diploid, tetraploid, and hexaploid forms with a basic chromosome number of $x = 7$.

Avena is a taxonomically complex genus, and the scientific community has not reached agreement regarding the number of species it contains. Numbers ranging from 7 (Ladizinsky and Zohary 1971) to 34 (Baum 1977) have been suggested, and a new species, *Avena insularis* Ladiz., was recently described (Ladizinsky 1998). Here, we use a classification based on ploidy level (table 13.1; Thomas 1995).

Hexaploid oats most likely have evolved from two distinct hybridization events: the first between two diploid species, with subsequent chromosome doubling; and the second between the resulting tetraploid and another diploid species, again followed by spontaneous chromosome doubling (Thomas 1995). The putative diploid ancestors are unknown (Ladizinsky 1995a). However, due to their cytogenetic affinities with *A. sativa*, the **AACC** genome species *A. murphyi* Ladiz., *A. maroccana* Gand., and *A. insularis* Ladiz. are possible tetraploid progenitors in the evolution of hexaploid oats (Ladizinsky and Zohary 1971; Ladizinsky 1998). The (diploid) **D**-genome donor of domesticated oats is unknown (Ladizinsky 1995a).

Hybridization

Wild oats are classified into primary, secondary, and tertiary gene pools according to their ploidy levels, genomic constitution, and, hence, their sexual compatibility with domesticated oats.

Primary Gene Pool (GP-1)

The primary gene pool of domesticated oats (table 13.1) is composed of the hexaploid species, all of which are interfertile. This group is often referred to as the *Avena sterilis-A. fatua* complex of wild and weedy forms, and it should theoretically constitute a single biological species (Bervillé et al. 2005), including the wild oats *A. occidentalis* Durieu and *A. ludoviciana* Durieu (which is often considered to be a subspecies or synonym of *A. sterilis*). Taxonomic considerations aside, domesticated oats are fully cross-compatible with hexaploid wild oats; the latter are sympatric with domesticated oats and are significant crop weeds in cereal fields throughout the world (Rajhathy and Thomas 1974; Thomas 1995). All are native to the Mediterranean region; *A. sterilis* and *A. fatua* are now widely naturalized and can be found in temperate regions on all continents (Baum 1977).

Hexaploid wild oats have been used extensively in breeding programs—in particular *A. sterilis*, whose desirable characteristics have been incorporated into domesticated oats, including disease resistance and higher protein and oil content in the grain (Frey 1977; Marshall and Shaner 1992; Wilson and McMullen 1997).

In areas where cultivated and hexaploid wild oats occur in sympatry, intermediate morphological forms can often be found in and around oat fields (Stace 1975; Baum 1977). These natural hybrids are distinct from the so-called fatuoids—*fatua*-like off-types that occur at rates of up to 1% in oat fields (Baum 1969, 1977)—which initially were thought to be of hybrid origin (Aamodt et al. 1934). Today, fatuoids are known to result from the chromosomal mutation of cultivated *A. sativa* (Nilsson-Ehle 1921; Bickelmann 1989; Hoekstra et al. 2003), although their genetic bases have still not been conclusively resolved. The natural hybrids must be either F2 or advanced hybrid generations, since F1 hybrids are morphologically similar to their crop parents and can therefore be overlooked in the field (Stace 1975; Barr and Tasker 1992).

In Australia and England under natural conditions, hybridization rates between domesticated oats and *A. fatua* ranged from 0.1% to 0.93% (Derrick 1933; Burdon et al. 1992). However, outcrossing rates of up to 8.7% have been reported between domesticated oats and *A.*

sterilis (Shorter et al. 1978). The fitness of F1 hybrids is low, and it declines further in subsequent generations, due to the loss of essential weedy traits and the dominance of disadvantageous domesticated traits, such as nondormancy (Barr and Tasker 1992). Hybrids are therefore not expected to persist for more than a few generations. Neutral and beneficial crop alleles, however, could enter and become established in natural populations.

Secondary Gene Pool (GP-2)

The secondary gene pool (table 13.1) includes the tetraploid species that share their genomes with domesticated oats—namely, *Avena maroccana*, *A. murphyi*, and *A. insularis* (all **AACC** genome) (Zohary and Hopf 1993), and *A. macrostachya* (**CCCC** genome) (I. Loskutov, pers. comm. 2008).

Crosses between the crop and *A. maroccana* or *A. murphyi* can be obtained relatively easily if common oats are used as the female parent (Ladizinsky and Fainstein 1977; Ladizinsky 1992, 1995b). Ladizinsky (1995b) found that reciprocal crosses do not yield viable hybrid seeds, although they sometimes produce a few seeds; the F1 plants are morphologically intermediate between their parents.

Ladizinsky also observed that backcrossing the pentaploid F1 hybrids with the tetraploid wild parent as the pollen donor results in plants with chromosome numbers varying from $2n = 28$ to $2n = 35$, nearly all of which are sterile. Some aneuploid backcrosses, however, can produce viable seeds. A portion of the resulting F2 plants are vigorous and fertile tetraploids that are morphologically almost indistinguishable from domesticated oats. *Avena maroccana*, *A. murphyi*, and *A. macrostachya* have all been used in breeding programs to introgress traits of agronomic interest for domesticated oats (Pohler and Hoppe 1991; Marshall and Sorrells 1992; Thomas 1995; Yu and Herrmann 2006).

In theory, spontaneous hybridization in the field is possible, albeit to a very low extent. The resulting pentaploid F1 hybrids are highly sterile, and the probability that they can set seed and produce viable progeny is very low.

Table 13.2   Isolation distances for oat (*Avena sativa* L.)

Threshold*	Minimum isolation distance (m)	Country	Reference	Comment
n.a. / 0.3%	25	EU	EC 2001	certified seed requirements (hybrids)
0.1% / 0.1%	50	OECD	OECD 2008	foundation (basic) seed requirements
0.1% / 0.3%	20	OECD	OECD 2008	certified seed requirements
0.03% / 0.2%	0	USA	USDA 2008	foundation (basic) seed requirements
0.05% / 0.3%	0	USA	USDA 2008	registered seed requirements
0.1% / 0.5%	0	USA	USDA 2008	certified seed requirements
0.01% / n.a.	3	Canada	CSGA 2009	foundation (basic) and registered seed requirements
0.05% / n.a.	3	Canada	CSGA 2009	certified seed requirements

*Maximum percentage of off-type plants permitted in the field / seed impurities of other varieties or off-types permitted. n.a. = not available.

Tertiary Gene Pool (GP-3)

The tertiary gene pool includes the remaining tetraploid *Avena* species and all of the diploid relatives. They are genetically very different from the crop complex, and crosses with domesticated oats can only be obtained through substantial human intervention, such as embryo rescue and the use of synthetic hexaploids, autotetraploids, and derived tetraploids (Sadanaga and Simons 1960; D. Sharma and Forsberg 1977; Knott 1987; Ohm and Shaner 1992; Zohary and Hopf 1993; Aung et al. 1996; Rines et al. 2007). Intergenomic hybrids between A- and C-genome species are difficult to obtain, even between diploid species. Failures in chromosome pairing show that there is a lack of affinity between the two genomes (Rajhathy and Thomas 1974).

Outcrossing Rates and Distances

Pollen Flow

In spite of the fact that oats are, for the most part, self-pollinating species, outcrossing is frequently observed and can be as high as 6% among cultivated varieties (Harrington 1932; Jensen 1961; Bickelmann and Leist 1988).

Outcrossing rates between domesticated (*Avena sativa*) and wild oats (*A. sterilis* or *A. fatua*) range from 0.1% to up to 12% in pure stands (Coffman and Wiebe 1930; Derrick 1933; Shorter et al. 1978; Murray et al. 2002).

In a pollen flow study in Canada under natural conditions, Murray et al. (2002) found that pollen flow was limited to rather short distances, with low levels of outcrossing up to 1 m away from the pollen source (i.e., wild oats, *A. fatua*).

Isolation Distances

Domesticated oats, like wheat, are mainly self-fertilized, and in the United States no separation distance is required between crops for certified seed production (USDA 2008). In OECD countries and Europe, however, minimum separation distances of 20 and 25 m, respectively, are mandated (table 13.2).

State of Development of GM Technology

In comparison to wheat and other major crops, thus far there has not been much research devoted to the development of GM oats. The first reports of oat genetic transformation were published in the early 1990s (Somers et al. 1992; Torbert et al. 1998b; Zhang et al. 1999). Nowadays, particle bombardment allows the stable transformation of oats (see, e.g., Svitashev et al. 2002; Cho et al. 2003; Perret et al. 2003), and fertile GM plants have been produced expressing genes for virus resistance (McGrath et al. 1997; Koev et al. 1998; Torbert et al. 1998a) and stress tolerance (Maqbool et al. 2002; Oraby et al. 2005). However, the technology is still not sufficiently developed to be considered routine for oat transformation, and thus far no GM lines have been released for field testing. The traits investigated to date for the development of GM oat varieties include virus and herbicide resistance, abiotic stress tolerance, and some nutritional and health attributes.

Agronomic Management Recommendations

In the genus *Avena*, there are three main avenues of gene introgression from domesticated oats to wild or weedy species:

1   through pollen flow from domesticated oats

2   through volunteers in fields previously planted with oats

3   through oat grains dispersed by humans during cropping, harvesting, handling, storage, and transport of the crop

The following agronomic management recommendations may help reduce the likelihood of gene introgression as much as possible:

- Consider land-use history—not only of the field to be planted, but also of neighboring fields—to avoid contamination through oat volunteers from earlier cropping cycles.

- Limit wild oat infestation by applying a combination of tillage, crop rotation, and chemical control.

- Since gene flow is inevitable from domesticated oats to wild oats that coexist in the same fields, in regions where wild oats are already a serious problem, GM oats should not be released if they carry genes that can confer high adaptive values to the weedy partners of the crop-wild-weed oat gene pool (e.g., virus resistance) or traits (such as herbicide resistance) that can significantly increase weediness in wild oats and protect them against weed control measures. Indeed, the frequency of hybrids resulting from gene flow could increase and spread even further through later generations (Bervillé et al. 2005). By and large, herbicide-resistant oats should not be used without effective biological confinement techniques and strict monitoring guidelines. Otherwise, gene flow could quickly limit the effectiveness of that herbicide for controlling weedy oats within the crop (Glover 2002).

- If flowering overlap is likely with any of the wild relatives identified above as being susceptible to gene introgression, isolation through physical separation is an effective way to confine pollen, and distances of only a few meters can radically reduce pollen flow to surrounding oat fields. No plants should be in the area at distances shorter than those specified in legal regulations for seed production.

- Take special measures when transporting oat grains, to avoid ferals being established through spilled seed. Monitor roads for feral oat plants and, if present, control them by mechanical or chemical means.

Crop Production Area

In addition to wheat and barley, domesticated oats are one of the major temperate-zone cereals, grown on at least 12 million ha. The crop is cultivated in moist, temperate latitudes and Mediterranean climates; the Russian Federation is the main producer, providing over 20% of the world's oats, followed by Canada (19%), the United States (5%), and Europe (mainly in Finland, Germany, Poland, Spain, and Sweden) (FAO 2009).

Research Gaps and Conclusions

Gene flow via pollen dispersal may occur within the oat crop-wild-weedy complex, which consists of domesticated oats and their two fully interfertile hexaploid wild relatives, *A. sterilis* and *A. fatua*. There is evidence of hybridization between the crop and these wild species under natural conditions. Although the levels of outcrossing are usually low, the potential for gene flow and introgression into these wild oat species is substantial, because they are very widespread. Oats can also hybridize with their tetraploid wild relatives—*A. maroccana*, *A. murphyi*, and *A. insularis*—but thus far no spontaneous hybrids between the crop and these geographically restricted wild oats have been reported. Furthermore, outcrossing is possible only when a domesticated oat plant is the female parent. The resulting pentaploid hybrid progeny (if any) are completely sterile, so the probability of introgression between domesticated oats and these three species is very low. Given the high rates of self-pollination in the crop (*Avena sativa*), and the lack of success in achieving hybridization with other *Avena* species, even with human intervention, it is highly unlikely that spontaneous hybridization will occur with other wild relatives.

Based on current knowledge about the relationships between various forms of oats, the likelihood of gene introgression from domesticated oats to their wild relatives—assuming physical proximity (less than 20 m) and flowering overlap, and being conditional on the nature of the respective trait(s)—is as follows:

- *high* for their cross-compatible weedy relatives *Avena fatua, A. ludoviciana, A. occidentalis,* and *A. sterilis* (N, E, and S Africa, N and S America, central Asia, Australia, Europe, and the Middle East) (map 13.1)
- *very low* for the four tetraploid wild relatives—*A. insularis, A. macrostachya, A. maroccana,* and *A. murphyi*—sharing their genomes with domesticated oats (Algeria, Morocco, Sicily, and S Spain)

REFERENCES
....

Aamodt OS, Johnson LPV, and Manson JM (1934) Natural and artificial hybridization of *Avena fatua* and its relation to the origin of the fatuoids. *Can. J. Res.* 11: 701–727.

Aung T, Chong J, and Leggett M (1996) The transfer of crown rust resistance gene *Pc94* from a wild diploid to cultivated hexaploid oat. In: Kema GHJ, Niks RE, and Daamen RA (eds.) *Proceedings of the 9th European and Mediterranean Cereal Rusts and Powdery Mildews Conference, Lunteren, Netherlands, September 2–6, 1996* (European and Mediterranean Cereal Rust Foundation: Wageningen, The Netherlands), pp. 167–171.

Banting JD (1965) Studies on the persistence of *Avena fatua. Can. J. Plant Sci.* 46: 129–140.

Barr AR and Tasker SD (1992) Breeding herbicide resistance in oats: Opportunities and risks. In: Barr AR and Medd RW (eds.) *Wild Oats in World Agriculture.* Vol. 2 in *Proceedings of the 4th International Oat Conference, Hilton International Hotel, Adelaide, South Australia, October 19–23, 1992* (Fourth International Oat Conference: Norwood, SA, Australia), pp. 51–56.

Baum BR (1969) The use of lodicule type in assessing the origin of *Avena* fatuoids. *Can. J. Bot.* 47: 931–944.

——— (1977) *Oats: Wild and Cultivated; A Monograph of the Genus Avena L. (Poaceae).* Canada Department of Agriculture, Research Branch, Monograph No. 14 (Biosystematics Research Institute, Research Branch, Canada Department of Agriculture: Ottawa, ON, Canada). 463 p.

Baum BR and Fedak G (1985) A new tetraploid species of *Avena* discovered in Morocco. *Can. J. Bot.* 63: 1379–1385.

Bervillé A, Breton C, Cunliffe K, Darmency H, Good AG, Gressel J, Hall LM, McPherson MA, Médail F, Pinatel C, Vaughan DA, and Warwick SI (2005) Issues of ferality or potential for ferality in oats, olives, the pigeon-pea group, ryegrass species, safflower, and sugarcane. In: Gressel J (ed.) *Crop Ferality and Volunteerism* (CRC Press: Boca Raton, FL, USA), pp. 231–255.

Bickelmann U (1989) The pedigree of German oat cultivars (*Avena sativa* L.) and the occurrence of fatuoids. *Plant Breed.* 103: 163–170.

Bickelmann U and Leist N (1988) Homogeneity of oat cultivars with respect to outcrossing. In: Mattsson B and Lyhagen R (eds.) *Proceedings of the 3rd International Oat Conference, Lund, Sweden, July 4–8, 1988* (Svalöf AB: Svalöf, Sweden), pp. 358–363.

Bonnett OT (1966) *Inflorescences of Maize, Wheat, Rye, Barley, and Oats: Their Initiation and Development.* University of Illinois, College of Agriculture, Agricultural Experiment Station Bulletin 721 (College of Agriculture, University of Illinois [at Urbana-Champaign]: Urbana, IL, USA). 105 p.

Brown CM (1980) Oat. In: Fehr W and Hadley HH (eds.) *Hybridization of Crop Plants* (American Society of Agronomy: Madison, WI, USA), pp. 427–441.

Brown CM and Shands HL (1956) Factors influencing seed set of oat crosses. *Agron. J.* 48: 173–177.

Burdon JJ, Marshall DR, and Oates JD (1992) Interactions between wild and cultivated oats in Australia. In: Barr AR and Medd RW (eds.) *Wild Oats in World Agriculture.* Vol. 2 in *Proceedings of the 4th International Oat Conference, Hilton International Hotel, Adelaide, South Australia, October 19–23, 1992* (Fourth International Oat Conference: Norwood, SA, Australia), pp. 82–87.

Cho MJ, Choi H, Okamoto D, Zhang S, and Lemaux P (2003) Expression of green fluorescent protein and its inheritance in transgenic oat plants generated from shoot meristematic cultures. *Plant Cell Rep.* 21: 467–474.

Coffman FA (1937) Factors influencing seed set in oat crossing. *J. Hered.* 28: 296–303.

——— (1946) Origin of cultivated oats. *J. Am. Soc. Agron.* 38: 983–992.

Coffman FA and Wiebe GA (1930) Unusual crossing in oats in Aberdeen, Idaho. *J. Am. Soc. Agron.* 22: 245–250.

CSGA [Canadian Seed Growers Association] (2009) Canadian Regulations and Procedures for Pedigreed Seed Crop Production. Circular 6–2005/Rev. 01.4–2009 (Canadian Seed Growers Association [CSGA]: Ottawa, Canada). www.seedgrowers.ca/cropcertification/circular.asp.

Derrick RA (1933) Natural crossing with wild oats, *Avena fatua*. *Sci. Agric.* 13: 458–459.

EC [European Commission] (2001) European Communities (Cereal Seed) Regulations, 2001: S.I. [Statutory Instrument] No. 640/2001, Amending the Council Directive No. 66/402/EEC of 14 June 1966 on the Marketing of Cereal Seed. OJ No. L125 2309–66, July 11, 1966. www.irishstatutebook.ie/2001/en/si/0640.html.

FAO [Food and Agriculture Organization of the United Nations] (2009) FAOSTAT data. http://faostat.fao.org/site/567/default.aspx.

Frey KJ (1977) Protein of oats. *Z. Pflanzenzucht.* 78: 185–215.

Glover J (2002) *Gene Flow Study: Implications for GM Crop Release in Australia* (Bureau of Rural Sciences: Canberra, ACT, Australia). 81 p.

Harrington JB (1932) Natural crossing in wheat, oats, and barley at Saskatoon, Saskatchewan. *Sci. Agric.* (Ottawa)12: 470–483.

Hoekstra GJ, Burrows VD, and Mather DE (2003) Inheritance and expression of the naked-grained and fatuoid characters in oat. *Crop Sci.* 43: 57–62.

Holm LG, Plucknett DL, Pancho JV, and Herberger JP (1977) *The World's Worst Weeds: Distribution and Biology* (University Press of Hawaii: Honolulu, HI, USA). 609 p.

Imam AG and Allard RW (1965) Population studies in predominantly self-pollinated species: VI. Genetic variability between and within natural populations of wild oats from differing habitats in California. *Genetics* 51: 49–62.

Jensen NF (1961) Genetics and inheritance in oats: Inheritance of morphological and other characters. In: Coffman FA (ed.) *Oats and Oats Improvement* (American Society of Agronomy: Madison, WI, USA), pp. 125–136.

Johnson LPV (1935) The inheritance of delayed germination in hybrids of *Avena fatua* and *A. sativa. Can. J. Res., Sect.* C 13: 367–387.

Knott DR (1987) Wheat breeding: 7E. Transferring alien genes to wheat. In: Heyne EG (ed.) *Wheat and Wheat Improvement.* 2nd ed. Agronomy Monograph 13 (American Society of Agronomy, Crop Science Society of America, and Soil Science Society of America: Madison, WI, USA), pp. 418–506.

Koev G, Mohan BR, Dinesh-Kumar SP, Torbert KA, Somers DA, and Miller WA (1998) Extreme reduction of disease in oats transformed with the 5' half of the barley yellow dwarf virus-PAV genome. *Phytopathol.* 88: 1013–1019.

Ladizinsky G (1992) Genetic resources of tetraploid wild oats and their utilization. In: Barr AR and Medd RW (eds.) *Wild Oats in World Agriculture.* Vol. 2 in *Proceedings of the 4th International Oat Conference, Hilton International Hotel, Adelaide, South Australia, October 19–23, 1992* (Fourth International Oat Conference: Norwood, SA, Australia), pp. 65–70.

—— (1995a) Characterization of the missing diploid progenitors of the common oat. *Genet. Resour. Crop Evol.* 142: 49–55.

—— (1995b) Domestication via hybridization of the wild tetraploid oats *Avena magna* and *A. murphyi. Theor. Appl. Genet.* 91: 639–646.

—— (1998) A new species of oat from Sicily, possibly the tetraploid progenitor of hexaploid oats. *Genet. Resour. Crop Evol.* 45: 263–269.

Ladizinsky G and Fainstein R (1977) Introgression between the cultivated hexaploid oat *Avena sativa* and the tetraploid wild oats *A. magna* and *A. murphyi*. *Can. J. Genet. Cytol.* 19: 59–66.

Ladizinsky G and Zohary D (1971) Notes on species delimitation, species relationships, and polyploidy in *Avena*. *Euphytica* 20: 380–395.

Loskutov IG (2007) *Oat (Avena L.): Distribution, Taxonomy, Evolution, and Breeding Value* (N.I. Vavilov Research Institute of Plant Industry [VIR]: St. Petersburg, Russia). 336 p.

────── (2008) On evolutionary pathways of *Avena* species. *Genet. Resour. Crop Evol.* 55: 211–220.

Maqbool S, Zhong H, El-Maghraby Y, Wang W, Ahmad A, Chai B, and Sticklen M (2002) Competence of oat (*Avena sativa* L.) shoot apical meristems for integrative transformation, inherited expression, and osmatic tolerance of *HVA1* transgene. *Theor. Appl. Genet.* 105: 201–208.

Marshall HG and Shaner GE (1992) Genetics and inheritance in oats. In: Marshall HG and Sorrells ME (eds.) *Oat Science and Technology.* Agronomy Monograph 33 (American Society of Agronomy: Madison, WI, USA), pp. 509–571.

Marshall HG and Sorrells ME (eds.) (1992) *Oat Science and Technology.* Agronomy Monograph 33 (American Society of Agronomy: Madison, WI, USA). 846 p.

McGrath PF, Vincent JR, Lei CH, Pawlowski WP, Torbert KA, Gu W, Kaeppler HF, Wan Y, Lemaux PG, Rines HR, Somers DA, Larkins BA, and Lister RM (1997) Coat protein-mediated resistance to barley yellow dwarf in oats and barley. *Eur. J. Plant Pathol.* 103: 695–710.

Murray BG, Morrison IA, and Friesen LF (2002) Pollen-mediated gene flow in wild oat. *Weed Sci.* 50: 321–325.

Nilsson-Ehle H (1921) Fortgesetzte Untersuchungen über Fatuoidmutationen beim Hafer. *Hereditas* 2: 401–409.

OECD [Organization for Economic Cooperation and Development] (2008) OECD Scheme for the Varietal Certification of Cereal Seed Moving in International Trade: Annex VIII to the Decision. C(2000)146/FINAL. OECD Seed Schemes, February 2008. www.oecd.org/document/0/0,3343,en_2649_33905_1933504 _1_1_1_1,00.html.

Ohm HW and Shaner G (1992) Breeding oat for resistance to diseases. In: Marshall HG and Sorrells ME (eds.) *Oat Science and Technology.* Agronomy Monograph 33 (American Society of Agronomy: Madison, WI, USA), pp. 657–698.

Oraby HF, Ransom CB, Kravchenko AN, and Sticklen MB (2005) Barley *HVA1* gene confers salt tolerance in R3 transgenic oat. *Crop Sci.* 45: 2218–2227.

Perret SJ, Valentine J, Leggett JM, and Morris P (2003) Integration, expression, and inheritance of transgenes in hexaploid oat (*Avena sativa* L.). *J. Plant Physiol.* 160: 931–943.

Pohler W and Hoppe HD (1991) *Avena macrostachya*—a potential gene source for oat breeding. *Vortr. Pflanzenzucht.* 20: 66–71.

Rajhathy T and Thomas H (1974) *Cytogenetics of Oats (Avena L.)*. Miscellaneous Publications of the Genetics Society of Canada No. 2 (Genetics Society of Canada: Ottawa, ON, Canada). 90 p.

Rines HW, Porter HL, Carson ML, and Ochocki GE (2007) Introgression of crown rust resistance from diploid oat *Avena strigosa* into hexaploid cultivated oat *A. sativa* by two methods: Direct crosses and through an initial $2x + 4x$ synthetic hexaploid. *Euphytica* 158: 67–79.

Sadanaga K and Simons MD (1960) Transfer of crown rust resistance of diploid and tetraploid species to hexaploid oat. *Agron. J.* 52: 285–288.

Scursoni J, Benech-Arnold R, and Hirchoren H (1999) Demography of wild oat in barley crops: Effect of crop, sowing rate, and herbicide treatment. *Agron. J.* 91: 478–485.

Sharma DC and Forsberg RA (1977) Spontaneous and induced interspecific gene transfer for crown rust resistance in *Avena*. *Crop Sci.* 17: 855–860.

Sharma MP and van den Born WH (1978) The biology of Canadian weeds: 27. *Avena fatua* L. *Can. J. Plant Sci.* 58: 141–157.

Shorter R, Gibson P, and Frey KJ (1978) Outcrossing rates in oat species crosses (*Avena sativa* L. × *A. sterilis* L.). *Crop Sci.* 18: 877–878.

Somers DA, Rines HW, Gu W, Kaeppler HF, and Bushnell WR (1992) Fertile, transgenic oat plants. *Bio/Technol.* 10: 1589–1594.

Stace CA (1975) *Avena* L. In: Stace CA (ed.) *Hybridization and the Flora of the British Isles* (Academic Press: London, UK), pp. 573–574.

Svitashev SK, Pawlowski WP, Makarevitch I, Plank DW, and Somers DA (2002) Complex transgene locus structures implicate multiple mechanisms for plant transgene rearrangement. *Plant J.* 32: 433–445.

Thomas H (1995) Oats. In: Simmonds S and Smartt J (eds.) *Evolution of Crop Plants*. 2nd ed. (Longman Scientific and Technical: Harlow, UK), pp. 132–137.

Torbert KA, Gopalraj M, Medberry SL, Olszewski NE, and Somers DA (1998a) Expression of the Commelina yellow mottle virus promoter in transgenic oat. *Plant Cell Rep.* 17: 284–287.

Torbert KA, Rines HW, Kaeppler HF, Menon GK, and Somers DA (1998b) Genetically engineering elite oat cultivars. *Crop Sci.* 38: 1685–1687.

USDA [United States Department of Agriculture] (2008) 7 CFR [Code of Federal Regulations] § 201.76. Federal Seed Act Regulations: Minimum Land, Isolation, Field, and Seed Standards. Source: 59 FR [Federal Register] 64516, Dec. 14, 1994, as amended at 65 FR 1710, Jan. 11, 2000 (Government Printing Office: Washington, DC, USA), pp. 372–378. www.access.gpo.gov/nara/cfr/wais idx_08/7cfr201_08.html.

Vavilov NI (1951) The origin, variation, immunity, and breeding of cultivated plants. *Chron. Bot.* 13: 1–366.

—— (1992) *Origin and Geography of Cultivated Plants*, trans. D Löve (Cambridge University Press: Cambridge, UK). 498 p.

Wilson WA and McMullen MS (1997) Dosage-dependent genetic suppression of oat crown rust resistance gene *Pc-62*. *Crop Sci.* 37: 1699–1705.

Yu J and Herrmann M (2006) Inheritance and mapping of a powdery mildew resistance gene introgressed from *Avena macrostachya* in cultivated oat. *Theor. Appl. Genet.* 113: 429–437.

Zeven AC and de Wet JMJ (1982) *Dictionary of Cultivated Plants and Their Regions of Diversity, Excluding Ornamentals, Forest Trees, and Lower Plants* (Pudoc: Wageningen, The Netherlands). 263 p.

Zhang S, Cho MJ, Koprek T, Yun R, Bregitzer P, and Lemaux PG (1999) Genetic transformation of commercial germplasm of oat (*Avena sativa* L.) and barley (*Hordeum vulgare* L.) using *in vitro* shoot meristematic cultures derived from germinated seedlings. *Plant Cell Rep.* 18: 959–966.

Zhou X, Jellen EN, and Murphy JP (1999) Progenitor germplasm of domesticated hexaploid oat. *Crop Sci.* 39: 1208–1214.

Zohary D and Hopf M (1993) *Domestication of Plants in the Old World: The Origin and Spread of Cultivated Plants in West Asia, Europe, and the Nile Valley* (Clarendon Press: Oxford, UK). 328 p.

FOURTEEN

Peanut, Groundnut
(*Arachis hypogaea* L.)

....

Center(s) of Origin and Diversity

Domesticated peanuts originated in S America, in particular from lowland Bolivia or Paraguay in the eastern foothills of the Andes in the southern Amazonian region (Krapovickas 1969; Gregory et al. 1980; Hammons 1994). The exact location, however, is uncertain and origins in NE Argentina (Stalker and Simpson 1995; Kochert et al. 1996), S Bolivia (Gregory et al. 1980; Piperno and Pearsall 1998), or NW Peru (Simpson et al. 2002) have also been suggested, based on morphological and genetic diversity, a multiplicity of uses, and vernacular names, among others.

The primary center of diversity for the genus *Arachis* L. is the Chaco region between S Bolivia and NW Argentina (Gregory and Gregory 1979). Several other regions in S America can be considered as secondary centers, including the SW Amazon region of Bolivia; the states of Goiás, Minas Geraís, Rondonia, and NW Mato Grosso in Brazil; NE Brazil; the geographic region between the Paraguay and Paraná rivers S of Paraguay; and the upper Amazon area and W coast of Peru (Singh and Simpson 1994). An important tertiary center of *Arachis* diversity is located in Africa (Gibbons et al. 1972; Singh 1995).

Biological Information

Domesticated peanuts (*Arachis hypogaea* L.) are thought to have originated from a single hybridization event between two diploids (Husted 1936; Gregory and Gregory 1976; Singh and Smartt 1998). This species

Table 14.1 Peanut (*Arachis hypogaea* L.) and its closest wild relatives in section *Arachis*

Species	Ploidy	Genome	Lifespan and life form	Origin and distribution
Primary gene pool (GP-1)				
A. hypogaea L. (domesticated peanut, groundnut)	$2n = 4x = 40$	AB	annual; known only from cultivation	tropical and subtropical regions worldwide
A. monticola Krapov. & Rigoni	$2n = 4x = 40$	AB	annual; wild progenitor or weedy form	NW Argentina
Secondary gene pool (GP-2)				
A. batizocoi Krapov. & W.C. Greg.	$2n = 2x = 20$	B	annual	Bolivia, Paraguay
A. benensis Krapov. et al.	$2n = 2x = 20$	A	annual	Bolivia
A. cardenasii Krapov. & W.C. Greg.	$2n = 2x = 20$	A	perennial	Bolivia, Paraguay
A. correntina (Burkart) Krapov. & W.C. Greg.	$2n = 2x = 20$	A	perennial	Argentina, Paraguay
A. cruziana Krapov. et al.	$2n = 2x = 20$	B	annual	Bolivia
A. decora Krapov. et al.	$2n = 2x = 18$	A	annual	Brazil
A. diogoi Hoehne	$2n = 2x = 20$	A	perennial	Bolivia, Brazil, Argentina, Paraguay
A. duranensis Krapov. & W.C. Greg.	$2n = 2x = 20$	A	annual	Bolivia, Argentina, Paraguay
A. glandulifera Stalker	$2n = 2x = 20$	D	annual	Bolivia, Brazil
A. gregoryi C.E. Simpson et al.*	$2n = 2x = 20$	B	annual	Brazil
A. helodes Mart. ex Krapov. & Rigoni	$2n = 2x = 20$	A	perennial	Brazil
A. herzogii Krapov. et al.	$2n = 2x = 20$	A	perennial	Bolivia
A. hoehnei Krapov. & W.C. Greg.	$2n = 2x = 20$	B	perennial	Brazil
A. ipaënsis Krapov. & W.C. Greg.	$2n = 2x = 20$	B	annual	Bolivia
A. kempff-mercadoi Krapov. et al.	$2n = 2x = 20$	A	perennial	Bolivia
A. krapovickasii C.E. Simpson et al.*	$2n = 2x = 20$	B	annual	Bolivia
A. kuhlmannii Krapov. & W.C. Greg.	$2n = 2x = 20$	A	perennial	Brazil

Species	Chromosome number	Genome	Habit	Distribution
A. *linearifolia* Valls et al.*	2n = 2x = 20	**A**	perennial	Brazil
A. *magna* Krapov. et al.	2n = 2x = 20	**B**	annual	Bolivia, Brazil
A. *microsperma* Krapov. et al.	2n = 2x = 20	**A**	perennial	Brazil
A. *palustris* Krapov. et al.	2n = 2x = 18	**A**	annual	Brazil
A. *praecox* Krapov. et al.	2n = 2x = 18	**A**	annual	Brazil
A. *schininii* Krapov. et al.*	2n = 2x = 20	**A**	annual	Paraguay
A. *simpsonii* Krapov. & W.C. Greg.	2n = 2x = 20	**A**	perennial	Bolivia, Brazil
A. *stenosperma* Krapov. & W.C. Greg.	2n = 2x = 20	**A**	long-lived annual (perennial)	Brazil
A. *trinitensis* Krapov. & W.C. Greg.	2n = 2x = 20	**A**	annual	Bolivia
A. *valida* Krapov. & W.C. Greg.	2n = 2x = 20	**A**	annual	Brazil
A. *villosa* Benth.	2n = 2x = 20	**A**	perennial	Argentina, Uruguay
A. *williamsii* Krapov. & W.C. Greg.	2n = 2x = 20	**B**	annual	Bolivia

Sources: Stalker and Simpson 1995; Krapovickas and Gregory 2007.

Note: The domesticated form is in bold. Chromosome numbers and genome designations follow Holbrook and Stalker 2003 and Peñaloza and Valls 2005. Gene pool delimitations are according to Smartt 1990.

*Species newly described by Valls and Simpson 2005.

and its putative wild progenitor, *A. monticola* Krapov. & Rigoni, are included in sect. *Arachis* (Gregory and Gregory 1979; Stalker 1990). Apart from these two tetraploid ($2n = 4x = 40$) species, the section contains approximately 29 cross-compatible diploid annual and perennial species (table 14.1). The basic chromosome number for most diploid species in this section is $x = 10$ ($2n = 2x = 20$), but three aneuploid species with $x = 9$ ($2n = 2x = 18$) have also been described (Lavia 1998; Krapovickas and Lavia 2000). Three genome types have been defined in sect. *Arachis*, with the A-type genome being the most common, the B-type genome appearing in six species (Lavia 1996; Burow et al. 1997; Tallury et al. 2001; Holbrook and Stalker 2003; Valls and Simpson 2005), and the D-type genome represented by *A. glandulifera* Stalker (Stalker 1991).

Flowering

Peanut flowers, similar to the flowers of most legumes, have petals that almost entirely enclose the male and female organs. The stigma usually becomes receptive a day or two before the flower opens (Hassan and Srivastava 1966; Sastri and Moss 1982), and pollen is shed from the anthers either the evening before or in the morning when the flower opens (Bolhuis et al. 1965; Lim and Hamdan 1984; Pattee et al. 1991).

Pollen Dispersal and Longevity

Self-pollination does not require insect visits. However, tripping by insects aids self-pollination and seed production in peanuts (Coolbear 1994) and can also facilitate cross-pollination from foreign pollen attached to the visiting insect (Culp et al. 1968; Gibbons and Tattersfield 1969; Knauft et al. 1992).

Peanut pollen can be stored artificially; otherwise it is only viable on the day the flower opens (C. Simpson, pers. comm. 2008).

Sexual Reproduction

The domesticated peanut plant is an annual allotetraploid ($2n = 4x = 40$, **AABB** genome), which—like all *Arachis* species—is predominantly autogamous. However, there can be up to 8% natural cross-pollination in *A. hypogaea* (Krapovickas and Rigoni 1960; Knauft et al. 1992; Krapovickas and Gregory 2007).

Seed Dispersal and Dormancy

Arachis species flower above the ground, but their fruits (pods) develop and mature below the soil surface (geocarpic), a characteristic that significantly hinders seed dispersal. Peanut seeds are primarily dispersed through soil movement and water (rivers) (Gregory et al. 1973; Smartt and Stalker 1982), but they can also be carried for distances of several kilometers by animals and birds (C. Simpson, pers. comm. 2008).

Seed dormancy is variable among domesticated peanuts. Seeds from some lines are dormant for a few months (e.g., in the 'Virginia' cultivar group), whereas others are completely nondormant (e.g., 'Spanish-Valencia' cultivar groups) (Toole et al. 1964; Bhapkar et al. 1986).

Volunteers, Ferals, and Their Persistence

Peanut volunteers frequently appear in cultivated fields in subsequent cropping cycles, regenerating from seed left in the ground from previous harvests.

Domesticated peanuts are known only from cultivation, as they are unable to become feral or sustain long-term natural populations outside of cultivation. They cannot survive longer than two to three generations in the wild without human intervention (Singh and Simpson 1994).

Weediness and Invasiveness Potential

Domesticated peanuts are not considered to be a significant weed, and they lack invasive potential.

Crop Wild Relatives

The genus *Arachis* includes about 80 diploid ($2n = 20$ or $2n = 18$) and tetraploid ($2n = 4x = 40$) species native to S America, most of them perennial (Valls 2000; Valls and Simpson 2005; Krapovickas and Gregory

2007). Wild *Arachis* species are confined to five S American countries (Argentina, Bolivia, Brazil, Paraguay, and Uruguay), where they are found from the eastern slopes of the Andes to the Atlantic (Valls et al. 1985; Singh 1995).

The genus is currently divided into nine sections, based on morphology, geographic distribution, and cross-compatibility (Krapovickas and Gregory 2007). *Arachis hypogaea* (domesticated peanuts) is the only widespread domesticated species. Another species, *A. villosulicarpa* Hoehne, is also considered to be domesticated, as it has never been found in the wild; it is used for human consumption by indigenous tribes in Brazil. Three other wild species are widely cultivated: *A. repens* Handro as groundcover in S America, and *A. glabrata* Benth. and *A. pintoi* Krapov. & W.C. Greg. as forages and groundcover in S America, Australia, and the United States.

Hybridization

Domesticated peanuts (*Arachis hypogaea*) probably arose from a single hybridization event between two thus far unidentified diploid species (Husted 1936; Singh and Simpson 1994), probably *A. duranensis* Krapov. & W.C. Greg. and *A. ipaënsis* Krapov. & W.C. Greg. (Fernández and Krapovickas 1994; Kochert et al. 1996; Seijo et al. 2004, 2007). The crop is isolated reproductively from both its genome donors and from other species in the genus (Kochert et al. 1996; Milla et al. 2005). *Arachis hypogaea* is particularly variable in Paraguay, which has been attributed to the natural hybrid swarms between different *A. hypogaea* genotypes that freely outcross with each other in that region (Krapovickas and Rigoni 1960).

Based on crossability studies, the wild relatives of domesticated peanuts can be grouped into three gene pools (Smartt 1990).

Primary Gene Pool (GP-1)

The primary gene pool contains the two tetraploid species, *Arachis hypogaea* (domesticated peanuts, including S American landraces and modern cultivars and breeding lines) and its closest wild relative, *A. monticola.*

Arachis monticola is morphologically and genetically very similar to domesticated peanuts (Halward et al. 1991; Kochert et al. 1991) and is either the direct ancestor or a weedy escape of domesticated peanut plants (Stalker and Simpson 1995). It is the only known wild species that is completely cross-compatible with domesticated peanuts, resulting in highly fertile progeny (Kirti et al. 1982; Singh and Simpson 1994; Singh 1995). Hybrids often have prolonged dormancy (Stalker and Simpson 1995). To our knowledge, there are no reports of spontaneous hybrids occurring in the wild.

Secondary Gene Pool (GP-2)

The secondary gene pool includes the remaining species of sect. *Arachis*, all of which are diploid. In spite of their ploidy difference with domesticated peanuts, many of the diploid species show limited compatibility with the crop (Smartt and Gregory 1967; Singh 1989; Holbrook and Stalker 2003). However, hybrids are not easy to obtain, due to pre- and post-zygotic hybridization barriers, genomic incompatibilities, and cryptic genetic differences (Stalker et al. 1991). When the wild relative is used as the pollen donor, hybridization has a greater chance of success (Holbrook and Stalker 2003). The resulting F1 progeny are viable, but they are triploid and thus usually sterile (Gregory and Gregory 1979; Stalker 1992; Holbrook and Stalker 2003). Fertility can sometimes be restored by simply propagating plants under field conditions for prolonged periods of time in a frost-free environment (Singh and Moss 1984). Gene introgression, however, is difficult, due to the sterility of hybrid progeny and limited genetic recombinations (Holbrook and Stalker 2003).

Some wild relatives have been used in peanut breeding programs, and several important agronomic traits, such as disease and pest resistance, have been successfully transferred to cultivated peanuts (e.g., Stalker and Beute 1993; Isleib et al. 1994; Garcia et al. 1995; Simpson and Starr 2001). However, introgression requires significant human intervention, including artificial techniques such as chromosome doubling after colchicine treatment (reviewed by Holbrook and Stalker 2003).

Intrasectional hybridization between most diploid species is possible, and hybrids between A-genome species show moderate to high (60%–91%) fertility levels (Singh and Moss 1984; Stalker and Moss

1987; Stalker et al. 1991). Intergenomic hybrids between **A**-genome and **B**- or **D**-genome species are sterile (Stalker 1991; Stalker et al. 1991).

Tertiary Gene Pool (GP-3)

The tertiary gene pool contains all species in the remaining eight *Arachis* sections. These wild relatives are either primarily or completely incompatible with domesticated peanuts, due to strong post-zygotic barriers (Gregory and Gregory 1979; Singh 1989, 1998). There are reports of viable intersectional hybrids between domesticated peanuts and several of their wild relatives: *A. chiquitana* Krapov. et al. (sect. *Procumbentes* Krapov. & W.C. Greg.; Mallikarjuna 2005), *A. glabrata* (sect. *Rhizomatosae* Krapov. & W.C. Greg.; Varisai Muhammad 1973) and *A. villosulicarpa* (sect. *Extranervosae* Krapov. & W.C. Greg.; Raman 1959, 1960). However, these reports are controversial (Gregory and Gregory 1979), and some hybrid crosses could not be repeated, even though material from the same sources was used (Pompeu 1977).

Other experiments crossing cultivated peanuts with species from sections other than *Arachis* have been unsuccessful thus far. Hence intersectional hybrids involving domesticated peanuts seem to be possible only with artificial hybridization techniques, such as hormone-aided pollination or embryo rescue (Gregory and Gregory 1979; Holbrook and Stalker 2003). The resulting hybrids are completely sterile.

Bridging

Bridge crosses with species in sects. *Rhizomatosae* and *Erectoides* have been attempted, but they failed. Therefore, sexual hybridization cannot be used to access germplasm outside of sect. *Arachis* for gene transfer to domesticated peanuts (Stalker 1985).

Outcrossing Rates and Distances

Pollen Flow

To our knowledge, no studies have been conducted to assess peanut pollen flow and outcrossing rates at various distances. The crop is mainly self-fertilizing, but natural outcrossing rates of up to 8% have been reported (Krapovickas and Rigoni 1960; Knauft et al. 1992; Krapovickas and Gregory 2007).

Isolation Distances

For peanut seed production in the United States, no separation distances have been established by regulatory authorities at the national level (USDA 2008). However, in Texas a distance of 50 feet (19.6 m) is required between cultivars for the Certified Seed Production Program (C. Simpson, pers. comm. 2008). In OECD countries, peanut varieties shall be "isolated from other crops by a definite barrier or a space sufficient to prevent mixture during harvest" (OECD 2008, p. 18). In Canada, a minimum separation distance of 3 m is suggested (CSGA 2009).

State of Development of GM Technology

Methods to genetically transform peanuts have been under development since the early 1990s (e.g., Ozias-Akins et al. 1993; Brar et al. 1994; Livingstone and Birch 1999; Sarker et al. 2000; Sharma and Anjaiah 2000; Venkatachalam et al. 2000), and efficient transformation protocols have recently become available (Deng et al. 2001; Ozias-Akins and Gill 2001; Swathi Anuradha et al. 2006).

In spite of research advances in Australia, Canada, China, India, and the United States, thus far no GM cultivars have been released for commercial production. However, field trials have been reported from China, India, and the United States. GM traits that have been researched include insect, fungus, and virus resistance. Other traits in the pipeline include higher oil content for biofuel production, higher yield, herbicide resistance, abiotic stress (drought) tolerance, and enhanced nutritional value (vitamin A). At this point, the main constraint for developing transgenic peanut cultivars is the lack of identified, agronomically useful genes (Holbrook and Stalker 2003).

Agronomic Management Recommendations

There are no specific recommendations.

Crop Production Area

Peanuts are grown in over 100 countries in tropical and subtropical regions throughout the world. More than 90% of the world's peanut production comes from Asia and Africa, with India and China being the most important producers (38% and 19%, respectively), followed by Nigeria (11%). The total crop production area is estimated at 23.4 million ha (FAO 2009).

Research Gaps and Conclusions

Gene flow and introgression between domesticated peanuts and their wild relatives is likely only in S America, where there are native wild *Arachis* species. In the five countries where wild peanut species occur (Argentina, Bolivia, Brazil, Paraguay, and Uruguay), there are currently more than 780,000 ha under production (FAO 2007). However, strong hybridization barriers between domesticated peanuts and most of their wild relatives greatly reduce the likelihood of naturally occurring hybrids. There is moderate cross-compatibility with several species in the secondary gene pool, but F1 hybrids are triploids and thus are completely sterile.

The only wild relative that is fully sexually compatible with domesticated peanuts—the tetraploid *A. monticola*—has a restricted natural range (Argentina). Hybridization between these two species results in fully fertile F1 progeny, but thus far no spontaneous hybrids have been reported in the wild.

Cultivated peanuts possess a narrow genetic base, and their genetic diversity is considerably lower than that of their wild relatives, due to their origin from a single hybridization event and their subsequent reproductive isolation (e.g., Kochert et al. 1991; Halward et al. 1992; He and Prakash 1997; Hopkins et al. 1999). If gene flow was asymmetric—that is, largely from the domesticated to the wild species—then this might result in reduced genetic diversity of the wild species.

For an adequate risk assessment of gene flow and introgression between cultivated peanuts and *A. monticola*, additional information about the frequency of outcrossing between these two compatible species would be useful. Natural outcrossing rates of domesticated peanuts can be as high as 8%, but only between plants in close proximity. Outcrossing rates with *A. monticola* have not yet been investigated, and no studies have been conducted to determine the predominant direction of gene flow, quantify the extent of introgression, or study the fate of introgressed alleles in later generations.

Based on the knowledge we currently have about the relationships between *Arachis* species, the likelihood of gene introgression from domesticated peanuts to their wild relatives—assuming physical proximity (less than 3 m) and flowering overlap, and being conditional on the nature of the respective trait(s)—is as follows (also see map 14.1):

- *high* for its cross-compatible close relative *Arachis monticola* (Argentina)

- *very low* for any other of the wild relatives belonging to the same genus (S Amazon in S America)

REFERENCES

Bhapkar DG, Patil PS, and Patil VA (1986) Dormancy in groundnut—a review. *J. Maharastra Agric. Univ.* 11: 68–71.

Bolhuis GG, Frinking HD, Leeuwaugh J, Rens RG, and Staritsky G (1965) Observations on the opening of flowers, dehiscence of anthers, and growth of pollen tubes in *Arachis hypogaea* L. *Neth. J. Agric. Sci.* 13: 361–365.

Brar GS, Cohen BA, Vick CL, and Johnson GW (1994) Recovery of transgenic peanut (*Arachis hypogaea* L.) plants from elite cultivars utilizing ACCELL(r) technology. *Plant J.* 5: 745–753.

Burow MD, Paterson AH, Starr JL, and Simpson CE (1997) Identification of additional B-genome peanut accessions by use of RFLP markers. *Proc. Am. Peanut Res. Educ. Soc.* 29: 55.

Coolbear P (1994) Reproductive biology and development. In: Smartt J (ed.) *The Groundnut Crop: A Scientific Basis for Improvement* (Chapman and Hall: London, UK), pp. 138–172.

CSGA [Canadian Seed Growers Association] (2009) Canadian Regulations and Procedures for Pedigreed Seed Crop Production. Circular 6–2005/Rev. 01.4–2009 (Canadian Seed Growers Association [CSGA]: Ottawa, ON, Canada). www.seedgrowers.ca/cropcertification/circular.asp.

Culp TW, Bailey WK, and Hammons RO (1968) Natural hybridization of peanut *Arachis hypogaea* L. in Virginia. *Crop Sci.* 8: 108–111.

Deng XY, Wei ZM, and An HL (2001) Transgenic peanut plants obtained by particle bombardment via somatic embryogenesis regeneration system. *Cell Res.* 11: 156–160.

FAO [Food and Agriculture Organization of the United Nations] (2007) FAOSTAT data. http://faostat.fao.org/site/567/default.aspx.

Fernández A and Krapovickas A (1994) Chromosomas y evolución en *Arachis* (Leguminosae). *Bonplandia* 8: 188–220.

García GM, Stalker HT, and Kochert GA (1995) Introgression analysis of an interspecific hybrid population in peanuts (*Arachis hypogaea* L.) using RFLP and RAPD markers. *Genome* 38: 166–176.

Gibbons RW and Tattersfield JR (1969) Out-crossing trials with groundnuts, *Arachis hypogaea* L. *Rhod. J. Agric. Res.* 7: 71–75.

Gibbons RW, Bunting AH, and Smartt J (1972) The classification of varieties of groundnut (*Arachis hypogaea* L.). *Euphytica* 21: 78–85.

Gregory WC and Gregory MP (1976) Groundnut. In: Simmonds (ed.) *Evolution of Crop Plants* (Longman Scientific and Technical: London, UK), pp. 151–154.

—— (1979) Exotic germplasm of *Arachis* L. interspecific hybrids. *J. Hered.* 70: 185–193.

Gregory WC, Gregory M, Krapovickas A, Smith BW, and Yarbrough JA (1973) Structure and genetic resources of peanuts. In: Pattee HE and Young CT (eds.) *Peanuts: Culture and Uses* (American Peanut Research and Education Association: Stillwater, OK, USA), pp. 47–133.

Gregory WC, Krapovickas A, and Gregory MP (1980) Structure, variation, evolution, and classification in *Arachis*. In: Summerfield RJ and Bunting AH (eds.) *Advances in Legume Sciences* (Royal Botanic Gardens, Kew: Richmond, Surrey, UK), pp. 469–481.

Halward TM, Stalker HT, LaRue E, and Kochert G (1991) Genetic variation detectable with molecular markers among unadapted germplasm resources of cultivated peanut and related wild species. *Genome* 34: 1013–1020.

—— (1992) Use of single-primer DNA amplifications in genetic studies of peanut (*Arachis hypogaea* L.). *Plant Mol. Biol.* 18: 315–325.

Hammons RO (1994) The origin and history of the groundnut. In: Smartt J (ed.) *The Groundnut Crop: A Scientific Basis for Improvement* (Chapman and Hall: London, UK), pp. 24–42.

Hassan MA and Srivastava DP (1966) Floral biology and pod development of peanut studied in India. *J. Indian Bot. Soc.* 45: 92–102.

He G and Prakash CS (1997) Identification of polymorphic DNA markers in cultivated peanuts (*Arachis hypogaea* L.). *Euphytica* 97: 143–149.

Holbrook CC and Stalker HT (2003) Peanut breeding and genetic resources. *Plant Breed. Rev.* 22: 297–355.

Hopkins MS, Casa AM, Wang T, Mitchell SE, Dean RE, Kochert GD, and Kresovich S (1999) Discovery and characterization of polymorphic simple sequence repeats (SSRs) in peanut. *Crop Sci.* 39: 1243–1247.

Husted L (1936) Cytological studies of the peanut *Arachis*: II. Chromosome number, morphology, and behavior and their application to the origin of cultivated forms. *Cytologia* 7: 396–423.

Isleib TG, Wynne JC, and Nigam SN (1994) Groundnut breeding. In: Smartt J (ed.) *The Groundnut Crop: A Scientific Basis for Improvement* (Chapman and Hall: London, UK), pp. 552–623.

Kirti PB, Murty UR, Bharati M, and Rao NGP (1982) Chromosome pairing in F1 hybrid *Arachis hypogaea* L. × *A. monticola* Krap. et. Rig. *Theor. Appl. Genet.* 62: 139–144.

Knauft DA, Chiyembekeza AJ, and Gorbet DW (1992) Possible reproductive factors contributing to outcrossing in peanut. *Peanut Sci.* 19: 29–31.

Kochert G, Halward T, Branch WD, and Simpson CE (1991) RFLP variability in peanut (*Arachis hypogaea* L.) cultivars and wild species. *Theor. Appl. Genet.* 81: 565–570.

Kochert G, Stalker HT, Gimenes M, Galgaro L, Romero Lopes C, and Moore K (1996) RFLP and cytogenetic evidence on the origin and evolution of allotetraploid domesticated peanut, *Arachis hypogaea* (Leguminosae). *Am. J. Bot.* 83: 1282–1291.

Krapovickas A (1969) The origin, variability, and spread of the groundnut (*Arachis hypogea*). In: Ucko PJ and Dimbleby GW (ed.) *The Domestication and Exploration of Plants and Animals* (Duckworth: London, UK), pp. 427–441.

Krapovickas A and Gregory WC (2007) Taxonomy of the genus *Arachis* (Leguminosae), trans. DE Williams and CE Simpson. *Bonplandia* 16 (Suppl.): 1–205.

Krapovickas A and Lavia G (2000) Advances in the taxonomy of the genus *Arachis*. *Proc. Am. Peanut Res. Educ. Soc.* 32: 46.

Krapovickas A and Rigoni VA (1960) La nomenclatura de las sub-espécies y variedades de *A. hypogaea* L. *Rev. Invest. Agric.* 14: 197–228.

Lavia GI (1996) Estudios cromosómicos en *Arachis* (Legiminosae). *Bonplandia* 9: 111–120.

——— (1998) Karyotypes of *Arachis palustris* and *A. praecox* (section *Arachis*), two species with basic chromosome number $x = 9$. *Cytologia* 63: 177–181.

Lim ES and Hamdan O (1984) The reproductive characters of four varieties of groundnuts (*Arachis hypogaea* L.). *Pertanika* 7: 61–66.

Livingstone DM and Birch RG (1999) Efficient transformation and regeneration of diverse cultivars of peanut (*Arachis hypogaea* L.) by particle bombardment into embryogenic callus produced from mature seeds. *Mol. Breed.* 5: 43–51.

Mallikarjuna N (2005) Production of hybrids between *Arachis hypogaea* and *A. chiquitana* (section *Procumbentes*). *Peanut Sci.* 32: 148–152.

Milla SR, Isleib TG, and Stalker HT (2005) Taxonomic relationships among *Arachis* sect. *Arachis* species as revealed by AFLP markers. *Genome* 48: 1–11.

OECD [Organization for Economic Cooperation and Development] (2008) OECD Scheme for the Varietal Certification of Grass and Legume Seed Moving in International Trade: Annex VI to the Decision. C(2000)146/FINAL. OECD Seed Schemes, February 2008. www.oecd.org/document/0/0,3343,en_2649_33905_1933504_1_1_1_1,00.html.

Ozias-Akins P and Gill R (2001) Genetic engineering of *Arachis. Peanut Sci.* 28: 123–131.

Ozias-Akins P, Schnall JA, Anderson WF, Singsit C, Clemente TE, Adang MJ, and Weissinger AK (1993) Regeneration of transgenic peanut plants from stably transformed embryogenic callus. *Plant Sci.* 93: 185–194.

Pattee HE, Stalker HT, and Giesbrecht FG (1991) Comparative peg, ovary, and ovule ontogeny of selected cultivated and wild-type *Arachis* species. *Bot. Gaz.* 152: 64–71.

Peñaloza APS and Valls JFM (2005) Chromosome number and satellited chromosome morphology of eleven species of *Arachis* (Legiminosae). *Bonplandia* 14: 65–72.

Piperno DR and Pearsall DM (1998) *The Origins of Agriculture in the Lowland Neotropics* (Academic Press: New York, NY, USA). 400 p.

Pompeu AS (1977) Cruzamentos entre *Arachis hypogaea* as espécies *A. villosa* var. *correntina*, *A. diogoi*, e *A. villosulicarpa. Cienc. Cult.* 29: 319–321.

Raman VS (1959) Studies in the genus *Arachis*: VI. Investigation on 30-chromosomed interspecific hybrids. *Ind. Oilseeds J.* 3: 157–161.

—— (1960) Studies in the genus *Arachis*: IX. A fertile synthetic tetraploid groundnut from the interspecific backcross *A. hypogaea* × (*A. hypogaea* × *A. villosa*). *Ind. Oilseeds J.* 4: 90–92.

Sarker RH, Islam MN, Islam A, and Seraj ZI (2000) *Agrobacterium*-mediated genetic transformation of peanut (*Arachis hypogaea* L.). *Plant Tissue Cult.* 10: 137–142.

Sastri DC and Moss JP (1982) Effects of growth regulators on incompatible crosses in the genus *Arachis* L. *J. Exp. Bot.* 53: 1293–1301.

Seijo JG, Lavia GI, Fernández A, Krapovickas A, Ducasse DA, and Moscone EA (2004) Physical mapping of the 5s and 18s-25s rRNA genes by FISH as evidence that *Arachis duranensis* and *A. ipaënsis* are the wild diploid progenitors of *A. hypogaea* (Leguminosae). *Am. J. Bot.* 91: 1294–1303.

Seijo JG, Lavia GI, Fernández A, Krapovickas A, Ducasse DA, Bertioli DJ, and Moscone EA (2007) Genomic relationships between the cultivated peanut (*Ara-*

chis hypogaea, Leguminosae) and its close relatives revealed by double GISH. *Am. J. Bot.* 94: 1963–1971.

Sharma KK and Anjaiah V (2000) An efficient method for the production of transgenic plants of peanut (*Arachis hypogaea* L.) through *Agrobacterium tumefaciens*-mediated genetic transformation. *Plant Sci.* 159: 7–19.

Simpson CE and Starr J (2001) Registration of 'COAN' peanut. *Crop Sci.* 41: 918.

Simpson CE, Krapovickas A, and Valls JFM (2002) History of *Arachis* including evidence of *A. hypogaea* L. progenitors. *Peanut Sci.* 28: 79–81.

Singh AK (1989) *Exploitation of* Arachis *Species for Improvement of Cultivated Groundnut*. Progress Report 1988 (International Crops Research Institute for the Semi-Arid Tropics [ICRISAT]: Patancheru, India). 73 p.

———— (1995) Groundnut. In: Simmonds S and Smartt J (eds.) *Evolution of Crop Plants*. 2nd ed. (Longman Scientific and Technical: Harlow, UK), pp. 246–250.

———— (1998) Hybridization barriers among the species of *Arachis* L., namely of the sections *Arachis* (including the groundnut) and *Erectoides*. *Genet. Resour. Crop Evol.* 45: 41–45.

Singh AK and Moss JP (1984) Utilization of wild relatives in genetic improvement of *Arachis hypogaea* L.: 5. Genome analysis in section *Arachis* and its implication in gene transfer. *Theor. Appl. Genet.* 68: 350–364.

Singh AK and Simpson CE (1994) Biosystematics and genetic resources. In: Smartt J (ed.) *The Groundnut Crop: A Scientific Basis for Improvement* (Chapman and Hall: London, UK), pp. 96–137.

Singh AK and Smartt J (1998) The genome donors of the groundnut/peanut (*Arachis hypogaea* L.) revisited. *Genet. Resour. Crop Evol.* 45: 113–118.

Smartt J (1990) The groundnut, *Arachis hypogaea* L. In: Smartt J (ed.) *Grain Legumes: Evolution and Genetic Resources* (Cambridge University Press: Cambridge, UK), pp. 30–84.

Smartt J and Gregory WC (1967) Interspecific cross-compatibility between the cultivated peanut *Arachis hypogaea* L. and other members of the genus *Arachis*. *Oléagineux* 22: 455–459.

Smartt J and Stalker HT (1982) Speciation and cytogenetics in *Arachis*. In: Pattee HE and Young CT (eds.) *Peanut Science and Technology* (American Peanut Research and Education Society: Yoakum, TX, USA), pp. 21–49.

Stalker HT (1985) Cytotaxonomy of *Arachis*. In: Moss JP (ed.) *Proceeding of an International Workshop on Cytogenetics of* Arachis, *ICRISAT Center, Patancheru, India, October 31–November 2, 1983* (International Crops Research Institute for the Semi-Arid Tropics [ICRISAT]: Patancheru, India), pp. 65–79.

———— (1990) A morphological appraisal of wild species in section *Arachis* of peanuts. *Peanut Sci.* 17: 117–122.

————— (1991) A new species in section *Arachis* of peanuts with a D genome. *Am. J. Bot.* 78: 630–637.

————— (1992) Utilizing *Arachis* germplasm resources. In: Nigam SN (ed.) *Groundnut, a Global Perspective: Proceedings of an International Workshop, ICRIST Center, November 25–29, 1991* (International Crops Research Institute for the Semi-Arid Tropics [ICRISAT]: Patancheru, India), pp. 281–295.

Stalker HT and Beute MK (1993) Registration of four interspecific peanut germplasm lines resistant to *Cercospora arachidicola. Crop Sci.* 33: 1117.

Stalker HT and Moss JP (1987) Speciation, cytogenetics, and utilization of *Arachis* species. *Adv. Agron.* 41: 1–40.

Stalker HT and Simpson CE (1995) Germplasm resources in *Arachis.* In: Pattee HE and Stalker HT (eds.) *Advances in Peanut Science* (American Peanut Research and Education Society: Stillwater, OK, USA), pp. 14–53.

Stalker HT, Dhesi JS, Parry DC, and Hahn JH (1991) Cytological and interfertility relationships of *Arachis* section *Arachis. Am. J. Bot.* 78: 238–246.

Swathi Anuradha T, Jami SK, Datla RS, and Kirti PB (2006) Genetic transformation of peanut (*Arachis hypogaea* L.) using cotyledonary node as explant and a promoterless *gus::npt*II fusion gene based vector. *J. Biosci.* 31: 235–246.

Tallury SP, Mila SR, Copeland SC, and Stalker HT (2001) Genome donors of *Arachis hypogaea* L. *Proc. Am. Peanut Res. Educ. Soc.* 33: 60.

Toole VK, Bailey WK, and Toole EH (1964) Factors influencing dormancy of peanut seeds. *Plant Physiol.* 39: 822–832.

USDA [United States Department of Agriculture] (2008) 7 CFR [Code of Federal Regulations] § 201.76. Federal Seed Act Regulations: Minimum Land, Isolation, Field, and Seed Standards. Source: 59 FR [Federal Register] 64516, Dec. 14, 1994, as amended at 65 FR 1710, Jan. 11, 2000 (Government Printing Office: Washington, DC, USA), pp. 372–378. www.access.gpo.gov/nara/cfr/wais idx_08/7cfr201_08.html.

Valls JFM (2000) Recent advances in the characterization of wild *Arachis* germplasm in Brazil. *Proc. Am. Peanut Res. Educ. Soc.* 32: 47.

Valls JFM and Simpson CE (2005) New species of *Arachis* (Leguminosae) from Brazil, Paraguay, and Bolivia. *Bonplandia* 14: 35–64.

Valls JFM, Ramanatha Rao V, Simpson CE, and Krapovickas A (1985) Current status of collection and conservation of South American groundnut germplasm with emphasis on wild species of *Arachis.* In: Moss JP (ed.) *Proceeding of an International Workshop on Cytogenetics of* Arachis, *ICRISAT Center, Patancheru, India, October 31–November 2, 1983* (International Crops Research Institute for the Semi-Arid Tropics [ICRISAT]: Patancheru, India), pp. 15–35.

Varisai Muhammad S (1973) Cytogenetical investigations in the genus *Arachis* L.: II. Triploid hybrids and their derivatives. *Madras Agric. J.* 60: 1414–1427.

Venkatachalam P, Geetha N, Khandelwal A, Shaila MS, and Sita GL (2000) *Agrobacterium*-mediated genetic transformation and regeneration of transgenic plants from cotyledon explants of groundnut (*Arachis hypogaea* L.) via somatic embryogenesis. *Curr. Sci.* 78: 1130–1136.

FIFTEEN

Pearl Millet
(*Pennisetum glaucum* (L.) R. Br.)

Center(s) of Origin and Diversity

Most authors concur that pearl millet originated in the Sahel in tropical W Africa, south of the Sahara from W Sudan to Senegal, but an Abyssinian (Ethiopian) origin has also been proposed (Rachie and Majmudar 1980).

Several centers of diversity are known for pearl millet (i.e., domesticated pearl millet and its progenitor), including semi-arid to arid W and E Africa, south of the Sahara Desert and north of the forest zone. India is considered to be a secondary center of diversity.

Biological Information

Flowering

Pearl millet is a protogynous species, that is, the stigma becomes receptive several days before pollen is released, resulting in high frequencies of cross-pollination (Burton 1980). However, inflorescences of the same plant flower in succession, so heads with exerted anthers can pollinate other heads on the same plant that are just exerting their stigmas, facilitating self-pollination at rates of 22% (Sandmeier 1993) or even up to 31% (Burton 1974). Fertilization occurs a few hours after pollination, and stylar branches usually wilt and dry out within 24 hours after pollination (Burton 1980).

In Niger, phenological studies showed that the flowering periods of domesticated (*Pennisetum glaucum*) and wild (*P. violaceum* (Lam.) Rich. *ex* Pers.) pearl millets only partly overlap, resulting in predominantly endogamic reproduction of the wild forms after the cultivated plants finished flowering (Renno and Winkel 1996).

Pollen Dispersal and Longevity

Pearl millet pollen is mainly wind dispersed (Burton 1974). However, insect pollinators—such as honeybees (*Apis mellifera*), bumblebees (*Bombus impatiens* Cresson), and halictid bees (*Dialictus pilosus* Smith)—are also important pollen vectors, and honeybees have transported pearl millet pollen for distances of up to 1.6 km or more (Leuck and Burton 1966).

Pearl millet pollen can remain viable between three to five days under field conditions in India (Kumar et al. 1995). Pollen longevity is negatively affected by high average temperature, increased periods of sunshine, and low average relative humidity (Kumar et al. 1995), as well as by low temperatures (Fussell et al. 1980; Mashingaidze and Muchena 1982).

Sexual Reproduction

Pearl millet is an annual, mainly allogamous diploid ($2n = 2x = 14$). Its rate of self-pollination can be as high as 31% (see above).

Seed Dispersal and Dormancy

While the spikelets and seeds of wild relatives of pearl millet are easily shattered when mature, the crop has lost this trait during domestication; most modern cultivars do not shatter at all (Poncet et al. 1998, citing studies by Niangado in 1981 and Joly in 1984). Birds are important depredators of pearl millet grains (Burton 1980), as are rodents (J. Wilson, pers. comm. 2008), but there is no data on how well seeds survive after their passage through digestive tracts. Wild pearl millet seeds are dispersed by wind, and their spikelets (containing the seed) can have abundant fine bristles that are easily blown across the ground for short distances (J. Wilson, pers. comm. 2008).

Seeds of most wild pearl millet genotypes are dormant for several weeks (Burton 1980), whereas dormancy has been lost in cultivated forms, which germinate within five days under favorable conditions (NRC 1996; Poncet et al. 1998). The minimum temperature for germination is 12°C, with germination rates increasing linearly to a clearly defined optimum between 30°C and 35°C (Ong 1983).

Volunteers, Ferals, and Their Persistence

Pearl millet volunteers, from leftover seeds after a harvest, frequently appear in subsequent cropping cycles and at disposal sites for mulch residue.

Pearl millet is very frost sensitive, so volunteers and ferals usually do not survive the winter in temperate regions. In tropical and subtropical regions, pearl millet can last outside of cultivation for several years. However, this very rarely happens, since birds and rodents often eat remnant seeds (J. Wilson, pers. comm. 2008).

Weediness and Invasiveness Potential

While pearl millet does not have weedy properties (J. Wilson, pers. comm. 2008), its weedy and wild relatives, *P. sieberianum* (Schltdl.) Stapf & C.E. Hubb. and *P. violaceum*, are aggressive colonizers of disturbed habitats in W Africa (de Wet 1995). Other wild relatives—such as *P. setaceum* (Forssk.) Chiov., *P. clandestinum* Hochst. *ex* Chiov., *P. ciliare* (L.) Link (= *Cenchrus ciliaris* L.), and *P. polystachion* (L.) Schult.—are invasive grasses (GISD 2007).

Crop Wild Relatives

The genus *Pennisetum* Rich. is widely distributed throughout the tropics and subtropics. The taxonomy and number of species included in the genus have been open to dispute (see, e.g., Nath et al. 1971; Purseglove 1972; Lebrun and Stork 1995), but currently the genus appears to have more than 100 species, divided into five sections: *Gymnothrix* (P. Beauv.) Benth. & Hook. f., *Brevivalvula* Döll, *Heterostachya* Stapf & C.E. Hubb., *Eupennisetum* Stapf, and *Penicillaria* (Willd.) Benth. & Hook. f. (= *Pennisetum*) (Stapf and Hubbard 1934).

Pearl millet (*Pennisetum glaucum*) and its closest wild relatives are native to W and E Africa and are now part of sect. *Penicillaria* (table 15.1).

Table 15.1 Domesticated pearl millet (*Pennisetum glaucum* (L.) R. Br.) and its wild relatives in section *Penicillaria* (Willd.) Benth & Hook. f.

Species	Common name	Ploidy	Genome	Origin and distribution
Primary gene pool (GP-1)				
P. glaucum (L.) R. Br.	pearl millet, cattail millet	$2n = 2x = 14$	**AA**	native to Africa; also cultivated in India, Pakistan, USA, Australia, and Brazil
P. sieberianum (Schltdl.) Stapf & C.E. Hubb.	weedy pearl millets, shibra	$2n = 2x = 14$	n.a.	weedy; native to Africa (ANG, BKN, CHA, GAM, GHA, MLI, NAM, NGA, NGR, SEN, SUD)
P. violaceum (Lam.) Rich. ex Pers.	wild pearl millet	$2n = 2x = 14$	n.a.	wild; native to Africa (CAF, MLI, MTN, NGA, NGR, SEN, SUD)
Secondary gene pool (GP-2)				
P. purpureum Schumach.	elephant grass, napier grass	$2n = 4x = 28$	**A'A'BB**	native to Africa; naturalized elsewhere (tropical Asia, Australia, USA, C and S America, West Indies)

Sources: Stapf and Hubbard 1934; Clayton and Renvoize 1982.

Note: Gene pool boundaries are according to Harlan and de Wet 1971. n.a. = not available. Abbreviations of African countries follow Brummitt 2001: ANG, Angola; BKN, Burkina; CAF, Central African Republic; CHA, Chad; GAM, Gambia; GHA, Ghana; MLI, Mali; MTN, Mauritania; NAM, Namibia; NGA, Nigeria, NGR, Niger; SEN, Senegal; SUD, Sudan.

These include the wild forms of pearl millet, classified as *P. violaceum*, and the weedy forms (called shibras), classified as *P. sieberanum* (Clayton and Renvoize 1982), all of which are diploids. Wild *P. violaceum* (also known as *P. glaucum* subsp. *monodii* (Lam.) van der Zon, *nom. illeg.*) is the putative wild progenitor of pearl millet.

The taxonomic status of *P. violaceum* and *P. sieberanum* is debatable, since they are fully sexually compatible with the crop (*P. glaucum*), and hybrids are fully fertile. The species designations for these three forms are based on morphology, not on sexual compatibility, fertility, and the viability of their progeny. They are thus often considered to be nondomesticated and domesticated forms of the same species.

The weedy forms (shibras) are morphologically intermediate between cultivated and wild pearl millet (Brunken et al. 1977) and are thought to be naturally occurring hybrids between the wild and domesticated types (J. Wilson, pers. comm. 2008). Shibras are obligate weeds that mimic pearl millet and are associated with pearl millet fields. They are very common in plantings of traditional pearl millet landraces in W Africa and N Namibia, although they are absent in Asia (where pearl millet is also cultivated). The early-maturing grains of shibras are harvested by women and play a role in food security (B. Haussmann, pers. comm. 2008). Shibras are dependent on humans, and they do not persist for more than one generation after cultivated fields are abandoned (de Wet 1995).

Besides the diploid annual *Pennisetum* species, sect. *Penicillaria* also includes the tetraploid ($2n = 4x = 28$) perennial wild relative elephant grass (*P. purpureum* Schumach.). This species can propagate vegetatively via rhizomes. While it is considered to be one of the world's worst weeds (Holm et al. 1977) in portions of its range, it is also a valuable forage grass in other regions.

Hybridization

The wild relatives of pearl millet can be grouped into primary, secondary, and tertiary gene pools, according to their cross-compatibility and level of genetic similarity with the domesticated crop (Harlan and de Wet 1971; Martel et al. 2004).

Primary Gene Pool (GP-1)

The primary gene pool includes the domesticated-weedy-wild pearl millet complex, comprising domesticated pearl millet (*Pennisetum glaucum*) and its weedy and wild forms *P. sieberianum* (shibras) and *P. violaceum* (table 15.1). All forms are cross-fertile and frequently hybridize with each other in Africa, where they occur in sympatry (Brunken 1977; Robert et al. 1991; Marchais and Tostain 1992). Hybridization between domesticated *P. glaucum* and wild *P. violaceum* may have been involved in the evolution of the weedy pearl millet form *P. sieberanum*, which is morphologically intermediate between the two species (Brunken et al. 1977).

Genetic studies have provided evidence of substantial rates of natural hybridization between wild and cultivated pearl millets and of introgression in both directions (Tostain 1992; Mariac et al. 2006a). In Niger, for instance, as many as 35% of the hybrid offspring have originated from crosses between domesticated pearl millet and *P. violaceum*, and up to 39% from crosses with *P. sieberanum* (Marchais 1994; Renno et al. 1997).

F1 hybrids resulting from crosses between cultivated and wild pearl millets are usually vigorous and fully fertile. They are morphologically intermediate and segregate in the F2 generation, where both wild and cultivated phenotypes are found, due to preferential associations or linkages between their characteristics at the genetic level (Pernès 1986; Poncet et al. 1998). A recent genetic study provides evidence for the presence of a single putative supergene that controls a series of differences between the crop and the weedy forms (Miura and Terauchi 2005).

In addition, moderate pre- and post-zygotic hybridization barriers have been identified that control gene flow between the wild and domesticated forms and thus contribute to maintaining a genetic differentiation between the species. Some of these barriers include pollen competition (Sarr et al. 1988; Robert et al. 1992) and reduced viability of hybrid grains (Amoukou and Marchais 1993).

Temporal isolation in particular regions, for instance in Niger (see above), is an important factor in restraining the extent of gene flow and introgression between cultivated and wild forms. Also, human and natural selection may reduce the frequency of hybrids between cultivated and wild pearl millets in the field (Couturon et al. 2003).

Secondary Gene Pool (GP-2)

The secondary gene pool contains a single species (table 15.1), the wild relative elephant grass (*P. purpureum*), which is a perennial tetraploid with $2n = 4x = 28$ (Harlan 1975). Domesticated pearl millet and elephant grass are sexually compatible and are easy to cross, but the resulting vigorous triploid F1 hybrids ($2n = 3x = 21$, **AAB** genome) are completely sterile (Burton 1944; Pantulu 1967; Jauhar 1968, 1981; Sethi et al. 1970; de Wet 1995; Techio et al. 2006). To our knowledge, no spontaneous hybrids between pearl millet and *P. purpureum* have been reported to occur in the wild. Hybrids, however, are used far and wide as forage and are propagated vegetatively by stem cuttings.

Because fertility in these hybrids can be restored after artificial chromosome doubling (Hanna 1981; Gonzales and Hanna 1984; Diz and Schank 1993), these ploidy manipulations have allowed the genes for early flowering, long inflorescences, leaf size, and restored male fertility to be transferred from wild to domesticated species despite their natural reproductive barriers (Hanna 1990, 1993).

Tertiary Gene Pool (GP-3)

The remaining *Pennisetum* species belong to the tertiary gene pool. Their ploidy levels (basic chromosome numbers $x = 5, 7, 8$, and 9) differ from that of cultivated pearl millet, and strong pre- and post-zygotic hybridization barriers impede natural crossing with the crop (Kaushal and Sidhu 2000). Hybridization is possible only through substantial human intervention (Burton and Powell 1968; Dujardin and Hanna 1984a; Schmelzer 1997), and the resulting hybrids tend to be anomalous, sterile, or lethal, which points out the difficulty of transferring genes between pearl millet and these species.

Breeding efforts involving wild relatives from GP-3 have concentrated on the transfer of apomixis genes from different apomictic wild species to pearl millet (Kaushal et al. 2005)—for example, from *P. squamulatum* Fresen. (Dujardin and Hanna 1983, 1984a, b, 1989a, b;

Hanna et al. 1993; Ozias-Akins et al. 1993; Roche et al. 2001; Kaushal et al. 2003), *P. orientale* Rich. (Patil and Singh 1964; Hanna and Dujardin 1982; Zadoo and Singh 1986; Dujardin and Hanna 1987, 1989a), *P. schweinfurthii* Pilg. (Hanna and Dujardin 1986), and *P. setaceum* (Hanna 1979; Dujardin and Hanna 1989a).

Pearl millet has also been crossed with buffelgrass, *Pennisetum ciliare* (L.) Link (Read and Bashaw 1974; Marchais and Tostain 1997). However, aside from significant human intervention, hybridization of these two species is highly unlikely, if not impossible, under natural conditions, due to strong pre- and post-zygotic barriers. In spite of the application of artificial techniques, gene transfer has not been achieved thus far.

Outcrossing Rates and Distances

Pollen Flow

Leuck and Burton (1966) measured pollen dispersal distances of 1.6 km or more, and, based on this observation, Burton (1980) recommended minimum isolation distances of 1.6 km for breeder seed. However, no information about outcrossing frequencies at this distance was provided.

Isolation Distances

The separation distance established by regulatory authorities in the United States for pearl millet seed production is 200 m for certified seed and 400 m for registered and foundation seed (USDA 2008). In Africa, isolation distances of 400 m and 1.6 km are required for certified and basic seed production, respectively (Singh 1995). For breeder and nucleus seed production, 1 and 2 km separations are recommended (Singh 1995).

State of Development of GM Technology

Genetic transformation methods for pearl millet have been under development for some time (Taylor and Vasil 1991; Taylor et al. 1993; Lambe et al. 1995; Girgi et al. 2002; Goldman et al. 2003; O'Kennedy et al.

2004), but an efficient and stable transformation protocol has only recently become available (Madhavi et al. 2006). The GM traits that have been researched are pest and herbicide resistances. To date, no experimental field trials of GM pearl millet have been reported.

Agronomic Management Recommendations

There are no specific recommendations.

Crop Production Area

Pearl millet is grown primarily as staple food crop in dryland areas of sub-Saharan Africa and S Asia—extending to Pakistan, India, and China—but farmers also grow it throughout the tropical and subtropical lowlands of the Americas, as well as in Australia, the Near East, and the United States. In nontraditional growing areas—such as Australia, Brazil, Canada, Europe, and the United States—pearl millet is grown as forage, fodder grain, or a cover crop.

The total crop production area of pearl millet is estimated in 26 million ha (CGIAR 2006). To date, there is no commercial production of GM pearl millet.

Research Gaps and Conclusions

Gene flow from cultivated pearl millet to its wild relatives is likely in Africa and Asia, where these plants occur in sympatry. Naturally occurring hybrid swarms can be observed in and near pearl millet fields, and since pearl millet is a predominantly allogamous crop, outcrossing rates through pollen flow under natural conditions are very likely to exceed the 0.9% threshold for allowable GM trait contamination.

Little is known about the role of pollinator vectors (wind and insects), the distances pearl millet pollen can travel, and expected outcrossing rates at different distances from the pollen source. Although most granivorous birds crack millet seeds before eating them, thus rendering them unviable, the possibility of pearl millet seed survival after passage through the digestive tract of granivores should be studied. Also, further studies are needed to investigate the rates of natural outcrossing between pearl millet and its closest wild relatives, hybrid vigor and fertility, the extent and main direction of introgression, and the fate of introgressed alleles in later generations. In particular, the probability of natural outcrossing between pearl millet and elephant grass, one of the world's worst weeds, should be assessed in hybridization experiments under natural and controlled conditions.

Gene flow and introgression between pearl millet and its wild relatives can occur via seed exchange. In many tropical and subtropical regions, such as in Africa and India, farmers exchange pearl millet seeds within and among villages (Christinck 2002; vom Brocke et al. 2003; Mariac et al. 2006b). These informal seed systems facilitate gene flow not only within the crop (landraces and modern cultivars), but also within the crop-weedy-wild complex.

Based on partial knowledge about the relationships between *Pennisetum* species, the likelihood of gene introgression from domesticated pearl millet to its wild relatives—assuming physical proximity (less than 200 m) and flowering overlap, and being conditional on the nature of the respective trait(s)—is as follows (map 15.1):

- *high* for its weedy relative *Pennisetum sieberianum* and its wild ancestor *P. violaceum*

- *low* for the perennial wild relative elephant grass (*P. purpureum*) (sub-Saharan Africa and S and SE Asia)

REFERENCES

Amoukou AI and Marchais L (1993) Evidence of a partial reproductive barrier between wild and cultivated pearl millets (*Pennisetum glaucum*). *Euphytica* 67: 19–26.

Brummitt RK (2001) *World Geographical Scheme for Recording Plant Distributions*. 2nd ed. Plant Taxonomic Database Standards No. 2 (published for the International Working Group on Taxonomic Databases for Plant Sciences [TDWG] by the Hunt Institute for Botanical Documentation: Carnegie Mellon University, Pittsburgh, PA, USA). 137 p. http://www.tdwg.org/TDWG_geo2.pdf.

Brunken JN (1977) A systematic study of *Pennisetum* sect. *Pennisetum* (Gramineae). *Am. J. Bot.* 64: 61–176.

Brunken JN, de Wet JMJ, and Harlan JR (1977) The morphology and domestication of pearl millet. *Econ. Bot.* 31: 163–174.

Burton GW (1944) Hybrids between napiergrass and cattail millet. *J. Hered.* 35: 726–732.

—— (1974) Factors affecting pollen movement and natural crossing in pearl millet. *Crop Sci.* 14: 802–805.

—— (1980) Pearl millet. In: Fehr W and Hadley HH (eds.) *Hybridization of Crop Plants* (American Society of Agronomy: Madison, WI, USA), pp. 457–469.

Burton GW and Powell JB (1968) Pearl millet breeding and cytogenetics. *Adv. Agron.* 20: 48–89.

CGIAR [Consultative Group on International Agricultural Research] (2006) *Pearl Millet: A Hardy Staple for the World's Drylands*. Story of the Month, June 2006. www.cgiar.org/monthlystory/june2006.html.

Christinck A (2002) *"This Seed is Like Ourselves": A Case Study from Rajasthan, India on the Social Aspects of Biodiversity and Farmers' Management of Pearl Millet Seed* (Margraf: Weikersheim, Germany). 190 p.

Clayton WD and Renvoize SA (1982) Gramineae. In: Polhill RM (ed.) *Flora of Tropical Africa, Part 3* (AA Balkema on behalf of the East African Governments: Rotterdam, The Netherlands), pp. 451–898.

Couturon E, Mariac C, Bezançon G, Lauga J, and Renno JF (2003) Impact of natural and human selection on the frequency of the F1 hybrid between cultivated and wild pearl millet (*Pennisetum glaucum* (L.) R. Br.). *Euphytica* 133: 329–337.

de Wet JMJ (1995) Pearl millet. In: Simmonds S and Smartt J (eds.) *Evolution of Crop Plants*. 2nd ed. (Longman Scientific and Technical: Harlow, UK), pp. 157–159.

Diz DA and Schank SC (1993) Characterization of seed-producing pearl millet × elephantgrass hexaploid hybrids. *Euphytica* 67: 143–149.

Dujardin M and Hanna WW (1983) Apomictic and sexual pearl millet × *Pennisetum squamulatum* hybrids. *J. Hered.* 74: 277–279.

—— (1984a) Cytogenetics of double cross hybrids between *Pennisetum americanum-P. purpureum* amphiploids and *P. americanum* × *P. squamulatum* interspecific hybrids. *Theor. Appl. Genet.* 69: 97–100.

—— (1984b) Pseudogamous parthenogenesis and fertilization of a pearl millet × *Pennisetum squamulatum* apomictic derivative. *J. Hered.* 75: 503–504.

—— (1987) Inducing male fertility in crosses between pearl millet and *Pennisetum orientale* Rich. *Crop Sci.* 27: 65–68.

—— (1989a) Crossability of pearl millet with wild *Pennisetum* species. *Crop Sci.* 29: 77–80.

—— (1989b) Developing apomictic pearl millet—characterisation of a BC3 plant. *J. Genet. Breed.* 43: 145–151.

Fussell LK, Pearson CJ, and Norman MJT (1980) Effect of temperature during various growth stages on grain development and yield of *Pennisetum americanum*. *J. Exp. Bot.* 31: 621–633.

Girgi M, O'Kennedy MM, Morgenstern A, Mayer G, Lorz H, and Oldach KH (2002) Transgenic and herbicide-resistant pearl millet (*Pennisetum glaucum* (L.) R. Br.) via microprojectile bombardment of scutellar tissue. *Mol. Breed.* 10: 243–252.

GISD [Global Invasive Species Database] (2007) *Pennisetum*. www.issg.org/data base.

Goldman JJ, Hanna WW, Fleming G, and Ozias-Akins P (2003) Fertile transgenic pearl millet [*Pennisetum glaucum* (L.) R. Br.] plants recovered through micro-projectile bombardment and phosphinothricin selection of apical meristem-, inflorescence-, and immature-embryo-derived embryogenic tissues. *Plant Cell Rep.* 21: 999–1009.

Gonzalez B and Hanna WW (1984) Morphological and fertility responses in isogenic triploid and hexaploid pearl millet × napier grass hybrids. *J. Hered.* 75: 317–318.

Hanna WW (1979) Interspecific hybrids between pearl millet and fountain grass. *J. Hered.* 70: 425–427.

—— (1981) Method of reproduction in napier grass and in the 3*x* and 6*x* alloploid hybrids with pearl millet. *Crop Sci.* 21: 123–126.

—— (1990) Transfer of germplasm from the secondary to the primary gene pool in *Pennisetum*. *Theor. Appl. Genet.* 80: 200–204.

—— (1993) Registration of pearl millet parental lines Tift 8677 and A1/B1 Tift 90D2E1. *Crop Sci.* 33: 1119.

Hanna WW and Dujardin M (1982) Apomictic interspecific hybrids between pearl millet and *Pennisetum orientale*. *Crop Sci.* 22: 857–859.

—— (1986) Cytogenetics of *Pennisetum schweinfurthii* Pilger and its hybrids with pearl millet. *Crop Sci.* 26: 449–453.

Hanna WW, Dujardin M, Ozias-Akins P, Lubbers E, and Arthur L (1993) Repro-duction cytology and fertility of pearl millet × *Pennisetum squamulatum* BC4 plants. *J. Hered.* 84: 213–216.

Harlan JR (1975) Geographic patterns of variation in some cultivated plants. *J. Hered.* 66: 184–191.

Harlan JR and de Wet JMJ (1971) Towards a rational classification of cultivated plants. *Taxon* 20: 509–517.

Holm LG, Plucknett DL, Pancho JV, and Herberger JP (1977) *The World's Worst Weeds: Distribution and Biology* (University Press of Hawaii: Honolulu, HI, USA). 609 p.

Jauhar PP (1968) Inter- and intragenomal chromosome pairing in an interspecific hybrid and its bearing on the basic chromosome number in *Pennisetum*. *Genetica* 39: 360–370.

———— (1981) Cytogenetics of pearl millet. *Adv. Agron.* 34: 407–479.

Kaushal P and Sidhu JS (2000) Pre-fertilization incompatibility barriers to interspecific hybridizations in *Pennisetum* species. *J. Agric. Sci.* 134: 199–206.

Kaushal P, Zadoo SN, Roy AK, and Choubey RN (2003) Interspecific hybrids between tetraploid pearl millet and *P. squamulatum*. *IGFRI [Indian Grassland And Fodder Research Institute] Newsl.* 9: 6.

Kaushal P, Zadoo SN, Malaviya DR, and Roy AK (2005) Apomixis research in India: Past efforts and future strategies. *Curr. Sci.* 89: 1092–1096.

Kumar A, Chowdhury RK, and Dahiya OS (1995) Pollen viability and stigma receptivity in relation to meteorological parameters in pearl millet. *Seed Sci. Technol.* 23: 147–156.

Lambe P, Dinant M, and Matagne RF (1995) Differential long-term expression and methylation of the hygromycin phosphotransferase (hph) and _-glucuronidase (GUS) genes in transgenic pearl millet (*Pennisetum glaucum*) callus. *Plant Sci.* 108: 51–62.

Lebrun JP and Stork AL (1995) *Monocotylédones: Limnocharitaceae à Poaceae.* Vol. 3 in *Énumération des Plantes à Fleurs d'Afrique Tropicale* (Conservatoire et jardin botaniques de la ville de Genève: Geneva, Switzerland). 342 p.

Leuck DB and Burton GW (1966) Pollination of pearl millet by insects. *J. Econ. Entomol.* 59: 1308–1309.

Madhavi AL, Rao KV, Reddy TP, and Reddy VD (2006) Development of transgenic pearl millet (*Pennisetum glaucum* (L.) R. Br.) plants resistant to downy mildew. *Plant Cell Rep.* 25: 927–935.

Marchais L (1994) Wild pearl millet population (*Pennisetum glaucum*, Poaceae) integrity in agricultural Sahelian areas: An example from Keita (Niger). *Plant Syst. Evol.* 189: 233–245.

Marchais L and Tostain S (1992) Bimodal phenotypic structure of two wild pearl millet samples collected in agricultural area. *Biodiv. Cons.* 1: 170–178.

———— (1997) Analysis of reproductive isolation between pearl millet (*Pennisetum glaucum* (L.) R. Br.) and *P. ramosum, P. schweinfurthii, P. squamulatum, Cenchrus ciliaris. Euphytica* 93: 97–105.

Mariac C, Luong V, Kapran I, Mamadou A, Sagnard F, Deu M, Chantereau J, Gérard B, Ndjeunga J, Bezançon G, Pham JL, and Vigouroux Y (2006a) Diversity of wild and cultivated pearl millet accessions (*Pennisetum glaucum* [L.] R. Br.) in Niger assessed by microsatellite markers. *Theor. Appl. Genet.* 114: 49–58.

Mariac C, Robert T, Allinne C, Remigereau MS, Luxereau A, Tidjani M, Seyni O, Bezançon G, Pham JL, and Sarr A (2006b) Genetic diversity and gene flow among pearl millet crop/weed complex: A case study. *Theor. Appl. Genet.* 113: 1003–1014.

Martel E, Poncet V, Lamy F, Siljak-Yakovlev S, Lejeune B, and Sarr A (2004) Chromosome evolution of *Pennisetum* species (Poaceae): Implications of ITS phylogeny. *Plant Syst. Evol.* 249: 139–149.

Mashingaidze K and Muchena SC (1982) The induction of floret sterility by low temperatures in pearl millet (*Pennisetum typhoides* (Burm.) Stapf & Hubbard). *Zimbabwe J. Agric. Res.* 20: 29–37.

Miura R and Terauchi R (2005) Genetic control of weediness traits and the maintenance of sympatric crop-weed polymorphism in pearl millet (*Pennisetum glaucum*). *Mol. Ecol.* 14: 1251–1261.

Nath J, Swaminathan MS, and Mehra KL (1971) Cytological studies in the tribe Paniceae, Gramineae. *Cytologia* 35: 111–131.

NRC [National Research Council] (1996) Pearl millet. In: National Research Council (ed.) *Grains*. Vol. 1 in *Lost Crops of Africa* (National Academy Press: Washington DC, USA), pp. 77–126.

O'Kennedy MM, Burger JT, and Botha FC (2004) Pearl millet transformation system using the positive selectable marker gene phosphomannose isomerase. *Plant Cell Rep.* 22: 684–690.

Ong CK (1983) Response to temperature in a stand of pearl millet (*Pennisetum typhoides* S[tapf] & H[ubb.]): I. Vegetative development. *J. Exp. Bot.* 34: 322–336.

Ozias-Akins P, Lubbers EL, Hanna WW, and McNay JW (1993) Transmission of the apomictic mode of reproduction in *Pennisetum*: Coinheritance of the trait and molecular markers. *Theor. Appl. Genet.* 85: 632–638.

Pantulu JV (1967) Pachytene pairing and meiosis in the F1 hybrid of *Pennisetum typhoides* and *P. purpureum*. *Cytologia* 32: 532–541.

Patil BD and Singh A (1964) An interspecific cross in the genus *Pennisetum* involving two basic numbers. *Curr. Sci.* 33: 255.

Pernès J (1986) L'allogamie et la domestication des céréales: L'exemple du maïs (*Zea mays* L.) et du mil (*Pennisetum americanum* (L.) K. Schum.). *Bull. Soc. Bot. Fr.* 133: 27–34.

Poncet V, Lamy F, Enjalbert J, Joly H, Sarr A, and Robert T (1998) Genetic analysis of the domestication syndrome in pearl millet (*Pennisetum glaucum* L., Poaceae): Inheritance of the major characters. *Heredity* 81: 648–658.

Purseglove JW (1972) *Tropical Crops: Monocotyledons*, vol. 2 (Longman Scientific and Technical: Harlow, UK). 607 p.

Rachie KO and Majmudar JV (1980) *Pearl Millet* (Pennsylvania State University Press: University Park, PA, USA). 307 p.

Read JC and Bashaw EC (1974) Intergeneric hybrid between pearl millet and buffelgrass. *Crop Sci.* 14: 401–403.

Renno JF and Winkel T (1996) Phenology and reproductive effort of cultivated and wild forms of *Pennisetum glaucum* under experimental conditions in the Sahel:

Implications for the maintenance of polymorphism in the species. *Can. J. Bot.* 74: 959–964.

Renno JF, Winkel T, Bonnefous F, and Bezançon G (1997) Experimental study of gene flow between wild and cultivated *Pennisetum glaucum. Can. J. Bot.* 75: 925–931.

Robert T, Lespinasse R, Pernès J, and Sarr A (1991) Gametophytic competition as influencing gene flow between wild and cultivated forms of pearl millet (*Pennisetum typhoides*). *Genome* 34: 195–200.

Robert T, Lamy F, and Sarr A (1992) Evolutionary role of gametophytic selection in the domestication of *Pennisetum typhoides* (pearl millet): A two-locus asymmetrical model. *Heredity* 69: 372–381.

Roche D, Chen Z, Hanna WW, and Ozias-Akins P (2001) Non-Mendelian transmission of an apospory-specific genomic region in a reciprocal cross between sexual pearl millet (*Pennisetum glaucum*) and an apomictic Fı (*P. glaucum* × *P. squamulatum*). *Sex. Plant Reprod.* 13: 217–223.

Sandmeier M (1993) Selfing rates of pearl millet (*Pennisetum typhoides* Stapf & Hubb.) under natural conditions. *Theor. Appl. Genet.* 86: 513–517.

Sarr A, Sandmeier M, and Pernès J (1988) Gametophytic competition in pearl millet, *Pennisetum typhoides* (Burm.) Stapf & Hubb. *Genome* 30: 924–929.

Schmelzer GH (1997) Review of *Pennisetum* section *Brevivalvula* (Poaceae). *Euphytica* 97: 1–20.

Sethi GS, Kalia HR, and Ghai BS (1970) Cytogenetical studies of three interspecific hybrids between *Pennisetum typhoides* Stapf and Hubb. and *P. purpureum* Schumach. *Cytologia* 35: 96–101.

Singh P (1995) Pearl millet (*Pennisetum americanum* L.). In: Muliokela SW (ed.) *Zambia Seed Technology Handbook* (Ministry of Agriculture, Food, and Fisheries: Lusaka, Zambia), pp. 171–182.

Stapf O and Hubbard CE (1934) *Pennisetum.* In: Prain D (ed.) *Flora of Tropical Africa, Part 9* (Crown Agents: London, UK), pp. 954–1070.

Taylor MG and Vasil IK (1991) Histology of, and physical factors affecting, transient GUS expression in pearl millet (*Pennisetum glaucum* (L.) R. Br.) embryos following microprojectile bombardment. *Plant Cell Rep.* 10: 120–125.

Taylor MG, Vasil V, and Vasil IK (1993) Enhanced GUS expression in cereal/grass cell suspensions and immature embryos using the maize ubiquitin-based plasmid pAHC25. *Plant Cell Rep.* 12: 491–495.

Techio VH, Davide LC, and Pereira AV (2006) Meiosis in elephant grass (*Pennisetum purpureum*), pearl millet (*Pennisetum glaucum*) (Poaceae, Poales), and their interspecific hybrids. *Genet. Mol. Biol.* 29: 353–362.

Tostain S (1992) Enzyme diversity in pearl millet (*Pennisetum glaucum* L.): 3. Wild millet. *Theor. Appl. Genet.* 83: 733–742.

USDA [United States Department of Agriculture] (2008) 7 CFR [Code of Federal Regulations] § 201.76. Federal Seed Act Regulations: Minimum Land, Isolation, Field, and Seed Standards. Source: 59 FR [Federal Register] 64516, Dec. 14, 1994, as amended at 65 FR 1710, Jan. 11, 2000 (Government Printing Office: Washington, DC, USA), pp. 372–378. www.access.gpo.gov/nara/cfr/wais idx_08/7cfr201_08.html.

vom Brocke K, Christinck A, Weltzien E, Presterl T, and Geiger HH (2003) Farmers' seed systems and management practices determine pearl millet genetic diversity patterns in semiarid regions of India. *Crop Sci.* 43: 1680–1689.

Zadoo SN and Singh A (1986) Recurrent addition of the *Pennisetum americanum* genome in a *P. americanum* × *P. orientale* hybrid. *Plant Breed.* 97: 187–189.

SIXTEEN

Pigeonpea
(*Cajanus cajan* (L.) Millsp.)

Center(s) of Origin and Diversity

The primary center of origin and diversity for domesticated pigeonpeas is India (De 1974; Maesen 1980, 1986; D. Sharma and Green 1980), while E Africa can be considered as a secondary center of domestication and crop diversity (Maesen 1995). The greatest diversity of species in the genus *Cajanus* DC. is found in N Australia, Yunnan province in China, and Myanmar (Maesen 1986).

Biological Information

Flowering

The inflorescence of the pigeonpea plant is a terminal or axillary raceme. Within the branches and inflorescences, individual flowers open progressively from the bottom to the top (Prasad et al. 1977). As with common beans, the flowering behavior of domesticated pigeonpeas is strongly influenced by photoperiod and temperature (D. Sharma and Green 1980; Wallis et al. 1981; Silim et al. 2006). In dry weather and with low soil-moisture conditions, high temperatures adversely affect the timing and extent of pollination, as well as that of setting pods (Howard et al. 1919; Mahta and Dave 1931).

The individual flowers are similar in structure to those of other legumes, consisting of a calyx with five sepals and a corolla with a standard (flag), two wings, and a keel. As is typical for the Papilionoideae, the anterior keel petals almost enclose the male and female organs. The stigmas become receptive a few days before anthesis, and the anthers usually dehisce one day before the flowers open, which favors high rates of self-pollination (Durga Prasad and Narasimha Murthy 1963; Prasad et al. 1977). The flowers open early in the morning during the summer, and by noon in winter. They remain open from several hours up to a day and a half, depending on weather conditions (Mahta and Dave 1931; D. Sharma and Green 1980). Fertilization occurs 48 to 58 hours after pollination (Datta and Deb 1970).

Pollen Dispersal and Longevity

Bees (*Apis* and *Megachile* Latreille species) are considered to be the main pollinators of pigeonpeas (D. Sharma and Green 1980; Saxena et al. 1990), but thrips (*Taeniothrips distalis* Karny) also seem to facilitate pollination (Sen and Sur 1964).

There appear to be no studies investigating the longevity of pigeonpea pollen under natural conditions in the field. In the laboratory, pigeonpea pollen was found to remain viable up to 42 hours at room temperature (Prasad et al. 1977).

Sexual Reproduction

The pigeonpea plant is a self-compatible, predominantly autogamous diploid ($2n = 2x = 22$) that is generally grown as an annual (Maesen 1995). Some degree of perenniality, however, is present in almost all cultivar types (D. Sharma and Green 1980). Domesticated pigeonpeas are mainly self-fertilizing, yet cross-pollination is frequent and outcrossing rates of 15% to 20% are often observed (Datta and Deb 1970; Veeraswamy et al. 1973; Saxena et al. 1990, 1992; Githiri 1991).

Vegetative Reproduction

The shrubby crop shows strong regrowth after cutting, and ratoons (new shoots from cut stem bases) have been used for experimental crosses and seed multiplication (Saxena et al. 1976; Smartt 1990). However, like most legume crops, as a rule pigeonpeas are propagated by seed.

Seed Dispersal and Dormancy

Most pigeonpea genotypes have been selected against for seed shatter-ing, but some still retain this primitive trait. The pods of wild *Cajanus* species dehisce explosively, and the seeds, which are quite heavy, are deposited no more than a few meters from the parent plant (Maesen 1995).

Most pigeonpea cultivars have lost seed dormancy and germinate right away, although hard-seededness can occasionally be observed. Most wild *Cajanus* species have hard seed coats and show seed dor-mancy for up to a few months (Smartt 1990).

Volunteers, Ferals, and Their Persistence

Volunteer plants and feral populations of domesticated pigeonpeas can establish themselves with relative ease from seeds dispersed through the dehiscence of pods that have been left on the plant after maturity (Smartt 1990).

Since pigeonpeas are shrubby plants, some degree of perenniality is found in most cultivars, which allows the crop to persist for several years in the wild (Ruthenberg 1980; Smartt 1990). However, because pigeonpeas are also highly susceptible to frost (D. Sharma and Green 1980), they do not survive in frost-prone regions.

Weediness and Invasiveness Potential

Neither domesticated pigeonpeas nor their wild relatives are competi-tive with weeds and other plant species in the wild, particularly during their establishment phase (Smartt 1990).

Crop Wild Relatives

Once the three interrelated genera *Cajanus, Atylosia* Wight & Arn., and *Endomallus* Gagnep. were merged (Maesen 1986, 2003), the genus *Cajanus* now consists of over 30 diploid ($2n = 2x = 22$) species (table 16.1). Six sections can be distinguished, based on morphological char-acters (Maesen 1986).

Table 16.1 Domesticated pigeonpea (*Cajanus cajan* (L.) Millsp.) and its wild relatives

Species	Origin and distribution
Primary gene pool (GP-1)	
C. cajan (L.) Millsp. (pigeonpea)	known only from cultivation; pantropical
Secondary gene pool (GP-2)	
C. acutifolius (F. Muell.) Maesen	Australia
C. albicans (Wight & Arn.) Maesen	S India, Sri Lanka
C. cajanifolius (Haines) Maesen	SE India
C. lanceolatus (W. Fitzg.) Maesen	Australia
C. latisepalus (S.T. Reynolds & Pedley) Maesen	Australia
C. lineatus (Wight & Arn.) Maesen	S India, Sri Lanka
C. reticulatus (Aiton) F. Muell.	Australia, New Guinea
C. scarabaeoides (L.) Graham *ex* Wall.	Australia, SE Asia, Pacific coastal Africa
C. sericeus (Benth. *ex* Baker) Maesen	S India
C. trinervius (DC.) Maesen	S India, Sri Lanka
Tertiary gene pool (GP-3)	
C. aromaticus Maesen	Australia
C. cinereus (F. Muell.) F. Muell.	Australia
C. confertiflorus F. Muell.	Australia
C. crassicaulis Maesen	Australia
C. crassus (Prain *ex* King) Maesen	SE Asia, China
C. elongatus (Benth.) Maesen	NE India, Vietnam
C. geminatus Pedley *ex* Maesen	N Australia
C. goensis Dalzell	India, SE Asia
C. grandiflorus (Benth. *ex* Baker) Maesen	NE India, S China
C. heynei (Wight & Arn.) Maesen	SW India, Sri Lanka
C. hirtopilosus Maesen	N Australia
C. kerstingii Harms	W Africa
C. lanuginosus Maesen	Australia
C. mareebensis (S.T. Reynolds & Pedley) Maesen	Australia
C. marmoratus (R. Br. *ex* Benth.) F. Muell.	Australia
C. membranifolius Maesen	Philippines, Indonesia
C. mollis (Benth.) Maesen	Himalayan foothills
C. niveus (Benth.) Maesen	Myanmar, S China
C. platycarpus (Benth.) Maesen	Indian subcontinent, Java
C. pubescens (Ewart & Morrison) Maesen	Australia
C. rugosus (Wight & Arn.) Maesen	S India, Sri Lanka
C. villosus (Benth. *ex* Baker) Maesen	NE India
C. viscidus Maesen	Australia
C. volubilis (Blanco) Blanco	Myanmar, India, SE Asia

Source: Maesen 1986.
Note: Gene pool designations follow Smartt 1990.

Most *Cajanus* species are native to S and SE Asia (18 species) and Australia (15), and one to W Africa (Fortunato 2000). Many of them are endemic, either to Australia (13 species) or to the Indian subcontinent and Myanmar (8). Only domesticated pigeonpeas (*C. cajan* (L.) Millsp.) and *C. scarabaeoides* (L.) Thouars have a widespread Asiatic-Australasian distribution (Smartt 1990).

Hybridization

Primary Gene Pool (GP-1)

The primary gene pool for pigeonpeas contains only the crop itself. Smartt (1990) classified the crop wild relatives of the genus *Cajanus* into secondary and tertiary gene pools, according to their crossability with the crop (table 16.1). In general, the Indian relatives are more closely related to domesticated pigeonpeas than the Australian species (Maesen 1995).

Secondary Gene Pool (GP-2)

The secondary gene pool is composed of 10 of the most closely related wild species (table 16.1). Of these, *C. cajanifolius* (Haines) Maesen is considered to be the species most closely related to the crop and its putative progenitor, based on morphological (Maesen 1980, 1986, 1990), cytological, and crossability studies (Pundir and Singh 1985a, b; Saxena 2005); isozymes (Krishna and Reddy 1982); seed protein (Panigrahi et al. 2007); and molecular markers (Nadimpalli et al. 1993; Parani et al. 2000). Spontaneous outcrossing between domesticated pigeonpeas and *C. cajanifolius* has been reported in greenhouse nurseries in India (Maesen 1980, 1990).

Most GP-2 relatives show a high degree of cross-compatibility with the crop, but some rather well-developed barriers to hybridization are present in most GP-2 species, and thus far no spontaneous hybrids have been reported in the wild (Maesen 1986).

Hybrid progeny between pigeonpeas and their GP-2 wild relatives can be obtained relatively easily up to F4 by hand pollination (see, e.g., Deodikar and Thakar 1956; Roy and De 1965; Sikdar and De 1967;

B. Reddy et al. 1977, 1978; Dundas 1990). Crosses are more likely to succeed if the wild relative is used as the pollen donor (De 1974). In general, the F1 progeny have reduced fertility, but seed setting is often observed (L. Kumar et al. 1958; L. Reddy 1973; De 1974; Pundir and Singh 1985c).

Interspecific gene transfer is possible at times through conventional hybridization, but more often the use of artificial techniques is required to overcome crossability barriers (Mallikarjuna 1998; Mallikarjuna 2003). For example, using gibberellic acid and embryo rescue, gene transfer to the crop has been achieved from *C. acutifolius* (F. Muell.) Maesen (Mallikarjuna and Saxena 2002), *C. sericeus* (Benth. *ex* Baker) Maesen (Ariyanayagam et al. 1995), and *C. scarabaeoides* (Tikka et al. 1997; Saxena and Kumar 2003; Aruna et al. 2005).

Seed protein studies provide some evidence that introgression of wild *Cajanus* alleles to domesticated pigeonpeas may have occurred (Smartt 1990). However, no molecular evidence has been provided thus far, and more detailed studies would be required to confirm or refute the possibility of permanent introgression between the crop and its wild relatives.

Tertiary Gene Pool (GP-3)

The remaining *Cajanus* species constitute the tertiary pigeonpea gene pool (Smartt 1990). Crosses with the crop can only be obtained through considerable human intervention, and F1 hybrids are, by and large, either male sterile or completely sterile (see, e.g., L. Reddy and Faris 1981; Mallikarjuna and Moss 1995; Tikka et al. 1997; Wanjari et al. 2001; Bajpai et al. 2003; Mallikarjuna 2003; Saxena and Kumar 2003; Mallikarujuna et al. 2006). Crosses with species outside of the genus (e.g., *Dunbaria ferruginea* Wight & Arn. and *Rhynchosia* spp.) have failed to produce hybrids thus far (McComb 1975; Maesen 1995).

Outcrossing Rates and Distances

The isolation distances established by regulatory authorities for the seed production of pigeonpeas range from 100 to 200 m (table 16.2). The recommended isolation distance for pigeonpea hybrid seed production

Table 16.2   Isolation distances for domesticated pigeonpea (*Cajanus cajan* (L.) Millsp.)

Threshold*	Minimum isolation distance (m)	Country	Reference	Comment
1 plant in 30 m²	200	OECD	OECD 2008	foundation (basic) seed requirements (fields < 2 ha)
1 plant in 10 m²	200	OECD	OECD 2008	certified seed requirements (fields < 2 ha)
1 plant in 30 m²	100	OECD	OECD 2008	foundation (basic) seed requirements (fields > 2 ha)
1 plant in 10 m²	100	OECD	OECD 2008	certified seed requirements (fields > 2 ha)

*Maximum number of off-type plants permitted in the field.

is 500 m (Saxena 2006). However, to our knowledge no systematic pollen flow studies have been conducted to measure pollen dispersal distances and rates of outcrossing.

State of Development of GM Technology

Success with the genetic modification of pigeonpeas is recent, as the lack of methods for efficiently regenerating pigeonpeas has long been an obstacle to the development of transformation protocols for the crop (see, e.g., Shiva Prakash et al. 1994; Franklin et al. 1998). Stable GM protocols for different explants using *Agrobacterium* or particle bombardment only have become available lately (see, e.g., Geetha et al. 1999; Dayal et al. 2003; Satyavathi et al. 2003; S. Kumar et al. 2004a, b; K. Sharma et al. 2006a, b). The GM traits researched to date are increased lysine content and pod-borer resistance (Thu et al. 2003, 2007). Field trials have been reported in India, but thus far there is no commercial production of GM pigeonpeas.

Agronomic Management Recommendations

There are no specific recommendations.

Crop Production Area

Pigeonpeas are grown in about 20, mostly tropical, countries and cover 4.6 million ha. India raises about 75% of the world's production, followed by Myanmar (16%) (FAO 2009). Other major producers are E Africa and the Caribbean. The crop has recently been reintroduced in China, and its production area has extended from only 50 ha in Guangxi and Yunnan provinces in 1999 to 100,000 ha in 12 provinces in 2006 (CGIAR 2007).

Research Gaps and Conclusions

An assessment of the likelihood of gene flow in pigeonpeas would benefit from more detailed work on pollen flow and outcrossing rates at different distances. Also, more studies are warranted to better understand the relationships between the crop and its sexually compatible wild relatives—especially with its putative progenitor, *C. cajanifolius*—including biosystematic and molecular studies. Additional collecting trips would be very helpful, given the scarcity of wild material in *ex situ* conservation genebanks. Questions awaiting answers include the likelihood and extent of outcrossing, and the fate of introgressed alleles in later generations.

Based on the limited knowledge currently at hand for pigeonpeas and their wild relatives, the likelihood of gene flow and introgression from the crop—assuming physical proximity (less than 100 m) and flowering overlap, and being conditional on the nature of the respective trait(s)—is as follows:

- *moderate* for its GP-2 wild relatives *Cajanus acutifolius*, *C. lanceolatus* (W. Fitzg.) Maesen, *C. latisepalus* (S.T. Reynolds & Pedley) Maesen, *C. reticulatus* (Aiton) F. Muell., and *C. scarabaeoides* (all occur in Australia, and some additionally in SE Asia and E Africa); and *C. albicans* (Wight & Arn.) Maesen, *C. cajanifolius*, *C. lineatus* (Wight & Arn.) Maesen, *C. sericeus*, and *C. trinervius* (DC.) Maesen (in India and Sri Lanka) (map 16.1)

REFERENCES

Ariyanayagam RP, Rao AN, and Zaveri PP (1995) Cytoplasmic-genic male sterility in interspecific matings of *Cajanus*. *Crop Sci.* 35: 981–985.

Aruna R, Manohar Rao D, Reddy LJ, Upadhyaya HD, and Sharma HC (2005) Inheritance of trichomes and resistance to pod borer (*Helicoverpa armigera*) and their association in interspecific crosses between cultivated pigeonpea (*Cajanus cajan*) and its wild relative C. *scarabaeoides*. *Euphytica* 145: 247–257.

Bajpai GC, Singh J, and Tewari SK (2003) Wide hybridization and its genetic and cytogenetic consequences in pigeonpea—a review. *Agric. Rev.* 24: 265–274.

CGIAR [Consultative Group on International Agricultural Research] (2007) Homing pigeonpea. *CGIAR News*, March 2007 (Consultative Group on International Agricultural Research [CGIAR] Secretariat: Washington, DC, USA). www.cgiar.org/enews/march2007/story_10.html.

Datta PC and Deb A (1970) Floral biology of *Cajanus cajan* (L.) Millsp. var. *bicolor* DC. (Papilionaceae). *Bull. Bot. Soc. Beng.* 24: 135–145.

Dayal S, Lavanya M, Devi P, and Sharma KK (2003) An efficient protocol for shoot regeneration and genetic transformation of pigeonpea [*Cajanus cajan* (L.) Millsp.] using leaf explants. *Plant Cell Rep.* 21: 1072–1079.

De DN (1974) Pigeon pea. In: Hutchinson JB (ed.) *Evolutionary Studies in World Crops: Diversity and Change in the Indian Subcontinent* (Cambridge University Press: Cambridge, UK), pp. 79–87.

Deodikar GB and Thakar CV (1956) Cytotaxonomic evidence for affinity between *Cajanus indicus* Spreng. and certain erect species of *Atylosia* W[ight] & A[rn]. *Proc. Indian Acad. Sci., B* 43: 37–45.

Dundas IS (1990) Pigeonpea: Cytology and cytogenetics—perspective and prospects. In: Nene YL, Hall SD, and Sheila VK (eds.) *The Pigeonpea* (CAB International and International Crops Research Institute for the Semi-Arid Tropics [ICRISAT]: Wallingford, UK), pp. 117–136.

Durga Prasad MMK and Narasimha Murthy BL (1963) Some observations on anthesis and pollination in redgram (*Cajanus cajan* L.). *Andhra Agric. J.* 10: 161–167.

FAO [Food and Agriculture Organization of the United Nations] (2009) FAOSTAT data. http://faostat.fao.org/site/567/default.aspx.

Fortunato RH (2000) Systematic relationships in *Rhynchosia* (Cajaninae-Phaseoleae-Papilionoideae-Fabaceae) from the neotropics. In: Herendeen PS and Bruneau A (eds.) *Advances in Legume Systematics, Part 9* (Royal Botanic Gardens, Kew: Richmond, Surrey, UK), pp. 339–354.

Franklin G, Jeyachandran R, Melchias G, and Ignacimuthu S (1998) Multiple shoot induction and regeneration of pigeon pea (*Cajanus cajan* (L.) Millsp.) cv. Vamban₁ from apical and axillary meristem. *Curr. Sci.* 74: 936–937.

Geetha N, Venkatachalam P, and Sita GL (1999) *Agrobacterium*-mediated genetic transformation of pigeonpea (*Cajanus cajan* L.) and development of transgenic plants via direct organoenesis. *Plant Biotech.* 16: 213–218.

Githiri SM (1991) Natural outcrossing in dwarf pigeonpea. *Euphytica* 53: 37–39.

Howard A, Howard GLC, and Khan R (1919) Studies of the pollination of Indian crops. *Mem. Dep. Agric. India, Bot.* 3: 195–225.

Krishna TG and Reddy LJ (1982) Species affinities between *Cajanus cajan* and some *Atylosia* species based on esterase isoenzymes. *Euphytica* 31: 709–713.

Kumar LSS, Thombre MV, and D'Cruz R (1958) Cytological studies of an intergeneric hybrid of *Cajanus cajan* (Linn.) Millsp. and *Atylosia lineata* W[ight] & A[rn]. *Proc. Indian Acad. Sci., B* 47: 252–261.

Kumar SM, Syamala D, Sharma KK, and Devi P (2004a) *Agrobacterium tumefaciens*-mediated genetic transformation of pigeonpea [*Cajanus cajan* (L.) Millsp.]. *J. Plant Biotech.* 6: 69–75.

Kumar SM, Kumar BK, Sharma KK, and Devi P (2004b) Genetic transformation of pigeonpea with rice chitinase gene. *Plant Breed.* 123: 485–489.

Maesen LJG van der (1980) India is the native home of the pigeon pea. In: Arends JC, Boelema G, de Groot CT, and Leeuwenberg AJM (eds.) *Liber Gratulatorius in Honorem HCD de Wit*. Landbouwhogeschool Miscellaneous Paper No. 19 (H Veenman and Zonen BV: Wageningen, The Netherlands), pp. 257–263.

———— (1986) *Cajanus DC. and Atylosia W[ight] & A[rn]. (Leguminosae): A Revision of All Taxa Closely Related to the Pigeonpea, with Notes on Other Related Genera within the Subtribe Cajaninae*. Agricultural University Wageningen Papers 85–4 (Department of Plant Taxonomy, Agricultural University Wageningen: Wageningen, The Netherlands). 225 p.

———— (1990) Pigeonpea: Origin, history, evolution, and taxonomy. In: Nene YL, Hall SD, and Sheila VK (eds.) *The Pigeonpea* (CAB International and International Crops Research Institute for the Semi-Arid Tropics [ICRISAT]: Wallingford, UK), pp. 15–45.

———— (1995) Pigeonpea. In: Simmonds S and Smartt J (eds.) *Evolution of Crop Plants*. 2nd ed. (Longman Scientific and Technical: Harlow, UK), pp. 251–254.

———— (2003) Cajaninae of Australia (Leguminosae: Papilionoideae). *Aust. Syst. Bot.* 16: 219–227.

Mahta DN and Dave BB (1931) Studies in *Cajanus indicus*. *Mem. Dep. Agric. India, Bot.* 19: 1–25.

Mallikarjuna N (1998) Ovule culture to rescue aborting embryos from pigeonpea (*Cajanus cajan* (L.) Millspaugh) wide crosses. *Indian J. Exp. Biol.* 36: 225–228.

————— (2003) Wide hybridization in important food legumes. In: Jaiwal PK and Singh RP (eds.) *Improvement Strategies of Leguminosae Biotechnology* (Kluwer Academic Publishers: Dordrecht, The Netherlands), pp. 155–170.

Mallikarjuna N and Moss JP (1995) Production of hybrids between *Cajanus platycarpus* and *Cajanus cajan*. *Euphytica* 83: 43–46.

Mallikarjuna N and Saxena KB (2002) Production of hybrids between *Cajanus acutifolius* and *C. cajan*. *Euphytica* 124: 107–110.

Mallikarjuna N, Jadhav D, and Reddy P (2006) Introgression of *Cajanus platycarpus* genome into cultivated pigeonpea, *C. cajan*. *Euphytica* 149: 161–167.

McComb JA (1975) Is intergeneric hybridization in the Leguminosae possible? *Euphytica* 24: 497–502.

Nadimpalli RJ, Jarret RL, Pathak SC, and Kochert G (1993) Phylogenetic relationships of the pigeonpea (*C. cajan*) based on nuclear restriction fragment length polymorphisms. *Genome* 36: 216–223.

OECD [Organization for Economic Cooperation and Development] (2008) OECD Scheme for the Varietal Certification of Grass and Legume Seed Moving in International Trade: Annex VI to the Decision. C(2000)146/FINAL. OECD Seed Schemes, February 2008. www.oecd.org/document/0/0,3343,en_2649_33905_1933504_1_1_1_1,00.html.

Panigrahi J, Kumar DR, Mishra M, Mishra RP, and Jena P (2007) Genomic relationships among 11 species in the genus *Cajanus* as revealed by seed protein (albumin and globulin) polymorphisms. *Plant Biotech. Rep.* 1: 109–116.

Parani M, Lakshmi M, Senthilkumar P, and Parida A (2000) Ribosomal DNA variation and phylogenetic relationships among *Cajanus cajan* (L.) Millsp. and its wild relatives. *Curr. Sci.* 78: 1235–1238.

Prasad S, Prakash R, and Haque MF (1977) Floral biology of pigeonpea. *Trop. Grain Leg. Bull.* 7: 12.

Pundir RPS and Singh RB (1985a) Biosystematic relationships among *Cajanus, Atylosia*, and *Rhynchosia* species and evolution of pigeonpea. *Theor. Appl. Genet.* 69: 531–534.

————— (1985b) Crossability relationships among *Cajanus, Atylosia*, and *Rhynchosia* species and detection of crossing barriers. *Euphytica* 34: 303–308.

————— (1985c) Cytogenetics of F1 hybrids between *Cajanus* and *Atylosia* species and its phylogenetic implications. *Theor. Appl. Genet.* 71: 216–220.

Reddy BVS, Reddy LJ, and Murthy AN (1977) Reproductive variants in *Cajanus cajan* (L.) Millsp. *Trop. Grain Leg. Bull.* 7: 11.

Reddy BVS, Green JM, and Bisen SS (1978) Genetic male sterility in pigeonpea. *Crop Sci.* 18: 362–364.

Reddy LJ (1973) Interrelationships of *Cajanus* and *Atylosia* species as revealed by hybridization and pachytene analysis. PhD dissertation, Indian Institute of Technology.

Reddy LJ and Faris DJ (1981) A cytoplasmic-genetic male-sterile line in pigeonpea. *Int. Pigeonpea Newsl.* 1: 16–17.

Roy A and De ND (1965) Intergeneric hybridization of *Cajanus* and *Atylosia. Sci. Cult.* 31: 93–95.

Ruthenberg H (1980) *Farming Systems in the Tropics.* 3rd ed. (Oxford University Press: London, UK). 424 p.

Satyavathi VV, Prasad V, Khandelwal A, Shaila MS, and Sita GL (2003) Expression of haemagglutinin protein of Rinderpest virus in transgenic pigeonpea [*Cajanus cajan* (L.) Millsp.] plant. *Plant Cell Rep.* 21: 651–658.

Saxena KB (2005) A cytoplasmic-nuclear male-sterility system derived from a cross between *Cajanus cajanifolius* and *Cajanus cajan. Euphytica* 145: 289–294.

——— (2006) *Hybrid Pigeonpea Seed Production Manual.* Information Bulletin No. 74 (International Crops Research Institute for the Semi-Arid Tropics [ICRISAT]: Patancheru, India). 32 p.

Saxena KB and Kumar RV (2003) Development of a cytoplasmic nuclear male-sterility system in pigeonpea using *C. scarabaeoides* (L.) Thouars. *Indian J. Genet. Plant Breed.* 63: 225–229.

Saxena KB, Sharma D, and Green JM (1976) Pigeonpea rattooning—an aid to breeders. *Trop. Grain Leg. Bull.* 4: 21.

Saxena KB, Singh L, and Gupta MD (1990) Variation for natural outcrossing in pigeonpea. *Euphytica* 46: 143–148.

Saxena KB, Ariyanayagam P, and Reddy LJ (1992) Genetics of a high-selfing trait in pigeonpea. *Euphytica* 59: 125–127.

Sen SK and Sur SC (1964) A study on vicinism in pigeonpea (*Cajanus cajan* (L.) Millsp.). *Agricultura* (Louvain) 12: 421–426.

Sharma D and Green JM (1980) Pigeonpea. In: Fehr W and Hadley HH (eds.) *Hybridization of Crop Plants* (American Society of Agronomy: Madison, WI, USA), pp. 471–481.

Sharma KK, Lavanya M, and Anjaiah V (2006a) *Agrobacterium*-mediated production of transgenic pigeonpea (*Cajanus cajan* (L.) Millsp.) expressing the synthetic *Bt cry1AB* gene. In Vitro *Cell. Dev. Biol., Pl.* 42: 165–173.

Sharma KK, Sreelatha G, and Dayal S (2006b) Pigeonpea (*Cajanus cajan* (L.) Millsp.). *Methods Mol. Biol.* 343: 359–367.

Shiva Prakash N, Pental D, and Bhalla-Sarin N (1994) Regeneration of pigeonpea (*Cajanus cajan*) from cotyledonary node via multiple shoot formation. *Plant Cell Rep.* 13: 623–627.

Sikdar AK and De DN (1967) Cytological studies of two species of *Atylosia. Bull. Bot. Soc. Bengal* 21: 25–28.

Silim SN, Coe R, Omanga PA, and Gwata ET (2006) The response of pigeonpea genotypes of different duration types to variation in temperature and photoperiod under field conditions in Kenya. *J. Food Agric. Environ.* 4: 209–214.

Smartt J (1990) The pigeonpea (*Cajanus cajan* (L.) Millsp.). In: Smartt J (ed.) *Grain Legumes: Evolution and Genetic Resources* (Cambridge University Press: Cambridge, UK), pp. 278–294.

Thu TT, Xuan Mai TT, Dewaele E, Farsi S, Tadesse Y, Angenon G, and Jacobs M (2003) *In vitro* regeneration and transformation of pigeonpea [*Cajanus cajan* (L.) Millsp]. *Mol. Breed.* 11: 159–168.

Thu TT, Dewaele E, Trung LQ, Claeys M, Jacobs M, and Angenon G (2007) Increasing lysine levels in pigeonpea (*Cajanus cajan* (L.) Millsp.) seeds through genetic engineering. *Plant Cell Tissue Organ Cult.* 91: 135–143.

Tikka SBS, Parmer LD, and Chauhan RM (1997) First record of cytoplasmic-genic male-sterility system in pigeonpea (*Cajanus cajan* (L.) Millsp.) through wide hybridization. *Gujarat Agric. Univ. Res. J.* 22: 160–162.

Veeraswamy R, Palaniswamy GA, and Rathnaswamy R (1973) Natural cross-pollination in *Cajanus cajan* (L.) Millsp. and *Lablab niger* Medikus. *Madras Agric. J.* 60: 1828.

Wallis ES, Byth DE, and Whiteman PC (1981) Mechanized dry seed production of pigeonpea. In: Nene YL and Kumble V (eds.) *Proceedings of the International Workshop on Pigeonpeas, ICRISAT [International Crops Research Institute for the Semi-Arid Tropics] Center, Patancheru, India, December 15–19, 1980*, vol. 1 (International Crops Research Institute for the Semi-Arid Tropics [ICRISAT]: Patancheru, India), pp. 51–60.

Wanjari KB, Patil AN, Manapure P, Manjayya JG, and Manish P (2001) Cytoplasmic male-sterility in pigeonpea with cytoplasm from *Cajanus volubilis*. *Ann. Plant Physiol.* 13: 170–174.

SEVENTEEN

Potato
(*Solanum tuberosum* L.)

Center(s) of Origin and Diversity

Domesticated potatoes originated in the Andes of S Peru (Ross 1986; Spooner et al. 2005; Spooner and Hetterscheid 2006).

The center of diversity for wild tuber-bearing potatoes (sect. *Petota* Dumort.) is in Latin America, which is also the center of origin for domesticated potatoes. Peru and Bolivia have the highest numbers of wild potato species (91 and 36, respectively), followed by N Argentina, central Bolivia, central Ecuador, and central Mexico (Hijmans and Spooner 2001).

Biological Information

Solanum tuberosum L. belongs to the taxonomic sect. *Petota* subsect. *Potatoe* ser. *Tuberosa* Bukasov. The domesticated potato (*S. tuberosum*) includes modern cultivars (accounting for most of the commercial potato production worldwide) and eight landrace cultivar groups that are mainly cultivated in S America. Prior to being included in *S. tuberosum*, these eight groups had previously been classified as seven separate species (Huamán and Spooner 2002).

This chapter will focus on the likelihood of gene flow from modern cultivars to landrace cultivars and wild relatives, since up until now, modern cultivars are the ones most frequently genetically modified. However, it is important to note that there is a growing demand for diploid potatoes (e.g., diploid 'Phureja' cultivars) on international markets and that protocols have been tested for the genetic modification of 'Phureja' cultivars (Ducreux et al. 2005).

Flowering

In general, the flowers of *Solanum* L. are actinomorphic, with a five-lobed calyx and corolla, five stamens, and a two-carpellate superior ovary. The anthers of most *Solanum* species, including potatoes, dehisce via terminal pores. This facilitates buzz-pollination by bees that cling to the anther cone and discharge pollen through the pores by vibrating their indirect flight muscles—a pollination syndrome found in about 200 genera of flowering plants (Buchmann 1983). These flowers have no nectar, and pollen is the sole floral reward.

Modern potato cultivars flower less profusely than landrace or wild material and often drop their blossoms after pollination. Successful flower development does not ensure fruit set, since reduced pollen fertility, or even pollen sterility, is very common among modern cultivars. Reduced female (i.e., ovule) fertility can also occur. Most modern cultivars seldom or never produce seeds (Burton 1989).

Several cultivars, however, do produce sexual seeds—also termed "true potato seeds" to distinguish them from asexual "seeds" (potato tubers) used by potato growers (Arndt et al. 1990). The domesticated potato is not a true breeder, in that seeds from the same fruit differ genetically among each other and from the parent plant. The extent and duration of flowering, and the amount of seed production, depend on the genotype and on climatic and environmental conditions, including light intensity and duration (day length), temperature, water supply, and available soil nutrients (Eastham and Sweet 2002).

Pollen Dispersal and Longevity

Potatoes do not produce nectar and are therefore not attractive to honeybees (*Apis mellifera*) and other pollinating insects, such as *Bombus fervidus* Fabr. (Sanford and Hanneman 1981). Some insects, though, do visit potato flowers and facilitate self- and cross-pollination. This is the case for certain bumblebee species, such as *Bombus funebris* Smith and *B. opifex* Smith in S America, and *B. impatiens* in the United States, all of which are considered to be good pollinators for potatoes (White 1983). In Peru, 12 native bee species were recorded visiting potato flowers (Celis et al. 2004). The potato beetle *Meligethes aeneus* Fabr. is also a pollinator of modern cultivars in Europe (Skogsmyr 1994; Petti et al. 2007).

Bumblebees typically forage over several hundred meters, and even kilometers (Osborne et al. 1999; Goulson and Stout 2001; Knight et al. 2005). Yet bumblebee cross-pollination is very localized around the pollen source, because pollen is often deposited on a limited number of nearby flowers visited directly afterwards (Free and Butler 1959; Heinrich 1979; Creswell et al. 2002).

The role that wind plays in potato outcrossing is not clear. Since potato pollen is rather large and dense, its wind-dispersal range seems to be limited to short distances of less than 10 m (White 1983; McPartlan and Dale 1994). Some authors, however, consider wind pollination to be more important than insect pollination (Eastham and Sweet 2002; Tolstrup et al. 2003).

Under natural conditions, potato pollen dies within a few (3–7) days of its release from the anthers (Howard 1958; Trognitz 1991).

Sexual Reproduction

Modern potato cultivars, grown in most parts of the world, are tetraploids ($2n = 4x = 48$). Among the landraces, which are grown almost entirely in S America, diploid ($2n = 2x = 24$), triploid ($3x = 36$), tetraploid, and pentaploid ($5x = 60$) types are found.

Domesticated potatoes are mainly self-pollinating, but cross-pollination (up to 20%) can occur (Plaisted 1980; Brown 1993). By and large, polyploids within sect. *Petota* are inbreeding, whereas diploids are self-incompatible, and thus depend on insect pollination (Kirch et al. 1989; Simmonds 1995).

Many cultivated and wild diploid potatoes produce unreduced gametes ($2n$) as well as normal haploid gametes (n), a phenomenon quite common in *Solanum* species (Hanneman 1995; Carputo et al. 2000). The triploid and pentaploid landrace cultivars seem to be fully sterile and strictly maintained by tuber propagation (clonal) by farmers.

Vegetative Reproduction

Because of the highly heterozygous nature of the domesticated potato, both modern and landrace cultivars are not maintained through seed propagation but instead reproduce vegetatively, through the planting of tubers or tuber sections having at least one eye.

Seed Dispersal and Dormancy

The number of true seeds produced in commercial fields of fertile modern potato cultivars depends on the cultivar grown, environmental conditions, and pest/pathogen activity, but it can be as high as 150–250 million seeds/ha (Accatino 1980; Lawson 1983). True potato seeds can form persistent seed banks in potato fields (Eastham and Sweet 2002) and remain viable in the soil for 7 to as many as 20 years, regardless of temperature (Lawson 1983; Love 1994; Tolstrup et al. 2003).

The entire potato plant contains toxic glycoalkaloids (Filadelfi 1982; Dalvi and Bowie 1983; Morris and Lee 1984; Sharma and Salunkhe 1989), and the concentrations in ripe fruits (berries) can be 10–20 times those of the tubers (Coxon 1981; Friedman 1992). The fruits of several *Solanum* species that are closely related to potatoes and have similar glycoalkaloid components and concentrations are dispersed by frugivore animals, including birds, bats, and rodents (see, e.g., Tamboia et al. 1996; Cipollini and Levey 1997a, b; Wahaj et al. 1998; Albuquerque et al. 2006). However, no scientific evidence thus far has either proven or refuted whether animals (birds and mammals) ingest—and thus possibly disperse—potato berries.

Potato tubers can remain in the field after harvest or be dispersed by handling and transportation of the crop from the fields to the stores. These tubers can germinate and result in volunteers and ferals that may then flower and eventually produce true potato seeds.

Tubers of domesticated potatoes have a fairly low frost tolerance (Palta et al. 1981). Shallow tubers are often destroyed by frost periods of about 50 hours at $-2°C$ or 5 hours at $-10°C$ (Lumkes and Sijtsma 1972). In temperate climates, up to 80% of the tubers left in the soil die even in mild winters (Lutman 1977). The remaining tubers show no dormancy and will sprout the next season (Conner 2007).

Tubers in deeper soil are seldom exposed to temperatures low enough to render them unviable, but they are also unlikely to sprout unless brought near the surface by tillage (Davies et al. 1999; Steiner et al. 2005). In frost-free regions, the tubers of modern potato cultivars can persist for several years (Hawkes 1990; Crawley et al. 2001).

Volunteers, Ferals, and Their Persistence

For the most part, volunteers are established from ground-keepers, that is, tubers left in the soil at harvest time that overwinter and emerge in the next and subsequent seasons. In potato-growing areas where potatoes are cultivated in rotation, volunteers occur to a greater or lesser extent in almost all subsequent planting seasons (Askew 1993). Volunteers established from tubers also occasionally occur near animal feedlots and at waste-disposal sites. The number of tubers left in the soil after a harvest depends on harvesting efficiency, which in turn reflects the effectiveness of either hand lifting or the quality of the machinery, as well as the weather and soil conditions at harvest. In Denmark, the number of residual tubers is estimated to range from 500 to 40,000/ha (Tolstrup et al. 2003).

Ferals often are established outside of cultivation from potato tubers lost during transportation. Ferals and volunteers act as sources of pollen flow outside the cultivation periods and area.

Although volunteers can survive over the winter and persist in rotation fields (Green et al. 2005), they usually last only one or just a few years, due to standard weed control practices and unfavorable environmental conditions, such as low temperatures (Makepeace et al. 1978). Ferals of modern commercial varieties do not become permanently established outside of cultivation, since they are at a competitive disadvantage and do not survive in the wild (Makepeace et al. 1978; Evenhuis and Zadoks 1991; Love 1994).

Weediness and Invasiveness Potential

Volunteers from modern cultivars constitute a short-term weed problem in areas of commercial potato cultivation (Lutman 1977; Lawson 1983), but they are often controlled by conventional agronomic prac-

Table 17.1 The wild relatives (*Solanum* sect. *Petota* Dumort. and sect. *Etuberosum* (Juz. ex Bukasov & Kameraz) A. Child) of domesticated potato (*Solanum tuberosum* L.), their ploidy and endosperm balance numbers (EBN), and countries where they occur

Wild species	Ploidy level and EBN number	Country
Primary gene pool (GP-1)		
Solanum albicans (Ochoa) Ochoa, *S. demissum* Lindl., *S. guerreroense* Correll, *S. bougasii* Correll, *S. iopetalum* (Bitter) Hawkes, *S. jaenense* Ochoa, *S. moscopanum* Hawkes, *S. nemorosum* Ochoa, *S. oplocense* Hawkes, *S. schenckii* Bitter, *S. tundalomense* Ochoa	6× (4EBN)	ARG, BOL, COL, ECU, GUA, MEX, PER
Solanum acaule subsp. *palmirense* Kardolus	6×	ECU
Solanum leptophyes Bitter, *S. oplocense* Hawkes, *S. ×sucrense* Hawkes	4× (4EBN)	ARG, BOL, PER
Solanum acaule Bitter, *S. bombycinum* Ochoa, *S. flahaultii* Bitter, *S. hoopesii* Hawkes & K.A. Okada, *S. longiconicum* Bitter, *S. neovalenzuelae* L.E. López, *S. pamplonense* L.E. López, *S. subpanduratum* Ochoa, *S. ugentii* Hawkes & K.A. Okada	4×	BOL, COL, CRI, MEX, PAN, VEN
Solanum acroglossum Juz., *S. albornozii* Correll, *S. amayanum* Ochoa, *S. ambosinum* Ochoa, *S. anamatophilum* Ochoa, *S. ancophilum* (Correll) Ochoa, *S. andreanum* Baker, *S. ariduphilum* Ochoa, *S. ayacuchense* Ochoa, *S. berthaultii* Hawkes, *S. ×blanco-galdosii* Ochoa, *S. boliviense* Dunal, *S. brevicaule* Bitter, *S. buesii* Vargas, *S. bukasovii* Juz., *S. candolleanum* P. Berthault, *S. cantense* Ochoa, *S. chacoense* Bitter, *S. chilliasense* Ochoa, *S. chiquidenum* Ochoa, *S. chomatophilum* Bitter, *S. coelestipetalum* Vargas, *S. contumazaense* Ochoa, *S. ×doddsii* Correll, *S. gandarillasii* Cárdenas, *S. hastiforme* Correll, *S. huancabambense* Ochoa, *S. huancavelicae* Ochoa, *S. huarochiriense* Ochoa, *S. incasicum* Ochoa, *S. infundibuliforme* Phil., *S. irosinum* Ochoa, *S. jalcae* Ochoa, *S. kurtzianum* Bitter & Wittm., *S. laxissimum* Bitter, *S. leptophyes* Bitter, *S. limbaniense* Ochoa, *S. ×litusinum* Ochoa, *S. marinasense* Vargas, *S. medians* Bitter, *S. megistacrolobum* Bitter, *S. microdontum* Bitter, *S. multiinterruptum, S. neovavilovii* Ochoa, *S. olmosense* Ochoa, *S. oplocense* Hawkes, *S. orophilum* Correll, *S. pampasense*	2× (2EBN)	ARG, BOL, BRA, CHL, COL, ECU, MEX, PER, PAR, URU

Hawkes, S. paucissectum Ochoa, S. peloquinianum Ochoa, S. pillahuatense
Vargas, S. piurae Bitter, S. raphanifolium Cárdenas & Hawkes, S.
rhomboideilanceolatum Ochoa, S. sanctae-rosae Hawkes, S. sandemanii Hawkes,
S. santolallae Vargas, S. sarasarae Ochoa, S. sauyeri Ochoa, S. saxatilis Ochoa,
S. sogarandinum Ochoa, S. sparsipilum (Bitter) Juz. & Bukasov, S. spegazzinii
Bitter, S. tacnaense Ochoa, S. tapojense Ochoa, S. tarijense Hawkes, S. taulisense
Ochoa, S. urubambae Juz., S. venturii Hawkes & Hjert., S. vernei Bitter & Wittm.,
S. verrucosum Schltdl., S. vidaurrei Cárdenas, S. violaceimarmoratum Bitter,
S. yungasense Hawkes

Secondary gene pool (GP-2)

Solanum acaule Bitter, S. agrimonifolium Rydb., S. colombianum Bitter, S. bjertingii Hawkes, S. lobbianum Bitter, S. nubicola Ochoa, S. oxycarpum Schiede, S. paucijugum Ochoa, S. stoloniferum Schltdl. & Bouchet*, S. tuquerrense Hawkes	4× (2EBN)	ARG, BOL, COL, ECU, GUA, HON, MEX, PER, VEN
Solanum augustii Ochoa, S. bulbocastanum Dunal, S. cajamarquense Ochoa, S. cardiophyllum Lindl., S. chancayense Ochoa, S. circaeifolium Bitter, S. commersonii Dunal, S. dolichocremastrum Bitter, S. ehrenbergii (Bitter) Rydb., S. guzmanguense Whalen & Sagást., S. humectophilum Ochoa, S. hypacrarthrum Bitter, S. immite Dunal, S. incahuasinum Ochoa, S. ingifolium Ochoa, S. jamesii Torr.*, S. lignicaule Vargas, S. lopez-camarenae Ochoa, S. minutifoliolum Correll, S. mochiquense Ochoa, S. pinnatisectum Dunal, S. raquialatum Ochoa, S. simplicissimum Ochoa, S. stenophyllidium Bitter, S. trifidum Correll, S. trinitense Ochoa, S. wittmackii Bitter	Tuber-bearing 2× (1EBN)	ARG, BRA, BOL, ECU, GUA, MEX, PER, PAR, URU

Tertiary gene pool (GP-3)

Solanum ×edinense P. Berthault	5×	MEX
Solanum burtonii Ochoa, S. calvescens Bitter, S. cardiophyllum Lindl., S. flavoviridens Ochoa, S. immite Dunal, S. ×indunii K.A. Okada & A.M. Clausen, S. maglia Schltdl., S. medians Bitter, S. microdontum Bitter, S. ×neouerberbaueri Wittm., S. ×rechei Hawkes & Hjert., S. ×vallis-mexici Juz., S. ×viirsoii K.A. Okada & A.M. Clausen	3×	ARG, BOL, BRA, CHL, ECU, MEX, PER
Solanum achacachense Cárdenas, S. acroscopicum Ochoa, S. alandiae Cárdenas, S. ancoripae Ochoa, S. ×arahuayum Ochoa, S. avilesii Hawkes & Hjert., S. aymaraesense Ochoa, S. billhookeri Ochoa, S. burkartii Ochoa,	2×	ARG, BOL, CHL, ECU, GUA, HON, MEX, PER

(continued)

Table 17.1 (continued)

Wild species	Ploidy level and EBN number	Country
S. calacalinum Ochoa, *S. chillonanum* Ochoa, *S. chiquidenum* Ochoa, *S. chomatophilum* Bitter, *S. clarum* Correll, *S. gracilifrons* Bitter, *S. incamayoense* K.A. Okada & A.M. Clausen, *S. irosinum* Ochoa, *S. lesteri* Hawkes & Hjert., *S. longiusculus* Ochoa, *S. maglia* Schltdl., *S. megistacrolobum* Bitter, *S. ×michoacanum* (Bitter) Rydb., *S. microdontum* Bitter, *S. morelliforme* Bitter & Münch, *S. multiinterruptum* Bitter, *S. neocardenasii* Hawkes & Hjert., *S. neorosii* Hawkes & Hjert., *S. neovargasii* Ochoa, *S. okadae* Hawkes & Hjert., *S. ortegae* Ochoa, *S. pascoense* Ochoa, *S. polyadenium* Greenm., *S. puchupuchense* Ochoa, *S. ×rechei* Hawkes & Hjert., *S. regularifolium* Correll, *S. salasianum* Ochoa, *S. ×sambucinum* Rydb., *S. scabrifolium* Ochoa, *S. ×setulosistylum* Bitter, *S. soestii* Hawkes & Hjert., *S. tacnaense* Ochoa, *S. tarapatanum* Ochoa, *S. tarnii* Hawkes & Hjert., *S. urubambae* Juz., *S. velardei* Ochoa, *S. virgultorum* (Bitter) Cárdenas & Hawkes, *S. yamobambense* Ochoa		
S. etuberosum Lindl., *S. fernandezianum* Phil., *S. palustre* Poepp.	Non-tuber-bearing 2× (1EBN)	ARG, CHL
Solanum arnezii Cárdenas, *S. ×bruecheri* Correll, *S. donachui* (Ochoa) Ochoa, *S. garcia-barrigae* Ochoa, *S. hintonii* Correll, *S. orocense* Ochoa, *S. otites* Dunal, *S. ×ruiz-lealii* Brücher, *S. solisii* Hawkes, *S. sucubunense* Ochoa, *S. woodsonii* Correll	unknown	ARG, BOL, COL, ECU, MEX, PAN, VEN

Source: Taxonomy follows Spooner and Salas 2006.

Note: Ploidy levels and EBN numbers are according to Spooner and Hijmans 2001 (table 4) and Spooner and Salas 2006; some species have more than one ploidy level and EBN number. Gene pool designations are according to Ortiz 1998. Country abbreviations are ARG, Argentina; BOL, Bolivia; BRA, Brazil; CHL, Chile; COL, Colombia; CRI, Costa Rica; ECU, Ecuador; GUA, Guatemala; HON, Honduras; MEX, Mexico; PAN, Panama; PER, Peru; PAR, Paraguay; URU, Uruguay; VEN, Venezuela.

*These species occur in Mexico and the USA.

tices and low temperatures or other adverse climatic conditions (Crawley et al. 2001).

In S America, however, several landrace cultivars are weedy and persistent. These weedy potatoes with large tubers—referred to as 'arakka' in the Quechua language—persist in cultivated fields and are common throughout much of Bolivia and Peru. Other landrace cultivars that escape and persist naturally outside of cultivated fields are referred to as 'siwwa' or 'sihua' potatoes (Spooner et al. 1999).

Crop Wild Relatives

The genus *Solanum* contains over 1000 species and is divided into several sections (D'Arcy 1991; Bohs 2005). The domesticated potato and its tuber-bearing wild relatives belong to *Solanum* sect. *Petota* (Hawkes 1990). There are few, if any, discrete morphological characters useful for cladistic analyses at the species or series level in this section (Spooner and van den Berg 1992; Giannattasio and Spooner 1994; Spooner and Castillo 1997), which is currently under revision (Knapp et al. 2004). At present, there are approximately 188 recognized tuber-bearing wild potato species in sect. *Petota* (Spooner and Salas 2006; table 17.1). Furthermore, three non-tuber-bearing wild species in section *Etuberosum* (Juz. *ex* Bukasov & Kameraz) A. Child (*S. etuberosum* Lindl., *S. fernandezianum* Phil., and *S. palustre* Poepp. *ex* Schltdl.) are also considered to be close relatives of domesticated potatoes. They have often been used in breeding programs, but in nature they cannot cross with domesticated potatoes (Hijmans 2002).

Wild potatoes are distributed from the SW United States to S Chile (between 38° N and 41° S), although they are more common in Mexico and in the Andean highlands of Argentina, Bolivia, and Peru (Burton 1989; Child 1990; Hawkes 1990). The potato gene pool includes diploid, tetraploid, and hexaploid species, with occasional sterile triploids and pentaploids.

Hybridization

Hybridization is widespread, and introgression is probably common, among many of the species in sect. *Petota* (Dale et al. 1992; Masuelli et al. 2009). About 27 wild and cultivated species are believed to have originated through hybridization (Spooner and van den Berg 1992; Clausen and Spooner 1998; see also Spooner and Hijmans 2001 for their putative parents). Hybridization and subsequent introgression are common among many other wild potato species (see, e.g., Hawkes 1962, 1990; Hawkes and Hjerting 1969, 1989; Ugent 1970; Watanabe and Peloquin 1989) as well as between domesticated and wild species (see, e.g., Debener et al. 1990; Rabinowitz et al. 1990). However, the species concept in the *Solanum* complex, its taxonomy, and the long domestication process for the genus make it difficult to assess true natural interspecific hybridizations.

Among tuber-bearing potatoes (sect. *Petota*), some species with similar ploidy cannot cross in nature, whereas other species with different ploidy cross successfully (Hawkes and Hjerting 1969, 1989; Hawkes 1990; Ochoa 1990). For example, there is evidence of natural hybridization between the wild potato species *Solanum acaule* Bitter (tetraploid) and *S. megistacrolobum* Bitter (diploid) in several localities in the Argentinean province of Jujuy. The hybrids, which are almost completely male sterile, are thriving colonizers in disturbed areas around farmers' dwellings (Okada and Clausen 1982).

Johnston et al. (1980) discovered that the main cause of failure in interspecific crosses in sect. *Petota* are post-zygotic barriers at the endosperm level, which prevent seed development after fertilization. Johnston and Hanneman (1982) introduced the concept of the Endosperm Balance Number (EBN), which they defined as a measure of the "effective ploidy" of a genome in the endosperm (p. 447). EBNs are independent of ploidy and are empirically determined, based on crossability with standard EBN test crossers or other species of known EBNs. Crosses between species with differing EBNs fail more often than not, whereas natural crosses between species with the same EBN are mostly successful (even if their ploidy differs) and their endosperm develops normally after fertilization (Ortiz and Ehlenfeldt 1992). Some hybrids, though, can show reduced fertility or vigor in later generations (Hawkes 1958, 1990). EBN is the major biological isolating mechanism in potatoes, but in some cases the pollen grain's failure to germinate on or grow through the style (pollen-pistil incompatibility), and seed abortion, can prevent crosses between EBN-compatible species (Hawkes and

Jackson 1992; Fritz and Hanneman 1989; Camadro et al. 2004). The degree of cross-compatibility between two species can be predicted, to a great extent, from their ploidy and EBNs, but this still needs to be tested experimentally (Hanneman 1994).

Jackson and Hanneman (1999) performed field and greenhouse crossing experiments with S. *tuberosum* and 134 wild potato species. In general, their results confirmed expectations based on ploidy level and EBNs, although—contrary to expectations from these EBNs— some 1EBN species produced seed from crosses with domesticated potatoes. However, no information is available yet about their viability and germinability. The best crossability was found with species in four *Solanum* series: ser. *Tuberosa*, ser. *Conicibaccata* Bitter, ser. *Piurana* Hawkes, and ser. *Acaulia* Juz. *ex* Bukasov & Kameraz. Likewise, hand- and open-pollination crossing experiments with an S. *tuberosum* cultivar group, the *Andigenum* landrace (4x, 4EBN), and six wild relatives (6x, 4EBN; 4x, 2EBN; 2x, 2EBN) agreed with predictions based on ploidy and EBNs (Celis et al. 2004).

Taking ploidy levels and EBNs into account, it is likely that the domesticated potato, S. *tuberosum* (4x, 4EBN), can hybridize readily with all tetraploid and hexaploid wild relatives in sect. *Petota* that share the same EBN (i.e., 4EBN). Furthermore, many wild potato species can produce unreduced (2n) gametes—both 2n pollen and 2n eggs—which provides an opportunity for gene flow among ploidy levels (den Nijs and Peloquin 1977a, b; Watanabe and Peloquin 1989, 1991; Werner and Peloquin 1991; Carputo et al. 2003). If a diploid or tetraploid 2EBN wild relative produces unreduced gametes, these gametes become 4EBN. *Solanum tuberosum* can therefore cross with wild potatoes that are 2x (2EBN) and 4x (2EBN) through unreduced gametes. There are an additional nine tetraploid and two hexaploid species with unknown EBNs that can cross with domesticated potatoes, because their EBNs will either be 4 or 6. This leads to a total of 109 wild species with a known ability to hybridize spontaneously with domesticated potatoes (table 17.2).

The ploidy levels and EBNs of 14 more species are unknown, and there are another 38 diploid species with unknown EBNs. Therefore, some of these species may have ploidy levels and EBNs that are compatible with domesticated potatoes, and this needs to be taken into ac-

Table 17.2 Hybridization potential of modern potato cultivars (*Solanum tuberosum* L., 4x, 4EBN) with wild relatives

Wild species group	F1 hybrid progeny	F1 hybrid fertility	Hybridization potential	Number of species*	Country
Primary gene pool (GP-1)					
Natural hybridization possible					
4x (4EBN)	4x (4EBN)	fertile	100	3	ARG, BOL, PER
4x	4x (4EBN) or 6x (4EBN)	potentially fertile	100 or 20	9	BOL, COL, CRI, MEX, PAN, VEN
2x (2EBN)†	4x (4EBN)	n.r.	4	74	ARG, BOL, BRA, CHL, COL, ECU, MEX, PER, PAR, URU
6x (4EBN)	5x (4EBN)	sterile	0	11	ARG, BOL, COL, ECU, GUA, MEX, PER
6x	5x	sterile	0	1	ARG, BOL, ECU, MEX, PER
Secondary gene pool (GP-2)					
Natural hybridization possible					
4x (2EBN)†	6x (4EBN)	n.r.	20	10	ARG, BOL, COL, ECU, GUA, HON, MEX, PER, VEN
Hybridization only possible artificially					
2x (1EBN)	n.r.	n.r.	2	27	ARG, BRA, BOL, ECU, GUA, MEX, PER, PAR, URU
Tertiary gene pool (GP-3)					
Hybridization only possible artificially					
3x	n.r.	n.r.	0	13	ARG, BOL, BRA, CHL, ECU, MEX, PER
5x	n.r.	n.r.	0	1	MEX
2x (1EBN)‡	n.r.	n.r.	0	3	ARG, CHL

Hybridization potential unknown $2x$	unknown	max. 4	46	ARG, BOL, CHL, ECU, GUA, HON, MEX, PER
Gene pool, ploidy level, and EBN unknown	unknown	100	11	ARG, BOL, COL, ECU, MEX, PAN, VEN

Source: adapted from Hijmans 2002.

Note: Ploidy level is listed first, and the endosperm balance number (EBN) is in parentheses. Hybridization potential refers to the risk values (o = no risk; 100 = high risk) extracted from Hijmans 2002, in line with ploidy and EBN. For species with unknown EBN, the assigned hybridization potential is the highest potential within the respective ploidy group. When the ploidy was also unknown, plants were assumed to be $4x$ (4EBN) (i.e., risk factor = 100). n.r. = not reported. Gene pool designations follow Ortiz 1998. Country abbreviations are ARG, Argentina; BOL, Bolivia; BRA, Brazil; CHL, Chile; COL, Colombia; CRI, Costa Rica; ECU, Ecuador; GUA, Guatemala; HON, Honduras; MEX, Mexico; PAN, Panama; PAR, Paraguay; PER, Peru; URU, Uruguay; VEN, Venezuela.

*See table 17.1. Some species have more than one ploidy level and EBN.

†Hybridization is possible only when unreduced ($2n$) gametes are produced.

‡The three non-tuber-bearing wild relatives are from sect. *Etuberosum* (*S. etuberosum*, *S. fernandezianum*, and *S. palustre*).

count. The triploid and pentaploid species are generally highly sterile (Hijmans 2002). No crosses are possible between domesticated potatoes and species with 1EBN (Hawkes and Jackson 1992; Dinu et al. 2005).

Primary Gene Pool (GP-1)

Among the species in the primary gene pool, hybridization between modern cultivars (4x, 4EBN) and 4EBN tetraploid or 2EBN diploid (through unreduced gametes) wild relatives often results in viable, fully or partly fertile F1 progeny. Outcrossing rates can reach up to 100% in the greenhouse and more than 40% in the field (Scurrah et al. 2008). The actual cross-compatibility of individual plants varies greatly within species and is genotype-dependent (Jansky 2006).

Thus far, it is not known how frequently unreduced gametes are produced in the wild, nor how often they lead to fertile offspring when crossed with a tetraploid 4EBN species. Unreduced gametes are observed quite often under controlled conditions in the laboratory (e.g., up to 33% in Jackson and Hanneman 1999). However, in the wild there are only a few tetraploids (which would be the common progeny), suggesting that F1 hybrids via 2n gametes either do not occur or do not survive for long in nature (Hawkes 1994; Hijmans 2002).

The hybridization of modern cultivars (4x, 4EBN) with hexaploid species is possible in theory, but no such hybrids have ever been observed in the wild. Even if this were the case, the likelihood of introgression would be very low, since F1 hybrids are pentaploids and thus are likely to be sterile.

Secondary Gene Pool (GP-2)

The secondary gene pool consists of 2EBN tetraploid and 1EBN diploid species in sect. *Petota*. The former species (2EBN tetraploid) can hybridize with modern cultivars through unreduced gametes. As is the case with 2EBN diploid species (see GP-1), crosses generally result in viable, fully or partly fertile F1 progeny.

In the case of 1EBN diploid wild relatives, there are strong barriers to hybridization with modern cultivars, due to differences in their EBNs and ploidy levels. These barriers, which can be overcome by using ploidy manipulations and bridge crosses (Johnston and Hanneman 1982; Jansky 2006), prevent spontaneous hybridization in the wild.

Tertiary Gene Pool (GP-3)

The tertiary gene pool contains the remaining diploid, triploid, and pentaploid tuber-bearing wild relatives; the three wild non-tuber bearing *Solanum* species (1EBN diploid) of sect. *Etuberosum*; and all remaining *Solanum* species. There are strong barriers to hybridization with modern cultivars, due to differences in EBNs and ploidy levels, and crosses succeed only via artificial chromosome doubling, unreduced gametes, embryo rescue, or intermediate bridging crosses that do not occur in nature (Ramanna and Hermsen 1981; Chavez et al. 1988; Watanabe et al. 1995).

For example, in Europe there are about 13 other *Solanum* species, most of which were introduced. Neither open- nor hand-pollination between domesticated potatoes and the two native and common *Solanum* species—*S. dulcamara* L. (bittersweet or woody nightshade) and *S. nigrum* L. (black nightshade)—produce any viable offspring (Dale et al. 1992; Conner 1994; Eijlander and Stiekema 1994; McPartlan and Dale 1994). Some viable but weak and male-sterile hybrids have been obtained, but only after applying artificial embryo rescue techniques (Eijlander and Stiekema 1994).

As mentioned above, there are 57 additional wild relatives with unknown ploidy levels and/or EBNs in sect. *Petota* (table 17.2). Some of these species likely belong to GP-1, and others to GP-2.

Outcrossing Rates and Distances

Pollen Flow

Although the role of wind has not been categorically determined in pollen dispersal, it appears that insects are the main pollinators of domesticated potatoes. Outcrossing would thus be restricted to within the distances flown by pollinating insects, such as bumblebees and bees. The foraging range of bumblebees is estimated to be less than 1 km

Table 17.3   A few representative studies of pollen dispersal distances and outcrossing rates of modern potato cultivars (*Solanum tuberosum* L.)

Pollen dispersal distance (m)	% outcrossing	Country	Reference
0, 1.5, 3, 4.5, 6, 10	1.14, 0.03, 0.05, 0.05, 0, 0	New Zealand	Tynan et al. 1990
0, 3, 9, 10	0.05, 0, 0.008, 0	New Zealand	Conner 1993
0, 3, 10, 20	24, 2, 0.017, 0	Scotland	Dale et al. 1992
0, 3, 10, 20	28.8, 2.6, 0.08, 0	UK	McPartlan and Dale 1994
10, 80	5.1, 0.2	Denmark	Schittenhelm and Hoekstra 1995
1, 2, 3, 10, 100, 1000	72, 32, 39, 34, 36, 31	Sweden	Skogsmyr 1994

(Osborne et al. 1999; Knight et al. 2005), and less than 3 km for other bees (Reheul 1987), although pollen transport over several kilometers has been reported (see, e.g., Eckert 1933; Goulson and Stout 2001). In the case of bumblebees, however, the effective cross-pollination distances are much shorter, since the majority of the pollen is deposited in the immediate surroundings to the pollen source (Free and Butler 1959; Skogsmyr 1994; Creswell et al. 2002). The extent of pollen flow between modern and landrace cultivars and their wild relatives varies, depending on the species composition of pollinators that are present in that locale, their density and foraging behavior, and local climatic conditions (Treu and Emberlin 2000).

Only a few results of field trials to experimentally determine outcrossing distances have been published (table 17.3). They indicate that—as in other crops—outcrossing rates are highest immediately around the pollen source and decrease rapidly with distance, following a leptocurtic curve (fig. 17.1). In general, outcrossing rates approach zero at distances greater than 20 m, as shown in several field experiments in Ireland, New Zealand, and Sweden (Conner and Dale 1996; Petti et al. 2007).

In one field experiment in Sweden, long-distance dispersal was detected up to 1 km away from the pollen source, and outcrossing frequencies (31% at 1 km) were much higher than reported in other studies (Skogsmyr 1994). These results were attributed to pollen beetle colonies

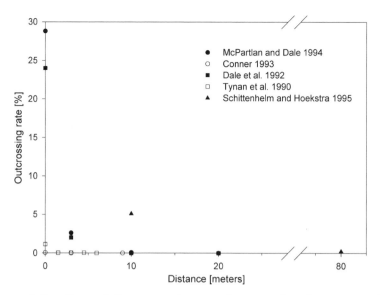

Fig. 17.1 Pollen dispersal distances and outcrossing rates of modern potato cultivars (*Solanum tuberosum* L.)

that emigrate from a (pollen) food source in large numbers and often fly
over large distances. However, this study was criticized for containing large
numbers of false-positive PCR results and no controls for the PCR tests on
which the presence of the GM trait was based (Conner and Dale 1996).

Isolation Distances

The isolation distances established by regulatory authorities vary greatly
among countries, ranging from 3 to 50 m between GM and conventional non-GM potatoes, and from 10 to 100 m between GM and organic non-GM potatoes (table 17.4). Based on the majority of published
studies, hybridization between modern and landrace cultivars, or between modern cultivars and wild potatoes, is likely to be effectively
contained with a 20-meter isolation distance. It should be acknowl-
edged, however, that the possibility of rare long-distance cross-pollination events—especially when pollen beetles are present—and the persistence of true potato seeds in the soil are uncertainties in the potato gene
flow equation.

Table 17.4   Isolation distances for modern potato cultivars (*Solanum tuberosum* L.)

Threshold*	Minimum isolation distance (m)	Country	Reference	Comment
n.a.	at least 1 skip row	USA	CCIA 2009	certified seed requirements
n.a.	20†	Denmark	CEC 2006	seed production
0.9%	20	Denmark	CEC 2006	between GM and organic non-GM potatoes
0.9%	20	Denmark	CEC 2006	between GM and conventional non-GM potatoes
n.a.	20	Latvia	CEC 2006	seed production
0.9%	100	Latvia	CEC 2006	between GM and organic non-GM potatoes
0.9%	20	Latvia	CEC 2006	between GM and conventional non-GM potatoes
n.a.	10	Netherlands	CEC 2006	seed production
0.9%	10	Netherlands	CEC 2006	between GM and organic non-GM potatoes
0.9%	3	Netherlands	CEC 2006	between GM and conventional non-GM potatoes
n.a.	50	Poland	CEC 2006	regardless of production scheme

*Maximum percentage of (GM) seed impurities permitted. n.a. = not available.
†If nonflowering or male sterile, then the isolation distance is 15 m.

Gene Flow through Seed (Tuber) Exchange

In Central and S America, the center of origin and diversity for potatoes, another important aspect of gene flow is the dynamic process of informal seed exchange (referring here to tubers) among farmers. Farmers often trade and sow tubers from different sources in the same or nearby fields, including occasional modern cultivars, and they even select and grow naturally occurring domesticated × wild potato hybrids when these plants have good agronomic characteristics (Quiros et al. 1992; Scurrah et al. 2008). Hence traditional tuber seed-exchanges among farmers are conducive to the introgression of GM traits into landrace cultivars if the hybrids are desirable to farmers.

State of Development of GM Technology

Potatoes were one of the first crops to be genetically modified in the early 1980s (An et al. 1986; Shahin and Simpson 1986). Since then, more efficient transformation systems have been developed, and field tests of potatoes expressing a wide range of GM traits have been conducted in several regions worldwide, including the crop's center of diversity in Central and S America (Goy and Duesing 1995; Jackson and Hanneman 1996). Today, several reliable and efficient transformation protocols for different potato cultivars are available (see, e.g., Romano et al. 2001; Trujillo et al. 2001; Banerjee et al. 2006). Genetically modified potatoes were in commercial cultivation in the United States and Canada during the mid 1990s. Later, production was abandoned due to market constraints.

Field trials have been reported from a wide range of countries, including Argentina, Australia, Austria, Belgium, Brazil, Canada, China, the Czech Republic, Denmark, Finland, France, Germany, Hungary, India, Ireland, Italy, Japan, Korea, the Netherlands, New Zealand, the Philippines, Poland, Portugal, the Russian Federation, Spain, Sweden, Switzerland, South Africa, the United Kingdom, and the United States.

The GM traits researched to date include insect, nematode, disease (virus, fungus, bacteria), and herbicide resistance; altered starch content and composition; abiotic stress tolerance; anti-bruise genes; amylose-free tubers; biofortification (e.g., enhanced protein, carotenoid, and beta-carotene [golden potato] content); and potatoes as bioreactors for the production of recombinant spider silk, vitamin E, and other pharmaceutical and technical products.

Agronomic Management Recommendations

There are four main ways genes introgress to landrace cultivars or wild species of potatoes:

1   through pollen flow from flower-producing modern cultivars

2   through volunteers from seed or from ground-keepers in fields previously planted with modern cultivars

3   through ferals from tubers of modern cultivars during cropping, harvesting, handling, storage, and transport of the crop

4   through hybrids resulting from modern-cultivar tuber exchanges among farmers via informal farmer seed systems (especially in Central and S America)

The following management recommendations can be used to reduce the likelihood of gene introgression as much as possible:

- Take land-use history into consideration when estimating the presence of crop-derived volunteer populations left by an earlier cropping cycle—not only that of the field to be planted, but also of neighboring fields.

- Apply crop rotation, preferably sowing cereals after potato production. Cereal herbicides can be applied to control emerging potato ground-keepers, when landrace cultivars are to be planted after GM modern cultivars have been grown.

- In areas where compatible landrace cultivars and wild relatives occur, isolate fields planted with GM modern cultivars from non-GM fields at the distances specified by legal regulations, but in no case less than 20 m.

- Take special measures when transporting potato tubers from the field to storage facilities (e.g., covering the tubers) to avoid establishing ferals through spilt tubers. Monitor roads for feral potato plants and, if present, control them by mechanical or chemical means.

To further minimize gene flow and introgression in the center of origin and diversity for potatoes, genetic modification efforts in this region might focus on completely sterile landrace cultivars, such as triploid cultivars from the *Solanum juzepczukii* Bukasov and *S. chaucha* Juz. & Bukasov cultivar groups (Hijmans 2002; Ghislain et al. 2003; Buijs et al. 2005), rather than on modern tetraploid or even diploid cultivars (Celis et al. 2004; Buijs et al. 2005; Green et al. 2005).

Crop Production Area

Potatoes are now produced in more than 150 countries worldwide— throughout the temperate regions (Asia, Europe, the Russian Federation, and the United States and Canada) as well as in the S and Central American Andes and, to a lesser extent, in Africa. China is the world's leading potato producer (22%), followed by the Russian Federation (11%) and India (8%) (FAO 2009). The total crop production area is estimated at 19.3 million ha, but none of this is currently planted with GM potatoes for commercial production.

Research Gaps and Conclusions

Future research efforts addressing the following issues would help greatly in reducing knowledge gaps and in creating more detailed and smaller-scale assessments of the likelihood of gene flow and introgression between modern cultivars and wild potatoes:

- the role of both wind and pollen beetles in potato outcrossing, particularly with regard to long-distance pollen dispersal
- the viability and longevity of pollen
- the viability in the soil of true potato seeds produced by modern cultivars
- the ploidy levels and EBNs of all species
- the development of an inventory of the wild potato species that occur in the areas where modern cultivars are grown, as well as noting their flowering times, pollinators, and seed set
- an estimation of the actual probabilities for successful crosses between modern cultivars and wild potatoes under real-life, small- and large-scale agronomic conditions (e.g., *Solanum stoloniferum* Schltdl. & Bouché in the United States)
- the viability and fitness of modern cultivar × wild potato hybrids over several generations

Tetraploid modern potato cultivars may only hybridize in nature with wild relatives of sect. *Petota*. These wild relatives are distributed from the SW United States to S Chile. Only two wild species are found in the United States (*S. jamesii* Torr. and *S. stoloniferum*), and no natural hybrids have been reported between these species and modern cultivars in the United States (Wozniak 2002). However, based on ploidy and EBNs, natural hybridization is possible in theory between modern cultivars and *S. stoloniferum* (4*x*, 2EBN) through unreduced gametes. In fact, Jackson and Hanneman (1999) produced hybrid seed through hand pollination of *S. stoloniferum* (= *S. fendleri* A. Gray), with *S. tuberosum* as the pollen donor. Still, this species is not expected to cross easily with modern cultivars, and Hijmans (2002) assigned a moderate risk factor (20 out of 100) to it. Similarly, Love (1994) considered that there was only a small chance of spontaneous outcrossing in the wild and, subsequently, of introgression. As expected, no fruit or seed set was observed with the diploid 1EBN species *S. jamesii* (Jackson and Hanneman 1999).

The likelihood of gene flow and introgression from modern cultivars to their landrace and wild relatives is therefore listed for the entire distribution range of the wild species. Here—and particularly in the Andes, where wild potato diversity is highest—species with compatible ploidy and EBNs occur almost everywhere that modern cultivars are grown, with only a few exceptions in Chile, Colombia, and S Argentina (map 17.1). As Hijmans (2002) noted, his estimation of crossing potential—based exclusively on ploidy levels and EBNs—is preliminary, and it may overestimate the real potential in a number of cases. For example, conservative (i.e., the highest possible) crossing potential factors are assigned to the many species with unknown ploidy levels and/or EBNs. Also, experimental data suggest that the actual outcrossing rates and the extent of viable seed production of many compatible species may be much lower than anticipated (Jackson and Hanneman 1999). The—also conservative—mapping approach presented here posits that the area with the highest likelihood of gene flow and introgression—stretching from SW United States to S Chile (along the Andean mountain range)—is locality and genotype dependent. The map therefore needs to be interpreted in light of the multiple conditions required for successful gene flow, including the fertility of the modern cultivar, the presence of a wild relative within less than 20 m, overlapping flowering time, the presence of pollinators, local climatic conditions, traditional tuber seed-exchange practices among farmers, and hybrids appreciated by farmers, among others.

Assuming physical proximity (less than 20 m) and flowering overlap, and being conditional on the nature of the respective trait(s), the likelihood of gene introgression from modern potato cultivars to landrace cultivars and wild relatives is as follows:

- *high* for tetraploid species with 4EBN, and probably also for some tetraploids with unknown EBNs (the mountainous area from Mexico to Argentina; map 17.1)

- *moderate* for diploid and tetraploid species with 2EBN, and probably also for some diploids and tetraploids with unknown EBNs (in map 17.1, part of this area is hidden by the areas in red, corresponding to high likelihood)

- *highly unlikely* for hexaploid species (since their F1 hybrids are sterile pentaploids), pentaploid and triploid species (since they are sterile), and diploid 1EBN species (due to genetic incompatibility)

REFERENCES

Accatino P (1980) Agronomic management in the utilization of true potato seed: Preliminary results. In: *Production of Potatoes from True Seed: Report of a Planning Conference, Manila, Philippines, September 13–15, 1979.* CIP [Centro Internacional de la Papa] Planning Conference Report 19 (International Potato Centre [CIP]: Lima, Peru), pp. 61–99.

Albuquerque LB, Velázquez A, and Mayorga-Saucedo R (2006) Solanaceae composition, pollination, and seed dispersal syndromes in Mexican mountain cloud forest. *Acta Bot. Bras.* 20: 599–613.

An G, Watson BD, and Chiang CC (1986) Transformation of tobacco, tomato, potato, and *Arabidopsis thaliana* using a binary Ti vector system. *Plant Physiol.* 81: 301–305.

Arndt GC, Rueda JL, Kidane-Mariam HM, and Peloquin SJ (1990) Pollen fertility in relation to open pollinated true seed production in potatoes. *Am. Potato J.* 67: 499–505.

Askew MF (1993) Volunteer potatoes from tubers and true potato seed. *Asp. Appl. Biol.* 35: 9–15.

Banerjee AK, Prat S, and Hannapel DJ (2006) Efficient production of transgenic potato (*S. tuberosum* L. subsp. *andigena*) plants via *Agrobacterium tumefaciens*-mediated transformation. *Plant Sci.* 170: 732–738.

Bohs L (2005) Major clades in *Solanum* based on *ndh*F sequence data. In: Keating RC, Hollowell VC, and Croat TB (eds.) *A Festschrift for William G. D'Arcy: The Legacy of a Taxonomist.* Monographs in Systematic Botany from the Missouri Botanical Garden 104 (Missouri Botanical Garden: St. Louis, MO, USA), pp. 27–49.

Brown CR (1993) Outcrossing rate in cultivated autotetraploid potato. *Am. Potato J.* 70: 725–734.

Buchmann SL (1983) Buzz pollination in angiosperms. In: Jones CE and Little RJ (eds.) *Handbook of Experimental Pollination Biology* (Van Nostrand Reinhold: New York, NY, USA), pp. 73–133.

Buijs J, Martinet M, de Mendiburu F, and Ghislain M (2005) Potential adoption and management of insect-resistant potato in Peru, and implications for genetically engineered potato. *Environ. Biosafety Res.* 4: 179–188.

Burton WG (1989) *The Potato.* 3rd ed. (John Wiley and Sons: New York, NY, USA). 742 p.

Camadro EL, Carputo D, and Peloquin SJ (2004) Substitutes for genome differentiation in tuber-bearing *Solanum*: Interspecific pollen-pistil incompatibility, nuclear-cytoplasmic male sterility, and endosperm. *Theor. Appl. Genet.* 109: 1369–1376.

Carputo D, Barone A, and Frusciante L (2000) 2*n* gametes in the potato: Essential ingredients for breeding and germplasm transfer. *Theor. Appl. Genet.* 101: 805–813.

Carputo D, Frusciante L, and Peloquin SJ (2003) The role of 2*n* gametes and endosperm balance number in the origin and evolution of polyploids in the tuber-bearing *Solanums. Genetics* 163: 287–294.

CCIA [California Crop Improvement Association] (2009) Potato Certification Standards. http://ccia.ucdavis.edu/seed_cert/seedcert_index.htm.

CEC [Commission of the European Communities] (2006) Annex to the Communication from the Commission to the Council and the European Parliament: Report on the Implementation of National Measures on the Coexistence of Genetically Modified Crops with Conventional and Organic Farming, 9 March 2006. Com(2006) 104, SEC(2006)313. http://ec.europa.eu/agriculture/coexistence/sec313_en.pdf.

Celis C, Scurrah M, Cowgill S, Chumbiauca S, Green J, Franco J, Main G, Kiezebrink D, Visser RG, and Atkinson HJ (2004) Environmental biosafety and transgenic potato in a centre of diversity for this crop. *Nature* 432: 222–225.

Chavez R, Brown CR, and Iwanaga M (1988) Application of interspecific sesquiploidy to introgression of PLRV resistance from non-tuber-bearing *Solanum etuberosum* to cultivated potato germplasm. *Theor. Appl. Genet.* 76: 497–500.

Child A (1990) A synopsis of *Solanum* subgenus *Potatoe* (G. Don) (D'Arcy) (*tuberarium*) (Dun.) Bitter (*s.l.*). *Feddes Repert.* 101: 209–235.

Cipollini M and Levey DJ (1997a) Antifungal activity of *Solanum* fruit glycoalkaloids: Implications for frugivory and seed dispersal. *Ecology* 78: 799–809.

—— (1997b) Secondary metabolites of fleshy vertebrate-dispersed fruits: Adaptive hypotheses and implications for seed dispersal. *Am. Nat.* 150: 346–372.

Clausen AM and Spooner DM (1998) Molecular support for the hybrid origin of the wild potato species *Solanum ×rechei*. *Crop Sci.* 38: 858–865.

Conner AJ (1993) Monitoring "escapes" from field trials of transgenic potatoes: A basis for assessing environmental risks. In: Organization for Economic Cooperation and Development [OECD] (ed.) *Report on the Seminar on Scientific Approaches for the Assessment of Research Trials with Genetically Modified Plants, Jouy-en-Josas, France, April 6–7, 1992*. OCDE/GD(93)118 (Organization for Economic Cooperation and Development [OECD]: Paris, France), pp. 34–40.

—— (1994) Analysis of containment and food safety issues associated with the release of transgenic potatoes. In: Belknap WR, Vayda ME, and Park WD (eds.) *The Molecular and Cellular Biology of the Potato.* 2nd ed. Biotechnology in Agriculture Series 12 (CAB International: Wallingford, UK), pp. 245–264.

—— (2007) Field testing of transgenic potatoes. In: Vreugdenhil D, Bradshaw J, Gebhardt C, Govers F, Taylor M, MacKerron DKL, and Ross H (eds.) *Potato Biology and Biotechnology: Advances and Perspectives* (Elsevier: Amsterdam, The Netherlands), pp. 685–701.

Conner AJ and Dale PJ (1996) Reconsideration of pollen dispersal data from field trials of transgenic potatoes. *Theor. Appl. Genet.* 92: 505–508.

Coxon DT (1981) The glycoalkaloid content of potato berries. *J. Sci. Food Agric.* 32: 412–414.

Crawley MJ, Brown SL, Hails RS, Kohn D, and Rees M (2001) Transgenic crops in natural habitats. *Nature* 409: 682–683.

Cresswell JE, Osborne JL, and Bell SA (2002) A model of pollinator-mediated gene flow between plant populations with numerical solutions for bumblebees pollinating oilseed rape. *Oikos* 98: 375–384.

Dale PJ, McPartlan HC, Parkinson R, MacKay GR, and Scheffler JA (1992) Gene dispersal from transgenic crops by pollen. In: Casper R and Landsmann J (eds.) *The Biosafety Results of Field Tests of Genetically Modified Plants and Microorganisms: Proceedings of the 2nd International Symposium, Goslar, Germany, May 11–14, 1992* (Biologische Bundesanstalt für Land- und Forstwirtschaft: Braunschweig, Germany), pp. 73–77.

Dalvi RR and Bowie WC (1983) Toxicology of solanine: An overview. *Vet. Hum. Toxicol.* 25: 13–15.

D'Arcy WG (1991) The *Solanaceae* since 1976, with a review of its biogeography. In: Hawkes JG, Lester RN, Nee M, and Estrada-R N (eds.) *Solanaceae III: Taxonomy, Chemistry, and Evolution* (Royal Botanic Gardens, Kew: Richmond, Surrey, UK), pp. 75–137.

Davies K, Milne F, and Golden A (1999) *Potato Volunteers and Their Management.* Technical Note 480 (Scottish Agricultural College [SAC]: Edinburgh, Scotland). www.sac.ac.uk/mainrep/pdfs/tn480potatovolunteers.pdf [limited access only].

Debener T, Salamini F, and Gebhardt C (1990) Phylogeny of wild and cultivated *Solanum* species based on nuclear restriction fragment length polymorphisms (RFLPs). *Theor. Appl. Genet.* 79: 360–368.

den Nijs TPM and Peloquin SJ (1977a) 2*n* gametes in potato species and their function in sexual polyploidization. *Euphytica* 26: 585–600.

——— (1977b) Polyploid evolution via 2*n* gametes. *Am. Potato J.* 54: 377–386.

Dinu II, Hayes RJ, Kynast RG, Phillips RL, and Thill CA (2005) Novel interseries hybrids in *Solanum*, section *Petota. Theor. Appl. Genet.* 110: 403–415.

Ducreux LJM, Morris WL, Taylor MA, and Millam S (2005) *Agrobacterium*-mediated transformation of *Solanum phureja. Plant Cell Rep.* 24: 10–14.

Eastham K and Sweet J (2002) *Genetically Modified Organisms (GMOs): The Significance of Gene Flow through Pollen Transfer.* Environmental Issue Report No. 28 (European Environment Agency [EEA]: Copenhagen, Denmark). 75 p. http://reports.eea.europa.eu/environmental_issue_report_2002_28/en/ GMOs%20for%20www.pdf.

Eckert JE (1933) The flight range of the honeybee. *J. Apicult. Res.* 47: 257–285.

Eijlander R and Stiekema WJ (1994) Biological containment of potato (*Solanum tuberosum*): Outcrossing to the related wild species black nightshade (*Solanum nigrum*) and bittersweet (*Solanum dulcamara*). *Sex. Plant Reprod.* 7: 29–40.

Evenhuis A and Zadoks JC (1991) Possible hazards to wild plants of growing transgenic plants: A contribution to risk analysis. *Euphytica* 55: 81–84.

FAO [Food and Agriculture Organization of the United Nations] (2009) FAOSTAT data. http://faostat.fao.org/site/567/default.aspx.

Filadelfi MA (1982) Naturally occurring toxicants in the potato. *Herbarist* 48: 21–23.

Free JB and Butler CG (1959) *The New Naturalist: Bumblebees* (Collins: London, UK). 208 p.

Friedman M (1992) Composition and safety evaluation of potato berries, potato and tomato seeds, potatoes, and potato alkaloids. In: Finley JW, Robinson S, and Armstrong A (eds.) *Food Safety Evaluation* (American Chemical Society: Washington, DC, USA), pp. 429–462.

Fritz NK and Hanneman RE Jr (1989) Interspecific incompatibility due to stylar barriers in tuber-bearing and closely related non-tuber-bearing *Solanums*. *Sex. Plant Reprod.* 2: 184–192.

Ghislain M, Lagnaoui A, and Walker T (2003) Fulfilling the promise of *Bt* potato in developing countries. *J. New Seeds* 5: 93–113.

Giannattasio R and Spooner DM (1994) A reexamination of species boundaries between *Solanum megistacrolobum* and *S. toralapanum* (*Solanum* sect. *Petota*, series *Megistacroloba*): Morphological data. *Syst. Bot.* 19: 89–105.

Goulson D and Stout JC (2001) Homing ability of the bumblebee *Bombus terrestris* (Hymenoptera: Apidea). *Apidol.* 32: 105–111.

Goy PA and Duesing JH (1995) From pots to plots: Genetically modified plants on trial. *Bio/Technol.* 13: 454–458.

Green J, Fearnehough MT, and Atkinson HJ (2005) Development of biosafe genetically modified *Solanum tuberosum* (potato) cultivars for growth within a centre of origin of the crop. *Mol. Breed.* 16: 285–293.

Hanneman RE Jr (1994) Assignment of endosperm balance numbers to the tuber-bearing *Solanums* and their close non-tuber-bearing relatives. *Euphytica* 74: 19–25.

——— (1995) Ecology and reproductive biology of potato: The potential for and the environmental implications of gene spread. In: Fredrick RJ, Virgin I, and Lindarte E (eds.) *Environmental Concerns with Transgenic Plants in Centers of Diversity: Potato as a Model; Proceedings from a Regional Workshop, Parque Nacional Iguazu, Argentina, June 2–3, 1995* (Biotechnology Advisory Commission, Stockholm Environmental Institute [SEI]: Stockholm, Sweden; Inter-American Institute for Cooperation on Agriculture [IICA]: San José, Costa Rica), pp. 19–38.

Hawkes JG (1958) Potato: I. Taxonomy, cytology, and crossability. In: Kappert H and Rudorf W (eds.) *Manual of Plant Breeding*, vol. 3. 2nd ed. (Paul Parey: Hamburg, Germany), pp. 1–43.

——— (1962) Introgression in certain wild potato species. *Euphytica* 11: 26–35.

——— (1990) *The Potato: Evolution, Biodiversity, and Genetic Resources* (Belhaven Press: London, UK). 259 p.

——— (1994) Origins of cultivated potatoes and species relationships. In: Bradshaw JE and Mackay GR (eds.) *Potato Genetics* (CAB International: Wallingford, UK), pp. 3–42.

Hawkes JG and Hjerting JP (1969) *The Potatoes of Argentina, Brazil, Paraguay, and Uruguay: A Biosystematic Study*. Annals of Botany Memoir No. 3 (Oxford University Press: London, UK). 525 p.

——— (1989) *The Potatoes of Bolivia: Their Breeding Value and Evolutionary Relationships* (Oxford University Press: Oxford, UK). 504 p.

Hawkes JG and Jackson MT (1992) Taxonomic and evolutionary implications of the Endosperm Balance Number hypothesis in potatoes. *Theor. Appl. Genet.* 84: 180–185.

Heinrich B (1979) "Majoring" and "minoring" by foraging bumblebees *Bombus vagans*: An experimental analysis. *Ecology* 60: 245–255.

Hijmans RJ (2002) Assessing the risk of natural geneflow between wild and genetically modified cultivated potatoes: Taxonomic and geographic considerations. Technical Report for the Servicio Nacional de Sanidad Agraria [SENASA], Ministerio de Agricultura, Peru, INCO-DEV project ICAA-1999–30106 (International Potato Center [CIP]: Lima, Peru). Unpublished ms, 45 p.

Hijmans RJ and Spooner DM (2001) Geography of wild potato species. *Am. J. Bot.* 88: 2101–2112.

Howard HW (1958) The storage of potato pollen. *Am. Potato J.* 35: 676–678.

Huamán Z and Spooner DM (2002) Reclassification of landrace populations of cultivated potatoes (*Solanum* sect. *Petota*). *Am. J. Bot.* 89: 947–965.

Jackson SA and Hanneman RE Jr (1996) Potential gene flow between cultivated potato and its wild tuber-bearing relatives: Implications for risk assessment of transgenic potatoes. *Proceedings and Papers from the 1996 Risk Assessment Research Symposium.* www.isb.vt.edu/brarg/brasym96/jackson96.htm.

—— (1999) Crossability between cultivated and wild tuber- and non-tuber-bearing *Solanums*. *Euphytica* 109: 51–67.

Jansky S (2006) Overcoming hybridization barriers in potato. *Plant Breed.* 125: 1–12.

Johnston SA and Hanneman RE Jr (1982) Manipulations of Endosperm Balance Number overcome crossing barriers between diploid *Solanum* species. *Science* 217: 446–448.

Johnston SA, den Nijs APM, Peloquin TPM, and Hanneman RE Jr (1980) The significance of genic balance to endosperm development in interspecific crosses. *Theor. Appl. Genet.* 57: 5–9.

Kirch HH, Uhrig H, Lottspeich F, Salamini F, and Thompson RD (1989) Characterisation of proteins associated with self-incompatibility in *Solanum tuberosum*. *Theor. Appl. Genet.* 78: 581–588.

Knapp S, Bohs L, Nee M, and Spooner DM (2004) Solanaceae—a model for linking genomics with biodiversity. *Comp. Funct. Genomics* 5: 285–291.

Knight ME, Martin AP, Bishop S, Osborne JL, Hale RJ, Sanderson RA, and Goulson D (2005) An interspecific comparison of foraging range and nest density of four bumblebee (*Bombus*) species. *Mol. Ecol.* 14: 1811–1820.

Lawson HM (1983) True potato seeds as arable weeds. *Potato Res.* 26: 237–46.

Love SL (1994) Ecological risk of growing transgenic potatoes in the United States and Canada. *Am. Potato J.* 71: 647–658.

Lumkes LM and Sijtsma R (1972) Mogelijkheden aardappelen als onkruid in volge-wassen te voorkomen en/of te bestrijden. *Landbouw Plantenziekten* 1: 17–36.

Lutman PJW (1977) Investigations into some aspects of the biology of potatoes as weeds. *Weed Res.* 17: 123–132.

Makepeace RJ, Cooper DC, and Holroyd J (1978) Weed control. In: Harris PM (ed.) *The Potato Crop* (Halsted Press: New York, NY, USA), pp. 402–405.

Masuelli RW, Camadro EL, Erazzú LE, Bedogni MC, and Marfil CF (2009) Homoploid hybridization in the origin and evolution of wild diploid potato species. *Plant Syst. Evol.* 277: 143–151.

McPartlan HC and Dale PJ (1994) An assessment of gene transfer by pollen from field-grown transgenic potatoes to non-transgenic potatoes and related species. *Transgen. Res.* 3: 216–225.

Morris SC and Lee TH (1984) The toxicity and teratogenicity of Solanaceae glycoalkaloids, particularly those of the potato (*Solanum tuberosum*). *Food Technol. Aust.* 36: 118–124.

Ochoa CM (1990) *The Potatoes of South America: Bolivia* (Cambridge University Press: Cambridge, UK). 570 p.

Okada KA and Clausen AM (1982) Natural hybridization between *Solanum acaule* Bitt. and *S. megistacrolobum* Bitt. in the province of Jujuy, Argentina. *Euphytica* 31: 817–835.

Ortiz R (1998) Potato breeding via ploidy manipulation. *Plant Breed. Rev.* 16: 1–86.

Ortiz R and Ehlenfeldt MK (1992) The importance of Endosperm Balance Number in potato breeding and the evolution of tuber-bearing *Solanum* species. *Euphytica* 60: 105–113.

Osborne JL, Clark SJ, Morris RJ, Williams IH, Riley JR, Smith AD, Reynolds DR, and Edwards AS (1999) A landscape-scale study of bumblebee foraging range and constancy, using harmonic radar. *J. Appl. Ecol.* 36: 519–533.

Palta JP, Chen HH, and Li PH (1981) Relationship between heat and frost resistance of tuber-bearing *Solanum* species: Effect of cold acclimation on heat resistance. *Bot. Gaz.* 142: 311–315.

Petti CA, Meade C, Downes M, and Mullins E (2007) Facilitating coexistence by tracking gene dispersal in conventional potato systems with microsatellite markers. *Environ. Biosafety Res.* 6: 223–235.

Plaisted RL (1980) Potato. In: Fehr WR and Hadley HH (eds.) *Hybridization of Crop Plants* (American Society of Agronomy: Madison, WI, USA), pp. 483–494.

Quiros CF, Ortega R, van Raamsdale L, Herrera-Montoya M, Cisneros P, Schmidt E, and Brush SB (1992) Increase of potato genetic resources in their center of diversity: The role of natural outcrossing and selection by the Andean farmer. *Genet. Resour. Crop Evol.* 39: 107–113.

Rabinowitz D, Linder CR, Ortega R, Begazo D, Murguia H, Douches DS, and Quiros CF (1990) High levels of interspecific hybridization between *Solanum sparsipilum* and *S. stenotomum* in experimental plots in the Andes. *Am. Potato J.* 67: 73–81.

Ramanna MS and Hermsen JGT (1981) Structural hybridity in the series *Etuberosa* of the genus *Solanum* and its bearing on crossability. *Euphytica* 30: 15–31.

Reheul D (1987) Ruimtelijke isolatie in de plantenveredeling: 2. Ruimtelijke isolatie bij insectenbestuivers. *Landbouwtijdschrift* 40: 15–23.

Romano A, Rarmakers K, Visser R, and Mooibroek H (2001) Transformation of potato (*Solanum tuberosum*) using particle bombardment. *Plant Cell Rep.* 20: 198–204.

Ross H (1986) Potato breeding—problems and perspectives. *J. Plant Breed.* 13 (Suppl.): 1–132.

Sanford JC and Hanneman RE Jr (1981) The use of bees for the purpose of intermating in potato. *Am. Potato J.* 58: 481–485.

Schittenhelm S and Hoekstra R (1995) Recommended isolation distances for the field multiplication of diploid tuber-bearing *Solanum* species. *Plant Breed.* 114: 369–371.

Scurrah M, Celis C, Chumbiauca S, Salas A, and Visser RGF (2008) Hybridization between wild and cultivated potato species in the Peruvian Andes and biosafety implications for deployment of GM potatoes. *Euphytica* 164: 881–892.

Shahin EA and Simpson RB (1986) Gene transfer system for potato. *Hort. Sci.* 21: 1199–1201.

Sharma RP and Salunkhe DK (1989) *Solanum* glycoalkaloids. In: Cheeke PR (ed.) *Alkaloids.* Vol. 1 in *Toxicants of Plant Origin* (CRC Press: Boca Raton, FL, USA), pp. 179–236.

Simmonds NW (1995) Potatoes. In: Simmonds S and Smartt J (eds.) *Evolution of Crop Plants.* 2nd ed. (Longman Scientific and Technical: Harlow, UK), pp. 466–471.

Skogsmyr I (1994) Gene dispersal from transgenic potatoes to conspecifics: A field trial. *Theor. Appl. Genet.* 88: 770–774.

Spooner DM and Castillo TR (1997) Reexamination of series relationships of South American wild potatoes (Solanaceae: *Solanum* sect. *Petota*): Evidence from chloroplast DNA restriction site variation. *Am. J. Bot.* 84: 671–685.

Spooner DM and Hetterscheid WLA (2006) Origins, evolution, and group classification of cultivated potatoes. In: Motley TJ, Zerega N, and Cross H (eds.) *Darwin's Harvest: New Approaches to the Origins, Evolution, and Conservation of Crops* (Columbia University Press: New York, NY, USA), pp. 285–307.

Spooner DM and Hijmans RJ (2001) Potato systematics and germplasm collecting, 1989–2000. *Am. J. Potato Res.* 78: 237–268.

Spooner DM and Salas A (2006) Structure, biosystematics, and genetic resources. In: Gopal J, and Khurana SMP (eds.) *Handbook of Potato Production, Improvement, and Post-Harvest Management* (Haworth's Press: Binghamton, NY, USA), pp. 1–39.

Spooner DM and van den Berg RG (1992) Species limits and hypotheses of hybridization of *Solanum berthaultii* Hawkes and *S. tarijense* Hawkes: Morphological data. *Taxon* 41: 685–700.

Spooner DM, Salas A, Huamán Z, and Hijmans RJ (1999) Potato germplasm collecting expedition in southern Peru (Departments of Apurímac, Arequipa, Cusco, Moquegua, Puno, Tacna) in 1998: Taxonomy and new genetic resources. *Am. J. Potato Res.* 76: 103–119.

Spooner DM, McLean K, Ramsay G, Waugh R, and Bryan GJ (2005) A single domestication for potato based on multilocus amplified fragment length polymorphism genotyping. *Proc. Natl. Acad. Sci. USA* 102: 14694–14699.

Steiner CM, Newberry G, Boydston R, Yenish J, and Thornton R (2005) *Volunteer Potato Management in the Pacific Northwest Rotational Crops*. Washington State University Extension Bulletin EB1993 (Washington State University: Pullman, WA, USA). http://cru.cahe.wsu.edu/CEPublications/eb1993/eb1993.pdf.

Tamboia T, Cipollini ML, and Levey DJ (1996) An evaluation of vertebrate seed dispersal syndromes in four species of black nightshade (*Solanum* sect. *Solanum*). *Oecologia* 107: 522–532.

Tolstrup K, Andersen SB, Boelt B, Buus M, Gylling M, Holm PB, Kjellsson G, Pedersen S, Ostergard H, and Mikkelsen SA (2003) *Report from the Danish Working Group on the Coexistence of Genetically Modified Crops with Conventional and Organic Crops*. Danish Institute for Agricultural Sciences [DIAS] Report, Plant Production, No. 94 (Danish Research Centre for Organic Farming [DARCOF]: Tjele, Denmark). 275 p.

Treu R and Emberlin J (2000) Pollen dispersal in the crops maize (*Zea mays*), oilseed rape (*Brassica napus* subsp. *oleifera*), potatoes (*Solanum tuberosum*), sugar beet (*Beta vulgaris* subsp. *vulgaris*), and wheat (*Triticum aestivum*). Report for the Soil Association from the National Pollen Research Unit (Soil Association: Bristol, UK). 54 p.

Trognitz BR (1991) Comparison of different pollen viability assays to evaluate pollen fertility of potato dihaploids. *Euphytica* 56: 143–148.

Trujillo C, Rodríguez E, Jaramillo S, Hoyos R, Orduz S, and Arango R (2001) One-step transformation of two Andean potato cultivars (*Solanum tuberosum* L. subsp. *andigena*). *Plant Cell Rep.* 20: 637–641.

Tynan JL, Williams MK, and Conner AJ (1990) Low frequency of pollen dispersal from a field trial of transgenic potatoes. *J. Genet. Breed.* 44: 303–306.

Ugent D (1970) The potato. *Science* 170: 1161–1166.

Wahaj SA, Levey DJ, Sanders AK, and Cipollini ML (1998) Control of gut retention time by secondary metabolites in ripe *Solanum* fruits. *Ecology* 79: 2309–2319.

Watanabe KN and Peloquin SJ (1989) Occurrence of 2*n* pollen and *ps* gene frequencies in cultivated groups and their related wild species in tuber-bearing *Solanums*. *Theor. Appl. Genet.* 78: 329–336.

———— (1991) The occurrence and frequency of 2*n* pollen in 2*x*, 4*x*, and 6*x* wild, tuber-bearing *Solanum* species from Mexico and Central and South America. *Theor. Appl. Genet.* 82: 621–626.

Watanabe KN, Orrillo M, Vega S, Valkonen JPT, Pehu E, Hurtado A, and Tanksley SD (1995) Overcoming crossing barriers between non-tuber-bearing and tuber-bearing *Solanum* species: Towards potato germplasm enhancement with a broad spectrum of solanaceous genetic resources. *Genome* 38: 27–35.

Werner JA and Peloquin SJ (1991) Occurrence and mechanisms of 2*n* egg formation in 2*x* potato. *Genome* 34: 975–982.

White JW (1983) Pollination of potatoes under natural conditions. *CIP [Centro Internacional de la Papa] Circular* 11: 1–2.

Wozniak C (2002) Gene flow assessment for plant-incorporated protectants by the biopesticide and Pollution Prevention Division, U.S. EPA. In: Snow AA (ed.) *Scientific Methods Workshop: Ecological and Agronomic Consequences of Gene Flow from Transgenic Crops to Wild Relatives; Meeting Proceedings, University Plaza Hotel and Conference Center, Ohio State University, Columbus, Ohio, March 5–6, 2002* (Ohio State University: Columbus, OH, USA), pp. 162–177. www.agbios.com/docroot/articles/02-280-003.pdf.

EIGHTEEN

Rice
(*Oryza sativa* L.)

There are two different species commonly referred to as rice, Asian cultivated rice (*Oryza sativa* L.) and African cultivated rice (*O. glaberrima* Steud.). African rice is only grown locally in W and central Africa and is now rapidly being replaced by Asian rice (Chang 1995; Lu 1996a). In this chapter, the focus will be on Asian cultivated rice, *O. sativa*, which is also the rice species most often used for genetic modification.

Center(s) of Origin and Diversity

The origin and domestication of Asian cultivated rice is a matter of continuing debate (Sweeney and McCouch 2007). The "where" seems to be relatively well understood; it is generally agreed that the crop originated in the region of the Yangtze River valley in China (Normile 2004; Fuller et al. 2007). However, the question about how domestication occurred is not yet conclusively resolved. In the past three decades, arguments have been brought forward for a single domestication event (Chang 1976; Oka 1988; Lu et al. 2002b; D. Vaughan et al. 2008) as well as for multiple (two or more) domestications (Second 1982; Cheng et al. 2003; Ma and Bennetzen 2004; Yamanaka et al. 2004; Q. Zhu and Ge 2005; Londo et al. 2006).

The primary center of diversity of Asian cultivated rice is in Indochina, which is also thought to be its principal domestication site (D. Vaughan et al. 2005). The mainland and islands of SE Asia are considered to be an additional primary center of diversity, associated with the intercrossing of the 'Indica' and 'Japonica' forms of *Oryza sativa*. A wide range of intermediate forms and tropical 'Japonica' are found in this region.

Secondary centers of diversity with distinctive *O. sativa* forms have been identified at locations in the Indian Ocean, W Asia, and Europe. More recently, other distinguishable secondary centers of diversity have been recognized in Africa and S America (Hamilton and Raymond 2005). The primary center of diversity for *Oryza* species in general is the island region ranging from SE Asia to the Pacific Ocean (Hamilton and Raymond 2005).

Biological Information

Flowering

The rice inflorescence is a panicle. In most cultivated rice varieties, the stigmas and stamens are short, only partially exerted, and largely protected from cross-pollination by remaining somewhat within the husk. Flowering lasts for about a week, until all the florets on the panicle have opened.

As a rule, the flowers open between 9 a.m. and 2 p.m. and remain open from as little as 6 minutes to up to 2.5 hours (Grist 1986; Moldenhauer and Gibbons 2003). Anthesis is dependent on humidity and temperature, and it occurs later and takes longer on cooler or cloudy days. Most pollen is shed before the anthers protrude from the flowers, and not later than 10 minutes after the florets open (Morishima 1984; Oka 1988). Fertilization is usually completed within six hours.

Most wild rice species have larger and longer stigmas and stamens, which expand outside the spikelet, generally leading to slightly higher outcrossing rates (Parmar et al. 1979).

Flowering Coincidence

The flowering habits of rice cultivars grown in different parts of the world vary greatly, depending on the variety and the differences in local cultivation time and seasons. Also, some varieties are photoperiod sensitive, while others are not. Likewise, the flowering times of wild rice vary both among species and among different populations of the same species, depending on geographic regions. Even if the flowering periods overlap, the time of the day when the flowers of different species open is an important factor, since rice flowers often remain open for less than three hours (Moldenhauer and Gibbons 2003).

In places where weedy rice (*Oryza sativa* f. *spontanea* Roshev.) occurs within fields of cultivated rice, the phenology, morphology, and adaptation of weedy rice can shift, due to cross-pollination and genetic recombination, to a point where they resemble (mimic) those of the crop (Delouche et al. 2007).

Pollen Dispersal and Longevity

Domesticated rice is mainly wind pollinated, but some varieties produce scented flowers that attract bees (Oka 1988).

The pollen of domesticated rice is very short lived and loses viability within a few minutes, even if temperature and moisture conditions are favorable. The longevity of wild rice pollen is a bit longer, up to 10 minutes (Oka and Morishima 1967). However, in a study of pollen longevity of three different rice types (wild rice, *O. rufipogon* Griff.; Asian rice, *O. sativa*; and their hybrids) pollen grains remained viable in the air for 70, 30, and 40 minutes, respectively (Song et al. 2001, 2004a).

Sexual Reproduction

Oryza sativa is a diploid ($2n = 2x = 24$, **AA** genome) annual crop. It is mostly autogamous and cleistogamous (Kolhe and Bhat 1981; Yoshida 1981). Outcrossing rates are typically less than 1%, to a maximum of 5% (Lord 1935; Beachell et al. 1938; Oka 1988; Gealy 2005). The degree of outcrossing tends to be higher in 'Indica' than in 'Japonica' cultivars, which are highly autogamous with very low rates of outcrossing (Oka 1988). The extent of outcrossing is variable between species and varieties, and even between populations within the same species. Actual outcrossing rates depend on several factors, including the morphology of the rice florets (which differs among varieties) and genotype. Also, the presence of honeybees leads at times to higher outcrossing rates (Gealy et al. 2003).

Like domesticated rice, most wild species are autogamous. However, some wild species (e.g., *O. longistaminata* A. Chev. & Roehr.) are self-incompatible and thus fully allogamous.

Vegetative Reproduction

Domesticated rice propagates sexually through seeds. Under favorable temperature and water conditions, some cultivars (particularly 'Indica' cultivars) and all perennial wild species can propagate vegetatively via tillers, or ratoons, that reshoot from subsoil nodes after the plant has ripened (Jia and Peng 2002). Ratoons can also detach from the parent plants and float down rivers, forming mats of regenerative material that can take root when reaching land (Morishima 1994; Lu 2006). Whole plants of tropical American wild rice, *Oryza grandiglumis* (Döll) Prod., can be dispersed by floating culms, and new plants can grow from loose culms lodged in newly exposed mud banks or attached to fallen trees in the river (Zamora et al. 2003).

Climatic and cultural conditions do not allow ratoon formation in most temperate rice-growing regions, such as Australia.

Seed Dispersal and Dormancy

Wild rice disperses its seeds freely at maturity to guarantee its propagation, while domesticated rice has almost completely lost this seed-shattering ability (Z. Lin et al. 2007). However, different levels of moderate seed shattering can still be found in some rice cultivars, since farmers in various regions have selected cultivars with different characteristics (from easy to hard grain shattering) for a variety of uses (e.g., for hand and machine harvesting). In general, 'Indica' cultivars have a wider range of grain shattering than 'Japonica' cultivars (D. Vaughan et al. 2008).

Rice seeds shattered before or during harvest can be buried in the soil or dispersed by water (e.g., by floating in irrigation canals or rivers), animals (rodents, ants, birds; also attached to passing animals), or humans.

Rodents, for example, mainly consume the seeds, but sometimes they carry rice grains away from fields. The distance for this form of seed dispersal depends on the size of the area in which the rodent roams.

The average area for mice ranges from 0.02 to 0.2 ha, that is, 40 × 50 m² (Caughley et al. 1998). Ants can also remove rice seeds and store them underground. Although there are differences across species in ant behavior and in the size of their foraging area (Peters et al. 2003), seed dispersal often occurs at a local scale, with a mean dispersal distance of less than 1 m and a maximum of 40 m (Beattie 1982; Gómez and Espadaler 1998).

Humans can also disperse viable seeds, either when the grains are threshed and dried in the open air or when they are handled and transported for milling. In rice intended for food consumption, the hulls of the rice seeds are frequently removed before shipping and exportation, rendering the seeds unviable. But rice can also be transported without dehulling, in the form of viable seeds intended for domestic seed sales.

Seeds of most modern rice varieties have a short dormancy period that persists until maturity and for a few days beyond. Yet some rice cultivars have seeds with high dormancy that can remain viable in the soil for several years (D. Vaughan 1994; Moldenhauer and Gibbons 2003). The seeds of weedy rice, and of hybrids of weedy and domesticated rice, form seed banks and can remain viable in the soil for more than 10 years (Small 1984; Diarra et al. 1985; Footitt and Cohn 1992). The dormancy of wild rice relatives, by and large, is strong.

Volunteers, Ferals, and Their Persistence

Under favorable water and temperature conditions, volunteer and feral plants of domesticated rice can establish themselves in paddies and upland fields. This can also happen outside of cultivation, from seeds shattered prior to, during, and after harvest (reviewed in Baki et al. 2000). In general, 'Indica' cultivars have a greater potential to become volunteers or ferals than 'Japonica' cultivars, because their grains shatter more readily.

Volunteer plants of domesticated rice do not persist for long in cultivated fields, since most are buried or killed by normal agronomic practices such as plowing, herbicide application, drainage, or flooding and rotation. Furthermore, irrigation encourages them to germinate, and the seedlings are subsequently destroyed.

Feral populations of *Oryza sativa* can persist and become natural-ized in the wild, as was reported in W Australia (Hussey et al. 1997). Weedy rice is more aggressive and competitive, and it has persisted in naturalized populations outside of cultivation in the United States (Ar-kansas) for more than 150 years (Sagers et al. 2002).

Weediness and Invasiveness Potential

Some rice cultivars possess many of the characteristics often associated with successful weeds, such as broad adaptation to different environ-ments, the ability to outcross as well as self-pollinate, seed dormancy, seed shattering, and vegetative regeneration. 'Japonica' rice tends to have a lower potential for weediness than 'Indica' cultivars. This is be-cause 'Japonica' rice is highly autogamous, does not regenerate vegeta-tively, has low seed shattering, and is not dormant if dried after harvest (OGTR 2005). Nonetheless, neither volunteer nor feral rice are consid-ered to be serious problems.

Weedy or red rice (*Oryza sativa* f. *spontanea*), on the other hand, is a weed that poses a major problem in commercial rice plantations worldwide. It shatters easily and has strong dormancy, forming seed banks in the soil that can persist for up to 12 years (Footitt and Cohn 1992; Delouche et al. 2007).

Crop Wild Relatives

The genus *Oryza* L. includes the two domesticated species *O. sativa* (Asian cultivated rice) and *O. glaberrima* (African cultivated rice), in addition to over 20 wild species widely distributed in the pantropics and subtropics (table 18.1; Chang 1976; Lu 1996a, 1998). Species in the genus *Oryza* include both diploids ($2n = 2x = 24$) and tetraploids ($2n = 4x = 48$), and 10 different genome types (the **AA, BB, CC, BBCC, CCDD, EE, FF, GG, HHJJ,** and **HHKK** genomes) (D. Vaughan 1994; Ge et al. 1999; Lu 1999).

Asian cultivated rice, *O. sativa*, is thought to have evolved from the domestication of its wild progenitor—variously considered to be *O. ru-fipogon* (Sampath and Govindaswami 1958; Oka 1974), *O. nivara* S.D. Sharma & Shastry (Chang 1976), or some intermediate type between both species (Sano et al. 1980). The three species are morphologically distinct, have somewhat different habitat preferences, and only partly overlap in their flowering times (Lu et al. 2002b; Kuroda et al. 2005). Even so, in large areas of SE Asia they form an interbreeding complex with extensive introgression and hybridization (Chang et al. 1982; Pental and Barnes 1985; Lu 1996a). Due to a lack of reproductive isolation between them, their classification into three independent species has been questioned. Instead, they are often considered to be one large species complex, with *O. rufipogon* and *O. nivara* as two ecotypes or subspecies of the same species (D. Vaughan 1994; Lu 1999; Lu et al. 2000, 2002b; Morishima 2001; D. Vaughan et al. 2003; Londo et al. 2006; Q. Zhu et al. 2007).

Weedy or red rice (*O. sativa* f. *spontanea*) is a highly variable intermediate between domesticated and wild rice that arose from natural hybridization and introgression. Weedy rice is autogamous, yet it has outcrossing rates of up to 52% (D. Vaughan and Morishima 2003). The multiple phenotypes, ecotypes, and biotypes observed and reported worldwide result from continuous natural crossings of weedy rice with cultivated varieties (Delouche et al. 2007). The taxonomic status of the highly diverse weedy rice complex and its origin are not yet fully clarified. For instance, in SE Asia and the United States, weedy rice has been reported as the result of hybridization between domesticated *O. sativa* and its two Asian wild relatives, *O. nivara* and *O. rufipogon* (see, e.g., Abdullah et al. 1996; Tang and Morishima 1996; Suh et al. 1997b; L. Vaughan et al. 2001; Chen et al. 2004). Other studies suggest that weedy rice could also include African *Oryza* species, such as *O. barthii* A. Chev., *O. glaberrima*, *O. longistaminata*, and *O. punctata* Kotschy *ex* Steud. (Holm et al. 1997). In NE China, weedy rice may have originated from segregants of intervarietal hybridization in *O. sativa*, without any direct involvement of wild *O. rufipogon* (Ishikawa et al. 2005; Cao et al. 2006). The tropical American weedy rice complex may be composed of several wild American *Oryza* species with varying degree of sexual compatibility, including the tetraploid *O. latifolia* Desv. (Lentini and Espinoza 2005). In Costa Rica, the weedy rice complex has been reported to involve *O. latifolia* and the Asian *O. rufipogon*, as well as an unidentified group of polymorphic plants, probably crosses between *O. sativa* cultivars and other *Oryza* species, such as *O. glu-*

Table 18.1 Domesticated rice (*Oryza sativa* L.) and its wild relatives

Species	Genome	Asia	Tropical America	N America	Africa	Oceania	Life form, weediness
Primary gene pool (GP-1)	AA						
sect. *Oryza*							
ser. *Sativae* S.D. Sharma & Shastry							
O. sativa L.		✓	✓*	✓*	✓*	✓*	'Asian rice'; annual; known only from cultivation
O. sativa f. spontanea Roshev.		✓	✓	✓*			'weedy,' 'red,' or 'black' rice; weedy annual ancestral, wild progenitor
O. rufipogon Griff.		✓	✓*	✓*		✓	weedy perennial; endangered in some regions
O. nivara S.D. Sharma & Shastry		✓					wild annual
O. glumaepatula Steud.			✓				wild perennial
O. glaberrima Steud.					✓		'African rice'; cultivated
O. barthii A. Chev.					✓		ancestral, wild annual
O. longistaminata A. Chev. & Roehr.					✓		ancestral, wild perennial
O. meridionalis N.Q. Ng		✓				✓	wild annual
Secondary gene pool (GP-2)	B, C, D, E, F						
sect. *Oryza*							
ser. *Latifoliae* S.D. Sharma & Shastry							
O. punctata Kotschy ex Steud.	BB				✓		
O. officinalis Wall. ex G. Watt	CC	✓					
O. rhizomatis Vaughan	CC	✓				✓	

Species / taxon	Genome				Notes
O. malampuzhaensis Krishnasw. & Chandras.	**BBCC**	✓			endemic to India (tetraploid form of O. officinalis)
O. minuta J. Presl	**BBCC**	✓			
O. eichingeri Peter	**BBCC**	✓	✓		tetraploid form of O.punctata
O. schweinfurthiana Prod.	**BBCC**		✓		
O. alta Swallen	**CCDD**	✓			
O. grandiglumis (Döll) Prod.	**CCDD**	✓			
O. latifolia Desv.	**CCDD**	✓			
ser. Australienses Tateoka ex S.D. Sharma & Shastry					
O. australiensis Domin	**EE**			✓	
sect. Brachyantha (S. Sampath ex S.D. Sharma & Shastry) B.Rong Lu	**FF**				
O. brachyantha A. Chev. & Roehr.			✓		
Tertiary gene pool (GP-3)	**G, H, J, K**				
sect. Padia (Moritzi) Baill.					
ser. Meyerianae S.D. Sharma & Shastry					
O. meyeriana (Zoll. & Moritzi) Baill.	**GG**	✓			
O. neocaledonica Morat†	**GG**			✓	
ser. Ridleyanae S.D. Sharma & Shastry					
O. ridleyi Hook. f.	**HHJJ**	✓			
O. longiglumis Jansen	**HHJJ**	✓			
ser. Schlechterianae S.D. Sharma & Shastry					
O. schlechteri Pilg.	**HHKK**	✓			
O. coarctata Roxb.	**HHKK**	✓			

Sources: Lu 1999; Lu and Ge 2005.
Note: Cultivated forms are in bold. Genome designations follow Chang 1995 and Ge et al. 1999. Gene pool classifications are according to Kush 1997.
*Introduced.
†Described by Morat et al. 1994.

maepatula Steud., *O. grandiglumis*, *O. latifolia*, and *O. rufipogon* (Arrieta-Espinoza et al. 2005). Thus the possibility that some of the diversity in weedy rice was derived from hybridization with entirely different species cannot be ignored and, as such, requires further study.

Hybridization

As mentioned above, in spite of the fact that domesticated rice is primarily a self-pollinated crop, outcrossing and hybridization with other *Oryza* species do occur (Nezu et al. 1960; Pental and Barnes 1985; Langevin et al. 1990), in particular with species sharing the same A genome (table 18.1). On the other hand, spontaneous hybridization between *O. sativa* and species with different genomes is very rare, due to significant reproductive barriers such as hybrid sterility, non-viability, or weakness (Oka 1974; Kubo and Yoshimura 1998).

Primary Gene Pool (GP-1)

The primary gene pool includes all species in sect. *Oryza* ser. *Sativae* S.D. Sharma & Shastry; they have the same ploidy level ($2n = 2x = 24$) and genome (**A** genome) as domesticated *O. sativa* (table 18.1). Hybridization of *O. sativa* with most of these diploid A-genome species occurs spontaneously in nature (Oka and Chang 1961). Hand crosses are relatively easy, and seed set can exceed 70% (Chu and Oka 1970b; Sitch et al. 1989; Morishima et al. 1992). The fitness of the resulting F1 hybrids is generally high (Naredo et al. 1997, 1998; Lu et al. 2002a), although reduced fertility or high sterility are sometimes observed (e.g., Nezu et al. 1960; Chu et al. 1969; Langevin et al. 1990). Backcrossing to one of the parents can restore fertility and stabilize the hybrids, thus facilitating gene flow and introgression between species (D. Vaughan and Morishima 2003).

There are eight A-genome rice species in GP-1 that co-occur with Asian cultivated rice (*O. sativa*) on different continents (Oka 1991; D. Vaughan 1994; D. Vaughan et al. 2003; Lu and Snow 2005): African cultivated rice (*O. glaberrima*), the ubiquitous weedy or red rice (*O. sativa* f. *spontanea*), and six annual and perennial wild relatives from Africa, tropical America, Asia, and Oceania (table 18.1). The six wild relatives are:

- perennial *Oryza rufipogon* and annual *O. nivara* from Asia
- perennial *O. longistaminata* and annual *O. barthii* from Africa
- perennial *O. glumaepatula* from tropical America
- annual *O. meridionalis* N.Q. Ng from N Australia and New Guinea

All of these species can hybridize naturally with domesticated rice and with each other, and they are therefore treated in more detail below.

Oryza sativa f. *spontanea* (weedy, red, or black rice)

Asian cultivated rice is completely cross-compatible with weedy rice (Langevin et al. 1990; Noldin et al. 1999; Dillon et al. 2002). Natural hybridization produces perennial, highly fertile F1 plants that exhibit the dominant traits of the parental weed and often outperform their parents in plant height, biomass, and seed production (Langevin et al. 1990; Oard et al. 2000; Snow et al. 2006). Rates of spontaneous outcrossing are similar to those for the outcrossing of domesticated rice and are usually below 1% (Gealy et al. 2003; N. Zhang et al. 2003; Chen et al. 2004), with maximum values up to 3% (Beachell et al. 1938; W. Zhang et al. 2004). They can increase greatly within only two years of contact with the crop, due to recurrent backcrossing (Langevin et al. 1990). Outcrossing depends on various factors, including the cultivar, the weedy rice ecotype, vertical and/or horizontal distances between panicles, synchronization of flowering periods, flower morphology, and seed production, as well as environmental conditions that can increase or reduce pollen viability and longevity (Delouche et al. 2007).

Outcrossing can occur with either plant type as the pollen donor, but there is controversy about preferential gene flow rates. In most studies, hybridization rates are higher when weedy rice is the male parent (pollen donor) and domesticated rice the pollen recipient (Gealy et al. 2005, 2007; Gealy and Estorninos 2007). This is explained by the fact that pollen flow—and consequently gene introgression—can be preferential from taller plants (usually the weedy rice) to shorter plants (the cultivated variety) (N. Zhang et al. 2003; Estorninos et al. 2004; Delouche et al. 2007).

Because of the dominance of most wild traits, hybridization between domesticated and weedy rice can lead to greater hybrid vigor, which is shown in vegetative traits such as plant height, flag leaf length, and panicle length (Cragmiles 1978; N. Zhang et al. 2003). Modeling studies by Madsen et al. (2002) estimate that herbicide resistance can become common in weedy rice populations within just three to eight years of continuous rice cropping. Other traits such as hybrid morphology, phenology, fitness, and fecundity are highly variable and depend on both the direction of gene flow and the parental genotypes involved (see, e.g., Langevin et al. 1990; Oard et al. 2000; N. Zhang et al. 2003; Noldin et al. 2004). In segregating F2 plants, for example, flowering behavior can range from extremely early to extremely late; in terms of plant height, some are shorter than the parental domesticated rice variety, while others are much taller than the red rice parent (Shivrain et al. 2006, 2007; Delouche et al. 2007). On the other hand, late flowering and seed set, and hybrid sterility, are fairly consistent traits in hybrid populations of domesticated and weedy rice (Jodon 1959; Gealy 2005). Knowledge about hybrid vigor in weedy rice and the fate of introgressed traits in later generations is still not sufficient, and more studies are needed to better understand the relationships between domesticated and weedy rice over several generations.

Oryza glaberrima (African cultivated rice)

The two domesticated rice species, *Oryza sativa* and *O. glaberrima*, are often grown (in various proportions) as mixtures in W African rice fields, and natural hybrid swarms between the two species—although rare—may be found in these areas (Chu et al. 1969; Pental and Barnes 1985; Messeguer et al. 2001; Linares 2002; Semon et al. 2005). Estimates of the outcrossing rate between the two species are low (between 2% and 5%) (Semon et al. 2005). In most cases, F1 hybrids exhibit normal vigor and intermediate morphology, leaning towards *O. glaberrima* (Bougerol and Pham 1989).

Although an average of 30%–56% seed set has been reported in interspecific crosses, there is no consensus on the extent of crossability. Hybrid sterility in *O. sativa* × *O. glaberrima* crosses has long been recognized (Morinaga and Kuriyama 1957; Morishima et al. 1962; Chu et al. 1969; Sano et al. 1979; Tao et al. 1997). Artificially produced F1 hybrids are often fully sterile or highly male sterile, with a few fertile embryo sacs allowing BC and fertility restoration that occasionally is up to 90% (Oka 1988; Bougerol and Pham 1989; Pham and Bougerol 1993, 1996; Jones et al. 1997; Heuer et al. 2003). Backcross populations are sometimes more sterile than their corresponding F1 hybrids (Bouharmont et al. 1985), and semi-sterile individuals are more and more frequent in subsequent BC generations (Tao et al. 2003). It has therefore been suggested that natural hybrids might disappear within a few generations, due to F1 weakness and hybrid breakdown (OECD 1999; OGTR 2005).

Still, cross-compatibility varies, depending on the *O. sativa* and *O. glaberrima* genotypes involved. For example, if compatible genotypes are used, hybridization via hand pollination is relatively easy and can produce hybrid progeny with good fertility to at least BC3 (Lorieux et al. 2000). Also, new, high-yielding varieties suitable for cultivation in the West African region (NERICA, or New Rice for Africa, varieties) have been developed from interspecific crosses between *O. glaberrima* and *O. sativa* (Jones et al. 1997; Gridley et al. 2002). Furthermore, in a genetic analysis of African *O. glaberrima* germplasm, two out of three accessions from Guinea Conakry, and every second accession from Sierra Leone, were identified as admixtures showing high levels of introgression from *O. sativa* (Semon et al. 2005). In general, *O. glaberrima* is more compatible with *O. sativa* 'Indica' cultivars than with 'Japonica' cultivars (Sarla and Mallikarjuna Swamy 2005). More research is needed to better understand the fitness and fertility of artificial and natural hybrid progeny from these species over several generations.

Oryza rufipogon

Asian cultivated rice and its progenitor, *Oryza rufipogon*, have a sympatric distribution in many regions of the world, and their flowering times overlap in broad sections of Asia. Genetically, the two species are closely related, and reproductive isolation is low (S. Lin and Yuan 1980; Chang et al. 1982; Pental and Barnes 1985; Lu 1996a). *Oryza rufipogon* and *O. sativa*, together with *O. nivara*, form an interbreeding complex and are often considered to be one large species complex (Lu et al. 2000; Morishima 2001; D. Vaughan et al. 2003; Londo et al. 2006; Q. Zhu et al. 2007).

Therefore, fertile hybrid swarms and gene introgression between

O. rufipogon and *O. sativa* are frequent in areas of sympatry (Oka and Chang 1961; Sano et al. 1980; Suh et al. 1997a; Lu 1999; Messeguer et al. 2001). Natural hybridization with domesticated rice has been implicated in the near extinction of an endemic Taiwanese subspecies, *O. perennis formosana* (= *Oryza sativa* var. *formosana* (Masam. & Suzuki) Yeh & Henderson, a synonym of *O. sativa* L.) (Kiang et al. 1979). Similarly, natural populations of wild *O. rufipogon* have decreased in size and number over the last decade or so, and their genetic diversity is in decline (Chitrakon 1994; Morishima 1994; Akimoto et al. 1999; Song et al. 2004b, 2005).

However, the degree of crossability varies with both the specific cultivar of rice and wild rice population (see, e.g., Oka and Morishima 1967; Chu and Oka 1970b; Oka 1988; Rao et al. 1997). Spontaneous outcrossing rates between domesticated rice and *O. rufipogon* in the wild usually do not exceed 3% (Chang 1995; Lu et al. 2002a; Song et al. 2003; Chen et al. 2004).

Outcrossing rates as high as 95% have been obtained using hand pollination (Song et al. 2002), and backcross selection has allowed the introduction of *O. rufipogon* traits into elite varieties of rice (IRRI 1998; Xiao et al. 1998; Moncada et al. 2001; Zhao et al. 2008). F1 hybrids are perennial and fully viable, but with varying levels of fitness in terms of fertility and seed set (Song et al. 2004a: 40% pollen viability; Lu et al. 2002a: > 11% spikelet fertility, 20% seed set). In general, hybrid progeny seem to be slightly inferior at the sexual reproduction stage—mainly due to lower seed set and pollen viability (Barrett 1983; Majumder et al. 1997; Song et al. 2001; Song et al. 2004a). During the growth phase, however, their hybrid vigor and clonal ability are usually higher than those of the parents (Song et al. 2004a; Schaal et al. 2006), suggesting that they may be more competitive, since their probability of survival is enhanced by vegetative reproduction. Depending on the cultivar type and the wild population, introgression between Asian cultivated rice and wild *O. rufipogon* is possible and probable (Majumder et al. 1997).

Oryza nivara

In addition to *Oryza rufipogon*, *O. nivara* is the other putative progenitor of Asian cultivated rice, *O. sativa*. This annual species is primarily inbreeding, with outcrossing rates ranging from 5% to 25% (Oka 1988). The wild relative is fully cross-compatible with the crop (Khush et al. 1977; Lu et al. 2000), and spontaneous hybridization between the two is common in areas of sympatry (Lu 1996b; Lu et al. 2002a, b; Chen et al. 2004; Kuroda et al. 2005, 2007). The resulting F1 plants are highly fertile and can form extensive hybrid swarms, often intermingled with *O. rufipogon* (Khush and Ling 1974; Khush 1977; Chang 1995; Naredo et al. 1997, 1998).

Genes of interest, such as grassy-stunt-virus resistance (Khush et al. 1977), have been introduced from *O. nivara* to domesticated rice, and an interspecific hybrid has a considerably increased protein content (Mahmoud et al. 2008).

Typical specimens of wild *O. nivara* are rarely found in Asia nowadays, and many wild populations have disappeared from their known habitats, which—among other factors—has been attributed to extensive hybridization with the crop (Chang 1995).

Oryza longistaminata

Oryza longistaminata is unusual in being a perennial, self-incompatible, outcrossing plant (Chu et al. 1969; Ghesquière 1986; D. Vaughan 1994). The species has two simultaneous modes of reproduction, rhizomes and seeds. There is a reproductive barrier between *O. sativa* and *O. longistaminata*, caused by the action of two complementary lethal genes—D1 (*O. longistaminata*) and D2 (*O. sativa*)—that leads to abortion of the embryo in subsequent hybrid generations (Chu and Oka 1970a; Ghesquière and Causse 1992). Albumen deterioration caused by these lethal genes results in the loss of at least 97% of the embryos formed, and 50% of the surviving F1 plants show hybrid weakness (Bezançon et al. 1977).

In spite of these reproductive barriers, naturally occurring hybrids between *O. longistaminata* and an African landrace (kalulu) have been observed in farmers' fields in Malawi (Bezançon et al. 1977; Kiambi et al. 2005), but there are no reports on the fertility and fitness of these hybrids. In Asia, *O. longistaminata* is thought to be involved in the formation of what are known as 'obake' off-types (see the paragraph on *O. sativa* × *O. barthii* below), which are complex interspecific hybrids involving crosses between Asian cultivated rice, *O. bar-*

thii, and *O. glaberrima*, among others (Ghesquière 1988; Bezançon et al. 1989). Artificially produced hybrids between domesticated rice and *O. longistaminata* using embryo rescue techniques yield F1 plants with reduced (< 35%) pollen fertility (Ghesquière et al. 1994; Tao and Sripichitt 2000; Tao et al. 2001). Seed set is highly variable and can reach from less than 10% to up to 43% (Kaushal and Ravi 1998; Tao and Sripichitt 2000). Artificial gene introgression is possible from *O. longistaminata* to domesticated rice (Khush et al. 1990; IRRI 1998; Schmit 1997).

Oryza barthii

Oryza barthii is a self-incompatible, cross-pollinated African wild rice that is isolated from its sympatric wild and domesticated AA-genome relatives by post-zygotic crossing barriers, hybrid weakness, and sterility (Chu et al. 1969; Chu and Oka 1970b). These hybridization barriers lead to the deterioration and death of F1 embryos and endosperms, resulting in less than 5% viable F1 seedlings (Chu and Oka 1970a).

As with *O. longistaminata*, the reproductive barrier between *O. barthii* and Asian cultivated rice is attributed to the action of a set of complementary lethal genes that cause abortion of the embryo in subsequent hybrid generations (Chu and Oka 1970a). Artificial *O. sativa* × *O. barthii* hybrids yield F1 plants with high pollen sterility (pollen fertility < 1%) and less than 5% (and often unviable) seed yield (Kaushal and Ravi 1998). In spite of these reproductive barriers, natural *O. sativa* × *O. barthii* hybrid swarms have been observed in W Africa (Chu and Oka 1970a). In Asia, so-called 'obakes' ('monster' in Japanese) occur naturally; they are complex hybrids swarms that involve crosses of *O. barthii* with *O. sativa*, and with *O. glaberrima* or *O. breviligulata* A. Chev. & Roehr (Chu and Oka 1970b). These hybrid populations show higher fertility than their wild parents, and may thus act as bridge species.

Oryza glumaepatula

The tropical American wild rice relative *Oryza glumaepatula* is predominantly autogamous, but it can act as a facultative allogamous species, with outcrossing rates of up to 35% (Buso et al. 1998). The species can spontaneously outcross with Asian cultivated rice, and natural hybrid swarms between the two have been observed in Cuba (Chu and Oka 1970b).

However, crossing experiments under controlled conditions have shown that *O. glumaepatula* is reproductively isolated from *O. sativa* and other **AA**-genome species by F1 sterility (Chu et al. 1969; Morishima 1969; Naredo et al. 1998; Brondani et al. 2001; Lentini and Espinoza 2005; Espinoza-Esquivel and Arrieta-Espinoza 2007). The spontaneous F1 plants from Cuba were also completely male sterile. Yet partial female fertility can be observed at times, and when *O. sativa* × *O. glumaepatula* hybrids are backcrossed with Asian cultivated rice, vigorous but sterile plants with different levels of hybrid weakness (poor growth stature, and small and completely sterile panicles) are recovered (Yang et al. 1991; Cavalheiro et al. 1996; Sobrizal et al. 1999; Brondani et al. 2001, 2002). A single gene governs hybrid weakness; its *O. glumaepatula* allele is dominant and can bring back hybrid weakness in plants having *O. glumaepatula* cytoplasm (Ikeda et al. 1999). Spontaneous gene flow and introgression between *O. glumaepatula* and *O. sativa* is therefore highly unlikely under natural conditions (Quesada et al. 2003; Espinoza-Esquivel and Arrieta-Espinoza 2007).

Oryza meridionalis

Oryza meridionalis is an annual wild rice from N Australia and New Guinea. The species is reproductively isolated from the Asian **AA**-genome species, including Asian cultivated rice, *O. sativa*. Still, hybridization between *O. meridionalis* and *O. sativa* is possible via hand pollination, although with differing levels of ease, depending on the genotypes (Lu et al. 1997; Naredo et al. 1997). Using compatible genotypes, crosses up to BC2 are relatively easy to obtain (M. Lorieux, pers. comm. 2008). Interspecific hybridization of *O. meridionalis* is also possible with other Asian **AA**-genome species (e.g., *O. rufipogon*, *O. nivara*) without embryo rescue intervention, but hybrid production is low, as many seeds fail to germinate (Chu et al. 1969; Naredo et al. 1997). There are no reports of spontaneous hybridization between *O. meridionalis* and Asian cultivated rice, or with any other Asian **AA**-genome species, under natural conditions.

Secondary Gene Pool (GP-2)

Species with **BB, CC, BBCC, CCDD, EE,** and **FF** genomes (sect. *Oryza* ser. *Latifoliae* Tateoka *ex* S.D. Sharma & Shastry and ser. *Australienses* Tateoka *ex* S.D. Sharma & Shastry; and sect. *Brachyantha* (S. Sampath *ex* S.D. Sharma & Shastry) B.Rong Lu) form the secondary gene pool of rice (table 18.1). Due to low homology between these genomes and the **AA** genome of *O. sativa*, only limited gene transfer is possible. Low crossability and the abortion of hybrid embryos are common features of such crosses (see, e.g., Wuu et al. 1963; Hayashi et al. 1988; Multani et al. 1994; Brar et al. 1996; Yan et al. 1997; Abbasi et al. 1999; Somantri 2001). Successful hybridization has only been achieved using artificial methods, such as embryo rescue and ovary culture, with the exception of one report of natural hybrids (fully sterile) between the diploid *O. glumaepatula* (**AA**-genome species) and a tetraploid species (*O. grandiglumis* or *O. latifolia*, both **CCDD**-genome species) from Costa Rica (Zamora et al. 2003).

F1 intergenomic hybrids are often male sterile (Khush and Brar 1988; Amante-Bordeos et al. 1992; Mariam et al. 1996; Multani et al. 2003; CIAT 2007), and their seeds have poorly developed endosperm, resulting in the embryos aborting during the early stages of development. Species with the **BB, CC, BBCC,** or **CCDD** genome are more crossable with *O. sativa* (0%–30% seed set) than the more distantly related **EE**- and **FF**-genome species (0.2%–3.8% seed set). However, all hybrids with *O. sativa* are highly male sterile and often also female sterile (Sitch 1990).

Tertiary Gene Pool (GP-3)

The tertiary gene pool of rice includes wild rice relatives with **GG, HHJJ,** and **HHKK** genomes (sect. *Padia* (Moritzi) Baill.; table 18.1). Crosses between *O. sativa* and GP-3 species are very difficult to accomplish, even with artificial hybridization techniques (see, e.g., Brar et al. 1991, 1997; Elloran et al. 1992; Khush 1997; Jelodar et al. 1999). Gene flow to these species is restricted to artificial breeding methods, such as embryo rescue and somatic hybridization in the laboratory, and the F1 hybrids are highly or completely sterile (see, e.g., Katayama and Onizuka 1979; Sitch et al. 1989; Jena 1994; Farooq et al. 1996; Lu and Snow 2005; Tan et al. 2006).

Bridging

Some hybrids between Asian cultivated rice and its compatible wild relatives can serve as a bridge for gene flow to other related wild species. In tropical America, for example, the likelihood of introgression to wild rice species is very low, due to the reproductive isolation of the only co-occurring wild species with the AA genome, *Oryza glumaepatula*. However, the situation is different if the introduced Asian *O. rufipogon* acts as a bridging species; then the probability of introgression rises from very low to low. Asian cultivated rice and *O. rufipogon* are very cross-compatible, and they have a sympatric distribution in tropical America. Furthermore, partly fertile hybrids between the latter and *O. glumaepatula* can occur with relative ease after hand pollination, despite the fact that no hybrids have been reported to occur in nature thus far. Therefore, it is possible that *O. rufipogon* × *O. sativa* hybrids might act as bridging species to *O. glumaepatula* in regions where all three species coexist. Also, *O. rufipogon* × *O. sativa* hybrids in Asia and Australia could serve as bridges for gene introgression from *O. sativa* to *O. meridionalis*, although there is a very low probability for these crosses to succeed, due to the reproductive isolation of the latter. In Africa, *O. glaberrima* × *O. barthii* hybrids could act as another catalyst for gene introgression to *O. sativa* (in addition to the fact that on their own, both species are cross-compatible with *O. sativa*).

Outcrossing Rates and Distances

Pollen Flow

Pollen-mediated gene flow in rice is influenced by several factors, including outcrossing rates, variation in flowering times, population sizes, the distance between populations, wind speed and direction, and humidity. Natural outcrossing rates of Asian cultivated rice are usually below 1% (Messeguer et al. 2001; Chen et al. 2004; Rong et al. 2004, 2005), and similar values have been measured in the field for gene flow

Table 18.2 Some representative studies on pollen dispersal distances and outcrossing rates of Asian cultivated rice (*Oryza sativa* L.)

Gene flow from *O. sativa* to	Pollen dispersal distance (m)	% outcrossing (max. values)	Country	Reference
O. sativa	0.2, 0.4, 0.8, 1.6, 2.4	0.08, 0.02, 0.00, 0.03, 0.00	Italy	Messeguer et al. 2001
O. sativa	0.25, 1, 2, 5, 10 (circular design)	0.05–0.11, 0.06, 0.05, 0.01, 0.01	Spain	Messeguer et al. 2004
O. sativa	1, 2, 3, 4, 5, 10, 30, 40, 50	11.6, 0.9, 0.8, 0.4, 0.4, 0.2, 0.03, 0.02, > 0.01	China	Jia et al. 2007
O. sativa	0.2, 1, 2.2, 5.4, 6.2, 8.2, 10.2	0.06, 0.09, 0.03, 0.04, 0.01, 0.01, > 0.01	China	Rong et al. 2007*
O. sativa	0, 7, 10, 15, 20, 40, 50	56, 0.1, 0.7, 0.7, 0.4, 0.06, 0.06	China	Jia 2002*
O. sativa	1, 2, 3, 4, 5, 10, 30, 40, 50	0.012, 0.01, 0.01, > 0.01, > 0.01, > 0.01, > 0.01, 0	China	Jia et al. 2007*
O. rufipogon	0.3, 1.2, 3.6, 10, 19, 28, 43, 58	1.02, 2.14, 2.94, 1.3, 1.2, 1.3, 0, 0	China	Song et al. 2003
O. rufipogon	1.2, 4.8, 9.6, 14.4, 19.2, 24, 28.8	170, 55, 22, 17, 10, 2, 0 pollen grains/cm²	China	Song et al. 2004b
O. rufipogon	1, 2, 5, 10, 20, 50, 70	3.9, 2.9, 1.4, 0.9, 0.1, 0.1, 0	China, Guangzhou	Wang et al. 2006
O. rufipogon	1, 2, 5, 10, 20, 50, 75, 250	11.2, 6.6, 5.0, 2.3, 1.1, 0.2, 0.06, 0.01	China, Sanya	Wang et al. 2006

*Using male-sterile bait plants.

from domesticated to weedy rice (Messeguer et al. 2001, 2004; Estorninos et al. 2002; N. Zhang et al. 2003; Chen et al. 2004). However, outcrossing rates in the field can be greatly enhanced within a few generations, due to recurrent backcrossing (Lentini and Espinoza 2005). In general, intraspecific outcrossing rates and gene flow frequencies are much higher to male-sterile lines than to common rice cultivars (see, e.g., Jia 2002; Jia et al. 2007).

The amount of pollen dispersal and the distance it travels are also influenced by the size of the pollen source. Larger fields generally produce more pollen, leading to greater outcrossing rates and larger dispersal distances than is the case for smaller fields (Song et al. 2004b; Wang et al. 2006).

In many rice-growing areas, the crop is cultivated in small-scale farming systems, with field sizes of 1 ha or less. In China, for example, rice cultivation in complex mixed-planting systems is a widespread practice, with traditional and hybrid rice varieties growing together in the same or adjacent fields (Y. Zhu et al. 2003). These intercropping systems have proved to be beneficial for both disease control and increased yield in rice, and their area is being increasingly extended (Y. Zhu et al. 2000). Gene flow frequencies from hybrid to traditional rice varieties in these production systems are extremely low (0.04%–0.18%), and they could be further minimized by sowing seeds of different varieties at different times, to avoid overlapping flowering periods (Rong et al. 2004).

Isolation Distances

Pollen dispersal distances are influenced by wind speed and direction, and outcrossing events are generally not detectable farther than 50 m from the pollen source (table 18.2). Based on regression analysis, pollen flow may be able to travel up to 110 m in a downwind direction with a fresh breeze, that is, at wind speeds under 10 m/s (Lu et al. 2002a; Song et al. 2004b). This is confirmed by molecular data from a field experiment in China, where outcrossing was measured at a maximum distance of 250 m (0.01%) along the prevalent wind direction (Wang et al. 2006).

The isolation distances established by regulatory authorities for the commercial production of rice seed are below 100 m (table 18.3). The International Rice Research Institute (IRRI) published a study

Table 18.3 Isolation distances for Asian cultivated rice (Oryza sativa L.)

Threshold*	Minimum isolation distance (m)	Country	Reference	Comment
0.0 / 0.1%	0	EU	EC 2001	foundation (basic) seed requirements (non-hybrids)
1 plant per 50m² / 0.3%	0	EU	EC 2001	certified seed requirements (non-hybrids)
n.a. / 0.3%	25	EU	EC 2001	certified seed requirements (hybrids)
n.a. / 0.1%	50	OECD	OECD 2008	foundation (basic) seed requirements (non-hybrids)
n.a. / 0.3%	20	OECD	OECD 2008	certified seed requirements (non-hybrids)
n.a. / 0.1%	600†	OECD	OECD 2008	foundation (basic) seed requirements (hybrids)
n.a. / 0.3%	500	OECD	OECD 2008	certified seed requirements (hybrids)
0.01 / 0.05%	3/15/30‡	USA	USDA 2008	foundation (basic) seed requirements
0.02 / 0.1%	3/15/30‡	USA	USDA 2008	registered seed requirements
0.1 / 0.2%	3/15/30‡	USA	USDA 2008	certified seed requirements
0.002% / 1 seed per pound	75	USA	CCIA 2009	foundation (basic) seed requirements
0.01% / 2 seeds per pound	75	USA	CCIA 2009	registered seed requirements
0.02% / 4 seeds per pound	75	USA	CCIA 2009	certified seed requirements

*Maximum percentage of off-type plants permitted in the field / seed impurities of other varieties or off-types permitted. n.a. = not available.
†If male-sterile varieties are used, the isolation distance is 1000 m.
‡Ground drilled, 3 m; ground broadcast, 15 m; aerial seeded, 30 m.

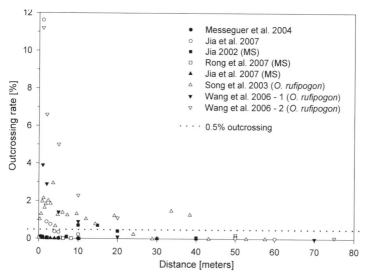

Fig. 18.1 Pollen dispersal distances and outcrossing rates of Asian cultivated rice (*Oryza sativa* L.) and from cultivated to wild rice (*O. rufipogon* Griff.) Wang et al. 2006–1: Guangzhou, Guangdong Province, China; Wang et al. 2006–2: Sanya, Hainan Province, China.

validating its practice of maintaining pure rice lines with a plot separation of 1.5 m and with the outer two rows of each plot not being harvested for seed. Using different field layouts and a range of rice varieties, crossing between adjacent plant rows was found to be below 0.25%, and even when the flowering panicles of different rice varieties were clipped together, outcrossing rates were less than 1% (Reano and Pham 1998).

However, an analysis of various pollen dispersal studies that measured pollen flow from cultivated to weedy and wild rice indicates that, for normal fertile varieties, there is a decline in outcrossing rates to less than 0.2% beyond a distance of 50 m, and to less than 0.06% beyond 75 m. Outcrossing rates of male-sterile varieties were similar to those for normal fertile varieties, but gene flow was considerably higher to a wild species, *Oryza rufipogon*, than to cultivated varieties (fig. 18.1). According to these data, in many cases separation distances of 50 m or less—as recommended for seed production by many authorities (table 18.3)—

seem to be insufficient to minimize seed contamination through pollen flow. Although the data suggest that these distances could be adequate to keep seeds under the maximum allowable amount of transgene contamination (0.9%), larger separation distances may be required to attain seed purity levels that comply with stipulated thresholds of, for example, 0.1% impurities or less.

State of Development of GM Technology

A comprehensive review of the state of development in GM technology for rice can be found in Bajaj and Mohanty (2005). Traits that have been used for the genetic modification of rice include herbicide resistance; disease and insect resistance (including *Bt*); abiotic stress tolerance (e.g., to salinity, drought, heat, cold, submergence, and mineral deficiency); increased yield; biofortification (e.g., enhanced protein, starch, iron, zinc, and beta-carotene [golden rice] content); other nutritional compounds (oil quality); male sterility; pharmaceutical proteins; and heavy-metal bioremediation.

The total rice production area is estimated at 157 million ha for paddy rice (FAO 2009), with 0.01% used for the production of GM rice. GM rice has been commercially produced in Iran since 2005. In the United States, two herbicide-resistant strains (LibertyLink(r)) have been deregulated since 1999, but thus far they have not been commercialized.

Nonetheless, field trials are being conducted in several countries (Argentina, Australia, Brazil, China, Colombia, Costa Rica, France, India, Italy, Japan, Korea, Pakistan, the Philippines, Spain, and the United States), and the adventitious presence of GM rice in the world's rice supply has been documented (EC 2007; USDA 2007).

Agronomic Management Recommendations

Although the survival and maintenance of hybrids is quite unlikely in the majority of cases, there are three main ways gene introgression to other cultivated or wild species can occur:

1 through pollen flow from cultivated rice

2 through volunteers in fields previously planted with rice

3   through rice grains shattered during cropping, harvesting, handling, storage, and transport of the crop

Rice seed, like maize, is also dispersed through informal seed-exchange systems among farmers (Dorward et al. 2007). Furthermore, vegetative propagation and competitive displacement among clones could promote (trans)gene persistence in perennial types, such as *Oryza glumaepatula*, *O. longistaminata*, and *O. rufipogon*.

To reduce the likelihood of gene introgression as much as possible, the following management recommendations could be adopted:

- Consider land-use history to avoid contamination through rice volunteers from earlier cropping cycles—not only of the field to be planted, but also of neighboring fields.

- Apply management practices to limit weedy rice infestation (see, e.g., Smith and Hill 1990; Moody 1994; Forner 1995; Askew et al. 1998; Delouche et al. 2007).

- Since gene flow from cultivated rice to weedy rice cannot be avoided when they occur in the same fields, care should be taken to avoid releasing rice with transgenes that may significantly increase weediness and protect against weed control (such as those for herbicide resistance). This is important in regions where weedy rice is already a serious problem, because the frequency of hybrids could accumulate and increase through subsequent generations. As a rule, herbicide-resistant rice should only be used with effective biological confinement techniques and strict monitoring guidelines (Gressel 2000, 2002; Olofsdotter et al. 2000).

- If flowering overlap is likely with any of the plant species identified above as being susceptible to gene introgression, then isolation by distance is an effective way of confining pollen, since distances of only a few meters can dramatically reduce pollen flow to surrounding rice fields. Plants should only be in the area at distances greater than those specified in legal regulations for seed production. Also, barriers with male-sterile bait plants (which are only fertilized by pollen from external sources) or other trap crops (e.g., maize or sugarcane) around the area planted with rice are an effective means to capture escaped pollen (den Nijs et al. 2004).

- At local scales and in mixed-rice cultivation systems (mainly used by small-scale farmers), sow different varieties at different times to avoid flowering overlap.

- Apply mechanical or chemical control measures after harvest to eliminate potential volunteer plants.

- Take special measures when transporting unhulled rice grains, to avoid the establishment of ferals through seed. Monitor roads for feral rice plants and, if present, control them by mechanical or chemical means.

Conservation Status and Protected Areas

Oryza rufipogon is listed as a noxious weed in the United States (Vandiver et al. 1992), but it is threatened with local extinction within parts of its native range—for example, in Bangladesh, China, Thailand, and Vietnam (Akimoto et al. 1999; Song et al. 2004b, 2005). In fact, in Asia many wild *O. nivara* and *O. rufipogon* populations have become extinct, and typical specimens are now found very rarely—due in part to habitat destruction, but also to extensive hybridization with the crop (Chang 1995; Gao et al. 2000; Zhou et al. 2003; Lu and Snow 2005). The Latin American *O. glumaepatula* also occurs in ecosystems that are fragile and vulnerable to extinction. Commercial rice plantations are often just 20 to 100 m distant from *O. glumaepatula* populations, and in some regions rice is cropped within these populations, such as in Costa Rica near the Nicaraguan border (Lentini and Espinoza 2005). *In situ* conservation of these species should be pursued assiduously.

In Africa (particularly in Kenya, Malawi, and Tanzania), habitat destruction by human activity poses a significant threat to the wild rice species. The presence of wild species indicates that conditions are suitable for growing rice, so farmers have cleared wild population areas and replaced them with cultivated rice (Kiambi et al. 2005). Many sites where wild rice populations had previously been documented have disappeared, either because they were displaced by settlements or disturbed by other factors, including agriculture, fishing, and brick making. Another serious threat is the reduction of plant populations through overgrazing by livestock (Emerton and Muramira 1999).

Crop Production Area

Oryza sativa (rice) is cultivated worldwide throughout the tropics, subtropics, and some temperate regions, from as far north as the border between the Russian Federation and China (55° N) to as far south as Argentina (35° S). Rice is grown under diverse conditions, and rice production systems can be classified as irrigated (55%), rain-fed lowland (25%), upland (12%), and flood-prone (8%) cultivation systems (Khush 1997). The vast majority (90%) of all rice is grown and consumed in Asia, with China and India alone producing more than 50% of the world's rice (FAO 2009).

Research Gaps and Conclusions

Little is known about how easily modern rice cultivars can revert to feral weedy rice, and thereby disperse as weedy biotypes (D. Vaughan 1994). The potential for weedy rice to act as a bridge species for gene flow to other wild rice relatives also deserves further attention. It is particularly vital to gain a better understanding of the evolution and population dynamics of weedy rice in different regions of the world in order to evaluate the consequences of gene flow from GM rice to weedy and wild rice populations.

A more thorough analysis of the ecological consequences of gene flow and introgression to wild rice species is urgently needed. Even the six species belonging to the primary gene pool of cultivated rice (**AA** genome) have not yet been studied in detail. The extent of backcrossing and introgression has not been sufficiently examined for any of these species, let alone the prevalent direction of gene flow or the fate of introgressed (domesticated or wild) traits over several generations. Longer-term studies are warranted to assess the ecological fitness of naturally occurring hybrids between cultivated and wild rice and their progeny, as well as the density and persistence of crop alleles in wild populations (Chen et al. 2004; Lu and Snow 2005). This particularly applies to Africa, tropical America, and Oceania (Australia and New Guinea), where hardly any information is available about the potential for hybridization between culti-vated rice and the native wild **AA**-genome species (*Oryza barthii*, *O. glu-maepatula*, *O. longistaminata*, *O. meridionalis*, and *O. nivara*). Furthermore, there is insufficient knowledge about the extent of possible fitness benefits from introgressed GM traits, or about whether and how these benefits affect the population dynamics of weedy and wild rice species. Virtually nothing is known about the degree to which insect and disease pressures regulate populations of weedy and wild rice relatives, and ecological studies of these populations should be con-ducted to address this question (Lu and Snow 2005).

Better information regarding all of these factors is crucial to ade-quately evaluate the likelihood of gene flow and introgression from GM rice cultivars. In the meantime, the following conclusions are based on current knowledge about gene flow in rice. First, Asian cultivated rice, *O. sativa*, is often considered to be a low-risk species with respect to in-trogression from GM rice varieties, due to its low outcrossing rates (< 5%) and short pollen dispersal distances (less than 50 m). However, spontaneous gene flow can occur in the field, due to outcrossing be-tween *O. sativa* and its relatives in GP-1, including fully cross-compati-ble, conspecific weedy rice, on the one hand, and cultivated African rice and the (at least) six wild *Oryza* species, on the other. In particular, hy-bridization with weedy rice could lead to increased weediness if the dominant wild traits are inherited (Sagers et al. 2002). Moreover, GM traits conferring fitness advantages (such as insect or disease resistance) may spread rapidly to weedy rice populations. In at least three cases (*O. glaberrima*, *O. nivara*, and *O. rufipogon*), hybridization with other *Oryza* species can produce partly or completely fertile progeny.

Second, hybrids between *O. sativa* and weedy or wild rice would probably be subjected to different selection pressures, as was shown to be the case for common beans (chapter 8). Their fate would be substan-tially influenced by (a) the direction of the cross and (b) whether their location was in either a cultivated or a wild environment.

In a crop situation, there would be an immediate selective disad-vantage for any cross of a weedy or wild species where the cultivated variety was the female parent, since the latter's non-shattering trait would lead to most of the F_1 hybrid seeds being harvested as grain for consumption, rather than remaining in the soil and propagating. Successful establishment of F1 hybrids would only be possible from seeds that were lost during harvest. Conversely, any cross with a weedy or wild rice as the female parent could produce hybrid seeds that would shatter onto the soil. If these seeds are not appropriately controlled, they could survive until the next growing season, infesting and backcrossing with the subsequent (rice) crop and other hybrids, and thereby creating seeds that would produce hybrid swarms in the following generation. The agronomic management recommendations given above may help minimize this possibility.

While the extent of outcrossing and hybridization may be small and localized, the likelihood of gene flow and introgression could increase, due to human seed dispersal during transportation of the harvested crop. Once seeds of cultivated rice germinate in wild environments, outcrossing with weedy or wild rice would be possible in both directions. The hybrid seeds resulting from a cross with the cultivated variety as the female parent may be less likely to survive for long periods in the soil, because of their reduced seed dormancy. However—independent of the direction of the cross—many surviving F1 hybrids would probably have higher fitness, due to the predominantly wild or weedy phenotype of the hybrids, thus facilitating gene introgression to F2 and later generations in wild populations. Even if it is minor, such an asymmetrical domesticated-to-wild (or weedy) gene flow could lead to a reduction in the genetic diversity of wild populations and even to local extinctions, as illustrated by the case of the now nearly extinct endemic Taiwanese wild rice, *O. rufipogon* subsp. *formosana* (Kiang et al. 1979). Gene introgression from cultivated rice—albeit at low rates—is therefore likely in regions where the crop occurs in sympatry with these other species.

Based on current knowledge, the likelihood of gene flow and introgression from Asian cultivated rice (*Oryza sativa*)—assuming physical proximity (less than 50 m) and flowering overlap, and being conditional on the nature of the respective trait(s)—is as follows:

- *high* for weedy rice (*Oryza sativa* f. *spontanea*; wherever it occurs in rice fields)

- *moderate* for African cultivated rice (*O. glaberrima*; Africa), wild *O. nivara* (Asia), and wild *O. rufipogon* (Asia, Australia, New Guinea, and—introduced—the Americas) (map. 18.1)

- *low* for wild *O. barthii* and *O. longistaminata* (both in Africa), *O. glumaepatula* (tropical America) and *O. meridionalis* (Australia) (map. 18.1).

Although these last four species do produce seeds with relative ease when hand pollinated with cultivated rice (without the use of sophisticated embryo rescue methods), significant genetic barriers to introgression exist, including F1 hybrid weakness and either reduced hybrid fertility or sterility. However, gene transfer—especially via bridge species such as introduced and sexually compatible wild *O. rufipogon*—cannot be completely ruled out.

REFERENCES

Abbasi FM, Brar DS, Carpena AL, Fukui K, and Khush GS (1999) Detection of autosyndetic and allosyndetic pairing among **A** and **E** genomes of *Oryza* through genomic *in situ* hybridization. *Rice Genet. Newsl.* 16: 24–25.

Abdullah M, Vaughan DA, Watanabe H, and Okuno K (1996) The origin of weedy rice in peninsular Malaysia. *MARDI [Malaysian Agricultural Research and Development Institute] Res. J.* 24: 160–174.

Akimoto M, Shimamoto Y, and Morishima H (1999) The extinction of genetic resources of Asian wild rice, *Oryza rufipogon* Griff.: A case study in Thailand. *Genet. Resour. Crop Evol.* 46: 419–425.

Amante-Bordeos AL, Sitch A, Nelson R, Dalmacio RD, Oliva NP, Aswidinnoor H, and Leung H (1992) Transfer of bacterial blight and blast resistance from the tetraploid wild rice *Oryza minuta* to cultivated rice, *Oryza sativa. Theor. Appl. Genet.* 84: 345–354.

Arrieta-Espinoza G, Sánchez E, Vargas S, Lobo J, Quesada T, and Espinoza AM (2005) The weedy rice complex in Costa Rica: I. Morphologic study and relationship with commercial rice varieties and wild *Oryza* relatives. *Genet. Resour. Crop Evol.* 52: 575–587.

Askew SD, Shaw DR, and Street JE (1998) Red rice (*Oryza sativa*) control and seedhead reduction with glyphosate. *Weed Technol.* 12: 504–506.

Bajaj S and Mohanty A (2005) Recent advances in rice biotechnology—towards genetically superior transgenic rice. *Plant Biotech. J.* 3: 275–307.

Baki BB, Chin DV, and Mortimer AM (eds.) (2000) *Wild and Weedy Rice in Rice Ecosystems in Asia: A Review.* Limited Proceedings No. 2 (International Rice

Research Institute [IRRI]: Los Baños, Manila, Philippines). 118 p. www.irri.
org/publications/limited/limited.asp?id=7.

Barrett SCH (1983) Crop mimicry in weeds. *Econ. Bot.* 37: 255–282.

Beachell HM, Adair CR, Jodon NE, Davis LL, and Jones JW (1938) Extent of natural crossing in rice. *J. Am. Soc. Agron.* 30: 743–753.

Beattie AJ (1982) Ants and gene dispersal in flowering plants. In: Armstrong JM,
Powell JM, and Richards AJ (eds.) *Pollination and Evolution: Based on the
Symposium on Pollination Biology, Held During the 13th International Botanical Congress (Section 6.7), at Sydney, Australia, August, 1981* (Royal Botanic Gardens: Sydney, NSW, Australia), pp. 1–8.

Bezançon G, Bozza JL, and Second G (1977) Variability of *Oryza longistaminata*
and *sativa* complex of *Oryza* in Africa: Ecological and evolutive aspects. In:
*Meeting on African Rice Species: Proceedings, IRAT [Institut de recherches
agronomiques tropicales et des cultures vivrieres]-ORSTOM [Office de la recherche scientifique et technique outre-mer], Paris, January 25–26, 1977* (Institut de recherches agronomiques tropicales et des cultures vivrieres [IRAT]
and Office de la recherche scientifique et technique outre-mer [ORSTOM]:
Paris, France), pp. 15–64.

Bezançon G, Causse M, Ghesquière A, de Kochko A, and Second G (1989) Les riz en
Afrique: Diversité génétique, relations interspécifiques, et évolution. *Bull. Soc.
Bot. Fr.* 136: 251–262. http://horizon.documentation.ird.fr/exl-doc/pleins_
textes/pleins_textes_5/b_fdi_23–25/30915.pdf.

Bougerol B and Pham JI (1989) Influence of the *Oryza sativa* genotype on the fertility and quantitative traits of F1 hybrids between the two cultivated rice species
O. sativa and *O. glaberrima*. *Genome* 32: 810–815.

Bouharmont J, Olivier M, and Dumont de Chassart M (1985) Cytological observations in some hybrids between the rice species *Oryza sativa* L. and *O. glaberrima* Steud. *Euphytica* 34: 75–81.

Brar DS, Elloran RM, and Khush GS (1991) Interspecific hybrids produced through
embryo rescue between cultivated and eight wild species of rice. *Rice Genet.
Newsl.* 8: 91–93.

Brar DS, Dalmacio RD, Elloran RM, Aggarwal R, Angeles R, and Khush GS (1996)
Gene transfer and molecular characterization of introgression from wild *Oryza*
species into rice. In: Khush GS (ed.) *Rice Genetics III* (International Rice Research Institute [IRRI]: Los Baños, Manila, Philippines), pp. 477–485.

Brar DS, Elloran RM, Talag JD, Abbasi F, and Khush GS (1997) Cytogenetic and
molecular characterization of an intergeneric hybrid between *Oryza sativa* L.
and *Porteresia coarctata* (Roxb.) Tateoka. *Rice Genet. Newsl.* 14: 43–45.

Brondani C, Brondani V, Pereira R, Rangel N, Hideo P, and Ferreira ME (2001)
Development and mapping of *Oryza glumaepatula*-derived microsatellite

markers in the interspecific cross *Oryza glumaepatula* × *O. sativa*. *Hereditas* 134: 59–71.

Brondani C, Rangel P, Brondani R, and Ferreira M (2002) QTL mapping and introgression of yield-related traits from *Oryza glumaepatula* to cultivated rice (*Oryza sativa*) using microsatellite markers. *Theor. Appl. Genet.* 104: 1192–1203.

Buso GSC, Rangel PH, and Ferreira ME (1998) Analysis of genetic variability of South American wild rice populations (*Oryza glumaepatula*) with isozymes and RAPD markers. *Mol. Ecol.* 7: 107–117.

Cao Q, Lu B-R, Xia H, Rong J, Sala F, Spada A, and Grassi F (2006) Genetic diversity and origin of weedy rice (*Oryza sativa* f. *spontanea*) populations found in northeastern China revealed by Simple Sequence Repeat (SSR) markers. *Ann. Bot.* (Oxford) 98: 1241–1252.

Caughley J, Bomford M, Parker B, Sinclair R, Griffiths J, and Kelly D (1998) *Managing Vertebrate Pests: Rodents* (Bureau of Resource Sciences and Grains Research and Development Corporation: Canberra, ACT, Australia). 130 p.

Cavalheiro ST, Brondani C, Rangel PHN, and Ferreira ME (1996) Paternity analysis of F1 interspecific progenies of crosses between *O. sativa* varieties and its wild relative *O. glumaepatula* using SSR and RAPD markers. *Rev. Bras. Genet.* 19: 225.

CCIA [California Crop Improvement Association] (2009) Rice Certification Standards. http://ccia.ucdavis.edu/seed_cert/seedcert_index.htm.

Chang TT (1976) The origin, evolution, cultivation, dissemination, and diversification of Asian and African rices. *Euphytica* 25: 425–441.

—— (1995) Rice. In: Smartt J and Simmonds NW (eds.) *Evolution of Crop Plants.* 2nd ed. (Longman Scientific and Technical: Harlow, UK), pp. 147–155.

Chang TT, Adair CR, and Johnston TH (1982) The conservation and use of rice genetic resources. *Adv. Agron.* 35: 37–91.

Chen LJ, Lee DS, Song ZP, Suh HS, and Lu B-R (2004) Gene flow from cultivated rice (*Oryza sativa*) to its weedy and wild relatives. *Ann. Bot.* (Oxford) 93: 67–73.

Cheng CY, Motohashi R, Tsuchimoto S, Fukuta Y, Ohtsubo H, and Ohtsubo E (2003) Polyphyletic origin of cultivated rice, based on the interspersion pattern of SINEs. *Mol. Biol. Evol.* 20: 67–75.

Chitrakon S (1994) Genetic erosion of rice in Thailand. *Tropics* 3: 223–225.

Chu YE and Oka HI (1970a) The genetic basis of crossing barriers between *Oryza perennis* subsp. *barthii* and its related taxa. *Evolution* 24: 135–144.

—— (1970b) Introgression across isolating barriers in wild and cultivated *Oryza* species. *Evolution* 24: 344–355.

Chu YE, Morishima H, and Oka HI (1969) Reproductive barriers distributed in cultivated rice species and their wild relatives. *Japan. J. Genet.* 4: 207–223.

CIAT [Centro Internacional de Agricultura Tropical, or International Center for Tropical Agriculture] (2007) *Annual Report 2006, Project IP4: Improved Rice*

Germplasm for Latin America and the Caribbean (Centro Internacional de Agricultura Tropical [CIAT]: Cali, Colombia). 157 pp.

Cragmiles JP (1978) Introduction. In: Eastin EF (ed.) *Red Rice: Research and Control; Proceedings of a Symposium Held at Texas A&M University Agricultural and Extension Center at Beaumont, Texas.* Texas Agricultural Experiment Station Bulletin B-1270 (Texas A&M University: College Station, TX, USA), pp. 5–6.

Delouche JC, Burgos NR, Gealy DR, San Martín GZ, and Labrada R (2007) *Weedy Rices—Origin, Biology, Ecology, and Control.* FAO Plant Production and Protection Series No. 188 (Food and Agriculture Organization of the United Nations [FAO]: Rome, Italy). 156 p.

den Nijs HCM, Bartsch D, and Sweet J (2004) *Introgression from Genetically Modified Plants into Wild Relatives* (CAB International: Wallingford, UK). 416 p.

Diarra A, Smith RJ Jr, and Talbert RE (1985) Interference of red rice (*Oryza sativa*) with rice (*O. sativa*). *Weed Sci.* 33: 644–649.

Dillon TL, Baldwin FL, Talbert RE, Estorninos LE Jr, and Gealy DR (2002) Gene flow from Clearfield rice to red rice. *Weed Sci. Soc. Am. Abst.* 42: 124.

Dorward P, Craufurd P, Marfo K, Dogbe W, and Bam R (2007) Improving participatory varietal selection processes: Participatory varietal selection and the role of informal seed diffusion mechanisms for upland rice in Ghana. *Euphytica* 155: 315–327.

EC [European Commission] (2001) European Communities (Cereal Seed) Regulations, 2001: S.I. [Statutory Instrument] No. 640/2001, Amending the Council Directive No. 66/402/EEC of 14 June 1966 on the Marketing of Cereal Seed, OJ No. L125 2309–66, July 11, 1966. www.irishstatutebook.ie/2001/en/si/0640.html.

—— (2007) *The Rapid Alert System for Food and Feed (RASFF): Annual Report 2006* (Office for Official Publications of the European Communities: Luxembourg). 72 p. http://ec.europa.eu/food/food/rapidalert/index_en.htm.

Elloran RM, Dalmacio RD, Brar DS, and Khush GS (1992) Production of backcross progenies from a cross of *Oryza sativa* × *O. granulata. Rice Genet. Newsl.* 9: 39.

Emerton L and Muramira E (1999) Uganda biodiversity: Economic assessment; Prepared with the National Environment Management Authority as part of the Uganda National Biodiversity Strategy and Action Plan (IUCN [International Union for Conservation of Nature]/National Environmental Authority [NEMA]: Kampala, Uganda). 57 p. Unpublished ms. www.cbd.int/doc/external/countries/uganda-eco-assessment-1999–en.pdf.

Espinoza-Esquivel AM and Arrieta-Espinoza G (2007) A multidisciplinary approach directed towards the commercial release of transgenic herbicide-tolerant rice in Costa Rica. *Transgen. Res.* 16: 541–555.

Estorninos LE Jr, Gealy DR, and Burgos NR (2004) Determining reciprocal out-crossing rates between non-herbicide rice and red rice using SSR markers. *WSSA [Weed Science Society of America] Ann. Mtg.* 44: 65–66.

Estorninos LE Jr, Gealy DR, Dillon TL, Baldwin EL, Burgos NR, and Tai TH (2002) Determination of hybridization between rice and red rice using microsatellite markers. *Proc. South. Weed Sci. Soc.* 55: 197–198.

FAO [Food and Agriculture Organization of the United Nations] (2009) FAOSTAT data. http://faostat.fao.org/site/567/default.aspx.

Farooq S, Iqbal N, Shah TM, and Asgher M (1996) Problems and prospects for utilizing *Porteresia coarctata* in rice breeding programs. *Cereal Res. Comm.* 24: 41–47.

Footitt S and Cohn MA (1992) Seed dormancy in red rice. *Plant Physiol.* 100: 1196–1202.

Forner MMC (1995) Chemical and cultural practices for red rice control in rice fields in Ebro Delta (Spain). *Crop Protect.* 14: 405–408.

Fuller DQ, Harvey E, and Qin L (2007) Presumed domestication? Evidence for wild rice cultivation and domestication in the fifth millennium BC of the Lower Yangtze region. *Antiquity* 81: 316–331.

Gao LZ, Ge S, and Hong DY (2000) A preliminary study on ecological differentiation within common wild rice *Oryza rufipogon* Griff. *Acta Agron. Sin.* 26: 210–216.

Ge S, Sang T, Lu B-R, and Hong DY (1999) Phylogeny of rice genomes with emphasis on origins of allotetraploid species. *Proc. Natl. Acad. Sci. USA* 96: 14400–14405.

Gealy DR (2005) Gene movement between rice (*Oryza sativa*) and weedy rice (*Oryza sativa*): A temperate rice perspective. In: Gressel J (ed.) *Crop Ferality and Volunteerism: A Threat to Food Security in the Transgenic Era?* (CRC Press: Boca Raton, FL, USA), pp. 323–354.

Gealy DR and Estorninos LE Jr (2007) SSR marker confirmation of reciprocal out-crossing rates between rice and red rice lines in Arkansas over a five-year period. *Proc. South. Weed Sci. Soc.* 60: 231.

Gealy DR, Mitten DH, and Rutger JN (2003) Gene flow between red rice (*Oryza sativa*) and herbicide-resistant rice (*O. sativa*): Implications for weed management. *Weed Technol.* 17: 627–645.

Gealy DR, Estorninos LE Jr, Wilson CE, and Agrama H (2005) Confirmation of hybridization between rice and phenotypically distinct red rice types in Arkansas rice fields. *Proc. North Central Weed Sci. Soc.* 60: 126.

Gealy DR, Estorninos LE Jr, and Burgos NR (2007) Multi-year evaluation of reciprocal outcrossing rates between selected rice cultivars and red rice types at Stuttgart, Arkansas. *Proc. South. Weed Sci. Soc.* 59: 186.

Ghesquière A (1986) Evolution of *Oryza longistaminata*. In: International Rice Research Institute (ed.) *Rice Genetics: Proceedings of the [1st] International Rice Genetics Symposium, May 27–31, 1985* (International Rice Research Institute [IRRI]: Los Baños, Manila, Philippines), pp. 15–25.

——— (1988) Diversité génétique de l'espèce sauvage de riz, *Oryza longistaminata* A. Chev. & Roehr. et dynamique des flux géniques au sein du groupe *sativa* en Afrique. PhD dissertation, University of Paris.

Ghesquière A and Causse M (1992) Linkage study between molecular markers and genes controlling the reproductive barrier in interspecific backcross between O. *sativa* and O. *longistaminata*. *Rice Genet. Newsl.* 9: 28–31.

Ghesquière A, Panaud O, Marmey P, Gavalda MC, Leblanc O, and Grimanelli D (1994) Suivi des introgressions dans les croisements interspécifiques chez le riz: Utilisation des marqueurs moléculaires. *Genet. Sel. Evol.* 26: 67s–80s. http://horizon.documentation.ird.fr/exl-doc/pleins_textes/pleins_textes_6/b_fdi_35–36/41177.pdf.

Gómez C and Espadaler X (1998) Myrmecochorous dispersal distances: A world survey. *J. Biogeogr.* 25: 573–580.

Gressel J (2000) Molecular biology of weed control. *Transgen. Res.* 9: 355–382.

——— (2002) Preventing, delaying, and mitigating gene flow from crops—rice as an example. In: *Proceedings of the 7th International Symposium on Biosafety of Genetically Modified Organisms (ISBGMO), Beijing, China, October 10–16, 2002* (International Society for Biosafety Research [ISBR], n.p.), pp. 59–77. www.isbr.info/symposia/docs/isbgmo.pdf.

Gridley HE, Jones MP, and Wopereis-Pura M (2002) Development of New Rice for Africa (NERICA) and participatory varietal selection. In: Witcombe JR, Parr LB, and Atlin GN (eds.) *Breeding Rainfed Rice for Drought-Prone Environments: Integrating Conventional and Participatory Plant Breeding in South and Southeast Asia; Proceedings of a DFID [Department for International Development] Plant Sciences Research Programme, Great Britain/IRRI Conference, International Rice Research Institute, Los Baños, Laguna, Philippines, March 12–15, 2002* (Centre for Arid Zone Studies [CAZS], University of Wales-Bangor: Bangor, Gwynedd, UK), pp 23–28.

Grist DH (1986) Weeds. In: *Rice.* 6th ed. (Longman Scientific and Technical: Harlow, UK). pp. 75–82.

Hamilton RS and Raymond R (2005) Toward a global strategy for the conservation of rice genetic resources. In: Toriyama K, Heong KL, and Hardy B (ed.) *Rice is Life: Scientific Perspectives for the 21st Century; Proceedings of the World Rice Research Conference Held in Tsukuba, Japan, November 4–7, 2004* (International Rice Research Institute [IRRI]: Los Baños, Manila, Philippines), pp. 47–49.

Hayashi Y, Kyozuka J, and Shimamoto K (1988) Hybrids of rice (*Oryza sativa* L.) and wild *Oryza* species obtained by cell fusion. *Mol. Gen. Genet.* 214: 6–10.

Heuer SM, Miezan K, and Gaye G (2003) Increasing biodiversity of irrigated rice in Africa by interspecific crosses of O. *glaberrima* Steud. × O. *sativa indica* L. *Euphytica* 132: 31–40.

Holm LG, Doll J, Holm E, Pancho JV, and Herbeger JP (1997) The wild rices. In: *World Weeds: Natural Histories and Distribution* (John Wiley and Sons: New York, NY, USA), pp. 531–547.

Hussey BMJ, Keighery GJ, Cousens RD, Dodd J, and LLoyd SG (1997) *Western Weeds: A Guide to the Weeds of Western Australia* (Plant Protection Society of Western Australia: Victoria Park, WA, Australia). 254 p. http://members.iinet. net.au/~weeds/western_weeds/poaceae_seven.htm.

Ikeda K, Sobrizal K, Sánchez PL, Yasui H, and Yoshimura A (1999) Hybrid weakness restoration gene (*Rhw*) for *Oryza glumaepatula* cytoplasm. *Rice Genet. Newsl.* 16: 62–64.

IRRI [International Rice Research Institute] (1998) *Program Report for 1997* (International Rice Research Institute [IRRI]: Los Baños, Manila, Philippines). www. irri.org/publications/program/prog1997.asp.

Ishikawa R, Toki N, Imai K, Sato YI, Yamagishi H, and Shinamoto Y (2005) Origin of weedy rice grown in Bhutan and the force of genetic diversity. *Genet. Resour. Crop Evol.* 52: 395–403.

Jelodar NB, Blackhall NW, Hartman TPV, Brar DS, Khush G, Davey MR, Cocking EC, and Power JB (1999) Intergeneric somatic hybrids of rice [*Oryza sativa* L. × *Porteresia coarctata* (Roxb.) Tateoka]. *Theor. Appl. Genet.* 99: 570–577.

Jena KK (1994) Development of intergeneric hybrid between O. *sativa* and *Porteresia coarctata*. *Rice Genet. Newsl.* 11: 78–79.

Jia SR (2002) Studies on gene flow in China—a review. In: In: *Proceedings of the 7th International Symposium on Biosafety of Genetically Modified Organisms (ISBGMO), Beijing, China, October 10–16, 2002* (International Society for Biosafety Research [ISBR], n.p.), pp. 92–96. www.isbr.info/symposia/docs/isb-gmo.pdf.

Jia SR and Peng YF (2002) GMO biosafety research in China—guest editorial. *Environ. Biosafety Res.* 1: 5–8.

Jia SR, Wang F, Shi L, Yuan Q, Liu W, Liao Y, Li S, Jin W, and Peng H (2007) Transgene flow to hybrid rice and its male-sterile lines. *Transgen. Res.* 16: 491–501.

Jodon NE (1959) Occurrence and importance of natural crossing in rice. *Rice J.* 62: 8–10.

Jones MP, Dingkuhn M, Aluko GK, and Semon M (1997) Interspecific O. *sativa* L. × O. *glaberrima* S[teud]. progenies in upland rice improvement. *Euphytica* 92: 237–246.

Katayama T and Onizuka W (1979) Intersectional F1 plants from *Oryza sativa* x *O. ridleyi* and *O. sativa* x *O. meyeriana*. *Japan. J. Genet.* 54: 43–46.

Kaushal P and Ravi (1998) and Pusa Basmati 1 for transfer of bacterial leaf blight resistance through interspecific hybridization. *J. Agric. Sci.* 130: 423–430.

Khush GS (1977) Disease and insect resistance in rice. *Adv. Agron.* 29: 265–341.

——— (1997) Origin, dispersal, cultivation, and variation in rice. *Plant Mol. Biol.* 35: 25–34.

Khush GS and Brar DS (1988) Wide hybridization in plant breeding. In: Zakri AH (ed.) *Plant Breeding and Genetic Engineering: Proceedings of the International Symposium and Workshop on Gene Manipulation for Plant Improvement in Developing Countries, Held at Kuala Lumpur, Malaysia, November 30–December 3, 1987* (Society for the Advancement of Breeding Researches in Asia and Oceania [SABRAO]: Bangi, Selangor Darul Ehsan, Malaysia), pp. 141–188.

Khush GS and Ling KC (1974) Inheritance of resistance to grassy stunt virus and its vectors in rice. *J. Hered.* 65: 134–136.

Khush GS, Ling KC, Aquino RC, and Aquiero VM (1977) Breeding for resistance to grassy stunt in rice. *Plant Breeding Papers.* Vol. 1 in *Proceedings of the 3rd International Congress of the Society for the Advancement of Breeding Researches in Asia and Oceania [SABRAO], in association with Australian Plant Breeding Conference, Canberra, Australia, February 1977* (Commonwealth Scientific and Industrial Research Organisation [CSIRO], Centre for Plant Biodiversity Research, CSIRO Plant Industry: Canberra, ACT, Australia), paper 4(b), pp. 3–9.

Khush GS, Bacalangco E, and Ogawa T (1990) A new gene for resistance to bacterial blight from *O. longistaminata*. *Rice Genet. Newsl.* 7: 121–122.

Kiambi DK, Ford-Lloyd BV, Jackson MT, Guarino L, Maxted N, and Newbury HJ (2005) Collection of wild rice (*Oryza* L.) in east and southern Africa in response to genetic erosion. *Plant Genet. Resour. Newsl.* 142: 10–20.

Kiang YT, Antonovics J, and Wu L (1979) The extinction of wild rice (*Oryza perennis formosana*) in Taiwan. *J. Asian Ecol.* 1: 1–9.

Kolhe GL and Bhat NR (1981) Genetic study of cleistogamy in rice (*Oryza sativa* L.). *Curr. Sci.* 50: 419–420.

Kubo T and Yoshimura A (1998) Linkage analysis of hybrid weakness in rice. *Breed. Sci.* 48 (Suppl. 1): 81 [English abstract].

Kuroda Y, Sato YI, Bounphanousay C, Kono Y, and Tanaka K (2005) Gene flow from cultivated rice (*Oryza sativa* L.) to wild *Oryza* species (*O. rufipogon* Griff. and *O. nivara* Sharma and Shastry) on the Vientiane plain of Laos. *Euphytica* 142: 75–83.

——— (2007) Genetic structure of three *Oryza* **AA** genome species (*O. rufipogon*,

O. nivara, and *O. sativa*) as assessed by SSR analysis on the Vientiane Plain of Laos. *Cons. Genet.* 8: 149–158.

Langevin SA, Clay K, and Grace JB (1990) The incidence and effects of hybridization between cultivated rice and its related weed red rice (*Oryza sativa* L.) *Evolution* 44: 1000–1008.

Lentini Z and Espinoza AM (2005) Coexistence of weedy rice and rice in tropical America: Gene flow and genetic diversity. In: Gressel J (ed.) *Crop Ferality and Volunteerism: A Threat to Food Security in the Transgenic Era?* (CRC Press: Boca Raton, FL, USA), pp. 303–319.

Lin SC and Yuan LP (1980) Hybrid rice breeding in China. In: International Rice Research Institute [IRRI] (ed.) *Innovative Approaches to Rice Breeding: Selected Papers from the 1979 International Rice Research Conference* (International Rice Research Institute [IRRI]: Los Baños, Manila, Philippines), pp. 35–51.

Lin Z, Griffith ME, Li X, Zhu Z, Tan L, Fu Y, Zhang W, Wang X, Xie D, and Sun C (2007) Origin of seed shattering in rice (*Oryza sativa* L.). *Planta* 226: 11–20.

Linares OF (2002) African rice (*Oryza glaberrima*): History and future potential. *Proc. Natl. Acad. Sci. USA* 99: 16360–16365.

Londo JP, Chiang YC, Hung KH, Chiang TY, and Schaal BA (2006) Phylogeography of Asian wild rice, *Oryza rufipogon*, reveals multiple independent domestications of cultivated rice, *Oryza sativa*. *Proc. Natl. Acad. Sci. USA* 103: 9578–9583.

Lord L (1935) The cultivation of rice in Ceylon. *J. Exp. Agric.* 3: 119–128.

Lorieux M, Ndjiondjop MN, and Ghesquière A (2000) A first interspecific *Oryza sativa* × *Oryza glaberrima* microsatellite-based genetic map. *Theor. Appl. Genet.* 100: 591–601.

Lu B-R (1996a) Diversity of the rice gene pool and its sustainable utilization. In: Zhang AL and Wu SG (eds.) *Floristic Characteristics and Diversity of East Asian Plants* (Higher Education Press: Beijing, China; Springer-Verlag: Berlin, Germany), pp. 454–460.

——— (1996b) *A [Trip] Report on Collecting of Wild Oryza Species in Lao PDR and Cambodia* (International Rice Research Institute [IRRI]: Los Baños, Manila, Philippines). www.irri.org/GRC/Biodiversity/Pdf%20files/CAM%20pdf/cam1.pdf.

——— (1998) Diversity of rice genetic resources and its utilization and conservation. *Chin. Biodiv.* 6: 63–72.

——— (1999) Taxonomy of the genus *Oryza* (Poaceae): A historical perspective and current status. *Int. Rice Res. Notes* 24(3): 4–8. www.irri.org/publications/irrn/irrn24-3.asp.

——— (2006) Identifying possible environmental hazard from GM rice in China to inform biosafety assessment. In: *Proceedings of the 9th International Symposium on Biosafety of Genetically Modified Organisms (ISBGMO), Jeju Island, South Korea, September 24–29, 2006* (International Society for Biosafety Research [ISBR], n.p.), pp. 108–112. www.isbr.info/isbgmo/docs/9th_isbgmo_program.pdf.

Lu B-R and Ge S (2005) *Oryza coarctata*: The name that best reflects the relationships of *Porteresia coarctata* (Poaceae: Oryzeae). *Nord. J. Bot.* 23: 555–558.

Lu B-R and Snow AA (2005) Gene flow from genetically modified rice and its environmental consequences. *BioScience* 55: 669–678.

Lu B-R, Naredo MEB, Juliano AB, and Jackson MT (1997) Hybridization of **AA** genome rice species from Asia and Australia: II. Meiotic analysis of *Oryza meridionalis* and its hybrids. *Genet. Resour. Crop Evol.* 44: 25–31.

——— (2000) Preliminary studies on taxonomy and biosystematics of the **AA** genome *Oryza* species (Poaceae). In: Jacobs SWL and Everett J (eds.) *Grasses: Systematics and Evolution* (Commonwealth Scientific and Industrial Research Organisation [CSIRO]: Collingwood, VIC, Australia), pp. 51–58.

Lu B-R, Song ZP, and Chen J (2002a) Gene flow from crops to wild relatives in Asia: Case studies and general expectations. In: *Proceedings of the 7th International Symposium on Biosafety of Genetically Modified Organisms (ISBGMO), Beijing, China, October 10–16, 2002* (International Society for Biosafety Research [ISBR], n.p.), pp. 29–36. www.isbr.info/symposia/docs/isbgmo.pdf.

Lu B-R, Zheng KL, Qian HR, and Zhuang JY (2002b) Genetic differentiation of wild relatives of rice as assessed by RFLP analysis. *Theor. Appl. Genet.* 106: 101–106.

Ma J and Bennetzen JL (2004) Rapid recent growth and divergence of rice nuclear genomes. *Proc. Natl. Acad. Sci. USA* 101: 12404–12410.

Madsen KH, Valverde BE, and Jensen JE (2002) Risk assessment of herbicide-resistant crops: A Latin American perspective using rice (*Oryza sativa*) as a model. *Weed Technol.* 16: 215–223.

Mahmoud AA, Sukumar S, and Krishnan HB (2008) Interspecific rice hybrid of *Oryza sativa* × *Oryza nivara* reveals a significant increase in seed protein content. *J. Agric. Food Chem.* 56: 476–482.

Majumder ND, Ram T, and Sharma AC (1997) Cytological and morphological variation in hybrid swarms and introgressed populations of interspecific hybrids (*Oryza rufipogon* Griff. × *Oryza sativa* L.) and its impact on the evolution of intermediate types. *Euphytica* 94: 295–302.

Mariam AL, Zakri AH, Mahani MC, and Normah MN (1996) Interspecific hybridization of cultivated rice, *Oryza sativa* L., with the wild rice *O. minuta* Presl. *Theor. Appl. Genet.* 93: 664–671.

Messeguer J, Fogher C, Guiderdoni E, Marfa V, Catala MM, Baldi G, and Melé E (2001) Field assessment of gene flow from transgenic to cultivated rices (*Oryza sativa* L.) using a herbicide resistance gene as tracer marker. *Theor. Appl. Genet.* 103: 1151–1159.

Messeguer J, Marfa V, Catala MM, Guiderdoni E, and Melé E (2004) A field study of pollen-mediated gene flow from Mediterranean GM rice to conventional rice and the red rice weed. *Mol. Breed.* 13: 103–112.

Moldenhauer KAK and Gibbons JH (2003) Rice morphology and development. In: Smith CW and Dilday RH (eds.) *Rice: Origin, History, Technology, and Production* (John Wiley and Sons: Hoboken, NJ, USA), pp. 103–127.

Moncada P, Martinez CP, Borrero J, Chatel M, Gauch H Jr, Guimaraes E, Tohme J, and McCouch SR (2001) Quantitative trait loci for yield and yield components in an *Oryza sativa* × *Oryza rufipogon* BC2F2 population evaluated in an upland environment. *Theor. Appl. Genet.* 102: 41–52.

Moody K (1994) Weed management in rice. In: Labrada R, Caseley JC, and Parker C (eds.) *Weed Management for Developing Countries*. FAO Plant Production and Protection Series No. 120 (Food and Agriculture Organization of the United Nations [FAO]: Rome, Italy), pp. 249–263.

Morat P, Deroin T, and Couderc H (1994) Présence en Nouvelle Calédonie d'une espèce endémique du genre *Oryza* L. (Gramineae). *Bull. Mus. Natl. Hist. Nat.* 16: 3–10.

Morinaga T and Kuriyama H (1957) Cytogenetical studies on *Oryza sativa* L.: IX. The F1 hybrid of O. *sativa* L. and O. *glaberrima* Steud. *Japan. J. Breed.* 7: 57–65.

Morishima H (1969) Variations in breeding system and numerical estimation of phylogeny in *Oryza perennis*. *Japan. J. Genet.* 44: 317–324.

——— (1984) Wild plants and domestication. In: Tsunoda S and Takahashi N (eds.) *Biology of Rice*. Developments in Crop Science No. 7 (Elsevier: Amsterdam, The Netherlands), pp. 3–30.

——— (1994) Observations at permanent study-site of wild rice in the suburb of Bangkok. *Tropics* 3: 227–233.

——— (2001) Evolution and domestication of rice. In: Khush GS, Brar DS, and Hardy B (eds.) *Rice Genetics IV: Proceedings of the 4th International Rice Genetics Symposium, Los Baños, Philippines, October 22–27, 2000* (Science Publishers: Enfield, NH, USA; International Rice Research Institute [IRRI]: Los Baños, Manila, Philippines), pp. 63–78.

Morishima H, Hinata K, and Oka HI (1962) Comparison between two cultivated rice species, *Oryza sativa* L. and O. *glaberrima* Steud. *Japan. J. Breed.* 12: 153–165.

Morishima H, Sano Y, and Oka HI (1992) Evolutionary studies in rice and its wild relatives. *Oxford Surv. Evol. Biol.* 8: 135–184.

Multani DS, Jena KK, Brar DS, de los Reyes BG, Angeles ER, and Khush GS (1994) Development of monosomic alien addition lines and introgression of genes from *Oryza australiensis* Domin to cultivated rice O. *sativa* L. *Theor. Appl. Genet.* 88: 102–109.

Multani DS, Khush GS, de los Reyes BG, and Brar DS (2003) Alien gene introgression and development of monosomic alien addition lines from *Oryza latifolia* Desv. to rice, O. *sativa* L. *Theor. Appl. Genet.* 107: 359–405.

Naredo MEB, Juliano AB, Lu B-R, and Jackson MT (1997) Hybridization of AA genome rice species from Asia and Australia: I. Crosses and development of hybrids. *Genet. Resour. Crop Evol.* 44: 17–24.

Naredo MEB, Amita BJ, Lu B-R, and Jackson MT (1998) Taxonomic status of *Oryza glumaepatula* Steud.: II. Hybridization between New World diploids and AA genome species from Asia and Australia. *Genet. Resour. Crop Evol.* 45: 205–214.

Nezu M, Katayama TC, and Kihara H (1960) Genetic study of the genus *Oryza*: I. Crossability and chromosomal affinity among 17 species. *Seiken Zihô* 11: 1–11.

Noldin JA, Chandler JM, and McCauley GN (1999) Red rice (*Oryza sativa*) biology: I. Characterization of red rice ecotypes. *Weed Technol.* 13: 12–18.

Noldin JA, Yokoyama S, Stuker H, Rampelotti FT, Gonçalves MIF, Eberhardt DS, Abreu A, Antunes P, and Vieira J (2004) Performance of F2 hybrid populations of red rice (*Oryza sativa*) with ammonium-glufosinate-resistant transgenic rice (*O. sativa*). *Planta Daninha* 22(3): 381–395.

Normile D (2004) Yangtze seen as earliest rice site. *Science* 275: 309.

Oard J, Cohn MA, Linscombe S, Gealy DR, and Gavois K (2000) Field evaluation of seed production, shattering, and dormancy in hybrid populations of transgenic rice (*Oryza sativa*) and the weed, red rice (*Oryza sativa*). *Plant Sci.* 157: 12–22.

OECD [Organization for Economic Cooperation and Development] (1999) Consensus Document on the Biology of *Oryza sativa* (rice). Consensus Documents for the Work on Harmonisation of Regulatory Oversight in Biotechnology No. 14, ENV/JM/MONO(99)26 (OECD: Paris, France). www.oecd.org/document/51/0,2340,en_2649_34391_1889395_1_1_1_1,00.html.

OECD [Organization for Economic Cooperation and Development] (2008) OECD Scheme for the Varietal Certification of Cereal Seed Moving in International Trade: Annex VIII to the Decision. C(2000)146/FINAL. OECD Seed Schemes, February 2008. www.oecd.org/document/0/0,3343,en_2649_33905_1933504_1_1_1_1,00.html.

OGTR [Office of the Gene Technology Regulator] (2005) *The Biology and Ecology of Rice (*Oryza sativa L.*) in Australia* (Office of the Gene Technology Regulator [OGTR]: Canberra, ACT, Australia). 28 p.

Oka HI (1974) Experimental studies on the origin of cultivated rice. *Genetics* 78: 475–486.

—— (1988) *Origin of Cultivated Rice* (Elsevier: Amsterdam, The Netherlands). 254 p.

—— (1991) Genetic diversity of wild and cultivated rice. In: Khush GS and Toenniessen GH (eds.) *Rice Biotechnology* (International Rice Research Institute [IRRI]: Los Baños, Manila, Philippines), pp. 55–81.

Oka HI and Chang WT (1961) Hybrid swarm between wild and cultivated species, *Oryza perennis* and *O. sativa*. *Evolution* 15: 418–430.

Oka HI and Morishima H (1967) Variations in the breeding systems of a wild rice, *Oryza perennis*. *Evolution* 21: 249–258.

Olofsdotter M, Valverde BE, and Madsen KH (2000) Herbicide resistant rice (*Oryza sativa* L.): Global implications for weedy rice and weed management. *Ann. Appl. Biol.* 137: 279–295.

Parmar KS, Siddiq EA, and Swaminathan MS (1979) Variation in anther and stigma characteristics in rice. *Indian J. Genet. Plant Breed.* 39: 551–559.

Pental D and Barnes SR (1985) Interrelationship of cultivated rices *Oryza sativa* and *O. glaberrima* with wild *O. perennis* complex. *Theor. Appl. Genet.* 70: 185–191.

Peters M, Oberrath R, and Bohning-Gaese K (2003) Seed dispersal by ants: Are seed preferences influenced by foraging strategies or historical constraints? *Flora* 198: 413–420.

Pham JL and Bougerol B (1993) Abnormal segregation in crosses between two cultivated rice species. *Heredity* 70: 447–466.

—— (1996) Variation in fertility and morphological traits in progenies of crosses between the two cultivated rice species. *Hereditas* 124: 179–183.

Quesada T, Trejos R, Lobo J, and Espinoza AM (2003) Reproductive biology of the wild rice species *Oryza glumaepatula* and its hybridization potential with *Oryza sativa*: First steps to evaluate gene flow. In: *Memorias de las Jornadas de Académicas de la Universidad de Costa Rica, August 18–24, 2003* (Universidad de Costa Rica: San Jose, Costa Rica).

Rao SA, Phetpaseut V, Bounphanousay C, and Jackson MT (1997) Spontaneous interspecific hybrids in *Oryza* in Lao PDR. *Int. Rice Res. Notes* 22(1): 4–5.

Reano R and Pham JL (1998) Does cross-pollination occur during seed regeneration at the International Rice Genebank? *Int. Rice Res. Notes* 23(3): 5–6.

Rong J, Xia H, Zhu YY, Wang YY, and Lu B-R (2004) Asymmetric gene flow between traditional and hybrid rice varieties (*Oryza sativa*) estimated by nuclear SSRs and its implication in germplasm conservation. *New Phytol.* 163: 439–445.

Rong J, Song ZP, Su J, Xia H, Lu B-R, and Wang F (2005) Low frequency of transgene flow from *Bt/CpTI* rice to its nontransgenic counterparts planted at close spacing. *New Phytol.* 168: 559–566.

Rong J, Lu B-R, Song ZP, Su J, Snow AA, Zhang X, Sun S, Chen R, and Wang F (2007) Dramatic reduction of crop-to-crop gene flow within a short distance from transgenic rice fields. *New Phytol.* 173: 346–353.

Sagers CL, Nigemann S, and Novak S (2002) Ecological risk assessment for the release of transgenic rice in southeastern Arkansas. In: Snow AA (ed.) *Scientific Methods Workshop: Ecological and Agronomic Consequences of Gene Flow from Transgenic Crops to Wild Relatives; Meeting Proceedings, University Plaza Hotel and Conference Center, Ohio State University, Columbus, Ohio, March 5–6, 2002* (Ohio State University: Columbus, OH, USA), pp. 94–105.

Sampath S and Govindaswami S (1958) Wild rices of *Oryza* and their relationships to the cultivated varieties. *Rice News Teller* 6: 17–20.

Sano Y, Chu Y, and Oka H (1979) Genetic studies of speciation in cultivated rice: 1. Genic analysis for the F1 sterility between *Oryza sativa* and *O. glaberrima*. *Japan. J. Genet.* 54: 121–132.

Sano Y, Morishima H, and Oka HI (1980) Intermediate perennial-annual populations of *Oryza perennis* found in Thailand and their evolutionary significance. *Bot. Mag.* (Tokyo) 93: 291–305.

Sarla N and Mallikarjuna Swamy BP (2005) *Oryza glaberrima*: A source for the improvement of *Oryza sativa. Curr. Sci.* 89: 955–963.

Schaal B , Bashir A, Jamjod S, Leverich W, Londo J, Maneechote C, Nirantraiyakun S, Rerkasem B, and Singhal S (2006) Introgression and gene flow in Asian rice. In: *Proceedings of the 9th International Symposium on Biosafety of Genetically Modified Organisms (ISBGMO), Jeju Island, South Korea, September 24–29, 2006* (International Society for Biosafety Research [ISBR], n.p.), pp. 127–130. www.isbr.info/isbgmo/docs/9th_isbgmo_program.pdf.

Schmit V (1997) Interspecific hybridization between *O. sativa* and *O. longistaminata* to develop a perennial upland rice. In: Jones MP, Dingkuhn M, Johnson DE, and Fagade SO (eds.) *Interspecific Hybridization Progress and Prospects: Proceedings of the Workshop; Africa/Asia Joint Research on Interspecific Hybridization between the African and Asian Rice Species (O. glaberrima and O. sativa), West Africa Rice Development Association (WARDA), M'bé, Bouake, Côte d'Ivoire, December 16–18, 1996* (West Africa Rice Development Association [WARDA]: Bouake, Côte d'Ivoire), pp. 141–158.

Second G (1982) Origin of the genic diversity of cultivated rice (*Oryza* spp.): Study of the polymorphism scored at 40 isozyme loci. *Japan. J. Genet.* 57: 25–57.

Semon M, Nielsen R, Jones MP, and McCouch SR (2005) The population structure of African cultivated rice *Oryza glaberrima* (Steud.): Evidence for elevated levels of linkage disequilibrium caused by admixture with *O. sativa* and ecological adaptation. *Genetics* 169: 1639–1647.

Shivrain VK, Burgos NR, Moldenhauer KAK, McNew RW, and Baldwin TL (2006)

Characterization of spontaneous crosses between Clearfield rice (*Oryza sativa*) and red rice (*Oryza sativa*). *Weed Technol.* 20: 576–584.

Shivrain VK, Burgos NR, Rajguru SN, Anders MM, Moore J, and Sales MA (2007) Gene flow between Clearfield and red rice. *Crop Prot.* 26: 349–356.

Sitch LA (1990) Incompatibility barriers operating in crosses of *Oryza sativa* with related species and genera. In: Gustafson JP (ed.) *Gene Manipulation in Plant Improvement II: 19th Stadler Genetics Symposium* (Plenum Press: New York, NY, USA), pp. 77–93.

Sitch LA, Dalmacio RD, and Romero GO (1989) Crossability of wild *Oryza* species and their potential use for improvement of cultivated rice. *Rice Genet. Newsl.* 6: 58–60.

Small E (1984) Hybridization in the domesticated-weed-wild complex. In Grant W (ed.) *Plant Biosystematics* (Academic Press: New York, NY, USA), pp. 195–210.

Smith RJ Jr and Hill JE (1990) Weed control in US rice. In: Copping LG, Grayson BT, and Green MB (eds.) *Pest Management in Rice* (Springer-Verlag: New York, NY, USA), pp. 314–327.

Snow AA, Sweeney PM, Lang N, and Buu B (2006) Hybrids between weedy and cultivated rice (*Oryza sativa*) in the Mekong Delta of Vietnam exhibit heterosis: Implications for rapid evolution in rice. *Proceedings of "Botany 2006— Looking to the Future, Conserving the Past"; Annual Meeting of the Botanical Society of America, California State University, Chico, California, USA, July 28–August 2, 2006.* www.2006.botanyconference.org/engine/search/index.php?func=.php?func=detail&aid=662.detail&aid=662.

Sobrizal K, Ikeda K, Sánchez PL, Doi K, Angeles ER, Khush GS, and Yoshimura A (1999) Development of *Oryza glumaepatula* introgression lines in rice, *O. sativa* L. *Rice Genet. Newsl.* 16: 107–108.

Somantri IH (2001) Wild rice (*Oryza* spp.): Their existence and research in Indonesia. *Bull. AgroBio* 5: 14–20.

Song ZP, Lu B-R, and Chen JK (2001) A study of pollen viability and longevity in *Oryza rufipogon, O. sativa,* and their hybrids. *Int. Rice Res. Notes* 26(2): 31–32.

Song ZP, Lu B-R, Zhu YG, and Chen JK (2002) Pollen competition between cultivated and wild rice species (*Oryza sativa* and *O. rufipogon*). *New Phytol.* 153: 289–296.

——— (2003) Gene flow from cultivated rice to the wild species *Oryza rufipogon* under experimental field conditions. *New Phytol.* 157: 657–665.

Song ZP, Lu B-R, Wang B, and Chen JK (2004a) Fitness estimation through performance comparison of F1 hybrids with their parental species *Oryza rufipogon* and *O. sativa. Ann. Bot.* (Oxford) 93: 311–316.

Song ZP, Lu B-R, and Chen JK (2004b) Pollen flow of cultivated rice measured under experimental conditions. *Biodiv. Cons.* 13: 579–590.

Song ZP, Li B, Chen JK, and Lu B-R (2005) Genetic diversity and conservation of common wild rice (*Oryza rufipogon*) in China. *Plant Spec. Biol.* 20: 83–92.

Suh HS, Sato YI, and Morishima H (1997a) Genetic characterization of weedy rice (*Oryza sativa* L.) based on morphophysiology, isozymes, and RAPD markers. *Theor. Appl. Genet.* 94: 316–321.

Suh HS, Back J, and Ha J (1997b) Weedy rice occurrence and position in transplanted and direct-seeded farmer's fields. *Korean J. Crop Sci.* 42: 352–356.

Sweeney M and McCouch S (2007) The complex history of the domestication of rice. *Ann. Bot.* (Oxford) 100: 951–957.

Tan GX, Xiong ZY, Jin HJ, Li G, Zhu LL, Shu LH, and He GC (2006) Characterization of interspecific hybrids between *Oryza sativa* L. and three wild rice species of China by genomic *in situ* hybridization. *J. Integr. Plant Biol.* 48: 1077–1083.

Tang LH and Morishima H (1996) Genetic characteristics and origin of weedy rice. In: Chau R, Gordon B, and Leir G (eds.) *Origin and Differentiation of Chinese Cultivated Rice* (China Agricultural University Press: Beijing, China), pp. 211–218. http://http-server.carleton.ca/~bgordon/Rice/papers/tang96.htm.

Tao D and Sripichitt P (2000) Preliminary report on transfer traits of vegetative propagation from wild rices to *O. sativa* via distant hybridization and embryo rescue. *Kasetsart J., Nat. Sci.* 34: 1–11.

Tao D, Hu F, Yang G, Yang J, and Tao H (1997) Exploitation and utilization of interspecific hybrid vigor between *Oryza sativa* and *O. glaberrima*. In: Jones MP, Dingkuhn M, Johnson DE, and Fagade SO (eds.) *Interspecific Hybridization Progress and Prospects: Proceedings of the Workshop; Africa/Asia Joint Research on Interspecific Hybridization between the African and Asian Rice Species (O. glaberrima and O. sativa), West Africa Rice Development Association (WARDA), M'bé, Bouake, Côte d'Ivoire, December 16–18, 1996* (West Africa Rice Development Association [WARDA]: Bouake, Côte d'Ivoire), pp. 103–112.

Tao D, Hu F, Yang Y, Xu P, Li J, Wen G, Sacks E, McNally K, and Sripichitt P (2001) Rhizomatous individual was obtained from interspecific BC2F1 progenies between *Oryza sativa* and *Oryza longistaminata*. *Rice Genet. Newsl.* 18: 11–13.

Tao D, Xu P, Yang Y, Hu F, Li J, and Zhou J (2003) Studies on fertility in interspecific hybrids between *O. sativa* × *O. glaberrima*. *Rice Genet. Newsl.* 20: 71–73.

USDA [United States Department of Agriculture] (2007) Report of LibertyLink Rice Incidents. *U.S. Department of Agriculture-Animal and Plant Health Inspection Service [USDA-APHIS] Report, October 4, 2007.* 8 p. www.aphis. usda.gov/newsroom/content/2007/10/index.shtml.

——— (2008) 7 CFR [Code of Federal Regulations] § 201.76. Federal Seed Act Regulations: Minimum Land, Isolation, Field, and Seed Standards. Source: 59 FR [Federal Register] 64516, Dec. 14, 1994, as amended at 65 FR 1710, Jan. 11,

2000 (Government Printing Office: Washington, DC, USA), pp. 372–378. www.access.gpo.gov/nara/cfr/waisidx_08/7cfr201_08.html.

Vandiver VV, Hall DW, and Westbrooks RG (1992) Discovery of *Oryza rufipogon* (Poaceae: Oryzeae) new to the United States, with its implications. *Sida* 15: 105–109.

Vaughan DA (1994) *The Wild Relatives of Rice: A Genetic Resources Guide Book* (International Rice Research Institute [IRRI]: Los Baños, Manila, Philippines). 137 p.

Vaughan DA and Morishima H (2003) Biosystematics of the genus *Oryza*. In: Smith CW and Dilday RH (eds.) *Rice: Origin, History, Technology, and Production* (John Wiley and Sons: Hoboken, NJ, USA), pp. 27–65.

Vaughan DA, Morishima H, and Kadowaki K (2003) Diversity in the genus *Oryza*. *Curr. Opin. Plant Biol.* 6: 139–146.

Vaughan DA, Sanchez PL, Ushiki J, Kago A, and Tomooka N (2005) Asian rice and weedy rice evolutionary perspectives. In: Gressel J (ed.) *Crop Ferality and Volunteerism: A Threat to Food Security in the Transgenic Era?* (CRC Press: Boca Raton, FL, USA), pp. 257–277.

Vaughan DA, Lu B-R, and Tomooka N (2008) The evolving story of rice evolution. *Plant Sci.* 174: 394–408.

Vaughan LK, Ottis BV, Prazak-Havey AM, Sneller C, Chandler J, and Park W (2001) Is all red rice found in commercial rice really *Oryza sativa*? *Weed Sci.* 49: 468–476.

Wang F, Yuan QH, Shi L, Qian Q, Liu WG, Kuang BG, Zeng DL, Liao YL, Cao B, and Jia SR (2006) A large-scale field study of transgene flow from cultivated rice (*Oryza sativa*) to common wild rice (*O. rufipogon*) and barnyard grass (*Echinochloa crusgalli*). *Plant Biotech. J.* 4: 667–676.

Wuu KD, Jui Y, Ly KCL, Chou C, and Li HW (1963) Cytogenetical studies of *Oryza sativa* L. and its related species: 3. Two intersectional hybrids, *O. sativa* L. × *brachyantha* A. Chev & Roehr. *Bot. Bull. Acad. Sin.* 4: 51–59.

Xiao J, Li J, Grandillo S, Nag Ahn S, Yuan L, Tanksley SD, and McCouch SR (1998) Identification of trait-improving quantitative trait loci alleles from a wild rice relative, *Oryza rufipogon. Genetics* 150: 899–909.

Yamanaka S, Nakamura I, Watanabe KN, and Sato YI (2004) Identification of SNPs in the waxy gene among glutinous rice cultivars and their evolutionary significance during the domestication process of rice. *Theor. Appl. Genet.* 108: 1200–1204.

Yan HH, Xiong ZM, Min SK, Hu HY, Zhang ZT, Tian SL, and Tang SX (1997) The transfer of brown planthopper resistance from *Oryza eichingeri* to *O. sativa*. *Chin. J. Genet.* 24: 277–284.

Yang SJ, Jin YD, Lee SK, and Chun GS (1991) Interspecific F1 hybrids resistant to rice black-streaked dwarf virus through embryo rescue. *Research Reports of the Rural Development Administration* (Suweon) 33: 1–5.

Yoshida S (1981) *Fundamentals of Rice Crop Science* (International Rice Research Institute [IRRI]: Los Baños, Manila, Philippines). 269 p.

Zamora A, Barboza C, Lobo J, and Espinoza AM (2003) Diversity of native rice (Poaceae: *Oryza*) species of Costa Rica. *Genet. Resour. Crop Evol.* 50: 855–870.

Zhang N, Linscombe S, and Oard J (2003) Outcrossing frequency and genetic analysis of hybrids between transgenic glufosinate herbicide-resistant rice and the weed, red rice. *Euphytica* 130: 35–45.

Zhang W, Linscombe S, Webster E, and Oard J (2004) Risk assessment and genetic analysis of natural outcrossing in Louisiana commercial fields between Clearfield rice and the weed, red rice. In: Rice Technical Working Group [RTWG] and U.S. Department of Agriculture [USDA] (eds.) *Proceedings of the 30th Rice Technical Working Group* (U.S. Department of Agriculture [USDA]: Washington, DC, USA), p.125.

Zhao M, Lafitte HR, Sacks E, Dimayuga G, and Botwright Acuña TL (2008) Perennial *O. sativa* × *O. rufipogon* interspecific hybrids: I. Photosynthetic characteristics and their inheritance. *Field Crops Res.* 204: 203–213.

Zhou H, Xie Z, and Ge S (2003) Microsatellite analysis of genetic diversity and population genetic structure of a wild rice (*Oryza rufipogon* Griff.) in China, *Theor. Appl. Genet.* 107: 322–339.

Zhu Q and Ge S (2005) Phylogenetic relationships among A-genome species of the genus *Oryza* revealed by intron sequences of four nuclear genes. *New Phytol.* 167: 249–267.

Zhu Q, Zheng X, Luo J, Gaut BS, and Ge S (2007) Multilocus analysis of nucleotide variation of *Oryza sativa* and its wild relatives: Severe bottleneck during domestication of rice. *Mol. Biol. Evol.* 24: 875–888.

Zhu YY, Chen HR, Fan JH, Wang YY, Li Y, Chen JB, Fan JX, Yang SS, Hu LP, Leung H, Mew TW, Teng PS, Wang ZH, and Mundt CC (2000) Genetic diversity and disease control in rice. *Nature* 406: 718–722.

Zhu YY, Wang YY, Chen HR, and Lu B-R (2003) Conserving traditional rice varieties through management for crop diversity. *Bioscience* 53: 158–162.

NINETEEN

Sorghum
(*Sorghum bicolor* (L.) Moench)

Center(s) of Origin and Diversity

Sorghum is thought to have originated in NE Africa, either in Ethiopia or Sudan (Vavilov 1951; Harlan and de Wet 1972; Mann et al. 1983; Dahlberg 1995; Doggett and Prasada Rao 1995), although the hypothesized number of domestication events varies from one to five (one for each major race) (Harlan 1975; Morden et al. 1990; Dahlberg 1995). The primary center of diversity of the genus *Sorghum* Moench is found in Ethiopia.

Biological Information

Flowering

Sorghum flowers bloom from the inflorescence apex downwards, and flowering can last for a period of 4 to 7 days. The sorghum stigma becomes receptive 1 to 2 days before blooming and remains in that state for 5 to 16 days after anthesis (Ayyangar and Rao 1931; Schertz and Dalton 1980). Fertilization usually occurs within two hours after pollination. The flowering times for domesticated sorghum (or grain sorghum) and its wild and weedy relatives overlap appreciably (Eastin and Lee 1985; Dogget 1988; Tesso et al. 2008).

Pollen Dispersal and Longevity

Sorghum is largely self-pollinated, but wind is an important vector for significant rates of outcrossing (Stephens and Quinby 1934). Insects (honeybees, wild bees, halictid bees, beetles) have been reported to visit sorghum flowers, and they may also contribute to cross-pollination (Stephens and Quinby 1934; Immelman and Eardley 2000; Schmidt and Bothma 2005).

Sorghum pollen is generally highly functional for about 30 minutes after the anthers dehisce, but its longevity is limited to two to four hours (Schertz and Dalton 1980; Lansac et al. 1994). Pollen viability and the ability to germinate are influenced by temperature, humidity, and cloud cover (Artschwager and McGuire 1949; Brooking 1979; Tuinstra and Wedel 2000).

Sexual Reproduction

Domesticated sorghum is a diploid ($2n = 2x = 20$) annual grass with a mixed mating system. The species is self-compatible and mainly autogamous, but outcrossing rates range from 2% to over 60%, depending on the genotype, and average about 6% (Schertz and Dalton 1980; Ellstrand and Foster 1983; Pedersen et al. 1998).

Vegetative Reproduction

Domesticated sorghum is not capable of vegetative reproduction, but several of its close wild relatives are, including the weed Johnsongrass, *Sorghum halepense* (L.) Pers. (Warwick et al. 1986).

Seed Dispersal and Dormancy

While seeds of wild and weedy sorghum relatives are easily shattered when mature, the crop lost this trait during domestication, and most modern cultivars completely lack this ability to shed seeds (Mann et al. 1983; Dahlberg 1995). In general, seeds of *Sorghum* species can spread by a number of means. They can be distributed via wind, water, animals, or humans (clothing, harvest machinery, vehicles) and could even travel for long distances when carried by water or in the excreta of birds or livestock (Holm et. al. 1977; Warwick and Black 1983).

Wild and sweet sorghum seeds (i.e., cultivars and landraces with low polyphenol content) are consumed by livestock, rodents, birds, and seed-eating ants (Dahlberg 1995), all of which can act as agents for seed dispersal. In most cases rodents probably consume the seeds, but they can also carry sorghum grains away from fields; the seed-dispersal distance depends on the size of the rodents' territory. The average territorial size for mice ranges from 0.02 to 0.2 ha, that is, 40×50 m^2 (Caughley et al. 1998).

Seeds of the noxious weed Johnsongrass have been shown to pass undamaged through the digestive tract of cattle and other livestock and subsequently be spread through manure fertilizer (Ball 1902). These seeds also attach themselves to the coats of passing animals (livestock).

Ants sometimes remove sorghum seeds and store them underground. Although there are differences in ant behavior and territory size across species (Peters et al. 2003), seed dispersal occurs at a local scale, with a mean dispersal distance of less than 1 m and a maximum of 40 m (Beattie 1982; Gómez and Espadaler 1998).

Seeds of domesticated sorghum may show dormancy during the first month after harvest, but in general they exhibit less dormancy than those of its wild relatives (Hoffman et al. 2002). The extent of dormancy, and the viability and survival of *Sorghum* seeds in the soil, depends on the species, the geographical region, the climatic and environmental conditions during seed development, the depth of seed burial in the soil, and the type of soil management (Fellows and Roeth 1992; Kegode and Pearce 1998). For instance, wild sorghum (*S. bicolor* subsp. *drummondii* (Nees *ex* Steud.) de Wet *ex* Davidse) seeds in the midwestern United States are rarely able to survive the winter, as more than 80% die between November and March and virtually none survive a second winter (Teo-Sherrell and Mortensen 2000). Likewise, seeds of the closely related noxious weed Johnsongrass (*S. halepense*) survived less than two years when buried in upper soil layers in Argentina (Leguizamón 1986).

Despite the rather short survival and persistence of *Sorghum* seeds in the upper layers of the soil, considerable seed banks can be built up by the frequent input of new seeds each year. Furthermore, Johnsongrass seeds are able to survive for more than six years at greater soil depths (> 22 cm), suggesting that substantial, persistent seed banks could be formed in these deeper layers (Leguizamón 1986).

Table 19.1 Domesticated sorghum (*Sorghum bicolor* (L.) Moench) and its closest wild relatives in section *Sorghum*

Species	Common name	Ploidy level	Origin and distribution	Life form and weediness
Primary gene pool (GP-1)				
S. *bicolor* (L.) Moench subsp. *bicolor*	grain sorghum, domesticated sorghum	diploid (2*n* = 2*x* = 20)	cultivated throughout tropic, subtropic, and warm-temperate regions	annual; cultivated grain sorghum
S. *bicolor* subsp. *verticilliflorum* (Steud.) de Wet *ex* Wiersema & J. Dahlb.	common wild sorghum	diploid (2*n* = 2*x* = 20)	native to tropical Africa, Madagascar, and perhaps the Mascarenes; naturalized in India, Australia, and the Americas	ancestor; wild annual
S. *bicolor* subsp. *drummondii* (Steud.) de Wet *ex* Davidse	chicken corn, Sudan grass	diploid (2*n* = 2*x* = 20)	native to Africa, but may be present wherever grain sorghum is cultivated	noxious annual crop weed, arising from hybridization of grain sorghum (subsp. *bicolor*) and its wild relative (subsp. *verticilliflorum*)
S. *propinquum* (Kunth.) Hitchc.		diploid (2*n* = 2*x* = 20)	SE Asia (Sri Lanka, S India, Myanmar, Thailand, Malaysia, and Philippines)	wild perennial; weedy
Secondary gene pool (GP-2)				
S. × *almum* Parodi	Columbus grass	allotetraploid (2*n* = 4*x* = 40)	natural hybrid; arose in S America (Argentina, Paraguay, and Uruguay)	wild perennial; noxious weed in USA
S. *halepense* (L.) Pers.	Johnson grass, weedy sorghum	allotetraploid (2*n* = 4*x* = 40)	native to W Asia and N Africa; widely naturalized in warm-temperate regions elsewhere, including N America	wild perennial; classified as a noxious weed in USA and Canada

Sources: Harlan and de Wet 1972; de Wet 1978; Dahlberg 2000; Wiersema and Dahlberg 2007.
Note: The cultivated form is in bold. Gene pool delimitations are according to Harlan and de Wet 1971.

Volunteers, Ferals, and Their Persistence

Spontaneous hybridization between different taxa of the cultivated-wild-weed complex results in intermediary, often weedy sorghum forms with varying levels of volunteerism and ferality (Ejeta and Grenier 2005). Consequently, domesticated sorghum plants frequently appear as volunteers in subsequent crops, or as ferals outside of cultivated areas, constituting a serious weed problem.

Sorghum is frost sensitive, so in temperate regions, volunteers and ferals normally do not survive the winter. In tropical and subtropical areas, rhizomatous sorghum plants can persist outside of cultivation for several years (W.L. Rooney, pers. comm. 2008). Nonrhizomatous plants most likely will die, either from drought or competition (A. Snow, pers. comm. 2008).

Weediness and Invasiveness Potential

Outcrossing among the taxa of the cultivated-wild-weed complex produces vigorous hybrids that are considered to be noxious weeds in many countries. In particular, weedy annual sorghum forms known as shattercane frequently infest sorghum fields (de Wet 1978; Dahlberg 1995, 2000; Ejeta and Grenier 2005). Weedy, volunteer, and feral sorghums constitute a serious problem, since they serve as hosts for pathogens that can be transferred to the crop (see, e.g., Bandyopadhyay et al. 1996; Isakeit et al. 1998).

Crop Wild Relatives

The genus *Sorghum* contains approximately 25 species and is divided into five subgenera, containing multiple sections (de Wet 1978). *Sorghum* sect. *Sorghum* is native to Africa and S Asia and is made up of three species: *Sorghum bicolor* (L.) Moench—including domesticated sorghum—and two perennial wild relatives, one of which (Johnsongrass, or *S. halepense*) is one of the world's worst weeds (McWhorter 1971; Holm et al. 1977). *Sorghum bicolor* refers to a crop-wild-weed complex with domesticated, wild, and weedy sorghum forms classified as three different subspecies (table 19.1): subsp. *bicolor* (L.) Moench (cultivated species, five races), subsp. *verticilliflorum* (Steud.) de Wet *ex* Wiersema & J.Dahlb. (= subsp. *arundinaceum* (Desv.) de Wet & J.R. Harlan *ex* Davidse, the wild progenitor of domesticated sorghum), and the annual weedy types (subsp. *drummondii* and shattercane).

Domesticated sorghum is native to Africa, where different sorghum types in the *Sorghum bicolor* crop-wild-weed complex occur both in and near cultivated sorghum fields (Ejeta and Grenier 2005; Tesso et al. 2008). In other areas of the world where sorghum is grown, it is sometimes sympatric with sexually compatible relatives such as Johnsongrass (*S. halepense*) and Columbus grass, *S.* ×*almum* Parodi (Arriola and Ellstrand 2002).

Hybridization

Domesticated sorghum is sexually compatible with all of its wild relatives that are included in sect. *Sorghum*, which can be divided into primary and secondary gene pools, according to the level of cross-compatibility with the crop (table 19.1).

Primary Gene Pool (GP-1)

The primary gene pool includes the *Sorghum bicolor* crop-wild-weed complex and the diploid perennial wild relative *S. propinquum* (Kunth) Hitchc. The crop-wild-weed complex consists of domesticated sorghum (*S. bicolor* subsp. *bicolor*), its wild ancestor *S. bicolor* subsp. *verticilliflorum*, and the annual weedy sorghums classified as *S. bicolor* subsp. *drummondii* and shattercane. All of these taxa are fully interfertile and can spontaneously outcross with each other in areas where their distributions overlap, leading to frequent introgression among them (Doggett and Majisu 1968; Baker 1972; Doggett and Prasada Rao 1995).

Sorghum bicolor subsp. verticilliflorum and subsp. drummondii

Wild and weedy sorghums, and their natural hybrid swarms with domesticated sorghum, are frequent weeds in African cereal fields (Doggett and Prasada Rao 1995; Snow et al. 2007). The S African sorghum race 'Kafir' might have arisen from introgression between domesticated and wild sorghum (Mann et al. 1983), and all other African sorghum races have been and still are diversified by introgression with wild sorghum (Doggett and Prasada Rao 1995). Molecular marker analysis of wild sorghum populations (subsp. *verticilliflorum*) growing in sympatry with domesticated sorghum in Africa revealed crop-specific alleles in wild sorghum plants, suggesting that intraspecific hybridization and introgression are common (Doggett and Majisu 1968; Aldrich and Doebley 1992; Aldrich et al. 1992).

The weed *S. bicolor* subsp. *drummondii* has evolved from natural hybrids between domesticated and wild sorghum (*S. bicolor* subsp. *bicolor* × subsp. *verticilliflorum*) and is fully cross-compatible with both species. It can be found in and near cultivated fields wherever sorghum is grown.

In field experiments in Africa and the United States, vigorous and fully fertile Fı hybrids were obtained via spontaneous outcrossing between local African sorghum cultivars and wild sorghum (Snow et al. 2007). The fitness of the Fı hybrids varied between genotypes; some combinations showed significantly greater seed production than their parents, while others had a similar or somewhat reduced fertility (Snow et al. 2007). At this point, there are no studies quantifying the extent of outcrossing and introgression between domesticated sorghum and its conspecifics.

Shattercane (*Sorghum bicolor*)

Shattercane is an annual weed that belongs to the same species as the crop (Harlan 1992). Shattercanes exist in a continuum of intermediary forms between wild and domesticated sorghums, and they are a noxious weed in areas of sorghum cultivation worldwide. The origins of shattercane have not been conclusively determined, but this taxon might have arisen from spontaneous hybridization between the crop and sudangrass (*S. bicolor* subsp. *drummondii*). Shattercanes easily out-compete domesticated sorghum, partly through their ability to disperse seeds through shattering and through their long seed dormancy (Ejeta and Grenier 2005).

To date, little is known about shattercane's role in facilitating gene flow in the sorghum crop-wild-weed complex. It is thought that shattercane can hybridize with the crop, but thus far this has not been confirmed.

Sorghum propinquum

The domesticated, wild, and weedy forms of *Sorghum bicolor* are fully cross-compatible with *S. propinquum* under controlled conditions (Doggett and Prasada Rao 1995). In the wild, both species are spatially isolated, due to their somewhat different adaptations (Dahlberg 1995). In the Philippines, however, where *S. propinquum* is native, introduced domesticated sorghum occurs in sympatry with the wild relative, and outcrossing can often be observed. Philippine farmers find the resulting hybrids to be noxious weeds (Ejeta and Grenier 2005).

Moreover, it is believed that the tetraploid noxious weed Johnsongrass, *S. halepense* (see secondary gene pool), arose from chromosome doubling of a natural hybrid between these two species (Doggett and Prasada Rao 1995; Paterson et al. 1995, 1998). Johnsongrass is native to the region between the distribution ranges of wild annual sorghum (*S. bicolor* subsp. *verticilliflorum*) and perennial *S. propinquum*.

Secondary Gene Pool (GP-2)

The secondary gene pool of domesticated sorghum includes Columbus grass (*Sorghum* ×*almum*), and one of the world's worst weeds, Johnsongrass (*S. halepense*).

Despite differences in ploidy level, domesticated sorghum ($2n = 2x = 20$) readily outcrosses with these tetraploid ($2n = 4x = 40$) wild relatives, both under controlled and natural conditions (e.g., Arriola and Ellstrand 1996, 1997; Morrell et al. 2005). In experimental field plantings in the United States, the rates of spontaneous outcrossing between *S. bicolor* and *S. halepense* can vary greatly, depending on the genotype and environmental conditions, and they have been shown to range from 0% to up to 100% (Arriola and Ellstrand 1996).

F1 hybrids are either completely sterile triploids or somewhat fertile tetraploids (Endrizzi 1957; Hadley 1958; Warwick and Black 1983; Godwin 2005). In triploid plants, partial self-fertility can be restored quite easily within two generations, via backcrossing to the diploid parent (Pritchard 1965; Arriola and Ellstrand 1996). F1 hybrids showed no differences in fitness from the weed (equal biomass and seed production, equal pollen fertility) under irrigated field conditions (Arriola and

Ellstrand 1997). However, nothing is known about the fitness of these hybrids under natural arid conditions, that is, without irrigation. Tetraploid hybrids are usually annual weeds, whereas sterile triploid hybrids develop extremely vigorous rhizomes that enable them to propagate vegetatively and persist over several years, even in cooler climates (Hadley 1958; Warwick et al. 1986; Yim and Bayer 1997).

Molecular marker analysis showed that introgression from domesticated sorghum has been involved in the evolution of very aggressive weedy *S. halepense* biotypes (Celarier 1958; Paterson et al. 1995; Morrell et al. 2005). Genetic evidence supports the hybrid origin of weedy Columbus grass (*S. ×almum*), which seems to have arisen from hybridization between domesticated sorghum and Johnsongrass in the Americas (Holm et al. 1977; Doggett 1988; Paterson et al. 1995).

Tertiary Gene Pool (GP-3)

The tertiary gene pool includes *Sorghum* species from other sections. Strong pre- and post-zygotic reproductive barriers exist between domesticated sorghum and wild relatives outside of sect. *Sorghum*, due to differences in chromosome morphology, genome size, pollen-pistil interactions, and embryo abortion (Garber 1950; Schertz and Dalton 1980; Doggett 1988; Sun et al. 1991; Price et al. 2006). Crosses with domesticated sorghum are only possible using artificial hybridization techniques, such as embryo rescue (Hodnett et al. 2005; Price et al. 2006).

Some intergeneric crosses of *Sorghum* have been obtained with sugarcane (*Saccharum* L.), but hybridization requires considerable human intervention and is highly unlikely to occur under natural conditions (Thomas and Venkatraman 1930; Janaki-Ammal and Singh 1936; Nair 1999; Nair et al. 2006).

Outcrossing Rates and Distances

Pollen Flow

As is the case for other crops, pollen-mediated gene flow in sorghum is influenced by such factors as variations in flowering times, outcrossing rates, population sizes, the distance between populations, wind speed,

Table 19.2 Isolation distances for sorghum (*Sorghum bicolor* (L.) Moench)

Threshold*	Minimum isolation distance (m)	Country	Reference	Comment
1 plant in 30 m²/ n.a.	400	OECD	OECD 2008	foundation (basic) seed requirements (non-hybrids)
1 plant in 10 m²/ n.a.	200	OECD	OECD 2008	certified seed requirements (non-hybrids)
1 plant in 30 m²/ 0.1%	400	OECD	OECD 2008	foundation (basic) seed requirements (hybrids)
1 plant in 10 m²/ 0.3%	200	OECD	OECD 2008	certified seed requirements (hybrids)
0.002†/ 0.005%	300	USA	CCIA 2009; USDA 2008a	foundation (basic) seed requirements (non-hybrids)
0.003†/ 0.01%	300	USA	CCIA 2009; USDA 2008a	registered seed requirements (non-hybrids)
0.005†/ 0.05%	200/300‡	USA	CCIA 2009; USDA 2008a	certified seed requirements (non-hybrids)
0.002†/ 0.005%	300	USA	CCIA 2009; USDA 2008a	foundation (basic) seed requirements (hybrids)
0.005†/ 0.1%	200/300‡	USA	CCIA 2009; USDA 2008a	certified seed requirements (hybrids)

*Maximum percentage of off-type plants permitted in the field / seed impurities of other varieties or off-types permitted. n.a. = not available.

†For definite varieties. For doubtful varieties: 0.005% (foundation seed), 0.01% (registered seed), and 0.1% (certified seed).

‡If the contaminating source does not genetically differ in height from the pollinator parent, or if it has a different chromosome number, the isolation distance is 200 m. If the contaminating source does differ genetically and has the same chromosome number, the distance is 300 m.

and humidity. Natural outcrossing rates of domesticated sorghum are highly variable and can reach up to 60% under optimal conditions (Schertz and Dalton 1980; Ellstrand and Foster 1983; Pedersen et al. 1998). Mean outcrossing frequencies of around 6% have been estimated (Schertz and Dalton 1980), and similar values have been reported from field trials in S Africa and the United States (Arriola 1995; Arriola and Ellstrand 1996; Schmidt and Bothma 2006).

In a field study conducted in S Africa with a male-sterile sorghum variety as the pollen receptor, natural outcrossing frequencies were highest at a distance of 13 m (2.5%) and decreased exponentially within 40 m of the pollen source (Schmidt and Bothma 2006). Low levels of outcrossing (< 0.5%) were detectable as far as 158 m, the farthest distance sampled. In a field study in the United States, outcrossing frequencies of up to 2% were detected at 100-meter distances between domesticated sorghum and Johnsongrass (Arriola and Ellstrand 1996).

Isolation Distances

The separation distances established for sorghum seed production in the United States and in OECD countries are 200, 300, and 400 m, depending on the seed category (table 19.2). Evidence from experimental field trials (see above) indicates that under certain climatic conditions, these distances may not be sufficient for maintaining seed purity standards or remaining under the maximum allowable amount of GM contamination (0.9%). To attain levels that comply with stipulated thresholds of 0.1% impurities or less, larger separation distances may be required. In practice, most breeders will use these greater distances to make sure there is complete isolation (W.L. Rooney, pers. comm. 2008). Further studies are needed to determine the frequency of outcrossing at different distances and at field scale.

State of Development of GM Technology

Sorghum is considered to be one of the most recalcitrant crops for tissue culture and plant regeneration, and, hence, for genetic modification. Transgenic technology in sorghum has been under development since the 1990s. Today, stable and quite effective protocols are available for

Agrobacterium-mediated transformation (see, e.g., Zhao et al. 2000; Gao et al. 2005; Howe et al. 2006) and transformation via particle bombardment (see, e.g., Jeoung et al. 2002; Tadesse et al. 2003; Girijashankar et al. 2005).

Traits currently being researched for genetic modification are enhanced protein content and quality; fungus resistance; insect tolerance (*Bt*); abiotic stress tolerance (to salinity, drought); enhanced nutritional quality (elevated lysine, vitamin A and E, and iron and zinc content); and enhanced dough quality for the baking industry. There has been no commercial production of genetically modified sorghum to date, but field trials are being conducted in India and the United States (USDA 2008b).

Agronomic Management Recommendations

There are no specific recommendations.

Crop Production Area

Sorghum is the world's fifth most important cereal crop. Its total production area is estimated to be 43.8 million ha, spread over 100 countries—primarily in hot, dry regions in Africa, SE Asia, Central and S America, and the United States (FAO 2009). In Asia and Africa, sorghum is mainly grown for human consumption, while in the rest of the world it is cultivated more for forage and for the production of ethanol. More than half of the area where sorghum is cultivated for food lies in three countries (India 19%, Nigeria 17%, Sudan 15%), which account for 11%, 16%, and 8%, respectively, of world sorghum grain production. The United States alone produce 20% of the world's sorghum grain on only 6% of the world production area. Together, Africa and Asia account for about 60% of world production, followed by S and Central America (18%) (FAO 2009).

Research Gaps and Conclusions

Gene flow unquestionably occurs between domesticated sorghum and its wild relatives, and introgression of crop alleles to wild species has been demonstrated. Pollen-mediated gene flow and introgression in sorghum are of particular concern, since one of its closest relatives—*Sorghum halepense*, or Johnsongrass—is one of the world's worst weeds and grows in sympatry with domesticated sorghum (Holm et al. 1977). Since this perennial wild relative (as well as the other two sexually compatible perennials, *S. propinquum* and *S. xalmum*) is capable of spreading vegetatively and persisting, because of its well-developed rhizomes, selectively neutral or advantageous crop alleles are expected to persist in wild Johnsongrass populations without any need for sexual reproduction. Moreover, the transfer of more fitness-enhancing alleles (e.g., herbicide tolerance) to *S. halepense* could increase the invasiveness of this weed.

In sorghum, there are four main ways gene introgression to other cultivated or wild species can occur:

1   through pollen flow from domesticated sorghum

2   through volunteers in fields previously planted with sorghum

3   through sorghum seeds dispersed by seed predators, such as birds and animals

4   through sorghum seeds dispersed by humans during cropping, harvesting, handling, storage, and transport of the crop

Sorghum seeds, like those of maize, are dispersed via informal seed exchange systems among farmers, particularly in Africa and Asia (see, e.g., Alvarez et al. 2005; Manzelli et al. 2007). These informal seed systems facilitate gene flow not only within the crop (between landraces and modern cultivars), but also within the crop-weedy-wild complex.

Despite the considerable amount of knowledge that has been gathered about the cross-compatibility of sorghum with its wild relatives, the origin of shattercanes still awaits clarification, as well as the role played by this annual weed in facilitating gene flow in the crop-wild-weed complex. Moreover, almost no research has been undertaken to quantify the extent of gene flow (through both pollen and seed) and introgression at the field scale in different agro-ecosystems. Surprisingly, the role of insects as pollen dispersal agents and the extent of pollen-mediated gene flow at varied distances from the pollen source, have received little attention. Also, no studies assessing the fate of introgressed alleles in later generations have been pinpointed, except for the study by Morrell et al. (2005), where molecular markers were used to show that introgression from domesticated sorghum has been involved in the evolution of particularly aggressive weedy *Sorghum halepense* biotypes.

Based on the knowledge currently available about the relationships between sorghum and its wild relatives, the risk of gene introgression from domesticated sorghum—assuming physical proximity (less than 400 m) and flowering overlap, and being conditional on the nature of the respective trait(s)—is as follows:

- *very high* for its conspecific annual wild and weedy relatives (*Sorghum bicolor* subsp. *drummondii*, subsp. *verticilliflorum*, and shattercane), as well as for its perennial weedy relatives Columbus grass (*S. xalmum*), Johnsongrass (*S. halepense*), and *S. propinquum* (widespread worldwide, map 19.1)

REFERENCES

Aldrich PR and Doebley J (1992) Restriction fragment variation in the nuclear and chloroplast genomes of cultivated and wild *Sorghum bicolor*. *Theor. Appl. Genet.* 85: 293–302.

Aldrich PR, Doebley J, Schertz KF, and Stec A (1992) Patterns of allozyme variation in cultivated and wild *Sorghum bicolor*. *Theor. Appl. Genet.* 85: 451–460.

Álvarez N, Garine E, Khasah C, Dounias E, Hossaert-McKey M, and McKey D (2005) Farmers' practices, metapopulation dynamics, and conservation of agricultural biodiversity on-farm: A case study of sorghum among the Duupa in sub-Sahelian Cameroon. *Biol. Cons.* 121: 533–543.

Arriola PE (1995) Crop to weed gene flow in *Sorghum*: Implications for transgenic release in Africa. *Afr. Crop Sci. J.* 3: 153–160.

Arriola PE and Ellstrand NC (1996) Crop-to-weed gene flow in the genus *Sorghum* (Poaceae): Spontaneous interspecific hybridization between Johnsongrass, *Sorghum halepense*, and crop sorghum, *S. bicolor*. *Am. J. Bot.* 83: 1153–1159.

—— (1997) Fitness of interspecific hybrids in the genus *Sorghum*: Persistence of crop genes in wild populations. *Ecol. Appl.* 7: 512–518.

—— (2002) Gene flow and hybrid fitness in the *Sorghum bicolor-Sorghum halepense* complex. In: Snow AA (ed.) *Scientific Methods Workshop: Ecological and Agronomic Consequences of Gene Flow from Transgenic Crops to*

Wild Relatives; Meeting Proceedings, University Plaza Hotel and Conference Center, Ohio State University, Columbus, Ohio, March 5–6, 2002 (Ohio State University: Columbus, OH, USA), pp. 25–30. www.agbios.com/docroot/articles/02-280-002.pdf.

Artschwager E and McGuire RC (1949) Cytology of reproduction in *Sorghum vulgare. J. Agric. Res.* 78: 659–673.

Ayyangar GNR and Rao VP (1931) Studies in sorghum: I. Anthesis and pollination. *Indian J. Agric. Sci.* 6: 1299–1322.

Ball CR (1902) Johnsongrass: Report of investigation made during the season of 1901. U.S. Dept. Agric., Bur. Plant Ind. Bull. 11: 1–24.

Baker HG (1972) Human influences on plant evolution. *Econ. Bot.* 26: 32–43.

Bandyopadhyay R, Frederickson DE, McLaren NW, and Odvody GN (1996) Ergot a global threat to sorghum. *Int. Sorghum Millet Newsl.* 37: 1–32.

Beattie AJ (1982) Ants and gene dispersal in flowering plants. In: Armstrong JM, Powell JM, and Richards AJ (eds.) *Pollination and Evolution: Based on the Symposium on Pollination Biology, Held During the 13th International Botanical Congress (Section 6.7), at Sydney, Australia, August, 1981* (Royal Botanic Gardens: Sydney, NSW, Australia), pp. 1–8.

Brooking IR (1979) Male sterility in *Sorghum bicolor* induced by low night temperature: II. Genotypic differences in sensitivity. *Aust. J. Plant Physiol.* 6: 143–147.

Caughley J, Bomford M, Parker B, Sinclair R, Griffiths J, and Kelly D (1998) *Managing Vertebrate Pests: Rodents* (Bureau of Resource Sciences and Grains Research and Development Corporation: Canberra, ACT, Australia). 130 p.

CCIA [California Crop Improvement Association] (2009) Sorghum Certification Standards. http://ccia.ucdavis.edu/seed_cert/seedcert_index.htm.

Celarier RP (1958) Cytotaxonomic notes on the subsection *Halepensia* of the genus *Sorghum. Bull. Torr. Bot. Club* 85: 49–62.

Dahlberg DJ (1995) Dispersal of *Sorghum* and the role of genetic drift. *Afr. Crop Sci. J.* 3: 143–151.

—— (2000) Classification and characterization of *Sorghum*: Weeds and their control in grain sorghum. In: Smith CW and Frederiksen RA (eds.) *Sorghum: Origin, History, Technology, and Production* (John Wiley and Sons: New York, NY, USA), pp. 99–130.

de Wet JMJ (1978) Systematics and evolution of *Sorghum* sect. *Sorghum* (Gramineae). *Am. J. Bot.* 65: 477–484.

Doggett H (1988) *Sorghum* (John Wiley and Sons: New York, NY, USA). 403 p.

Doggett H and Majisu BN (1968) Disruptive selection in crop development. *Heredity* 23: 1–23.

Doggett H and Prasada Rao KE (1995) Sorghum. In: Smartt J and Simmonds NW (eds.) *Evolution of Crop Plants.* 2nd ed. (Longman Scientific and Technical: Harlow, UK), pp. 173–180.

Eastin JD and Lee K (1985) *Sorghum bicolor.* In: Halevy AH (ed.) *CRC Handbook of Flowering,* vol. 4 (CRC Press: Boca Raton, FL, USA), pp. 367–375.

Ejeta G and Grenier C (2005) Sorghum and its weedy hybrids. In: Gressel J (ed.) *Ferality and Volunteerism: A Threat to Food Security in the Transgenic Era?* (CRC Press: Boca Raton, FL, USA), pp. 123–135.

Ellstrand NC and Foster KW (1983) Impact of population structure on the apparent outcrossing rate of grain sorghum (*Sorghum bicolor*). *Theor. Appl. Genet.* 66: 323–327.

Endrizzi JE (1957) Cytological studies of some species and hybrids in the *Eu-Sorghums. Bot. Gaz.* 119: 1–10.

FAO [Food and Agriculture Organization of the United Nations] (2009) FAOSTAT data. http://faostat.fao.org/site/567/default.aspx.

Fellows GM and Roeth FW (1992) Factors influencing shattercane (*Sorghum bicolor*) seed survival. *Weed Sci.* 40: 434–440.

Gao Z, Xie X, Ling Y, Muthukrishnan S, and Liang GH (2005) *Agrobacterium tumefaciens*-mediated sorghum transformation using a mannose selection system. *Plant Biotech. J.* 3: 591–599.

Garber ED (1950) Cytotaxonomic studies in the genus *Sorghum. Univ. Calif. Pub. Bot.* 23: 283–361.

Girijashankar V, Sharma HC, Sharma KK, Swathisree V, Prasad LS, Bhat BV, Royer M, Secundo BS, Narasu ML, Altosaar I, and Seetharama N (2005) Development of transgenic sorghum for insect resistance against the spotted stem borer (*Chilo partellus*). *Plant Cell Rep.* 24: 513–522.

Godwin ID (2005) Sorghum genetic engineering: Current status and prospectus. In: Seetharama N and Godwin ID (eds.) *Sorghum Tissue Culture and Transformation* (Oxford & IBH: New Delhi, India), pp. 1–8.

Gómez C and Espadaler X (1998) Myrmecochorous dispersal distances: A world survey. *J. Biogeogr.* 25: 573–580.

Hadley HH (1958) Chromosome numbers, fertility, and rhizome expression of hybrids between grain sorghum and Johnsongrass. *J. Agron.* 50: 278–282.

Harlan JR (1975) Geographic patterns of variation in some cultivated plants. *J. Hered.* 66: 184–191.

——— (1992) *Crops and Man.* 2nd ed. (American Society of Agronomy: Madison, WI, USA). 284 p.

Harlan JR and de Wet JMJ (1971) Towards a rational classification of cultivated plants. *Taxon* 20: 509–517.

——— (1972) A simplified classification of cultivated sorghum. *Crop Sci.* 12: 172–176.

Hodnett GL, Burson BL, Rooney WL, Dillon SL, and Price HJ (2005) Pollen-pistil interactions result in reproductive isolation between *Sorghum bicolor* and divergent *Sorghum* species. *Crop Sci.* 45: 1403–1409.

Hoffman ML, Buhler DD, and Regnier EE (2002) Utilizing *Sorghum* as a functional model of crop-weed competition: II. Effects of manipulating emergence time or rate. *Weed Sci.* 50: 473–478.

Holm LG, Plucknett DL, Pancho JV, and Herberger JP (1977) *The World's Worst Weeds: Distribution and Biology* (University Press of Hawaii: Honolulu, HI, USA). 609 p.

Howe A, Sato S, Dweikat I, Fromm M, and Clemente T (2006) Rapid and reproducible *Agrobacterium*-mediated transformation of sorghum. *Plant Cell Rep.* 25: 751–758.

Immelman K and Eardley C (2000) Gathering of grass pollen by solitary bees (Halictidae, Lipotriches) in South Africa. *Mitt. Mus. Naturk. Berlin, Zool. Reihe* 76: 263–268.

Isakeit T, Odvody GN, and Shelby RA (1998) First report of sorghum ergot caused by *Claviceps africana* in the United States. *Plant Dis.* 82: 592.

Janaki-Ammal EK and Singh TSN (1936) A preliminary note on a new *Saccharum-Sorghum* hybrid. *Indian J. Agric. Sci.* 6: 1105–1106.

Jeoung JM, Krishnaveni S, Muthukrishnan S, Trich HN, and Liang GH (2002) Optimization of sorghum transformation parameters using genes for green fluorescent protein and β-glucuronidase as visual markers. *Hereditas* 137: 20–28.

Kegode GO and Pearce RB (1998) Influence of environment during maternal plant growth on dormancy of shattercane (*Sorghum bicolor*) and giant foxtail (*Setaria faberi*) seed. *Weed Sci.* 46: 322–329.

Lansac AR, Sullivan CY, Johnson BE, and Lee KW (1994) Viability and germination of the pollen of sorghum (*Sorghum bicolor* (L.) Moench). *Ann. Bot.* (Oxford) 74: 27–33.

Leguizamón ES (1986) Seed survival and patterns of seedling emergence in *Sorghum halepense* (L.) Pers. *Weed Res.* 26: 397–404.

Mann JA, Kimber CT, and Miller FR (1983) *The Origin and Early Cultivation of Sorghums in Africa*. Texas Agricultural Experimental Station Bulletin No. 1454 (Texas Agricultural Experimental Station: College Station, TX, USA).

Manzelli M, Pileri L, Lacerenza N, Benedettelli S, and Vecchio V (2007) Genetic diversity assessment in Somali sorghum (*Sorghum bicolor* (L.) Moench) accessions using microsatellite markers. *Biodiv. Cons.* 16: 1715–1730.

McWhorter CG (1971) Introduction and spread of Johnsongrass in the United States. *Weed Sci.* 19: 496–500.

Morden CW, Doebley J, and Schertz KF (1990) Allozyme variation among the spontaneous species of *Sorghum* section *Sorghum* (Poaceae). *Theor. Appl. Genet.* 80: 296–304.

Morrell PL, Williams-Coplin TD, Lattu AL, Bowers JE, Chandler JM, and Paterson AH (2005) Crop-to-weed introgression has impacted allelic composition of Johnsongrass populations with and without recent exposure to cultivated sorghum. *Mol. Ecol.* 14: 2143–2154.

Nair NV (1999) Production and cyto-morphological analysis of intergeneric hybrids of *Sorghum* × *Saccharum*. *Euphytica* 108: 187–191.

Nair NV, Selvi A, Sreenivasan TV, Pushpalatha KN, and Sheji M (2006) Characterization of intergeneric hybrids of *Saccharum* using molecular markers. *Genet. Resour. Crop Evol.* 53: 163–169.

OECD [Organization for Economic Cooperation and Development] (2008) OECD Scheme for the Varietal Certification of Maize and Sorghum Seed Moving in International Trade: Annex XI to the Decision. C(2000)146/FINAL. OECD Seed Schemes. February 2008. www.oecd.org/document/0/0,3343,en_2649_3 3905_1933504_1_1_1_1,00.html.

Paterson AH, Schertz KF, Lin YR, Liu SC, and Chang YL (1995) The weediness of wild plants: Molecular analysis of genes influencing dispersal and persistence of Johnsongrass, *Sorghum halepense* (L.) Pers. *Proc. Natl. Acad. Sci. USA* 92: 6127–6131.

Paterson AH, Schertz KF, Lin YR, and Li Z (1998) Case history in plant domestication: Sorghum, an example of cereal evolution. In: Paterson A (ed.) *Molecular Dissection of Complex Traits* (CRC Press: Boca Raton, FL, USA), pp. 187–195.

Pedersen J, Toy J, and Johnson B (1998) Natural outcrossing of sorghum and sudangrass in the central Great Plains. *Crop Sci.* 38: 937–939.

Peters M, Oberrath R, and Bohning-Gaese K (2003) Seed dispersal by ants: Are seed preferences influenced by foraging strategies or historical constraints? *Flora* 198: 413–420.

Price HJ, Hodnett GL, Burson BL, Dillon SL, Stelly DM, and Rooney WL (2006) Genotype-dependent interspecific hybridization of *Sorghum bicolor*. *Crop Sci.* 46: 2617–2622.

Pritchard AJ (1965) Cytological and genetical studies on hybrids between *Sorghum almum* Parodi ($2n = 40$) and some diploid ($2n = 20$) species of sorghum. *Euphytica* 14: 307–314.

Schertz KF and Dalton LG (1980) Sorghum. In: Fehr W and Hadley HH (eds.) *Hybridization of Crop Plants* (American Society of Agronomy: Madison, WI, USA), pp. 577–488.

Schmidt MR and Bothma GC (2005) Indications of bee pollination in *Sorghum* and its implications in transgenic biosafety. *Int. Sorghum Millet Newsl. (ICRISAT)* 46: 72–75.

————— (2006) Risk assessment for transgenic sorghum in Africa: Crop-to-crop gene flow in *Sorghum bicolor* (L.) Moench. *Crop Sci.* 46: 790–798.

Snow AA, Sweeney PM, Grenier C, Kapran I, Tesso T, Bothma G, Ejeta G, and Pedersen JF (2007) Lifetime fecundity of F1 crop-wild sorghum hybrids: Implications for gene flow from transgenic sorghum in Africa. *Conference Program and Abstract Book: Plant Biology and Botany 2007 Joint Congress, Chicago, Illinois, July 7–11, 2007* (Botanical Society of America: St. Louis, MO, USA). http://2007.botanyconference.org/engine/search/index.php?func=detail&aid=1946.

Stephens JC and Quinby JR (1934) Anthesis, pollination, and fertilization in sorghum. *J. Agric. Res.* 49: 123–136.

Sun Y, Suksayretrup K, Kirkham MB, and Liang GH (1991) Pollen tube growth in reciprocal interspecific pollinations of *Sorghum bicolor* and *S. versicolor*. *Plant Breed.* 107: 197–202.

Tadesse Y, Sagi L, Swennen R, and Jacobs M (2003) Optimisation of transformation conditions and production of transgenic sorghum (*Sorghum bicolor*) via microparticle bombardment. *Plant Cell Tissue Organ Cult.* 75: 1–18.

Teo-Sherrell CPA and Mortensen DA (2000) Fates of buried *Sorghum bicolor* subsp. *drummondii* seed. *Weed Sci.* 48: 549–554.

Tesso TT, Kapran I, Grenier C, Snow AA, Sweeney PM, Pedersen JF, Marx D, Bothma G, and Ejeta G (2008) The potential for crop-to-wild gene flow in sorghum in Ethiopia and Niger: A geographic survey. *Crop Sci.* 48: 1425–1431.

Thomas R and Venkatraman TS (1930) Sugarcane-sorghum hybrids. *Agric. J. India* 25: 164.

Tuinstra MR and Wedel J (2000) Estimation of pollen viability in grain sorghum. *Crop Sci.* 40: 968–970.

USDA [United States Department of Agriculture] (2008a) 7 CFR [Code of Federal Regulations] § 201.76. Federal Seed Act Regulations: Minimum Land, Isolation, Field, and Seed Standards. Source: 59 FR [Federal Register] 64516, Dec. 14, 1994, as amended at 65 FR 1710, Jan. 11, 2000 (Government Printing Office: Washington, DC, USA), pp. 372–378. www.access.gpo.gov/nara/cfr/wais idx_08/7cfr201_08.html.

————— (2008b) Field test release applications in the U.S. Information Systems for Biotechnology [ISB] environmental releases database. www.isb.vt.edu/cfdocs/fieldtests1.cfm.

Vavilov NI (1951) The origin, variation, immunity, and breeding of cultivated plants. *Chron. Bot.* 13: 1–366.

Warwick SI and Black LD (1983) The biology of Canadian weeds: 61. *Sorghum halepense* (L.) Pers. *Can. J. Plant Sci.* 63: 997–1014.

Warwick SI, Phillips D, and Andrews C (1986) Rhizome depth: The critical factor in winter survival of *Sorghum halepense* (L.) Pers. (Johnsongrass). *Weed Res.* 26: 381–388.

Wiersema JH and Dahlberg J (2007) The nomenclature of *Sorghum bicolor* (L.) Moench (Gramineae). *Taxon* 56: 941–946.

Yim KO and Bayer DE (1997) Rhizome expression in a selected cross in the *Sorghum* genus. *Euphytica* 94: 253–256.

Zhao ZY, Cai T, Tagliani L, Miller M, Wang N, Pang H, Rudert M, Schroeder S, Hondred D, Seltzer J, and Pierce D (2000) *Agrobacterium*-mediated sorghum transformation. *Plant Mol. Biol.* 44: 789–798.

TWENTY

Soybean
(*Glycine max* (L.) Merr.)

Center(s) of Origin and Diversity

Soybeans originated in N and central China (Fukuda 1933; Hymowitz 1970; Zhuang 1999). Their primary center of diversity is in China, and it extends to Korea, Japan, and the Far East region of the Russian Federation. Many other Asian countries (such as Afghanistan, Cambodia, Laos, Myanmar, and Vietnam) also harbor a rich diversity of this crop (Lu 2004).

Biological Information

Flowering

Soybean flowers—like those of most legumes—have petals that almost entirely enclose the male and female organs. The stigma usually becomes receptive a day or two before the flower opens, and pollen is shed from the anthers either the evening before or in the morning when the flower opens (Williams 1950; H. Erickson 1975; Palmer et al. 1978; Carlson and Lersten 1987). Fertilization occurs within 10 hours from the time the flower opens (Fehr 1980).

The flowering periods for domesticated soybeans fluctuate with genotype, and some soybean cultivars have been found to overlap with wild soybeans for approximately 30 days in Japan (Fukui and Kaizuma 1971; Nakayama and Yamaguchi 2002). Since wild soybeans tend to flower later than cultivated soybeans, a flowering overlap is more likely to occur in regions where late-flowering soybean cultivars are grown.

Pollen Dispersal and Longevity

The main pollinators of domesticated soybeans are bee species, such as honeybees, carpenter bees, leaf-cutter bees (*Megachile turugensis* Cockerell) and halictid bees (*Halictus* spp.) (Juliano 1977; E. Erickson 1984; Fujita et al. 1997; Nakayama and Yamaguchi 2002; Chiari et al. 2005). Some cultivars are frequently visited by pollen-feeding thrips and predatory *Hemiptera* species, which can also act as pollinators (Miyazaki and Kudo 1988; Namai 1989; Nakayama and Yamaguchi 2002; Yoshimura et al. 2006).

As with most self-pollinating legumes, the longevity of soybean pollen is rather short, and it usually does not last for more than two to four hours.

Sexual Reproduction

The domesticated soybean plant is an annual allotetraploid ($2n = 4x = 40$; GG genome) (Gurley et al. 1979; Skorupska et al. 1989). The plants are self-fertile and highly autogamous, with natural outcrossing rates below 1% (Woodworth 1922; Poehlmann 1959; Caviness 1966; McGregor 1976; Carlson and Lersten 1987). Wild soybeans (*G. soja* Siebold & Zucc.), like their domesticated relatives, are self-pollinating, with mostly cleistogamous flowers (i.e., self-fertilization occurs before the flowers open). However, natural outcrossing rates can vary from less than 5% (Kiang et al. 1992; Kuroda et al. 2006) to up to 20% in some wild populations in Japan (Fujita et al. 1997; Ohara and Shimamoto 2002).

Vegetative Reproduction

There are no reports of vegetative reproduction in soybeans.

Seed Dispersal and Dormancy

Soybean seeds are dispersed mechanically, via explosive dehiscent pods (Oka 1983); they can then be transported by water and, occasionally, by birds. They can also be dispersed during mechanical planting, harvesting, handling, storage, and transportation operations.

The seeds of domesticated soybeans usually do not display dormancy characteristics, and they germinate quickly if soil temperatures are above 10°C (TeKrony et al. 1987).

Volunteers, Ferals, and Their Persistence

Volunteer and feral soybean plants sprout now and then from shattered seed. The crop is frost intolerant, so volunteers and ferals are likely to be frost killed during autumn or early winter in the year they were produced (Raper and Kramer 1987).

Weediness and Invasiveness Potential

Domesticated soybeans do not evince weedy traits, and they are not found outside of cultivation. They do not compete well with other cultivated or wild plants.

Crop Wild Relatives

The genus *Glycine* Willd. includes about 20 annual and perennial species, distributed for the most part in Australia and Asia (Lu 2004). It is divided into two subgenera, subg. *Soja* (Moench) F.J. Herm. (annual) and subg. *Glycine* (perennial).

Domesticated soybeans (*Glycine max* (L.) Merr.) are included in subg. *Soja*, together with their closest wild relatives, which are found in E Asia (China, Korea, Japan, and the Far East region of the Russian Federation) (Skvortzow 1927). *Glycine* subg. *Glycine* is comprised of the roughly 15 perennial wild relatives of soybeans. The majority are both native to and restricted to Australia, but the geographic range of some species extends to SE Asia, and even as far north as Japan (table 20.1).

The taxonomic boundaries of the wild soybean (*Glycine soja*) complex have not been conclusively resolved, and one to three closely related, interfertile wild relatives (*G. soja*, *G. gracilis* Skvortsov, and *G. formosana* Hosok.) have been delineated (Hymowitz et al. 1998).

Cytological, morphological, and molecular evidence suggest that the annual wild soybean, *G. soja*, is the immediate ancestor of the domesticated soybean *G. max* (Hymowitz 1970; Singh et al. 2001). The

Table 20.1 Domesticated soybean (*Glycine max* (L.) Merr.) and its crop wild relatives

Species	Ploidy	Genome	Origin and distribution
Primary gene pool (GP-1)			
subg. *Soja* (Moench) F.J. Herm.			
G. *max* (L.) Merr.	$2n = 4x = 40$	GG	annual; native to Asia; known only from cultivation
G. *soja* Siebold & Zucc. complex (including G. *gracilis* Skvortsov and G. *formosana* Hosok.)	$2n = 4x = 40$	GG	annual; wild ancestor; native to China, the Russian Federation, Taiwan, Japan, Korea
Tertiary gene pool (GP-3)			
subg. *Glycine*			
G. *albicans* Tindale & Craven	$2n = 40$	II	Australia
G. *aphyonota* B.E. Pfeil*	$2n = 40$	n.a.	Australia
G. *arenaria* Tindale	$2n = 40$	HH	Australia
G. *argyrea* Tindale	$2n = 40$	A_2A_2	Australia
G. *canescens* F.J. Herm.	$2n = 40$	AA	Australia
G. *clandestine* J.C. Wendl.	$2n = 40$	A_1A_1	Australia, S Pacific Islands
G. *curvata* Tindale	$2n = 40$	C_1C_1	Australia
G. *cyrtoloba* Tindale	$2n = 40$	CC	Australia
G. *falcata* Benth.	$2n = 40$	FF	Australia
G. *hirticaulis* Tindale & Craven	$2n = 40, 80$	H_1H_1	Australia
G. *lactovirens* Tindale & Craven	$2n = 40$	I_1I_1	Australia
G. *latifolia* (Benth.) C. Newell & Hymowitz	$2n = 40$	B_1B_1	Australia
G. *latrobeana* (Meisn.) Benth.	$2n = 40$	A_3A_3	Australia
G. *microphylla* (Benth.) Tindale	$2n = 40$	BB	Australia
G. *peratosa* B.E. Pfeil & Tindale*	$2n = 40$	n.r.	Australia
G. *pindanica* Tindale & Craven	$2n = 40$	H_2H_2	Australia
G. *pullenii* B.E. Pfeil et al.*	$2n = 40$	n.a.	Australia

G. *rubiginosa* Tindale & B.E. Pfeil*	$2n = 40$	n.a.	Australia
G. *tabacina* (Labill.) Benth.	$2n = 40, 80$	**AA, B$_2$B$_2$**	Australia, Pacific Islands, Philippines, Miyako Island (Japan), S China (including Fujian), Taiwan
G. *tomentella* Hayata	$2n = 38, 40, 78, 80$	**DD, EE, AADD**	Australia, Papua New Guinea, Philippines, Taiwan

Sources: Hymowitz et al. 1998; Hymowitz 2004.

Note: The domesticated form is in bold. Genome designations are according to Hymowitz et al. 1998 and Hymowitz 2004. Gene pool delimitations are according to Hymowitz 2004. n.a. = not available. n.r. = not recorded.

*Species newly described by Pfeil et al. 2001 and Pfeil and Craven 2002.

weedy or semi-wild form, *G. gracilis*, shows intermediate phenotypic traits between those of *G. max* and *G. soja*, and its taxonomic status has thus far not been categorically determined. Some authors consider *G. gracilis* to be an intermediate in the speciation of *G. max* from *G. soja* (Fukuda 1933), while others see it as the weedy form of domesticated soybeans that arose from hybridization between *G. soja* and *G. max* (Hymowitz 1970; Broich and Palmer 1980; Lackey 1981; Lu 2004). Sometimes a third wild species, the Taiwanese endemic *G. formosana*, is mentioned (Shimamoto 2000).

This chapter will only distinguish between domesticated soybeans (*G. max*) and the annual wild soybean (*G. soja*) complex.

Hybridization

Under natural conditions, domesticated soybeans can only hybridize with wild relatives within the same subgenus, that is, subg. *Soja*. These species comprise the primary gene pool of soybeans (table 20.1). A secondary gene pool does not exist (Hymowitz 2004).

Primary Gene Pool (GP-1)

Domesticated soybeans and their annual wild relatives in the *Glycine soja* complex are fully cross-compatible and hybridize with ease (Oka 1983; Singh and Hymowitz 1989; Hymowitz 1995; Yuichiro and Hirofumi 2002). Under controlled greenhouse conditions, mean hybridization rates of more than 3% have been reported between domesticated and wild soybeans (Dorokhov et al. 2004), but they drop (< 1%) under natural conditions (Manabe et al. 1996; Nakayama and Yamaguchi 2002; Yuichiro and Hirofumi 2002). Hybridization is less successful (0.7% versus 12.8%) when domesticated soybeans are used as the pollen source, rather than vice versa (Dorokhov et al. 2004).

Viable, fully fertile F1, F2, and BC hybrids can be obtained with relative ease from crosses between *G. max* and *G. soja* (see, e.g., Karasawa 1936; Tang and Chen 1959; Tang and Tai 1962; Ahmad et al. 1977; Newell and Hymowitz 1983; Singh and Hymowitz 1989; Zhuang 1999), and between *G. max* and *G. gracilis* (Karasawa 1952; Hadley and Hymowitz 1973; Singh and Hymowitz 1989; Zhuang 1999). Traits of agronomic interest have been effectively transferred from *G. soja* to domesticated soybeans (Leroy et al. 1991; Concibido et al. 2003), and there is molecular evidence for natural introgression of domesticated soybean alleles into wild populations (Abe et al. 1999; Kuroda et al. 2006).

Tertiary Gene Pool (GP-3)

The tertiary gene pool includes the perennial wild soybean relatives in subg. *Glycine*. Crosses between these species and domesticated soybeans are nearly impossible without the use of artificial hybridization techniques, due either to the pods' failure to develop or to pod abscission in the early stages of growth (Ahmad et al. 1977; Ladizinsky et al. 1979; Singh et al. 1987).

Hybrids can be obtained *in vitro* via embryo rescue—for instance between *G. max* and *G. clandestina* J.C. Wendl. (Singh et al. 1987); between *G. max* and *G. tomentella* Hayata (Newell and Hymowitz 1982; Singh and Hymowitz 1985; Newell et al. 1987; Singh et al. 1987, 1990); and between *G. max* and *G. canescens* F.J. Herm., using transplanted endosperm as a nurse layer (Broué et al. 1982). In all cases, it is very difficult to obtain progeny from such intersubgeneric hybrids; moreover, they are completely sterile, due to poor chromosome pairing.

Hybridization between domesticated soybeans and GP-3 species—which results in sterile hybrid progeny—only seems to be possible with considerable human intervention. Research has also demonstrated that soybeans do not hybridize with any species outside of the genus *Glycine* (Hymowitz and Singh 1987).

Outcrossing Rates and Distances

Pollen Flow

Outcrossing rates in the mainly autogamous domesticated soybean plant have been shown to be below 1% (see, e.g., Garber and Odland 1926; Weber and Hanson 1961; Caviness 1966; Sediyama et al. 1970; Beard and Knowles 1971; Vernetti et al. 1972; Carlson and Lersten 1987; Chiang and Kiang 1987). However, pollen flow and outcrossing

Table 20.2 Studies on pollen dispersal distances and outcrossing rates of domesticated soybean (*Glycine max* (L.) Merr.)

Pollen flow from	Pollen dispersal distance (m)	% outcrossing (max. values)	Country	Reference
G. max	0.9, 5.4	0.41, 0.05	USA	Ray et al. 2003
G. max	0.5, 1.0, > 6.5	0.45, 1.4, 0.0	Brazil	Abud et al. 2003
G. max	0.7, 2.1, > 7.0	0.19, 0.052, 0.0	Japan	Yoshimura et al. 2006
G. max	0.9, 4.6, 8.2, 15.5	0.44, 0.04, 0.02, 0.01	USA	Caviness 1966

frequency are influenced by intrinsic and external factors, such as genotype (cultivar), climatic conditions, and the presence and abundance of pollinators (Beard and Knowles 1971; E. Erickson et al. 1978; Fujita et al. 1997; Palmer et al. 2001). Under favorable conditions, outcrossing rates often exceed 2.5% (see, e.g., Cutler 1934; Beard and Knowles 1971; Gumisiriza and Rubaihayo 1978; Ahrent and Caviness 1994) and, when an abundant number of pollinators are present, they can climb to 7% (Abrams et al. 1978; Ray et al. 2003) or even 20% (Palmer et al. 2001).

The outcrossing levels found in wild soybeans are also highly variable. They tend to be higher than those for domesticated soybeans and can reach up to 20% (Fujita et al. 1997; Ohara and Shimamoto 2002).

Pollen flow in soybeans, like that in other crops, follows a leptokurtic distribution, characterized by rare pollen dispersal events over long distances. Outcrossing rates decrease quickly the greater the distance from the pollen source becomes, dropping to less than 1.5% beyond 1 m and less than 0.1% beyond 2 m (table 20.2). In most studies, no outcrossing could be detected beyond 7 m (Boerma and Moradshahi 1975; Abud et al. 2003; Yoshimura et al. 2006), and the longest distance at which pollen flow was detected was 15 m (Caviness 1966).

Isolation Distances

The separation distances established by regulatory authorities for soybean seed production range from 0 to 3 m in Canada, Europe, and the United States (EC 2002; USDA 2008; CSGA 2009). In OECD countries, soybean varieties for seed production "shall be isolated from other crops by a definite barrier or a space sufficient to prevent mixture during harvest" (OECD 2008, p. 18). These recommendations were created primarily to avoid mechanical mixtures of adjacent varieties or crops in the field. Although they are not intended to prevent occasional insect-vectored cross-pollination, the few pollen-flow studies available thus far (table 20.2) suggest that these distances could be sufficient to maintain respective seed purity standards and remain under the maximum allowable amount of GM contamination (0.9%). Still, more studies at field scale would be useful to better understand the dynamics of soybean pollen flow and the frequencies of outcrossing at different distances.

State of Development of GM Technology

Genetically modified soybeans are commercially grown on approximately 65.8 million ha in Argentina, Bolivia, Brazil, Canada, Chile, Mexico, Paraguay, Romania, South Africa, Uruguay, and the United States (James 2008). This represents nearly 70% of the global soybean production area. In addition, there are field trials in many other countries, such as China, France, Germany, Italy, Japan, the Russian Federation, and Spain.

All commercially planted GM soybean cultivars are herbicide-tolerant RR(r) soybeans (James 2008). Other experimentally tested GM traits are insect resistance (see, e.g., Dang and Wei 2007), enhanced nutritional quality, and abiotic stress tolerance.

Agronomic Management Recommendations

There are three main ways gene introgression to wild relatives of cultivated soybeans can occur:

1 through pollen flow from soybean fields

2 through volunteers in fields previously planted with soybeans

3 through soybean seed shattered during cropping, harvesting, handling, storage, and transportation of the crop

The following management recommendations should greatly reduce the risk of gene introgression:

- Consider land-use history, to avoid contamination with crop-derived volunteer populations left by an earlier cropping cycle—not only of the field to be planted, but also of neighboring fields.

- Consider establishing pollen-trapping barriers around the area planted with soybeans. Two or three border rows should effectively contain the majority of the pollen-mediated gene flow.

- In regions where wild annual soybeans occur, and if flowering overlap is likely, planting in the area should not be pursued at distances of less than 10 m.

- Although ferals are not likely to persist outside of cultivation, special measures should be taken when transporting soybean seeds to avoid the seeds falling from harvest vehicles. Monitor roads for volunteer soybean plants and, if present, control them by mechanical or chemical means.

Crop Production Area

Domesticated soybeans are grown in more than 90 countries worldwide, in an estimated area of about 95 million ha. The world's leading soybean producers are Argentina, Brazil, China, India, and the United States, which together account for 90% of total soybean production. In addition, Canada, Bolivia, and Paraguay grow significant amounts (FAO 2009).

Conservation Status of Soybean Wild Relatives and Protected Areas

Wild soybeans could once be found over nearly all of China's Yellow River delta and Sanjiang Plain, but now they are scattered in just a few sites in these areas (NEPA 1994).

Research Gaps and Conclusions

Gene flow from soybeans to their wild relatives is likely only in Asia, where wild annual soybeans occur in allopatry and sympatry with cultivated soybeans. Although the rates of and distances for pollen-mediated outcrossing of the crop are quite low, studies at the field scale would be useful to better understand the dynamics of soybean pollen flow and the outcrossing frequencies at diverse distances.

The likelihood of gene introgression from cultivated soybeans (*Glycine max*) to their wild relatives—assuming physical proximity (less than 3 m) and flowering overlap, and being conditional on the nature of the respective trait(s)—is as follows:

- *high* for annual wild soybeans belonging to the *Glycine soja* species complex (E Asia, map 20.1)

- *highly unlikely* for perennial wild soybeans belonging to subg. *Glycine*

REFERENCES

Abe J, Hasegawa A, Fukushi H, Mikami T, Ohara M, and Shimamoto Y (1999) Introgression between wild and cultivated soybeans of Japan revealed by RFLP analysis for chloroplast DNAs. *Econ. Bot.* 53: 285–291.

Abrams RI, Edwards CR, and Harris T (1978) Yields and cross-pollination of soybeans as affected by honey bees and alfalfa leaf-cutting bees. *Am. Bee J.* 118: 555–560.

Abud S, Souza PIM, Moreira CT, Andrade SRM, Ulbrich AV, Vianna GR, Rech EL, and Aragão FJL (2003) Gene flow in transgenic soybean in the Cerrado region, Brazil. *Pesq. Agropec. Bras.* (Brasilia) 38: 1229–1235.

Ahmad QN, Britten EJ, and Byth DE (1977) Inversion bridges and meiotic behaviour in species hybrids of soybeans. *J. Hered.* 68: 360–364.

Ahrent DK and Caviness CE (1994) Natural cross-pollination of twelve soybean cultivars in Arkansas. *Crop Sci.* 34: 376–378.

Beard BH and Knowles PF (1971) Frequency of cross-pollination of soybeans after seed irradiation. *Crop Sci.* 11: 489–492.

Boerma HR and Moradshahi A (1975) Pollen movement within and between rows to male-sterile soybean. *Crop Sci.* 15: 858–861.

Broich SL and Palmer R (1980) A cluster analysis of wild and domesticated soybean phenotypes. *Euphytica* 29: 23–32.

Broué P, Gouglass J, Grace JP, and Marshall DR (1982) Interspecific hybridization of soybeans and perennial *Glycine* species indigenous to Australia via embryo culture. *Euphytica* 31: 715–724.

Carlson JB and Lersten NR (1987) Reproductive morphology. In: Wilcox JR (ed.)

Soybeans: Improvement, Production, and Uses. 2nd ed. (American Society of Agronomy: Madison, WI, USA), pp. 95–134.

Caviness CE (1966) Estimates of natural cross-pollination in Jackson soybeans in Arkansas. *Crop Sci.* 6: 211.

Chiang YC and Kiang YT (1987) Geometric position of genotypes, honeybee foraging patterns, and outcrossing in soybean. *Bot. Bull. Acad. Sin.* 28: 1–11.

Chiari WC, Toledo VAA, Ruvolo-Takasusuki MCC, Oliveira AJB, Sakaguti ES, Attencia VM, Costa FM, and Mitsui MH (2005) Pollination of soybean (*Glycine max* (L.) Merrill) by honeybees (*Apis mellifera* L.). *Braz. Arch. Biol. Technol.* 48: 31–36.

Concibido VC, La Vallee B, Mclaird P, Pineda N, Meyer J, Hummel L, Yang J, Wu K, and Delannay W (2003) Introgression of a quantitative trait locus for yield from *Glycine soja* into commercial soybean cultivars. *Theor. Appl. Genet.* 106: 575–582.

CSGA [Canadian Seed Growers Association] (2009) Canadian Regulations and Procedures for Pedigreed Seed Crop Production. Circular 6–2005/Rev. 01.4–2009 (Canadian Seed Growers Association [CSGA]: Ottawa, ON, Canada). www .seedgrowers.ca/cropcertification/circular.asp.

Cutler GH (1934) A simple method for making soybean hybrids. *J. Am. Soc. Agron.* 26: 252–254.

Dang W and Wei ZM (2007) An optimized *Agrobacterium*-mediated transformation for soybean for expression of binary insect resistance genes. *Plant Sci.* 173: 381–389.

Dorokhov D, Ignatov A, Deineko E, Serjapin A, Ala A, and Skryabin K (2004) The chance for gene flow from herbicide-resistant GM soybean to wild soy in its natural inhabitation at Russian Far East region. In: den Nijs HCM, Bartsch D, and Sweet J (eds.) *Introgression from Genetically Modified Plants into Wild Relatives* (CAB International: Wallingford, UK), pp. 151–161.

EC [European Commission] (2002) Council Directive 2002/57/EC of 13 June 2002 on the Marketing of Seed of Oil and Fibre Plants. OJ No. L 193, 20.7.2002. www.legaltext.ee/text/en/U70003.htm.

Erickson EH (1984) Soybean pollination and honey production—a research progress report. *Am. Bee J.* 124: 775–779.

Erickson EH, Berger GA, Shannon JG, and Robbins JM (1978) Honey bee pollination increases soybean yields in the Mississippi Delta region of Arkansas and Missouri. *J. Econ. Entomol.* 71: 601–603.

Erickson HT (1975) Variability of floral characteristics influences honey bee visitation to soybean blossoms. *Crop Sci.* 15: 767–771.

FAO [Food and Agriculture Organization of the United Nations] (2009) FAOSTAT Data. http://faostat.fao.org/site/567/default.aspx.

Fehr WR (1980) Soybean. In: Fehr W and Hadley HH (eds.) *Hybridization of Crop Plants* (American Society of Agronomy: Madison, WI, USA), pp. 589–599.

Fujita R, Ohara M, Okazaki K, and Shimamoto Y (1997) The extent of natural pollination in wild soybean (*Glycine soja*). *J. Hered.* 76: 373–374.

Fukuda Y (1933) Cytological studies on the wild and cultivated Manchurian soybeans (*Glycine* L.). *Japan. J. Bot.* 6: 489–506.

Fukui J and Kaizuma N (1971) Morphological and ecological differentiation of various characters of the Japanese wild soybean, *Glycine soja*. *J. Fac. Agr. Iwate Univ.* 10: 195–208 [English abstract].

Garber RJ and Odland TE (1926) Natural crossing in soybeans. *J. Am. Soc. Agron.* 18: 967–970.

Gumisiriza G and Rubaihayo PR (1978) Factors that influence outcrossing in soybean. *Acker und Pflanzenbau* 147: 129–133.

Gurley WB, Hepburn AG, and Key JL (1979) Sequences organization of the soybean genome. *Biochem. Biophys. Acta* 561: 167–183.

Hadley HH and Hymowitz T (1973) Speciation and cytogenetics. In: Caldwell BE, Howell RW, Judd RW, and Johnson HW (eds.) *Soybeans: Improvement, Production, and Uses* (American Society of Agronomy: Madison, WI, USA), pp. 97–116.

Hymowitz T (1970) On the domestication of the soybean. *Econ. Bot.* 24: 408–421.

——— (1995) Soybean. In: Simmonds S and Smartt J (eds.) *Evolution of Crop Plants.* 2nd ed. (Longman Scientific and Technical: Harlow, UK), pp. 261–266.

——— (2004) Speciation and cytogenetics. In: Boerma HR and Specht JE (eds.) *Soybeans: Improvement, Production, and Uses.* 3rd ed. Agronomy Monograph 16 (American Society of Agronomy: Madison, WI, USA), pp. 97–136.

Hymowitz T and Singh RJ (1987) Taxonomy and speciation. In: Wilcox JR (ed.) *Soybeans: Improvement, Production, and Uses.* 2nd ed. (American Society of Agronomy: Madison, WI, USA), pp. 23–48.

Hymowitz T, Singh RJ, and Kollipara KP (1998) The genomes of the *Glycine. Plant Breed. Rev.* 16: 289–317.

James C (2008) *Global Status of Commercialized Biotech/GM Crops 2008.* ISAAA Brief 39 (The International Service for the Acquisition of Agri-biotech Applications [ISAAA]: Ithaca, NY, USA). 243 p.

Juliano JC (1977) *Entomophilus* pollination of soybean. In: Gonçalves LS (ed.) *Anais do 40. Congresso Brasileiro de Apicultura, Realizado em Curitiba, PR, no Centro de Treinamento do Magistério do Estado do Paraná (CETEPAR), [September] 9–11, 1976* (s.n.: Riberão Preto, SP, Brazil), pp. 235–239.

Karasawa K (1936) Crossing experiments with *Glycine soja* and *G. ussuriensis. Japan. J. Bot.* 8: 113–118.

——— (1952) Crossing experiments with *Glycine soja* and *G. gracilis. Genetica* 26: 357–358.

Kiang YT, Chiang YC, and Kaizuma N (1992) Genetic diversity in natural populations of wild soybean in Iwate Prefecture, Japan. *J. Hered.* 83: 325–329.

Kuroda Y, Kaga A, Tomooka N, and Vaughan DA (2006) Population genetic structure of Japanese wild soybean (*Glycine soja*) based on microsatellite variation. *Mol. Ecol.* 15: 959–974.

Lackey JA (1981) Phaseoleae. In: Polhill RM and Raven PH (eds.) *Advances in Legume Systematics, Part 1* (Royal Botanic Gardens, Kew: Richmond, Surrey, UK), pp. 301–327.

Ladizinsky G, Newell CA, and Hymowitz T (1979) Wild crosses in soybeans: Prospects and limitations. *Euphytica* 28: 421–423.

Leroy AR, Fehr WR, and Cianzio SR (1991) Introgression of genes for small seed size from *Glycine soja* into *G. max. Crop Sci.* 31: 693–697.

Lu B-R (2004) Conserving biodiversity of soybean gene pool in the biotechnology era. *Plant Spec. Biol.* 19: 115–125.

Manabe T, Kashiwara Y, Yamada S, Harada J, and Matsumura T (1996) Environmental safety evaluation of transgenic soybean with glyphosate-tolerance in the environmentally isolated field. *Breed. Sci.* 46: 261 [in Japanese].

McGregor SE (1976) *Insect Pollination of Cultivated Crop Plants.* U.S. Department of Agriculture, Agricultural Research Service, Agriculture Handbook 496 (Government Printing Office: Washington, DC, USA). 411 p.

Miyazaki M and Kudo I (1988) Feeding habit of thrips. In: Umeya K, Kudo I, and Miyazaki M (eds.) *Pest Thrips in Japan* (Zenkoku Noson Kyoiku Kyokai Publishing: Tokyo, Japan), pp. 62–76 [in Japanese].

Nakayama Y and Yamaguchi H (2002) Natural hybridization in wild soybean (*Glycine max* subsp. *soja*) by pollen flow from cultivated soybean (*Glycine max* subsp. *max*) in a designed population. *Weed Biol. Managem.* 2: 25–30.

Namai H (1989) Pollinators and pollination, fertilization, and fructification. In: Matsuo T (ed.) *Collected Data of Plant Genetic Resources*, vol. 1 (Kodansha Scientific: Tokyo, Japan), pp. 297–302 [in Japanese].

NEPA [National Environmental Protection Agency] (1994) *China: Biodiversity Conservation Action Plan* (China Environmental Science Press: Beijing, China), final version edited by C Maxey and J Lutz. 106 p. http://bpsp-neca.brim.ac.cn/books/actpln_cn/index.html.

Newell CA and Hymowitz T (1982) Successful wide hybridization between the soybean and a wild perennial relative, *G. tomentella* Hayata. *Crop Sci.* 22: 1062–1065.

———— (1983) Hybridization in the genus *Glycine* subgenus *Glycine* Willd. (Leguminosae, Papilionoideae). *Am. J. Bot.* 70: 334–348.

Newell CA, Delannay X, and Edge ME (1987) Interspecific hybrids between the soybean and wild perennial relatives. *J. Hered.* 78: 301–306.

OECD [Organization for Economic Cooperation and Development] (2008) OECD Scheme for the Varietal Certification of Grass and Legume Seed Moving in International Trade: Annex VI to the Decision. C(2000)146/FINAL. OECD Seed Schemes, February 2008. www.oecd.org/document/0/0,3343,en_2649_33905 _1933504_1_1_1_1,00.html.

Ohara M and Shimamoto Y (2002) Importance of genetic characterization and conservation of plant genetic resources: The breeding system and genetic diversity of wild soybean (*Glycine soja*). *Plant Spec. Biol.* 17: 51–58.

Oka HI (1983) Genetic control of regenerating success in semi-natural conditions observed among lines derived from a cultivated × wild soybean hybrid. *J. Appl. Ecol.* 20: 937–949.

Palmer RG, Albertsen MC, and Heer H (1978) Pollen production in soybeans with respect to genotype, environment, and stamen position. *Euphytica* 27: 427–433.

Palmer RG, Gai J, Sun H, and Burton JW (2001) Production and evaluation of hybrid soybean. *Plant Breed. Rev.* 21: 263–307.

Pfeil BE and Craven LA (2002) New taxa in *Glycine* (Fabaceae: Phaseolae) from north-western Australia. *Aust. Syst. Bot.* 15: 565–573.

Pfeil BE, Tindale MD, and Craven LA (2001) A review of the *Glycine clandestina* species complex (Fabaceae: Phaseolae) reveals two new species. *Aust. Syst. Bot.* 14: 891–900.

Poehlmann JM (1959) *Breeding of Field Crops* (Henry Holt: New York, NY, USA). 427 p.

Raper CD Jr and Kramer PJ (1987) Stress physiology. In: Wilcox JR (ed.) *Soybeans: Improvement, Production, and Uses.* 2nd ed. (American Society of Agronomy: Madison, WI, USA), pp. 589–641.

Ray JD, Kilen TC, Abel CA, and Paris RL (2003) Soybean natural cross-pollination rates under field conditions. *Environ. Biosafety Res.* 2: 133–138.

Sediyama T, Cardoso AA, Vieira C, and Andrade D (1970) Taxa de hibridação natural em soja, em Viçosa e em Capinópolis, Minas Gerais. *Revista Ceres* 2: 329–331.

Shimamoto Y (2000) Research on wild legumes with an emphasis on soybean germplasm. In: Oono K (ed.) *Wild Legumes: Proceedings of the 7th Ministry of Agriculture, Forestry, and Fisheries [MAFF], Japan, International Workshop on Genetic Resources, October 13–15, 1999* (National Institute of Agro-Biological Resources [NIAR]: Tsukuba, Japan), pp. 5–17.

Singh RJ and Hymowitz T (1985) An intersubgeneric hybrid between *Glycine tomentella* Hayata and the soybean, *G. max* (L.) Merr. *Euphytica* 34: 187–192.

―――― (1989) The genomic relationships between *Glycine soja* Sieb. and Zucc., *G. max* (L.) Merr., and '*G. gracilis*' Skvortz. *Plant Breed.* 103: 171–173.

Singh RJ, Kollipara KP, and Hymowitz T (1987) Intersubgeneric hybridization of soybeans with a wild perennial species *Glycine clandestina* Wendl. *Theor. Appl. Genet.* 74: 391–396.

——— (1990) Backcross-derived progeny from soybean and *Glycine tomentella* Hayata intersubgeneric hybrids. *Crop Sci.* 30: 871–874.

Singh RJ, Kim HH, and Hymowitz T (2001) Distribution of rDNA loci in the genus *Glycine* Willd. *Theor. Appl. Genet.* 103: 212–218.

Skorupska H, Albertsen MC, Langholz KD, and Palmer RG (1989) Detection of ribosomal RNA genes in soybean by *in situ* hybridization. *Genome* 32: 1091–1095.

Skvortzow BW (1927) The soy bean—wild and cultivated in Eastern Asia. *Proc. Manchurian Res. Soc., Nat. Hist. Sec. Publ. Ser. A* 22: 1–22.

Tang WT and Chen CH (1959) Preliminary studies on the hybridization of the cultivated and wild bean (*Glycine max* Merrill and *G. formosana* [Hosokawa]). *J. Agr. Assoc. China* 28: 7–23.

Tang WT and Tai G (1962) Studies on the qualitative and quantitative inheritance of an interspecific cross of soybean, *Glycine max* × *G. formosana*. *Bot. Bull. Acad. Sin.* 3: 39–54.

TeKrony DM, Egli DB, and White GM (1987) Seed production and technology. In: Wilcox JR (ed.) *Soybeans: Improvement, Production, and Uses.* 2nd ed. (American Society of Agronomy: Madison, WI, USA), pp. 295–354.

USDA [United States Department of Agriculture] (2008) 7 CFR [Code of Federal Regulations] § 201.76. Federal Seed Act Regulations: Minimum Land, Isolation, Field, and Seed Standards. Source: 59 FR [Federal Register] 64516, Dec. 14, 1994, as amended at 65 FR 1710, Jan. 11, 2000 (Government Printing Office: Washington, DC, USA), pp. 372–378. www.access.gpo.gov/nara/cfr/wais idx_08/7cfr201_08.html.

Vernetti FJ, Bonatto ER, Terazawa F, and Gastal MF (1972) Observações sobre a taxa de cruzamentos naturais em soja, em Pelotas e Sertão, RS e Ponta Grossa, PR. *Ciência e Cultura* 1: 36–41.

Weber CR and Hanson WD (1961) Natural hybridization with and without ionizing radiation in soybeans. *Crop Sci.* 1: 389–392.

Williams LF (1950) Structure and genetic characteristics of the soybean. In: Markley KS (ed.) *Soybeans and Soybean Products*, vol. 1 (Interscience: New York, NY, USA), pp. 111–134.

Woodworth CM (1922) The extent of natural cross-pollination in soybeans. *Agron. J.* 14: 278–283.

Yoshimura Y, Matsuo K, and Yasuda K (2006) Gene flow from GM glyphosate-tolerant to conventional soybeans under field conditions in Japan. *Environ. Biosafety Res.* 5: 169–173.

Yuichiro N and Hirofumi Y (2002) Natural hybridization in wild soybean (*Glycine max* subsp. *soja*) by pollen flow from cultivated soybean (*Glycine max* subsp. *max*) in a designed population. *Weed Biol. Managem.* 2: 25–30.

Zhuang BC (ed.) (1999) *Biological Studies of Wild Soybeans in China* (Science Press: Beijing, China) [in Chinese]. 323 p.

TWENTY ONE

Sweetpotato, Batata, Camote
(Ipomoea batatas (L.) Lam.)

Center(s) of Origin and Diversity

Cultivated sweetpotatoes (*I. batatas* (L.) Lam. var. *batatas*) are thought to have originated in Mesoamerica or northern S America, somewhere between Mexico's Yucatan peninsula and the mouth of the Orinoco River (Austin 1988b; McDonald and Austin 1990; Zhang et al. 2000).

The most important primary centers of diversity are in Central America (Guatemala, Mexico, and Nicaragua), northern South America (Colombia, Ecuador, and Peru), and Brazil (Austin 1988b; Bohac et al. 1995; Zhang et al. 2000). Secondary centers of sweetpotato diversity are found outside the Americas in China, E Africa, New Guinea, and SE Asia (Yen 1974, 1982; Austin 1983, 1988b).

Biological Information

Flowering

The flowering time and duration, flowering intensity, and seed set of sweetpotato cultivars are strongly influenced by genotype, photoperiod, and environmental stresses (Montelaro and Miller 1951; Jones 1980; Eguchi 1996).

The sweetpotato inflorescence is a compound cyme with up to 25 individual flowers opening singly or in groups (Jones 1966; Purseglove 1968). The bisexual flowers open at dawn and close and wilt by early afternoon of the same day (Onwueme 1978). The stigma is receptive for only a few hours in the morning (Jones 1980).

Pollen Dispersal and Viability

Insects, particularly bees, are the primary pollinators of sweetpotatoes (Purseglove 1968; F. Martin and Jones 1973; Jones 1980).

Not much is known about the survival of sweetpotato pollen, except that germination can continue for three to four hours after pollination (F. Martin and Cabanillas 1966). More systematic studies on the viability of sweetpotato pollen would be useful, especially under field conditions.

Sexual Reproduction

The cultivated sweetpotato plant is a hexaploid ($2n = 6x = 90$, **AB** genome), although wild tetraploid ($2n = 4x = 60$) forms have been found in Ecuador (Bohac et al. 1993). Thus far, no consensus has been reached as to whether the species is an autopolyploid (Shiotani and Kawase 1987, 1989; Shiotani 1988; Ukoskit and Thompson 1997) or an allopolyploid (Magoon et al. 1970; Yen 1976). The crop is allogamous and mostly self-incompatible, but some varieties show self-fertility (e.g., Stout 1924; Togari and Kawahara 1946; F. Martin and Ortiz 1967; F. Martin 1968).

Vegetative Reproduction

Sweetpotatoes are predominantly an outcrossing species, yet they can also reproduce vegetatively, either by stem or vine cuttings in the tropics or through adventitious root buds in temperate regions (Austin 1988b; Bohac et al. 1995).

Seed Dispersal and Dormancy

Despite flowering abundantly, most sweetpotato cultivars set seed poorly, because of their complex self-incompatibility system (F. Martin 1968, 1988; Onwueme 1978; Murata and Matsuda 2003). The main dispersal agents for sweetpotato seeds are birds and water (Yen 1960; Purseglove 1965; Bulmer 1966; Zhang et al. 2004).

Sweetpotatoes can show mechanical dormancy (J. Martin 1946), and the hard testa of its seeds requires scarification prior to germinating. The seeds remain viable for many years and can form seed banks in the soil, as has happened in Brazil (Martins 2001).

Volunteers, Ferals, and Their Persistence

Feral (4x) sweetpotatoes occur regularly in areas where herbicides are not used, particularly near rural farmers' fields (D.F. Austin, pers. comm. 2008).

Austin et al. (1993) located wild (4x) sweetpotato plants in Ecuador in the same areas where they had been collected decades before. This suggests that sweetpotatoes could survive for several years, propagating via seed or vegetatively.

Weediness and Invasiveness Potential

Under favorable conditions, sweetpotato vines grow vigorously and are able to compete with weeds, even in their early stages of development (Harris 1958). According to rural farmers, the wild tetraploid form is a serious weed that competes (and eventually outcrosses) with their cultivated sweetpotato plants (Austin et al. 1993). However, the crop itself is not considered to be a noxious weed.

Crop Wild Relatives

Ipomoea L. is a large genus that includes over 600 species worldwide (McDonald and Mabry 1992; Austin and Huáman 1996; Austin and Bianchini 1998). Cultivated sweetpotatoes (*I. batatas*) and their closest wild relatives form sect. *Eriospermum* ser. *Batatas* (Choisy) D.F. Austin, which is comprised of 13 species and 2 naturally occurring hybrids (table 21.1). Except for *I. littoralis* (L.) Blume (Austin 1991), all species are native to the New World, where they extend from the southern United States, throughout Central America and the Caribbean, to S America.

Most of the wild relatives of sweetpotatoes are diploids ($2x = 30$) and two are tetraploids. The crop *Ipomoea batatas* is a hexaploid, although a (wild) tetraploid cytotype ($2n = 4x = 60$) has been collected in Ecuador (Bohac et al. 1993). Two genomic subgroups (**A** and **B**) can be identified

Table 21.1 Sweetpotato (*Ipomoea batatas* (L.) Lam.) and its crop wild relatives in series *Batatas* (Choisy) D.F. Austin

Species	Genome group	Ploidy	Origin and distribution
I. batatas (L.) Lam. var. batatas (sweetpotato)	**B**	hexaploid (2*n* = 6*x* = 90)	S and C America; known only from cultivation
I. batatas var. apiculata (M. Martens & Galeotti) J.A. McDonald & D.F. Austin	**B**	allotetraploid (2*n* = 4*x* = 60)	wild; endemic to Mexico, spread along a small area of beach near Veracruz
I. cordatotriloba Dennst.	**A**	diploid (2*n* = 30)	native to USA, Mexico, Bolivia, and Argentina; naturalized throughout C America
I. cynanchifolia Meisn.	**A**	diploid (2*n* = 30)	native to Guyana and Brazil
I. × grandifolia (Dammer) O'Donell	**A**	diploid (2*n* = 30)	native to Brazil, Argentina, Paraguay, and Uruguay
I. lacunosa L.	**A**	diploid (2*n* = 30)	native to USA; naturalized elsewhere
I. × leucantha Jacq. (I. cordatotriloba × I. lacunosa)	**A**	diploid (2*n* = 30)	native to SE USA; naturalized in N Mexico, S America, Hawaii, and Philippines
I. littoralis (L.) Blume	**B**	diploid (2*n* = 30)	native to Pacific and Indian oceans
I. ramosissima (Poir.) Choisy	**A**	diploid (2*n* = 30)	C and S America
I. splendor-sylvae House	**A**	diploid (2*n* = 30)	C America
I. tabascana J.A. McDonald & D.F. Austin	**B**	allotetraploid (2*n* = 4*x* = 60)	endemic to Mexico
I. tenuissima Choisy	**A**	diploid (2*n* = 30)	Greater Antilles, S Florida, and Caribbean; probably extirpated in all formerly known sites except S Florida (D.F. Austin, pers. comm. 2008)
I. tiliacea (Willd.) Choisy	**A**	allotetraploid (2*n* = 4*x* = 60)	native to C and S America; naturalized elsewhere
I. trifida (Kunth) G. Don	**B**	diploid (2*n* = 30)	Caribbean
I. triloba L.	**A**	diploid (2*n* = 30)	native to Caribbean; pantropical distribution

Sources: The taxonomic classification follows Austin 1978,1979,1988a; McDonald and Austin 1990; Austin and Huáman 1996; and Austin and Bianchini 1998.

Note: The cultivated form is in bold. Genome designations are according to Jarret et al. 1992 and Rajapakse et al. 2004. Ploidy levels follow Jones 1974; Jarret et al. 1992, and Huáman and Zhang 1997.

in ser. *Batatas*, corresponding to their selfing abilities, interspecific crossing capabilities, and morphological and cytogenetic characteristics (Ting et al. 1957; Jones 1965; Jones and Deonier 1965; F. Martin and Jones 1973; Nishiyama and Sakamoto 1975; Oración et al. 1990). The species in group **A** are self-fertile, whereas those in group **B** (including sweetpotatoes) are self-incompatible (Nishiyama and Sakamoto 1975).

The genomic origin of sweetpotatoes is controversial. There is evidence that *I. trifida* (Kunth) G. Don (2*x*) and *I. tabascana* J.A. McDonald & D.F. Austin (4*x*) are the wild species most closely related to the crop (Jarret and Austin 1994; Hu et al. 2003; Rajapakse et al. 2004; Srisuwan et al. 2006). The tetraploid (4*x*) form may have originated from hybridization of the two diploid wild relatives *I. triloba* L. (**A** genome) and *I. trifida* (**B** genome), and the hexaploid (6*x*) form through hybridization of a diploid species with an unknown tetraploid, with subsequent spontaneous chromosome doubling (Jones and Deonier 1965; Magoon et al. 1970; Yen 1976). Another hypothesis suggests an autopolyploid origin with *I. trifida*, or an as yet unknown species, as the ancestor (Nishiyama 1971; Nishiyama et al. 1975; Shiotani and Kawase 1987, 1989; Shiotani 1988).

Hybridization

Cultivated sweetpotatoes are genetically isolated from most of their wild relatives, due, on the one hand, to the difference in ploidy level, and to complex compatibility and sterility systems, on the other (reviewed by F. Martin 1965; Williams and Cope 1967). Different compatibility factors—such as low seed germination, low seedling vigor, abnormal plant types, reduced flowering, ovule abortion, ineffective pollen tube growth, embryo abortion, and poor seed set—can express themselves at different stages in the sexual process (F. Martin 1968, 1970; Magoon et al. 1970).

However, unreduced pollen has been produced in tetraploid and hexaploid forms of *Ipomoea batatas* (Bohac et al. 1992), as well as in the diploid *I. trifida* (Jones 1990; Orjeda et al. 1990). Some researchers, therefore, do not rule out the existence of natural outcrossings and recurrent gene flows among related *Ipomoea* species through unreduced pollen (Austin 1978; Shiotani 1988).

In fact, laboratory and greenhouse hybridization experiments between the crop and its wild relatives (ser. *Batatas*) have shown that only crosses with *I. trifida* and the natural hybrid *I. ×leucantha* Jacq. (both diploids) produce fertile progeny (Wedderburn 1967; Kobayashi and Miyazaki 1976; Oración et al. 1990; Diaz et al. 1996). Crosses between *I. batatas* (4x and 6x) and *I. trifida* (2x) might require the use of artificial ploidy-level manipulations, such as colchicine-induced chromosome doubling (Freyre et al. 1991; Iwanaga et al. 1991; Mont et al. 1993), but hybridization is also possible through hand pollination (Orjeda et al. 1991). The extent of crossability is genotype dependent, and the average number of viable hybrid seeds that are produced (approximately one viable seed per 100 pollinations, with a sweetpotato plant as the female parent) is low (Freyre et al. 1991; Orjeda et al. 1991). Tetraploid F1 hybrids show varying degrees of reduced fertility (Oración et al. 1990).

There are no reports thus far on attempts to cross sweetpotatoes with their other two close relatives, *I. batatas* var. *apiculata* (M. Martens & Galeotti) J.A. McDonald & D.F. Austin and *I. tabascana*. However, both species are rare in the wild—*I. tabascana* is even considered to be endangered (Austin et al. 1991)—and probably no more seeds are available in *ex situ* collections (D.F. Austin, pers. comm. 2008).

Hybridization with any of the other wild relatives in ser. *Batata* is only possible with considerable human intervention, using artificial hybridization techniques such as embryo rescue and ovule culture (see, e.g., Kobayashi et al. 1994).

Based on these results, in theory cultivated sweetpotatoes could spontaneously hybridize with *I. trifida*. They perhaps might also outcross with *I. batatas* var. *apiculata* and *I. tabascana* under natural conditions, but the scarcity of these latter two taxa probably precludes such a possibility. In fact, it has been suggested that certain weedy tetraploids have resulted from spontaneous hybridization between sweetpotatoes and *I. trifida* (Austin 1977). To date, no molecular studies have been conducted on spontaneous or putative natural hybrids to either sustain or refute this hypothesis. Further research is needed to clarify the presence of natural hybrids involving cultivated sweetpotatoes.

Outcrossing Rates and Distances

We have found no reports of pollen flow studies that measure pollen dispersal distances and outcrossing rates.

State of Development of GM Technology

Given that it is difficult to regenerate sweetpotatoes from cultured tissues or cells, and that there are no efficient protocols to do so, for a long time there were few biotechnological applications for sweetpotatoes, as compared to other crops (Prakash et al. 1991; Prakash 1994; Qaim 1999). In the 1990s, attempts at sweetpotato transformation progressed slowly (Prakash and Varadarajan 1992; Gama et al. 1996; Otani et al. 1998; Saito et al. 1998; Kimura et al. 1999), but that situation has rapidly changed in recent years. Today, efficient and reliable transformation protocols are available (Okada et al. 2002a, b; Song et al. 2004; Lim et al. 2007; Yi et al. 2007), and the first transgenic cultivars are being evaluated in field trials in the United States and Kenya. The GM traits tested thus far include pest and virus resistance, decreased amylase content, tolerances for oxidative and chilling stresses, and herbicide resistance. No commercial production of GM sweetpotatoes has yet been reported.

Agronomic Management Recommendations

There are no specific recommendations.

Crop Production Area

Sweetpotatoes, also called the potatoes of the tropics, are the world's second most important root crop (Simpson and Ogorzaly 2000). The total crop area is estimated to be 9 million ha (FAO 2009). More than 95% of its global production area is in developing countries. China grows more than 80% of the world's sweetpotatoes, followed by two countries in sub-Saharan Africa (Nigeria and Uganda). The crop is grown as an annual, and it is also important in Africa, the Caribbean

Basin, Polynesia and other regions in Asia, and S America. While not a staple, it is also cultivated as a high-quality alternative vegetable in some temperate countries, such as Japan, New Zealand, and the United States.

Research Gaps and Conclusions

In spite of extensive literature on experimental crosses of sweetpotatoes with most of their closest wild relatives, several issues related to the likelihood of gene flow and introgression under natural conditions await further research. For example, it would help to have additional knowledge about the viability and duration of sweetpotato pollen under field conditions, as well as on pollen flow and outcrossing rates at different distances. More research is needed on the extent of cross-compatibility between sweetpotatoes and their two closest wild relatives, *Ipomoea trifida* and *I. tabascana*. No studies have been found that assessed the rates of outcrossing between these species under field conditions or analyzed the fertility and fitness of their hybrid progeny. Molecular studies to estimate the extent of introgression between sweetpotatoes and their wild relatives would also be useful.

Based on available knowledge about the relationships between *Ipomoea batatas* and its wild relatives, the likelihood of gene introgression from cultivated sweetpotatoes—assuming physical proximity and flowering overlap, and being conditional on the nature of the respective trait(s)—is as follows (map 21.1):

- *moderate* for its closest wild relative, *Ipomoea trifida* (Central and S America), and potentially for *I. batatas* var. *apiculata* and *I. tabascana*

- *low* for the natural hybrid *I. ×leucantha* (S United States and wherever it is naturalized)

REFERENCES

....

Austin DF (1977) Hybrid polyploids in *Ipomoea* section *batatas*. *J. Hered.* 68: 259–260.

——— (1978) The *Ipomoea batatas* complex: I. Taxonomy. *Bull. Torrey Bot. Club* 105: 114–129.

——— (1979) An infrageneric classification for *Ipomoea* (Convolvulaceae). *Taxon* 28: 359–361.

——— (1983) Variability in sweetpotato in America. *Proc. Am. Soc. Hort. Sci. Tropical Region* 27 (Part B): 15–26.

——— (1988a) Nomenclature changes in the *Ipomoea batatas* complex (Convolvulaceae). *Taxon* 37: 184–185.

——— (1988b) The taxonomy, evolution, and genetic diversity of sweet potatoes and related wild species. In: International Potato Centre (ed.) *Exploration, Maintenance, and Utilization of Sweet Potato Genetic Resources: Report of the First CIP [Centro Internacional de la Papa] Sweet Potato Planning Conference, Held at CIP Headquarters, Lima, Peru, February 23–27, 1987* (Centro Internacional de la Papa [CIP], or International Potato Centre: Lima, Peru), pp. 27–59.

——— (1991) *Ipomoea littoralis* (Convolvulaceae)—Taxonomy, distribution, and ethnobotany. *Econ. Bot.* 45: 251–256.

Austin DF and Bianchini RS (1998) Additions and corrections in American *Ipomoea* (Convolvulaceae). *Taxon* 47: 833–838.

Austin DF and Huáman Z (1996) A synopsis of *Ipomoea* (Convolvulaceae) in the Americas. *Taxon* 45: 3–38.

Austin DF, de la Puente F, and Contreras J (1991) *Ipomoea tabascana*, an endangered tropical species. *Econ. Bot.* 45: 435.

Austin DF, Jarret RL, Tapia C, and de la Puente F (1993) Collecting tetraploid *I. batatas* (Linnaeus) Lamarck in Ecuador. *Plant Genet. Resour. Newsl.* 91/92: 33–35.

Bohac JR, Jones A, and Austin DF (1992) Unreduced pollen: Proposed mechanism of polyploidization of sweet potato (*Ipomoea batatas*). *Hort. Sci.* 27: 611.

Bohac JR, Austin DF, and Jones A (1993) Discovery of wild tetraploid sweetpotatoes. *Econ. Bot.* 47: 193–201.

Bohac JR, Dukes PD, and Austin DF (1995) Sweet potato. In: Smartt J and Simmonds NW (eds.) *Evolution of Crop Plants.* 2nd ed. (Longman Scientific and Technical: Harlow, UK), pp. 57–62.

Bulmer R (1966) Birds as possible agents in the propagation of the sweet potato. *Emu* 65: 165–182.

Diaz J, Schmiediche P, and Austin DF (1996) Polygon of crossability between eleven species of *Ipomoea*: Section *Batatas*. *Euphytica* 88: 189–200.

Eguchi Y (1996) Flowering and seed production in the sweetpotato. In: Ramanatha Rao V (ed.) *Proceedings of the Workshop on the Formation of a Network for the Conservation of Sweet Potato Biodiversity in Asia, Bogor, Indonesia, May*

1–5, 1996 (International Plant Genetic Resources Institute [IPGRI]: Rome, Italy), pp. 86–92.

FAO [Food and Agriculture Organization of the United Nations] (2009) FAOSTAT data. http://faostat.fao.org/site/567/default.aspx.

Freyre R, Iwanga M, and Orjeda G (1991) Use of *Ipomoea trifida* (H.B.K.) G. Don germplasm for sweetpotato improvement: 2. Fertility of synthetic hexaploids and triploids with 2n gametes of *I. trifida,* and their interspecific crossability with sweet potato. *Genome* 34: 209–214.

Gama MICS, Leite RP, Cordeiro AR, and Cantliffe DJ (1996) Transgenic sweetpotato plants obtained by *Agrobacterium tumefaciens*-mediated transformation. *Plant Cell Tissue Organ Cult.* 46: 237–244.

Harris VC (1958) Nut grass controlled by competition. *Miss. Farm Res.* 22: 1–5.

Hu J, Nakatani M, Garcia Lalusin A, Kuranouchi T, and Fujimura T (2003) Genetic analysis of sweetpotato and wild relatives using inter-simple sequence repeats (ISSRs). *Breed. Sci.* 53: 297–304.

Huáman Z and Zhang D (1997) Sweetpotato. In: Fuccillo DA, Sears L, and Stapleton P (eds.) *Biodiversity in Trust: Conservation and Use of Plant Genetic Resources in CGIAR [Consultative Group on International Agricultural Research] Centres* (Cambridge University Press: Cambridge, UK), pp. 29–38.

Iwanaga M, Freyre R and Orjeda G (1991) Use of *Ipomoea trifida* (H.B.K.) G. Don germplasm for sweetpotato improvement: 1. Development of synthetic hexaploids of *I. trifida* by ploidy-level manipulations. *Genome* 34: 201–208.

Jarret RL and Austin DF (1994) Genetic diversity and systematic relationships in sweetpotato (*Ipomoea batatas* (L.) Lam.) and related species as revealed by RAPD analysis. *Genet. Resour. Crop Evol.* 41: 165–173.

Jarret RL, Gawel N, and Whittemore A (1992) Phylogenetic relationships of the sweetpotato *Ipomoea batatas* (L.) Lam. *J. Am. Soc. Hort. Sci.* 117: 633–637.

Jones A (1965) Cytological observations and fertility measurements of sweetpotato (*Ipomoea batatas* (L.) Lam.). *Proc. Am. Soc. Hort. Sci.* 86: 527–537.

——— (1966) *Morphological Variability in Early Generations of a Randomly Intermating Population of Sweet Potatoes (*Ipomoea batatas *(L.) Lam.).* University of Georgia Agricultural Experiment Station Technical Bulletin, n.s., 56 (College of Agriculture, University of Georgia: Athens, GA, USA). 31 p.

——— (1974) Chromosome numbers in genus *Ipomoea. J. Hered.* 55: 216–219.

——— (1980) Sweet potato. In: Fehr WR and Hadley HH (eds.) *Hybridization of Crop Plants* (American Society of Agronomy: Madison, WI, USA), pp. 645–655.

——— (1990) Unreduced pollen in a wild tetraploid relative of sweet potato. *J. Am. Soc. Hort. Sci.* 115: 512–516.

Jones A and Deonier MT (1965) Interspecific crosses among *Ipomoea lacunosa, I. ramoni, I. trichocarpa,* and *I. triloba. Bot. Gaz.* 126: 226–232.

Kimura T, Otani M, Noda T, Ideta O, Shimada T, and Saito A (1999) Decrease of amylase content in transgenic sweetpotato. *Breed. Res.* 1: 142.

Kobayashi RS and Miyazaki T (1976) Sweet potato breeding using wild related species. In: Cock J, MacIntyre R, and Graham M (eds.) *Proceedings of the 4th Symposium of the International Society for Tropical Root Crops, held at CIAT [Centro Internacional de Agricultura Tropical], Cali, Colombia, August 1–7, 1976* (International Development Research Centre [IDRC]: Ottawa, ON, Canada), pp. 53–57.

Kobayashi RS, Bouwkamp JC, and Sinden SL (1994) Interspecific hybrids from cross incompatible relatives of sweetpotato. *Euphytica* 80: 159–164.

Lim S, Kim JH, Kim SH, Kwon SY, Lee HS, Kim JS, Cho KY, Paek KY, and Kwak SS (2007) Enhanced tolerance of transgenic sweetpotato plants that express both CuZnSOD and APX in chloroplasts to methyl viologen-mediated oxidative stress and chilling. *Mol. Breed.* 19: 227–239.

Magoon ML, Krishnan R, and Baii KV (1970) Cytological evidence on the origin of sweetpotato. *Theor. Appl. Genet.* 40: 360–366.

Martin FW (1965) Incompatibility in the sweet potato: A review. *Econ. Bot.* 19: 406–415.

—— (1968) The system of self-incompatibility in *Ipomoea. J. Hered.* 59: 263–267.

—— (1970) Self- and interspecific incompatibility in the Convolvulaceae. *Bot. Gaz.* 131: 139–144.

—— (1988) Genetic and physiological basis for breeding and improving the sweet potato. In: Institut national de recherche agronomique [INRA] (ed.) *Proceedings of the 7th Symposium of the International Society for Tropical Root Crops, Gosier, Guadeloupe, July 1–6, 1985* (Institut national de recherche agronomique [INRA]: Paris, France), pp. 749–761.

Martin FW and Cabanillas E (1966) Post-pollen germination barriers to seed set in sweet potato. *Euphytica* 15: 404–411.

Martin FW and Jones A (1973) The species of *Ipomoea* closely related to the sweetpotato. *Econ. Bot.* 26: 201–215.

Martin FW and Ortiz S (1967) Anatomy of the stigma and style of sweet potato. *New Phytol.* 66: 109–113.

Martin JA Jr (1946) Germination of sweet potato seed as affected by different methods of scarification. *Proc. Am. Soc. Hort. Sci.* 47: 387–390.

Martins PS (2001) Dinâmica evolutiva em roças de caboclos amazônicos. In: Vieira ICG, Silva JMC, Oren DC, and d'Incao MA (eds.) *Diversidade Biológica e Cultural da Amazônia* (Museu Paraense Emílio Goeldi: Belém, Brazil), pp. 369–384.

McDonald JA and Austin DF (1990) Changes and additions in *Ipomoea* section *Batatas* (Convolvulaceae). *Brittonia* 42: 116–120.

McDonald JA and Mabry TJ (1992) Phylogenetic systematics of New World *Ipomoea* (Convolvulaceae) based on chloroplast DNA restriction site variation. *Plant Syst. Evol.* 180: 243–259.

Mont J, Iwanaga M, Orjeda G, and Watanabe K (1993) Abortion and determination of stages for embryo rescue in crosses between sweetpotato, *Ipomoea batatas* Lam. ($2n = 6x = 90$) and its wild relative, *I. trifida* (H.B.K.) G. Don ($2n = 2x = 30$). *Sex. Plant Reprod.* 6: 176–182.

Montelaro J and Miller JC (1951) A study of some factors affecting seed setting in the sweet potato. *Proc. Am. Soc. Hort. Sci.* 57: 329–334.

Murata T and Matsuda Y (2003) Histological studies on the relationship between the process from fertilization to embryogenesis and the low seed set of sweet potato, *Ipomoea batatas* (L.) Lam. *Breed. Sci.* 53: 41–50.

Nishiyama I (1971) Evolution and domestication of the sweet potato. *Bot. Mag.* (Tokyo) 84: 377–387.

Nishiyama I and Sakamoto S (1975) Evolutionary autoploidy in the sweet potato (*Ipomoea batatas*) and its progenitors. *Euphytica* 24: 197–208.

Nishiyama I, Niyazaki T, and Sakamoto S (1975) Evolutionary autoploidy in the sweet potato (*Ipomoea batatas* (L.) Lam.) and its progenitors. *Euphytica* 24: 197–208.

Okada Y, Nishiguchi M, Saito A, Kimura T, Mori M, Hanada K, Sakai J, Matsuda Y, and Murata T (2002a) Inheritance and stability of the virus-resistant gene in the progeny of transgenic sweet potato. *Plant Breed.* 121: 249–253.

Okada Y, Saito A, Nishiguchi M, Kimura T, Mori M, Matsuda Y, and Murata T (2002b) Microprojectile bombardment-mediated transformation of sweet potato (*Ipomoea batatas* (L.) Lam). *SABRAO [Society for the Advancement of Breeding Researches in Asia and Oceania] J. Breed. Genet.* 34: 1–8.

Onwueme IC (1978) *The Tropical Tuber Crops: Yams, Cassava, Sweet Potato, and Cocoyams* (John Wiley and Sons: New York, NY, USA). 234 p.

Oración MZ, Niwa K, and Shiotani I (1990) Cytological analysis of tetraploid hybrids between sweet potato and diploid *Ipomoea trifida* (H.B.K.) Don. *Theor. Appl. Genet.* 80: 617–624.

Orjeda G, Freyre R, and Iwanaga M (1990) Production of $2n$ pollen in diploid *Ipomoea trifida*, a putative wild ancestor of sweet potato. *J. Hered.* 81: 462–467.

——— (1991) Use of *Ipomoea trifida* (H.B.K.) G. Don germplasm for sweetpotato improvement: 3. Development of $4x$ interspecific hybrids between *Ipomoea batatas* (L.) Lam. ($2n = 6x = 90$) and *Ipomoea trifida* (H.B.K.) G. Don ($2n = 2x = 30$) as storage-root indicators for wild species. *Theor. Appl. Genet.* 81: 462–467.

Otani M, Shimada T, Kimura T, and Saito A (1998) Transgenic plants production from embryogenic callus of sweetpotato (*Ipomoea batatas* (L.) Lam.) using *Agrobacterium tumefaciens*. *Plant Biotech*. 15: 11–16.

Prakash CS (1994) Sweet potato biotechnology: Progress and potential. *Biotech. Dev. Monit.* 18: 18–22.

Prakash CS and Varadarajan U (1992) Genetic transformation of sweet potato by particle bombardment. *Plant Cell Rep.* 11: 53–57.

Prakash CS, Varadarajan U, and Kumar AS (1991) Foreign gene transfer to sweet potato (*Ipomoea batatas*). *Hort. Sci.* 26: 492.

Purseglove JW (1965) The spread of tropical crops. In: Baker HK and Stebbins GL (eds.) *The Genetics of Colonizing Species* (Academic Press: New York, NY, USA), pp. 375–386.

—— (1968) *Tropical Crops: Dicotyledons*, vol. 1 (Longman Scientific and Technical: Harlow, UK). 719 p.

Qaim M (1999) *The Economic Effects of Genetically Modified Orphan Commodities: Projections for Sweetpotato in Kenya* (International Service for the Acquisition of Agri-biotech Applications [ISAAA] and Center for Development Research [Zentrum für Entwicklungsforschung, or ZEF]: Bonn, Germany). 47 p.

Rajapakse S, Nilmalgoda SD, Molnara M, Ballarda RE, Austin DF, and Bohac JR (2004) Phylogenetic relationships of the sweetpotato in *Ipomoea* series *Batatas* (Convolvulaceae) based on nuclear-amylase gene sequences. *Mol. Phylogenet. Evol.* 30: 623–632.

Saito A, Kimura T, Ideta O, Mori M, and Nishiguchi M (1998) Transgenic sweet potato (*Ipomoea batatas* (L.) Lam.) exhibiting resistance to sweet potato feather mottle potyvirus. *Sweet Potato Res.* 7: 4.

Shiotani I (1988) Genomic structure and the gene flow in sweet potato and related species. In: International Potato Centre (ed.) *Exploration, Maintenance, and Utilization of Sweet Potato Genetic Resources: Report of the First CIP [Centro Internacional de la Papa] Sweet Potato Planning Conference, Held at CIP headquarters, Lima, Peru, February 23–27, 1987* (Centro Internacional de la Papa [CIP], or International Potato Centre: Lima, Peru), pp. 61–73.

Shiotani I and Kawase T (1987) Synthetic hexaploids derived from wild species related to sweet potato. *Japan. J. Breed.* 37: 367–376.

—— (1989) Genomic structure of the sweet potato and hexaploids in *Ipomoea trifida* (H.B.K) Don. *Japan. J. Breed.* 39: 57–66.

Simpson BB and Ogorzaly MC (2000) *Economic Botany: Plants in Our World*. 3rd ed. (McGraw-Hill: New York, NY, USA). 544 p.

Song GQ, Honda H, and Yamaguchi KI (2004) Efficient *Agrobacterium tumefaciens*-mediated transformation of sweet potato (*Ipomoea batatas* (L.) Lam.)

from stem explants using a two-step kanamycin-hygromycin selection method. In Vitro *Cell. Dev. Biol., Pl.* 40: 359–365.

Srisuwan S, Sihachakra D, and Siljak-Yakovlev S (2006) The origin and evolution of sweet potato (*Ipomoea batatas* Lam.) and its wild relatives through the cytogenetic approaches. *Plant Sci.* 171: 424–433.

Stout AB (1924) The flowers and seed of sweetpotatoes. *J. New York Bot. Gard.* 25: 153–168.

Ting YC, Kehr AE, and Miller JC (1957) A cytological study of the sweetpotato plant *Ipomoea batatas* (L.) Lam. and its related species. *Am. Nat.* 91: 197–203.

Togari Y and Kawahara U (1946) Studies on the self- and cross-incompatibility in sweetpotato (a preliminary report): I. On the different grades of compatibility among the compatible matings. *Imperial Agr. Exp. Sta. Tokyo Bull.* 52: 1–19 [in Japanese].

Ukoskit K and Thompson PG (1997) Autopolyploidy versus allopolyoloidy and low-density randomly amplified polymorphic DNA linkage maps of sweetpotato. *J. Am. Soc. Hort. Sci.* 122: 822–828.

Wedderburn MM (1967) A study of hybridisation involving the sweet potato and related species. *J. Plant Growth Regul.* 16: 69–75.

Williams DB and Cope FW (1967) Notes of self-incompatibility in the genus *Ipomoea* L. In: Tai EA (ed.) *Proceedings of the International Symposium on Tropical Root Crops, Held at the University of the West Indies, St. Augustine, Trinidad, April 2–8,1967*, vol. 1 (Department of Crop Science, University of the West Indies: St. Augustine, Trinidad, West Indies), pp. 16–30.

Yen DE (1960) The sweet potato in the Pacific: The propagation of the plant in relation to its distribution. *J. Polynesian Soc.* 69: 368–375.

——— (1974) *The Sweet Potato and Oceania: An Essay in Ethnobotany*. Bernice Pauahi Bishop Museum Bulletin No. 236 (Bishop Museum Press: Honolulu, HI, USA). 389 p.

——— (1976) Sweet potato *Ipomoea batatas* (Convolvulaceae). In: Simmonds NW (ed.) *Evolution of Crop Plants* (Longman Scientific and Technical: London, UK), pp. 42–45.

——— (1982) Sweetpotato in historical perspective. In: Villareal RL and Griggs TD (eds.) *Sweetpotato: Proceedings of the First International Symposium on Sweet Potato, Taiwan, ROC, March 23–27, 1981*. AVRDC [Asian Vegetable Research and Development Center] Publication No. 82–172 (Asian Vegetable Research and Development Center [AVRDC]: Shanhua, Tainan, Taiwan), pp. 17–30.

Yi G, Shin YM, Choe G, Shin B, Kim YS, and Kim KM (2007) Production of herbicide-resistant sweet potato plants transformed with the bar gene. *Biotech. Lett.* 29: 669–675.

Zhang D, Cervantes J, Huamán Z, Carey E, and Ghislain M (2000) Assessing genetic diversity of sweet potato (*Ipomoea batatas* (L.) Lam.) cultivars from tropical America using AFLP. *Genet. Resour. Crop Evol.* 47: 659–665.

Zhang D, Rossel G, Kriegner A, and Hijmans R (2004) AFLP assessment of diversity
in sweetpotato from Latin America and the Pacific Region: Its implications on
the dispersal of the crop. *Genet. Resour. Crop Evol.* 51: 115–120.

TWENTY TWO

Wheat, Bread Wheat
(*Triticum aestivum* L.)

Apart from bread (or common) wheat (*Triticum aestivum* L. subsp. *aestivum*) there are several other domesticated species within the genus *Triticum* L., including the widely known einkorn (*T. monococcum* L. subsp. *monococcum*, 2*x*), emmer (*T. turgidum* subsp. *dicoccum* (Schrank *ex* Schübl.) Thell., 4*x*) and durum wheat (*T. turgidum* subsp. *durum* (Desf.) Husn., 4*x*). This chapter will focus on hexaploid (6*x*) bread wheat (or common wheat) only, since it is both the most widely cultivated wheat crop and the one most frequently used for genetic transformation.

Center(s) of Origin and Diversity

Hexaploid bread wheat originated in the Fertile Crescent in the Middle East (McFadden and Sears 1946; Harlan 1992; Cauderon 1994; Feldman et al. 1995; Zohary and Hopf 2000; Feldman 2001; Özkan et al. 2002).

The primary center of diversity for wheat (i.e., cultivated wheat species and their progenitors) is the Levant, a rather limited area along the eastern Mediterranean coast (Mac Key 2005). Secondary centers of diversity developed in China and in the Hindu Kush-Himalayan region (Yamashita 1980).

Biological Information

Flowering

The wheat inflorescence is a spike that flowers when temperatures reach 14°C and higher (Percival 1921; Mandy 1970; Lersten 1987). Flowering begins in the middle and spreads towards the top and bottom of the spike (Mandy 1970; Evans et al. 1972; Allan 1980). Individual flowers open before sunrise in the early morning and close by 7 p.m. In central Europe, wheat plants flower for 4 to 15 days (Mandy 1970).

Self-fertilization rates are around 96% and higher (Heyne and Smith 1967; J. Martin et al. 1976), although the majority of pollen-shedding anthers may protrude from the spikelet and shed their pollen into the air (Leighty and Sando 1924; d'Souza 1970; Beri and Anand 1971; Waines and Hegde 2003). This is because wheat is slightly protandrous, that is, the anthers mature one to three days before the stigma becomes receptive. In particular, self-fertilization rates are high under biotic and abiotic stresses (e.g., drought, frost, pest attacks, microdeficiencies), since the anthers tend to be retained in the spikelets (Leighty and Sando 1924; Ueno and Itoh 1997). These rates are also higher in closed-flowering cultivars than in open-flowering types (Ueno and Itoh 1997; Kirby 2002). Unfertilized spikelets remain open and receptive to pollen for up to 13 days (Mandy 1970; de Vries 1971). Once a pollen grain is deposited on the stigma, it will germinate between 15 minutes and 1.5 hours (d'Souza 1970; de Vries 1971; Bennett et al. 1973).

Pollen Dispersal and Longevity

The pollen of bread wheat is quite heavy, a characteristic associated with a high ($6x$) ploidy level (de Vries 1971). Nonetheless, wind is the main pollen dispersal vector. The flowers of cereal crops do not produce nectar, so visiting insects do not contribute much to cross-fertilization (Treu and Emberlin 2000).

Wheat pollen has a very limited longevity. Under field conditions (20°C-30°C), it rarely remains viable for more than 30 minutes after dehiscence (d'Souza 1970; Fritz and Lukaszewski 1989; Treu and Emberlin 2000). Even under optimum conditions (16°C-20°C, 60%–75% relative atmospheric humidity), longevity often does not exceed three hours (d'Souza 1970; de Vries 1971).

Sexual Reproduction

Bread wheat is a hexaploid ($2n = 6x = 42$, **AABBDD**), primarily self-pollinating crop. In field conditions, the outcrossing rates are minimal

(0.1%) with high humidity but can exceed 10% in some genotypes (Poehlmann 1959; Heyne and Smith 1967; Wiese 1991; Hucl 1996; Enjalbert et al. 1998). On average, they are below 2% (J. Martin et al. 1976; Prakash and Singhal 2003).

Likewise, most wheat landraces and wild *Triticum* and *Aegilops* L. (goatgrass) relatives are mainly autogamous (Tsegaye 1996), except for *Ae. speltoides* Tausch, which is typically allogamous.

Seed Dispersal and Dormancy

Most wild wheat species have a brittle rachis, that is, when a ripe head of wheat is touched or blown by the wind, the main axis (rachis) of the inflorescence (or spike) splits and falls apart into numerous dispersal units, with each unit consisting of a single spikelet attached to a short segment of the rachis. Other wild species have heads with grains that shatter easily, so that when the ripe plant is touched or blown by the wind, the seeds are loosened and fall out of the spikelets onto the ground. Long awns further promote efficient seed dispersal, either by wind or by being attached to animals (Harper et al. 1970; Peart 1981; Chambers and MacMahon 1994; Elbaum et al. 2007).

In contrast, most modern bread wheat cultivars have completely lost their shattering ability, due to selection for a tough rachis during domestication. However, seeds may be dispersed by animals feeding on wheat grains (e.g., birds and small mammals such as rats and mice) or by water (e.g., in irrigation channels).

Seed dormancy is present in many wild wheat species, but the trait was under strong selection pressure during domestication. Still, modern wheat cultivars have a wide range of seed dormancy levels that are influenced by environmental factors, such as temperature and humidity (Pickett 1989; Anderson and Soper 2003; Biddulph et al. 2007).

Cereal seeds in the soil usually survive for periods of one to three years (Harker et al. 2005; Gaines et al. 2007a). However, seeds plowed into the soil after harvest may survive longer. Field studies indicate that wheat seeds may persist for up to five years in the soil (Pickett 1993; Anderson and Soper 2003), while seeds of *Aegilops* and other wild wheat species may remain viable for three to five years (Marañon 1987; Morishita 1996).

Volunteers, Ferals, and Their Persistence

Volunteer wheat plants are frequently found in subsequent cropping cycles, due to seeds being lost during cropping and harvesting. Under continuous monocropping, this can lead to an accumulation of seeds in the soil and the generation of significant populations of wheat volunteers over time (Komatsuzaki and Endo 1996; Wicks et al. 2000). Occasionally, individual wheat ferals and small naturalized feral populations are found outside of cultivation (see, e.g., von der Lippe and Kowarik 2007).

Winter wheat is fairly cold tolerant, and seedlings may survive temperatures as low as −25°C, whereas spring wheat seedlings are only cold tolerant to about −5°C, depending on the genotype and humidity (Hömmö and Pulli 1993). In western Canada, volunteer wheat species have been shown to persist for up to five years on up to 9% of the cultivated fields (Beckie et al. 2001). However, there are no indications that feral wheat plants are vigorous enough to establish persistent populations outside of cultivation (Sukopp and Sukopp 1993). In any case, volunteers and feral populations are easily controlled. For example, shallowly tilling the soil surface for several days after harvest encourages seed germination, and the resulting volunteer plants can then be destroyed, either by mechanical or chemical treatments.

Weediness and Invasiveness Potential

Hexaploid wheat species are not able to compete well and lack characteristic weediness traits—such as prolonged seed dormancy, extended persistence in the soil, germination under a broad range of environmental conditions, rapid vegetative growth, a short life cycle, very high seed output, a high rate of seed dispersal, and long-distance seed dispersal (Keeler 1989; Keeler et al. 1996). Therefore, they are not likely to become invasive weeds.

In contrast, many wild relatives are colonizers, quickly able to invade new territories, particularly the wild relatives of *Aegilops* (van Slageren 1994). For example, some *Aegilops* species introduced in the United States (*Ae. cylindrica* Host, *Ae. triuncialis* L., and *Ae. geniculata* Roth) are now troublesome weeds (Hitchcock 1951; Donald and

Ogg 1991). On the other hand, in Europe *Aegilops* species, although they are invasive, are not considered to be weeds (Hanf 1982).

Crop Wild Relatives

The gene pool for bread wheat comprises various domesticated (mainly *Triticum*) species and a large number of wild relatives. The latter belong to several genera in the tribe Triticeae, including *Aegilops*, *Agropyron* Gaertn., *Dasypyrum* (Coss. & Durieu) T. Durand, *Elymus* L., *Hordeum* L., *Secale* L., and *Triticum*. Here, our focus will be only on those species that are sexually compatible with bread wheat (tables 22.1 and 22.2).

The taxonomy of wheat species is controversial; thus, wheat researchers around the world follow different classifications. Those currently in use vary in their genus as well as their species concepts. Some treat all wheat species as one large genus (*Triticum sensu lato*) while others divide them into two separate genera (*Triticum* and *Aegilops*). An overview of the different classification systems and their synonymy is provided by the Wheat Genetic and Genomic Resources Center of the University of Kansas (www.k-state.edu/wgrc/Taxonomy/taxintro.html).

In this book, we consider *Triticum* and *Aegilops* to be two different genera. The classification of *Triticum* L. follows Mac Key (2005), who recognizes 10 species in four sections, with ploidy levels of 14, 28, 42 and 56 chromosomes. He also included triticale (formerly genus ×*Triticosecale* Wittm. *ex* A. Camus)—a man-made cross between wheat and rye—as a new section (table 22.1).

The classification of *Aegilops* follows van Slageren (1994), except that *Amblyopyrum muticum* (Boiss.) Eig is included in the genus *Aegilops* (Hammer 1980). This genus contains 11 diploid, 10 tetraploid, and 2 hexaploid species and is highly variable, as there is often natural hybridization and gene flow among different *Aegilops* species. Genetic exchange is especially common among most of the tetraploid *Aegilops* species within sect. *Aegilops* (= sect. *Pleionathera* Eig), resulting in the formation of complex intermediates and highly introgressed types, such as those in Greece, Israel, and Turkey (Feldman 1965; Feldman et al. 1995).

Wheat species (*Triticum* and *Aegilops*) can be grouped into three genome clusters (**A, D,** and **U**), according to their pivotal genomes (Zo-

Table 22.1 The primary gene pool of bread wheat (*Triticum aestivum* L.)

Species	Common name	Genome	Origin and distribution
sect. **Triticum** L. (= sect. *Speltoidea* Flaksb.)		hexaploid (2n = 6x = 42)	known only from cultivation
T. aestivum L.*	bread wheat or common wheat	AABBDD	cultivated worldwide in temperate zones
T. zhukovskyi Menabde & Ericzjan (= *T. timopheevii* × *T. monococcum*)	Zanduri wheat	AAAAGG	cultivated in Georgia
sect. **Dicoccoidea** Flaksb.		tetraploid (2n = 4x = 28)	wild and cultivated
T. turgidum subsp. **durum** (Desf.) Husn.†	durum wheat, macaroni wheat	AABB	cultivated throughout W Asia, N Africa, and Mediterranean region
T. turgidum subsp. **dicoccum** (Schrank ex Schübl.) Thell.	emmer wheat	AABB	relict crop grown in mountainous areas of Europe, Morocco, and S India
T. turgidum subsp. *dicoccoides* (Körn. ex Asch. & Graebn.) Thell.	wild emmer	AABB	wild; native to Fertile Crescent (Iran, Iraq, Israel, Lebanon, Syria, Turkey)
T. timopheevii (Zhuk.) Zhuk. subsp. **timopheevii**	part of the Zanduri wheat complex	AAGG	cultivated in Caucasus region (Transcaucasia, Armenia, Iraq, Iran)
T. timopheevii subsp. *armeniacum* (Jakubz.) van Slageren	wild emmer	AAGG	wild; native to Iran, Turkey, Armenia, Azerbaijan (NE Fertile Crescent)
sect. **Monococcon** Dumort.		diploid (2n = 2x = 14)	wild and cultivated
T. monococcum L. subsp. **monococcum**	cultivated einkorn, small spelt wheat	AA	cultivated in mountainous villages in Italy, Spain, Turkey, and elsewhere
T. monococcum subsp. *aegilopoides* (Link) Thell.	wild einkorn	AA	wild; origin in S Turkey, native to SE Europe, Caucasus, and W Asia
Triticum urartu Thumanjan ex Gandilyan		AA	wild; native to Caucasus (Armenia)

		polyploid	
sect. **Triticosecale** (Wittm. *ex* A. Camus) **Mac Key, sect. nov.**			known only from cultivation; cultivated worldwide
T. semisecale Mac Key, spec. nov.	subtriticale	**AARR** $(2n = 2x = 28)$	
T. neoblaringhemii (Wittm. ex A. Camus) **Mac Key, comb. nov.**	triticale	**AABBRR** $(2n = 6x = 42)$	
T. rimpaui (Wittm.) **Mac Key, comb. nov.**	eutriticale	**AABBDDRR** $(2n = 8x = 56)$	

Source: The taxonomic classification follows Mac Key 2005.

Note: Cultivated forms are in bold. Artificial species and mutants such as *T.* × *timococcum* Kostov, *T.* × *fungicidum* Zhuk., *T.* × *borisovii* Zhebrak, and *T.* × *kiharae* Dorof. & Migush. are excluded, as they have been developed during genetic experiments and only occur in artificial environments. Genome designations follow Kihara 1954 and Feldman et al. 1995.

*Apart from bread wheat (*T. aestivum* (L.) subsp. *aestivum*), this also includes the following cultivated subspecies: spelt wheat (subsp. *spelta* (L.) Thell.), macha wheat (subsp. *macha* (Dekapr. & Menabde) Mac Key), dwarf/club wheat (subsp. *compactum* (Host) Mac Key), and cake wheat (subsp. *sphaerococcum* (Percival) Mac Key).

†This includes the cultivated subspecies Persian wheat (*T. turgidum* subsp. *carthlicum* (Nevski) Á. Löve & D. Löve), kolchis wheat (subsp. *georgicum* (Dekapr. & A.M. Menabde) Mac Key), Georgian emmer (subsp. *paleocolchicum* Á. Löve & D. Löve), Polish wheat (subsp. *polonicum* (L.) Thell.), Oriental wheat (subsp. *turanicum* (Jakubz.) Á. Löve & D. Löve), and rivet/cone wheat (subsp. *turgidum*).

Table 22.2 The secondary gene pool of bread wheat (*Triticum aestivum* L.)

Species	Common name	Ploidy	Genome	Origin and distribution
sect. *Aegilops* (= sect. *Pleionathera* Eig)			U-genome	
Ae. biuncialis Vis.	Lorent's goatgrass	tetraploid ($2n = 4x = 28$)	UUMM	Mediterranean region, the Russian Federation, Turkey, Palestine, Iraq, Iran
Ae. columnaris Zhuk.	pillar or columnar Aegilops		UUXX	from Turkey to Iraq, Iran, Caucasia; weed
Ae. geniculata Roth.	haver (oat) grass, ovate goatgrass		UUMM	Mediterranean region, Palestine, Syria, Lebanon, Turkey, Iraq, Iran, Afghanistan; weed
Ae. kotschyi Boiss.	mouse barley		SSUU	from N Africa across Palestine, Iraq, Iran, Afghanistan, Caucasia
Ae. neglecta Req. ex Bertol.	three-awned ovate goatgrass		UUXX / UUXXNN	Mediterranean region, West Asia, Iraq, Iran, S parts of the Russian Federation
Ae. peregrina (Hack.) Eig			SSUU	W Asia, E Europe, N Africa (Egypt); widely naturalized elsewhere
Ae. triuncialis L.	barbed goatgrass		UUCC	Mediterranean region, Turkey, Palestine, Syria, Lebanon, Iraq, Iran, Turkmenistan, Afghanistan
Ae. umbellulata Zhuk.		diploid ($2n = 14$)	UU	Greek islands, Turkey, N Syria, Iraq, NW Iran, Transcaucasia; weed
sect. *Comopyrum* (Jaub.) Zhuk.				
Ae. comosa Sm. in Sibth. & Sm.	thick Aegilops	diploid ($2n = 14$)	MM	W Asia and E Europe
Ae. uniaristata Vis.	one-awned Aegilops		NN	native to SE Europe

sect. *Cylindropyrum* (Jaub.) Zhuk.			C-genome	
Ae. caudata L.	Cretan hardgrass	diploid (2n = 14)	CC	W Asia and SE Europe
Ae. cylindrica Host	cylindrical or jointed goatgrass	tetraploid (2n = 4x = 28)	CCDD	native to W Asia and E Europe; naturalized in W Europe and N America; weed
sect. *Vertebrata* Zhuk., emend. Kihara			D-genome	
Ae. crassa Boiss.		tetra- and hexaploid	XXDD / XXDDDD	Turkey, Syria, Palestine, Iraq, Iran, Afghanistan
Ae. juvenalis (Thell.) Eig.		hexaploid	XXDDUU	Iraq, Syria, Azerbaijan, Turkmenistan, Uzbekistan
Ae. tauschii Coss.*	goatgrass, rough-spiked hardgrass	diploid (2n = 14)	DD	Native to Mediterranean region (Armenia), C Asia (China), Iran, Iraq, Transcaucasia; widely naturalized elsewhere (pantropical weed)
Ae. vavilovii (Zhuk.) Chennav.		hexaploid	XXDDSS	W Asia and Arabia
Ae. ventricosa Tausch	bulbed goatgrass	tetraploid (2n = 4x = 28)	DDNN	S Europe and N Africa
sect. *Sitopsis* (Jaub. & Spach) Zhuk.			S^b-genome	
Ae. bicornis (Forssk.) Jaub. & Spach	mouse barley, two-horned *Aegilops*	diploid (2n = 14)	SS	N Africa and W Asia
Ae. longissima Schweinf. & Muschl.			SS	Egypt, Israel, Jordan
Ae. searsii Feldman & Kislev			SS	Israel, Jordan, Lebanon, Syria
Ae. sharonensis Eig			SS	Israel, Lebanon
Ae. speltoides Tausch	goatgrass		SS	S Turkey, N Syria, Iraq

(continued)

Table 22.2 (continued)

Species	Common name	Ploidy	Genome	Origin and distribution
sect. *Amblyopyrum* (Jaub. & Spach) Zhuk.				
Ae. mutica Boiss.	slim wheatgrass, hairgrass	diploid (2*n* = 14)	TT	restricted to Armenia, Turkey, NW Iran
Other genera				
Secale cereale L.	cultivated rye	2*n* = 2*x* = 14		known only from cultivation
Secale iranicum Kobyl.		2*n* = 2*x* = 14		Near East, Iran, Afghanistan, Transcaspia; wild
Secale montanum Guss.		2*n* = 2*x* = 14		S Europe, N Africa, Caucasus, Middle East, Pakistan; introduced to N USA; wild
Secale sylvestre Host		2*n* = 2*x* = 14		E and SE Europe, the Russian Federation, Middle East; wild
Dasypyrum breviaristatum (H. Lindb.) Fred.				Algeria, Morocco, Greece
Dasypyrum villosum (L.) P. Candargy				S and E Europe, Armenia, Azerbaijan, Turkmenistan
Thinopyrum elongatum (Host) D.R. Dewey	tall wheatgrass			S and E Europe, N Africa, Caucasus, S America (Bolivia)
Thinopyrum intermedium (Host) Barkworth & D.R. Dewey	intermediate wheatgrass			S and E Europe, Caucasus, C Asia, Middle East, Pakistan, China

Sources: The taxonomic classification of *Aegilops* L. follows van Slageren 1994; that of *Secale* L. is according to Sencer and Hawkes 1980 and Kobyljanskii 1989.
Note: The cultivated species is in bold. Genome designations follow Kihara 1954; Waines and Barnhart 1992; and Badaeva et al. 2002, 2004.
*Primary gene pool.

hary and Feldman 1962; Morris and Sears 1967; Kimber and Zhao 1983; Kimber and Feldman 1987):

1 the A-genome cluster includes all diploid and polyploid *Triticum* forms

2 the D-genome cluster is formed by diploid *Aegilops tauschii* Coss. (= *Ae. squarrosa* L.) and five polyploid species of *Aegilops* (sects. *Cylindropyrum* (Jaub.) Zhuk. and *Vertebrata* Zhuk., emend. Kihara)

3 the U-genome cluster contains the diploid *Ae. umbellulata* Zhuk. and seven polyploid species of *Aegilops* (sect. *Aegilops* = sect. *Pleionathera*)

Bread wheat belongs to the A-genome cluster and is a hexaploid ($2n = 6x = 42$, **AABBDD**) that resulted from two unique, spontaneous hybridization events between diploid strains (Allan 1980; Matsuoka et al. 2007). The first event involved a diploid A-genome and a diploid B-genome species. The A-genome donor was likely wild *T. urartu* Thumanjan *ex* Gandilyan (Dvorak 1976, 1988; Dvorak et al. 1993; Miyashita et al. 1994). The B-genome comes from *Aegilops* sect. *Sitopsis* (Jaub. & Spach) Zhuk., possibly from *Ae. speltoides* (or its early ancestors), as the B genome in this species most closely resembles that of bread wheat (Sarkar and Stebbins 1956; Riley et al. 1958). Hybridization between these two diploid species led to the evolution of the tetraploid wheat progenitor wild emmer (*T. turgidum* subsp. *dicoccoides* (Körn. *ex* Asch. & Graebn.) Thell.). A second hybridization event took place between domesticated tetraploid emmer (*T. turgidum* subsp. *dicoccum*), donor of the AB genome, and diploid *Ae. tauschii*, donor of the D genome (Kihara 1944; McFadden and Sears 1946; Dvorak et al. 1998).

Hybridization

From an evolutionary standpoint, one important feature is the complex polyploidy encountered in wheat and many of its wild relatives. The combination of various diploid genomes that were originally isolated later facilitated repeated recombinations and genetic exchanges (Feldman et al. 1995).

Likewise, due to its high (6*x*) ploidy level, crossing experiments with bread wheat in general show greater rates of success than crosses with other cereals (e.g., barley). Hybrids can be obtained relatively easily between bread wheat and species that share a common (**A, B,** or **D**) genome (Kimber and Sears 1987; Jauhar et al. 1991). In contrast, hybridization with species that have nonhomologous genomes is more difficult and requires considerable human intervention.

Overall, hybridization is more successful when the plant with the higher chromosome number is the female parent. For F1 hybrids, the probability of intergenomic recombination at the meiotic stage is influenced by the presence or absence of certain regulator genes (*Ph* locus and others) in the bread-wheat parent that suppress or induce homoeologous chromosome pairing (Riley and Chapman 1958; Mello-Sampayo 1971; Sears 1976). Plant breeders have successfully used *Ph*-deficient or appropriate *Ph*-mutant lines to improve the extent of genomic recombination (see, e.g., Sears 1972, 1981; Darvey 1984). Also, unreduced gametes may sometimes be produced, which may lead to fertile amphiploids (Mac Key 2005; Zaharieva and Monneveux 2006). Wide crosses between bread wheat and a considerable number of less-closely related species have been produced using artificial methods. However, these crosses are, on the whole, relevant for the breeder but not for estimating actual spontaneous outcrossing under natural conditions.

According to their level of sexual compatibility with the crop, the domesticated and wild relatives of bread wheat are divided into primary, secondary, and tertiary gene pools (Allan 1980; Mujeeb-Kazi 2005).

Primary Gene Pool (GP-1)

The primary gene pool for bread wheat contains all cultivated and wild *Triticum* forms (including the man-made cereal triticale; table 22.1) and the D-genome donor, *Aegilops tauschii*. It thus includes diploid, tetraploid, and hexaploid forms. Most of the cultivated and wild GP-1 forms are limited to geographically restricted areas of Europe and Asia. Even so, several of the diploid and tetraploid wild relatives in these regions occur in sympatry with bread wheat and are common weeds in cultivated fields (Jia 2002; Zaharieva and Monneveux 2006).

Hexaploid *Triticum* Forms

The hexaploid forms (sect. *Triticum*) are fully cross-compatible with bread wheat, because their genomes are homologous. Hybridization occurs relatively easily and results in viable and fertile progeny (Körber-Grohne 1988; Feldman et al. 1995).

Other *Triticum* Forms (Diploid and Tetraploid)

The diploid (sect. *Monococcon* Dumort.) and tetraploid (sect. *Dicoccoidea* Flaksb.) forms are interfertile with bread wheat, although to a lesser extent, as the fertility of the F1 hybrids is often substantially reduced. For example, successful manual crosses between bread wheat and tetraploid forms have been reported for *T. timopheevii* (Zhuk.) Zhuk. and for *T. turgidum* subsp. *carthlicum* (Nevski) Á. Löve & D. Löve, subsp. *durum* (Desf.) Husn., and subsp. *turgidum* (Mandy 1970; Sharma and Gill 1983a), but only the bread wheat × *T. turgidum* crosses were fertile. Spontaneous hybrids of bread wheat and *T. turgidum* relatives have also been reported in the field (Feldman et al. 1995). In general, hybridization of bread wheat is much more difficult with diploid than with tetraploid *Triticum* forms, and the resulting hybrids are usually sterile. Fertility can sometimes be restored by backcrossing or by chromosome doubling. However, the extent of cross-compatibility is also influenced by the direction of the cross (Gill and Waines 1978). For example, manual crosses with *T. monococcum* as the female parent produced F1 hybrids with grains that germinated, whereas grains of the reciprocal hybrid did not germinate.

Aegilops tauschii

Likewise, crosses between bread wheat and its diploid D-genome donor *Aegilops tauschii* (= *Ae. squarrosa*) may only succeed as a result of manipulation and artificial hybridization techniques. Certain genes have been transferred from the wild species to bread wheat using artificial hybridization techniques, such as embryo rescue or an initial combination of the diploid genome with that of a tetraploid species, forming a synthetic hexaploid that is then crossed to a hexaploid cultivar (see, e.g., Raupp et al. 1983, 1993; Gill and Raupp 1987; Cox et al. 1994; Innes and Kerber 1994; A. Lu et al. 2002, reported in Jia 2002). To date, there are no reports of spontaneous outcrossing between bread wheat and *Ae. tauschii* under natural conditions.

Triticale

Triticale (formerly genus ×*Triticosecale*, now included as a new section within the genus *Triticum*) is a man-made cereal resulting from hybridization between (bread or durum) wheat and rye (*Secale cereale* L.) (Lukaszewski and Gustafson 1987; Oettler 2005), with tetraploid ($2n = 4x = 28$), hexaploid ($2n = 6x = 42$), and octoploid forms ($2n = 8x = 56$). Triticale shows a certain extent of cross-compatibility with both wheat and rye (Chaubey and Khanna 1986; Lelley 1992; Guedes-Pinto et al. 2001), and spontaneous hybridization between wheat and triticale may occur in the field (Ammar et al. 2004). Hybridization success, however, is influenced by several factors, such as the triticale genotype and gender (Nkongolo et al. 1991). Although hybridization rates are higher when triticale is the male parent, the resulting hybrid seed is only very rarely viable (Bizimungu et al. 1998; Hills et al. 2007). In contrast, crosses where triticale is the female parent generally produce viable and fertile F1 seed (Chaubey and Khanna 1986; Khanna 1990; Nkongolo et al. 1991; Hills et al. 2007). The resulting F1 hybrids are usually completely sterile, but occasionally some fertile pollen grains can be found (Hills et al. 2007). With the use of sophisticated artificial techniques (e.g., the generation of 1B/1R translocation chromosomes; Rabinovich 1998), triticale can be exploited as a bridge for the introgression of genes from *Secale cereale*, but there are no reports of triticale being used as a bridge for hybridization with other species.

Wheat can thus be crossed artificially with most of its cultivated and wild primary genepool (GP-1) relatives. Gene transfer usually requires human intervention, such as hand pollination, embryo rescue, or the use of male-sterile recipient plants, and in general F1 hybrids are sterile or have significantly reduced fertility. However, for some combinations (particularly with hexaploid forms), the compatibility may be higher and hybridization and gene transfer may occur spontaneously in the field, albeit at low rates. Yet as a rule, natural introgression between homologous genomes in GP-1 is relatively unlikely.

Secondary Gene Pool (GP-2)

The secondary gene pool of bread wheat is rather large and includes all *Aegilops* species (except *Ae. tauschii*), as well as all *Secale* and *Dasypyrum* species, *Thinopyrum elongatum* (Host) D.R. Dewey (= *Elymus elongatus* (Host) Runemark = *Elytrigia elongata* (Host) Nevski = *Agropyron elongatum* (Host) P. Beauv.), and *Thinopyrum intermedium* (Host) Barkworth & D.R. Dewey (= *Elymus hispidus* (Opiz) Melderis = *Elytrigia intermedia* (Host) Nevski = *Agropyron intermedium* (Host) P. Beauv.) (table 22.2). The majority of the secondary gene pool is formed by the goatgrasses (*Aegilops* spp.), which are widely compatible with bread wheat. As mentioned above, the B and D genomes of bread wheat originated from *Aegilops* species.

Tetraploid *Aegilops* Species

Although *Aegilops* species are predominantly autogamous (as is bread wheat), spontaneous in-the-field bread wheat hybrids with most of the tetraploid *Aegilops* species have been reported for a long time in Europe and the Russian Federation. Hybrids are commonest with *Aegilops* species that share the D genome with bread wheat, as this genome has a substantial and well-documented affinity for pairing with the A genome of bread wheat (Lukas and Jahier 1988; Jauhar et al. 1991). Spontaneous hybrids also occur with *Aegilops* species having the U genome (sect. *Aegilops* = sect. *Pleionathera*), which is widely homoeologous with the D genome of bread wheat (Yang et al. 1996; Zhang et al. 1998).

Spontaneous hybrids have been documented for the following tetraploid *Aegilops* species:

- *Ae. biuncialis* Vis. (**UUMM**): Boguslavski 1978; Loureiro et al. 2007a

- *Ae. crassa* Boiss. (**XXDD**): van Slageren 1994

- *Ae. cylindrica* (**CCDD**): Johnston and Parker 1929; Rajhathy 1960; Zohary and Feldman 1962; Gotsov and Panayotov 1972; Donald and Ogg 1991; Zemetra et al. 1998; Guadagnuolo et al. 2001; A. Lu et al. 2002, reported in Jia 2002; Gaines et al. 2008

- *Ae. geniculata* (**UUMM**): Loureiro et al. 2007
- *Ae. neglecta* Req. *ex* Bertol. (**UUXX**): Godron 1854; Boguslavski 1978
- *Ae. triuncialis* (**UUCC**): Godron 1854; Boguslavski 1978
- *Ae. ventricosa* Tausch (**DDNN**): Dosba and Cauderon 1972; Zhao and Kimber 1984

Natural hybrids have also been reported with hexaploid *Ae. juvenalis* (Thell.) Eig (**XXDDUU**), which shares the **D** genome with bread wheat (van Slageren 1994).

In field trials, outcrossing rates of 0.3% were reported for spontaneous hybrids of bread wheat with *Ae. biuncialis* and *Ae. geniculata* (Loureiro et al. 2007a). Also, experimental crosses of bread wheat with most tetraploid *Aegilops* species can be obtained relatively easily (see, e.g., Kimber and Abu Bakar 1979; Claesson et al. 1990; Fernández-Calvín and Orellana 1992; Ye et al. 1997; Logojan and Molnár-Láng 2000; Jacot et al. 2004). The F1 progeny classically have very low fertility or are completely sterile, as they are pentaploids ($5x$) and lack chromosome pairing during meiosis, except for the chromosomes of the **D** genome (Zemetra et al. 1998). Obtaining backcross derivatives for many of these hybrids is a major problem (Popova 1923; Boguslavski 1978; Sharma and Gill 1986; van Slageren 1994). At times, partially fertile plants can be found that may backcross with the wheat parent, as well as self-pollinate at very low rates (Feldman et al. 1995; Loureiro et al. 2006, 2007a; Zaharieva and Monneveux 2006).

For example, hybrids with jointed goatgrass (*Ae. cylindrica*) can spontaneously backcross with both parents to produce viable and partly fertile BC1 progeny. This has been observed both in the greenhouse (Farooq et al. 1990; Zemetra et al. 1998; Wang et al. 2000, 2001, 2002; Lin 2001) and in the field—the latter in the United States (Mayfield 1927; Mallory-Smith et al. 1996; Seefeldt et al. 1998; Mallory-Smith and Snyder 1999; Snyder et al. 2000; Morrison et al. 2002a, b), Switzerland (Guadagnuolo et al. 2001; Schoenenberger et al. 2005, 2006), and the Russian Federation (Gotsov and Panayotov 1972; Boguslavski 1978). The frequency of *Ae. cylindrica* hybrids can accumulate and then increase over generations (Guadagnuolo et al. 2001). There is also evidence that the same may apply for *Ae. geniculata* (Schoenenberger 2005).

In general, the extent of genomic recombination of the F1 hybrids is influenced by pairing regulator genes in wheat, and this can be improved using appropriate mutant or *Ph*-deficient wheat lines (Sears 1972, 1981; Sharma and Gill 1986). There is an even greater probability of gene introgression when the gene of interest is located on a shared genome, such as the **D** genome of *Ae. cylindrica* and *Ae. ventricosa* (Zaharieva and Monneveux 2006). In breeding programs, several genes of interest from tetraploid *Aegilops* species have been introduced to bread wheat, for example, from *Ae. geniculata* (= *Ae. ovata* L.) and *Ae. peregrina* (Hack.) Eig (= *Ae. variabilis* Eig) (Ter-Kuile et al. 1988; Meissner 1991; Friebe et al. 1996a, b; Spetsov et al. 1997).

Diploid *Aegilops* Species

As mentioned above for diploid *Triticum* species, direct introgression from diploid *Aegilops* species into hexaploid wheat is often difficult and may require specialized techniques. The crossability of wheat cultivars, hybrid seed abortion, hybrid lethality, and high male and female sterility of F1 hybrids are major hurdles (The 1973). In addition, diploid *Aegilops* species are isolated from other wheat species and from each other, as they all contain different genomes. Their chromosomes show little affinity for each other and do not pair regularly in interspecific crosses, leading to complete sterility of the F1 hybrids (Feldman et al. 1995). However, the diploid species in sect. *Sitopsis* may be an exception, as their genome (S^b) is closely related to the **B** genome of bread wheat (Singh and Sharma 1997; Maestra and Naranjo 1998; Bálint et al. 2000). For example, *Ae. speltoides* is known to form spontaneous hybrids with bread wheat under natural (Boguslavski 1978) as well as laboratory conditions (Friebe et al. 2000; Mujeeb-Kazi 2005). The extent of genomic recombination between these two species is influenced by the presence of genes in *Ae. speltoides* that suppress the effect of the wheat *Ph1* locus and thereby allow homoeologous chromosome pairing (Feldman and Mello-Sampayo 1967; Dover and Riley 1972). A similar phenomenon has been observed in two other S-genome diploids, *Ae. longissima* Schweinf. & Muschl. and *Ae. sharonensis* Eig, as well as in other diploids (*Ae. caudata* L., **C** genome; *Ae. mutica* Boiss., **T** genome) (Dover and Riley 1972; van Slageren 1994). More information is needed on the S-genome species and on other diploid species to determine the extent of their cross-compatibility with bread wheat. Gene transfer from the only U-genome diploid species (*Ae. umbellulata* Zhuk.) has been obtained only through considerable human intervention (Sears 1956; Zhang et al. 1998).

Secale cereale (Rye)

Wheat and rye ($2n = 2x = 14$) have successfully been crossed in the laboratory, using embryo rescue and chromosome doubling (Lukaszewski und Gustafson 1987). The resulting artificial hybrid, triticale, represents the first crop synthesized by man.

Intergeneric hybridization between wheat and rye is also known to occur in the wild under natural conditions, documented by reports dating back a long way (see, e.g., Leighty 1915; Meister 1921; Dorofeev 1969; Minasyan 1969; Müntzing 1979; Larter 1995). The first such report was by Rimpau (1891), who described a mainly sterile plant with some partially fertile tillers. Nonetheless, the frequency with which wheat × rye hybrids occur in nature is very low, influenced by crossability genes (*Kr* and *Skr*) in the wheat parent (Lein 1943; Tanner and Falk 1981; Gale and Miller 1987; Zeven 1987; Tixier et al. 1998; Tikhenko et al. 2008). The extent of genomic recombination in the F1 hybrids is influenced by the presence or absence of pairing regulator genes *Ph1* and *Ph2* in the wheat parent (Naranjo and Fernández-Rueda 1996; Benavente et al. 1998). F1 progeny are typically completely male sterile. Female fertility is highly variable, but it rarely exceeds 10% (Riley and Chapman 1967; Oettler 1985; Maan 1987). Natural gene flow and introgression in the wild is theoretically possible but very unlikely.

Other Genera

Other intergeneric hybrids of bread wheat can be obtained relatively easily by hand pollination with the following species: *Dasypyrum breviaristatum* (H. Lindb.) Fred. (= *Triticum breviaristatum* H. Lindb.), *D. villosum* (L.) P. Candargy (= *Agropyron villosum* (L.) Link), *Thinopyrum elongatum*, *Thinopyrum intermedium*, *Secale iranicum* Kobyl., *S. montanum* Guss., and *S. sylvestre* Host (table 22.2). The resulting hybrids are occasionally fertile (Smith 1942; Tsvelev 1984). However, all hybrids obtained thus far were produced by manual cross-pollination under greenhouse conditions (see, e.g., Smith 1942; Poehlmann 1959;

Knott 1960; Sharma and Gill 1983a, b; Mujeeb-Kazi et al. 1984; Maan 1987; Cauderon 1994; Jiang et al. 1994; Sharma 1995; Cai and Jones 1997; Chen et al. 2001). To date, there are no reports of naturally occurring hybrids between bread wheat and any of these species.

For GP-2, the literature thus suggests that spontaneous hybridization in the wild is only likely between bread wheat and *Aegilops* species, and between bread wheat and rye. For the former, the extent of cross-compatibility depends on the degree of relatedness between the *Aegilops* and bread wheat species. Cross-compatibility is also influenced by the wheat cultivar and by the genotype of the wild species (Sharma and Gill 1983a; Farooq et al. 1989). In most cases, the pentaploid F1 hybrids are completely sterile, only very occasionally producing viable hybrid seed (Hegde and Waines 2004; Zaharieva and Monneveux 2006). However, partially female-fertile hybrids can produce seeds, and fertility can be restored in as little as two BC generations. Introgression is thus possible under natural conditions, albeit at low rates. The extent of gene flow and introgression depends largely on the ploidy and genomic compatibility of the parents, the potential for restored fertility, and the viability and fitness of the hybrid progeny (reviewed by Mujeeb-Kazi and Hettel 1995; Jauhar and Chibbar 1999; Schneider et al. 2008). Furthermore, depending on their location in either a cultivated or a wild environment, hybrids would probably be subjected to different selection pressures, as has been shown to be the case for other crops, such as common beans (chapter 8) and rice (chapter 18). Fertile hybrids between wheat and its other wild GP-2 relatives may be produced using hand pollination (Feldman et al. 1995; Jiang et al. 1994), but thus far no naturally occurring hybrids have been reported for these combinations.

Tertiary Gene Pool (GP-3)

The tertiary gene pool for bread wheat includes the remaining diploid and polyploid species of the genus *Elymus*, as well as species belonging to genera such as *Agropyron, Avena* L., *Hordeum* L., *Leymus* Hochst., and *Thinopyrum*. The genomes of these species are nonhomologous to those of bread wheat. Wide crossings and gene transfer between bread wheat and these species therefore require considerable human intervention, as well as the use of artificial hybridization techniques such as irradiation, gibberellic acid treatment, callus culture-mediated translocation induction, and embryo culture. They are only possible under laboratory conditions (see, e.g. Smith 1942; Cauderon and Cauderon 1956; Kruse 1969, 1973; Snape et al. 1979; Thomas et al. 1981; Sharma and Gill 1983a, b; Dewey 1984; Forster and Miller 1985; Blanco et al. 1986; Sharma and Baenziger 1986; Maan 1987; Mujeeb-Kazi et al. 1987, 1989; Stich and Snape 1987; Plourde et al. 1989; Ahmad and Comeau 1991; Fedak 1991; Koba et al. 1991; B-R. Lu and Bothmer 1991; Jauhar 1995; Koebner et al. 1995; A. Martín et al. 1999). The hybridization success of these wide crosses is often restricted and depends on the parental genotype, the ploidy level, the crossing procedure, and the direction of the cross (e.g., Mujeeb-Kazi and Kimber 1985; Fedak 1998). Therefore, no spontaneous hybrids between bread wheat and any of these species have been reported.

Outcrossing Rates and Distances

Pollen Flow

As is the case for most other plants, pollen-mediated gene flow in wheat is influenced by a variety of factors, including variation in flowering time, outcrossing rates, the sizes of and distance between populations, wind speed, and humidity (e.g., Heyne and Smith 1967; Jain 1975; Loureiro et al. 2007b). Bread wheat is mainly self-pollinating, and natural outcrossing rates tend to be below 2% (Poehlmann 1959; J. Martin et al. 1976; Redden 1977). Outcrossing rates can be up to 10%, depending on the population density, genotype, and environmental conditions (Harrington 1932; J. Wilson 1968; Griffin 1987; T. Martin 1990; Wiese 1991; Hucl 1996; Enjalbert et al. 1998; Hucl and Matus-Cádiz 2001; Hanson et al. 2005). Wheat-specific crop models have been developed to simulate pollen dispersal and predict cross-pollination rates (Gustafson et al. 2005; Brûlé-Babel et al. 2006; Gaines et al. 2007b).

Long-distance pollen dispersal is also influenced by pollen longevity. In the case of wheat, pollen grains loose their viability quite quickly (within a few hours). Bread wheat pollen is also heavy and very sensitive to environmental conditions (D'Souza 1970; de Vries 1971), which further reduces the possibilities for long-distance cross-pollination. This is in agreement with several pollen flow studies, where no cross-pollination events were recorded beyond 30 m from the source (J. Wilson 1968; Kertesz et al. 1995; Hucl and Matus-Cádiz 2001; A. Lu et al. 2002, reported in Jia 2002; Gatford et al. 2006; Loureiro et al. 2007b).

In some field experiments, however, wheat pollen dispersal and cross-fertilization were detected at distances exceeding 40 m (de Vries 1974; Hanson et al. 2005), 60 m (Jensen 1968; Khan et al. 1973), 300 m (Matus-Cádiz et al. 2004), and even 2.75 km from the pollen source, albeit at very low (< 0.01%) rates (for fully fertile recipient plants, at a commercial-scale level, in the dominant wind direction; Matus-Cádiz et al. 2007). In a pollen flow experiment in the laboratory (moderate vertical mass convection of 10 g/cm per second and a moderate wind speed of 3 m/sec), pollen traveled about 60 m at a height of 1 m (D'Souza 1970).

Isolation Distances

The isolation distances established by regulatory authorities for bread-wheat seed production vary greatly among countries. They range from 3 to 50 m for seed production from normally fertile varieties and from 25 to 200 m for hybrid seed production (table 22.3). Taking into account data from the majority of published studies, separation distances of 30 m are likely to be effective both for maintaining seed-purity standards and with respect to the maximum allowances (0.9%) for GM trait admixtures. Nevertheless, environmental conditions play a major role, and long-distance gene flow may occur under certain circumstances, albeit at low rates.

State of Development of GM Technology

Genetic transformation technology in wheat began in the 1990s (V. Vasil et al. 1990; I. Vasil and Vasil 1992; Becker et al. 1994; Nehra et al. 1995; Cheng et al. 1997). However, the development of a stable and efficient

Table 22.3 Isolation distances for bread wheat (*Triticum aestivum* L.)

Threshold*	Minimum isolation distance (m)	Country	Reference	Comment
n.a. / 0.3%	25	EU	EC 2001	certified seed requirements (hybrid)
n.a. / 0.1%	50	OECD	OECD 2008	foundation (basic) seed requirements
n.a. / 0.3%	20	OECD	OECD 2008	certified seed requirements
0.03% / 0.05%	0	USA	USDA 2008	foundation (basic) seed requirements (non-hybrid)
0.05% / 0.1%	0	USA	USDA 2008	registered seed requirements (non-hybrid)
0.1% / 0.2%	0	USA	USDA 2008	certified seed requirements (non-hybrid)
0.03% / 0.05%	200	USA	USDA 2008	foundation (basic) seed requirements (hybrid)
0.05% / 0.1%	200	USA	USDA 2008	registered seed requirements (hybrid)
0.1% / 0.2%	100	USA	USDA 2008	certified seed requirements (hybrid)
0.01% / 1 seed per pound	3	USA	CCIA 2009	foundation (basic) and registered seed requirements
0.02% / 1 seed per pound	3	USA	CCIA 2009	certified seed requirements
0.01% / n.a.	3	Canada	CSGA 2009	foundation (basic) and registered seed requirements
0.05% / n.a.	3	Canada	CSGA 2009	certified seed requirements

*Maximum percentage of off-type plants permitted in the field / seed impurities of other varieties or off-types permitted. n.a. = not available.

high-throughput methodology for wheat transformation has lagged behind that for most other crops, due to a variety of factors, including (1) the fact that this crop is considered to be outside of the host range of *Agrobacterium*; (2) the effects of genotype on plant regeneration; (3) low transformation efficiencies; and (4) problems with transgene regulation, inheritance, and expression. Today, reliable transformation protocols are available, and field trials of (mainly herbicide resistant) GM wheat plants have been conducted in various countries—Argentina, Australia, Belgium, Canada, Chile, China, Egypt, Germany, Italy, Japan, Lithuania, Spain, Switzerland, the United Kingdom, and the United States (Patnaik and Khurana 2001; Sahrawat et al. 2003; Jones 2005; Jones et al. 2007; I. Vasil 2007). The GM traits studied thus far include fungal and virus resistance, pest and herbicide tolerance, drought tolerance, male sterility and the restoration of fertility, alterations in baking quality, and improved starch quality. Other GM traits might be incorporated into wheat cultivars in the future (W. Wilson et al. 2003). Commercial production, however, has thus far been stopped, as in the case of herbicide-tolerant GM wheat in Canada (Huygen et al. 2003).

Agronomic Management Recommendations

In bread wheat, there are four main ways gene introgression to other cultivated or wild species can occur:

1 through pollen flow from cultivated wheat

2 through volunteers in fields previously planted with wheat

3 through wheat seeds dispersed by seed predators, such as birds and rodents

4 through wheat seeds dispersed by humans during cropping, harvesting, handling, storage, and transport of the crop

To reduce the likelihood of gene introgression, the following management recommendations are suggested:

- Take land-use history into consideration to avoid contamination through wheat volunteers from earlier cropping cycles—not only of the field to be planted, but also of neighboring fields, as the frequency of hybrids resulting from gene flow can accumulate and then increase within a few generations (Guadagnuolo et al. 2001; van Acker et al. 2003, 2004; Brûlé-Babel et al. 2006).

- Shallow tilling of the soil surface several days after harvesting encourages the germination of seeds, and the resulting volunteer plants can then be destroyed, either by mechanical or chemical treatments. In general, volunteer cereals can easily be controlled through crop rotation, the use of herbicides, and/or through mechanical means.

- If flowering overlap with any of the wild relatives identified above is known, spatial isolation is a very effective way to confine pollen, and distances of only a few meters can greatly reduce pollen flow to surrounding wheat fields. No plants should be in the area at distances shorter than those specified in legal regulations for seed production. Temporal isolation, by choosing varieties that do not flower simultaneously with their wild relatives, is another option.

- Take special measures when transporting wheat grains, to avoid the establishment of ferals through seed. Monitor roads for feral wheat plants and, if present, control them by mechanical or chemical means.

Crop Production Area

Bread wheat is the most widely adapted of the cereals, due to its high ($6x$) ploidy level. The total crop area for bread wheat is estimated at 217 million ha (FAO 2009).

Bread wheat is grown in a wide range of temperate environments, from 67° N (Finland, Norway, and the Russian Federation) to 45° S (Argentina), but it is primarily cultivated in more temperate areas (Miller 1987). There are different wheat varieties, including winter, facultative, and spring wheat, each of which is adapted to certain ecoclimatic conditions (Curtis 2002). Bread wheats are grown in N and S America; central, W, and S Asia; Australia; W, central, and E Europe; the Middle East; the Russian Federation; and the Ukraine (Allan 1980; FAO 2009).

Research Gaps and Conclusions

Although some general conclusions may be drawn about the likelihood of gene flow and introgression between wheat and several of its wild relatives, more information on individual wheat wild relatives would be useful. The emphasis should be on tetraploid *Aegilops* species—as they are the wild relatives most prone to spontaneous outcrossing with bread wheat—as well as on diploid species in sect. *Sitopsis* other than *Ae. speltoides*. Studies concerning pollen dispersal and outcrossing rates, and crop × wild relative hybridization as a function of distance, would be very useful. Also, there is very little information available with respect to hybrid vigor and fertility, the extent and primary direction of introgression, and the fate of introgressed alleles in later generations.

Based on the knowledge available about the relationships between the crop and its wild relatives, the likelihood of gene introgression from bread wheat—assuming physical proximity (less than 30 m) and flowering overlap, and being conditional on the nature of the respective trait(s)—is as follows:

- *high* for its cultivated hexaploid *Triticum* subspecies, and for hexaploid *T. zhukovskyi* Menabde & Ericzjan

- *moderate* for its tetraploid relatives *T. timopheevii* and *T. turgidum* (both cultivated and wild), and for the tetraploid **D**-genome species *Aegilops cylindrica* and *Ae. tauschii*

- *low* for the diploids *T. monococcum* and *T. urartu* Thumanjan *ex* Gandilyan; for the wild and weedy *Aegilops* species that have homoeologous genomes (i.e., **D**, **U**, and **S** genomes) to bread wheat (*Ae. biuncialis*, *Ae. columnaris* Zhuk., *Ae. crassa*, *Ae. geniculata*, *Ae. juvenalis*, *Ae. kotschyi* Boiss., *Ae. neglecta*, *Ae. peregrina*, *Ae. speltoides*, *Ae. triuncialis*, *Ae. ventricosa*); for cultivated rye (*Secale cereale*); and for the triticales (*T. neoblaringhemii* (Wittm. *ex* A. Camus) Mac Key, *T. rimpaui* (Wittm.) Mac Key, *T. semisecale* Mac Key) (note that more information is required on the extent of cross-compatibility between bread wheat and the other **S**-genome diploids in *Aegilops* sect. *Sitopsis* (*Ae. bicornis* (Forssk.) Jaub. & Spach, *Ae. longissima*, *Ae. sear-*

sii Schweinf. & Muschl., and *Ae. sharonensis*) to determine whether they need to be added to this group)

The geographical areas of overlap between bread wheat and its sexually compatible wild species are illustrated in map 22.1.

The classification of the wild relatives into these three categories is preliminary. While more studies are being conducted and information is becoming available about actual outcrossing rates and the extent of introgression under natural conditions, additional species may need to be added, and the likelihood category for some species should be reconsidered.

REFERENCES

Ahmad F and Comeau A (1991) A new intergeneric hybrid between *Triticum aestivum* L. and *Agropyron fragile* (Roth.) Candargy: Variation in *A. fragile* for suppression of the wheat *Ph*-locus activity. *Plant Breed.* 106: 275–283.

Allan RE (1980) Wheat. In: Fehr W and Hadley HH (eds.) *Hybridization of Crop Plants* (American Society of Agronomy: Madison, WI, USA), pp. 709–720.

Ammar K, Mergoum M, and Rajaram S (2004) The history and evolution of triticale. In: Mergoum M and Gómez-Macpherson H (eds.) *Triticale: Improvement and Production*. FAO Plant Production and Protection Series No. 179 (Food and Agriculture Organization of the United Nations [FAO]: Rome, Italy), pp. 1–11. ftp: //ftp.fao.org/docrep/fao/009/y5553e/y5553e00.pdf.

Anderson RL and Soper G (2003) Review of volunteer wheat (*Triticum aestivum*) seedling emergence and seed longevity in soil. *Weed Technol.* 17: 620–626.

Badaeva ED, Amosova AV, Muravenko OV, Samatadze TE, Chikida NN, Zelenin AV, Friebe B, and Gill BS (2002) Genome differentiation in *Aegilops*: 3. Evolution of the D-genome cluster. *Plant Syst. Evol.* 231: 163–190.

Badaeva ED, Amosova AV, Samatadze TE, Zoshchuk SA, Shostak NG, Chikida NN, Zelenin AV, Raupp WJ, Friebe B, and Gill BS (2004) Genome differentiation in *Aegilops*: 4. Evolution of the U-genome cluster. *Plant Syst. Evol.* 246: 45–76.

Bálint AF, Kovács G, and Sutka J (2000) Origin and taxonomy of wheat in the light of recent research. *Acta Agron. Hung.* 48: 301–313.

Becker D, Brettschneider R, and Lörz H (1994) Fertile transgenic wheat from microprojectile bombardment of scutellar tissue. *Plant J.* 5: 299–307.

Beckie HJ, Hall LM, and Warwick SI (2001) Impact of herbicide resistant crops as weeds in Canada. In: British Crop Protection Council (ed.) *Weeds 2001: The BCPC Conference Proceedings of an International Conference, Brighton, UK, November 12–15, 2001* (British Crop Protection Council [BCPC]: Farnham, UK) pp. 135–142.

Benavente E, Orellana J, and Fernández-Calvín B (1998) Comparative analysis of the meiotic effects of wheat *ph1b* and *ph2b* mutations in wheat × rye hybrids. *Theor. Appl. Genet.* 96: 1200–1204.

Bennett MD, Rao MK, Smith JB, and Bayliss MW (1973) Cell development in the anther, the ovule, and the young seed of *Triticum aestivum* L. var. 'Chinese Spring'. *Phil. Trans. Roy. Soc. Lond., Ser. B* 66: 39–81.

Beri SM and Anand SC (1971) Factors affecting pollen shedding capacity in wheat. *Euphytica* 20: 327–332.

Biddulph TB, Plummer JA, Setter TL, and Mares DJ (2007) Influence of high temperature and terminal moisture stress on dormancy in wheat (*Triticum aestivum* L.). *Field Crop Res.* 103: 139–153.

Bizimungu B, Collin J, Comeau A, and St-Pierre CA (1998) Hybrid necrosis as a barrier to gene transfer in hexaploid winter wheat × triticale crosses. *Can. J. Plant Sci.* 78: 239–244.

Blanco AC, Fracchiolla V, and Creco B (1986) Intergeneric wheat × barley hybrid. *J. Hered.* 77: 98–100.

Boguslavski RL (1978) Spontannaia guibridizatzia u vidov roda *Aegilops* L. v usloviyah yujnogo Dagestana. *Bull. VIR* 84: 65–68.

Brûlé-Babel AL, Willenborg CJ, Friesen LF, and van Acker RC (2006) Modeling the influence of gene flow and selection pressure on the frequency of a GE herbicide-tolerant trait in non-GE wheat and wheat volunteers. *Crop Sci.* 46: 1704–1710.

Cai X and Jones S (1997) Direct evidence for high level of autosyndetic pairing in hybrids of *Thinopyrum intermedium* and *Th. ponticum* with *Triticum aestivum*. *Theor. Appl. Genet.* 95: 568–572.

Cauderon Y (1994) Cytogénétique et amélioration des plantes: L'exemple des hybrides entre *Triticum* et *Elytrigia*. *C. R. Soc. Biol.* 188: 93–107.

Cauderon Y and Cauderon A (1956) Étude des hybrides F1 entre *Hordeum bulbosum* et *Hordeum secalinum. Ann. Inst. Rech. Agron.* 6: 307–317.

CCIA [California Crop Improvement Association] (2009) Grain Certification Standards. http://ccia.ucdavis.edu/seed_cert/seedcert_index.htm.

Chambers JC and MacMahon JA (1994) A day in the life of a seed: Movements and fates of seeds and their implications for natural and managed systems. *Annu. Rev. Ecol. Syst.* 25: 263–292.

Chaubey NK and Khanna VK (1986) A study of crossability between wheat, triticale, and rye. *Curr. Sci.* 55: 744–745.

Chen Q, Conner RL, Laroche A, and Ahmad F (2001) Molecular cytogenetic evidence for a high level of chromosome pairing among different genomes in *Triticum aestivum-Thinopyrum intermedium* hybrids. *Theor. Appl. Genet.* 102: 847–852.

Cheng M, Fry JE, Pang S, Zhou H, Hironaka CM, Duncan D, Conner TW, and Wan Y (1997) Genetic transformation of wheat mediated by *Agrobacterium tumefaciens*. *Plant Physiol.* 115: 971–980.

Claesson L, Kotimaki M, and Bothmer R von (1990) Crossability and chromosome pairing in some interspecific *Triticum* hybrids. *Hereditas* 112: 49–55.

Cox TS, Raupp WJ, and Gill BS (1994) Leaf rust-resistance genes Lr41, Lr42, and Lr43 transferred from *Triticum tauschii* to common wheat. *Crop Sci.* 34: 339–343.

CSGA [Canadian Seed Growers Association] (2009) Canadian Regulations and Procedures for Pedigreed Seed Crop Production. Circular 6–2005/Rev. 01.4–2009 (Canadian Seed Growers Association [CSGA]: Ottawa, ON, Canada). www.seedgrowers.ca/cropcertification/circular.asp.

Curtis BC (2002) Wheat in the world. In: Curtis BC, Rajaram S, and Gómez-Macpherson H (eds.) *Bread Wheat: Improvement and Production*. FAO Plant Production and Protection Series No. 30 (Food and Agriculture Organization of the United Nations [FAO]: Rome, Italy). www.fao.org/DOCREP/006/Y4011E/y4011e04.htm#bm04.

Darvey NL (1984) Alien wheat bank. *Genetics* 107 (Suppl.): 24.

de Vries AP (1971) Flowering biology of wheat, particularly in view of hybrid seed production—a review. *Euphytica* 20: 152–170.

——— (1974) Some aspects of cross-pollination in wheat (*Triticum aestivum* L.). *Euphytica* 23: 601–622.

Dewey DR (1984) The genomic system of classification as a guide to intergeneric hybridization with the perennial Triticeae. In: Gustafson JP (ed.) *Gene Manipulation in Plant Improvement: Proceedings of the 16th Stadler Genetics Symposium, Columbia, Missouri, USA* (Plenum Press: New York, NY, USA), pp. 209–279.

Donald WW and Ogg AG (1991) Biology and control of jointed goatgrass (*Aegilops cylindrica*), a review. *Weed Technol.* 53: 3–17.

Dorofeev VF (1969) Die Weizen Transkaukasiens und ihre Bedeutung in der Evolution der Gattung *Triticum* L.: II. Formenbildung in Populationen der Weizen Transkaukasiens. *Z. Pflanzenzucht.* 62: 201–230.

Dosba F and Cauderon Y (1972) A new interspecific hybrid: *Triticum aestivum* subsp. *vulgare* × *Aegilops ventricosa*. *Wheat Info. Serv.* 35: 22–24.

Dover GA and Riley R (1972) Prevention of pairing of homoeologous meiotic chromosomes of wheat by an activity of supernumerary chromosomes of *Aegilops*. *Nature* 240: 159–161.

D'Souza VL (1970) Investigations concerning the suitability of wheat as pollen donor for cross-pollination by wind as compared to rye, triticale, and Secalotricum. *Plant Breed.* 63: 246–269.

Dvorak J (1976) The relationship between the genome of *Triticum urartu* and the **A** and **B** genomes of *Triticum aestivum. Can. J. Genet.* 18: 371–377.

———— (1988) Cytogenetical and molecular inferences about the evolution of wheat. In: Miller TE and Koebner RMD (eds.) *Proceedings of the 7th International Wheat Genetics Symposium, Cambridge, UK, July 13–18, 1988,* vol. 1 (Institute of Plant Science Research: Cambridge, UK), pp. 187–192.

Dvorak J, di Terlizzi P, Zhang HB, and Resta P (1993) The evolution of polyploid wheats: Identification of the **A** genome donor species. *Genome* 36: 21–31.

Dvorak J, Luo MC, Yang ZL, and Zhang HB (1998) The structure of the *Aegilops tauschii* genepool and the evolution of hexaploid wheat. *Theor. Appl. Genet.* 67: 657–670.

EC [European Commission] (2001) European Communities (Cereal Seed) Regulations, 2001: S.I. [Statutory Instrument] No. 640/2001, Amending the Council Directive No. 66/402/EEC of 14 June 1966 on the Marketing of Cereal Seed. OJ No. L125 2309–66, July 11, 1966. www.irishstatutebook.ie/2001/en/si/0640.html.

Elbaum R, Zaltzman L, Burgert I, and Fratzl P (2007) The role of wheat awns in the seed dispersal unit. *Science* 316: 884–886.

Enjalbert J, Goldringer I, David J, and Brabant P (1998) The relevance of outcrossing for the dynamic management of genetic resources in predominantly selfing *Triticum aestivum* L. (bread wheat). *Genet. Sel. Evol.* 30: 197–211.

Evans LT, Bingham J, and Roskams MA (1972) The pattern of grain set within ears of wheat. *Aust. J. Biol. Sci.* 25: 1–8.

FAO [Food and Agriculture Organization of the United Nations] (2009) FAOSTAT data. http://faostat.fao.org/site/567/default.aspx.

Farooq S, Iqbal N, and Shah TM (1989) Intergeneric hybridization for wheat improvement: I. Influence of maternal and paternal genotypes on hybrid production. *Cereal Res. Commun.* 17: 17–22.

———— (1990) Intergeneric hybridization for wheat improvement: II. Utilization of the *ph1b* mutant for direct alien introgression into cultivated wheat and production of backcross seed. *Cereal Res. Commun.* 18: 21–26.

———— (1991) Intergeneric hybrids involving the genus *Hordeum.* In: Gupta PK and Tsuschiya T (eds.) *Chromosome Engineering in Plants: Genetics, Breeding, Evolution.* Developments in Plant Genetics and Breeding, 2A (Elsevier: Amsterdam, The Netherlands), pp. 433–438.

———— (1998) Procedures for transferring agronomic traits from alien species to crop plants. In: Slinkard AE (ed.) *Cytogenetics and Evolution: Proceedings of the 9th International Wheat Genetics Symposium, Saskatoon, Saskatchewan, Canada, August 2–7, 1998,* vol. 1 (University of Saskatchewan, University Extension Press: Saskatoon, SK, Canada), pp. 1–7.

Feldman M (1965) Fertility of interspecific F1 hybrids and hybrid derivatives involving tetraploid species of *Aegilops* section *Pleionathera. Evolution* 19: 562–568.

———— (2001) The origin of cultivated wheat. In: Bonjean AP and Angus WJ (eds.) *The World Wheat Book: A History of Wheat Breeding* (Intercept: London, UK), pp. 3–56.

Feldman M and Mello-Sampayo T (1967) Suppression of homoeologous pairing in hybrids of polyploid wheats × *Triticum speltoides. Can. J. Genet. Cytol.* 9: 307–313.

Feldman M, Lipton FGH, and Miller TE (1995) Wheats. In: Simmonds S and Smartt J (eds.) *Evolution of Crop Plants.* 2nd ed. (Longman Scientific and Technical: Harlow, UK), pp. 184–192.

Fernández-Calvín B and Orellana J (1992) Relationship between pairing frequencies and genome affinity estimations in *Aegilops ovata* × *Triticum aestivum* hybrid plants. *Heredity* 68: 165–172.

Forster B and Miller TE (1985) A 5B deficient hybrid between *Triticum aestivum* and *Agropyron junceum. Cereal Res. Commun.* 13: 93–95.

Friebe B, Jiang J, Raupp WJ, McIntosh RA, and Gill BS (1996a) Characterization of wheat alien translocations conferring resistance to diseases and pests: Current status. *Euphytica* 71: 59–87.

Friebe B, Tuleen NA, Badaeva ED, and Gill BS (1996b) Cytogenetic identification of *Triticum peregrinum* chromosomes added to common wheat. *Genome* 39: 272–276.

Friebe B, Qi LL, Nasuda S, Zhang P, Tuleen NA, and Gill BS (2000) Development of a complete set of *Triticum aestivum-Aegilops speltoides* chromosome addition lines. *Theor. Appl. Genet.* 101: 51–58.

Fritz SE and Lukaszewski AJ (1989) Pollen longevity in wheat, rye, and triticale. *Plant Breed.* 102: 31–34.

Gaines TA, Preston C, Byrne PF, Henry WB, and Westra P (2007a) Adventitious presence of herbicide-resistant wheat in certified and farm-saved seed lots. *Crop Sci.* 47: 751–756.

Gaines TA, Byrne PF, Westra P, Nissen SJ, Henry WB, Shaner DL, and Chapman PL (2007b) An empirically derived model of field-scale gene flow in winter wheat. *Crop Sci.* 47: 2308–2316.

Gaines TA, Henry WB, Byrne PF, Westra P, Nissen SJ, and Shaner DL (2008) Jointed goatgrass (*Aegilops cylindrica*) by imidazolinone-resistant wheat hybridization under field conditions. *Weed Sci.* 56: 32–36.

Gale MD and Miller TE (1987) The introduction of genetic variation into wheat. In:

Lupton FGH (ed.) *Wheat Breeding: Its Scientific Basis* (Chapman and Hall: London, UK), pp. 173–210.

Gatford K, Basri Z, Edlington J, Lloyd J, Qureshi J, Brettell R, and Fincher G (2006) Gene flow from transgenic wheat and barley under field conditions. *Euphytica* 151: 383–391.

Gill BS and Raupp WJ (1987) Direct genetic transfers from *Aegilops squarrosa* L. to hexaploid wheat. *Crop Sci.* 27: 445–450.

Gill BS and Waines JG (1978) Paternal regulation of seed development in wheat hybrids. *Theor. Appl. Genet.* 51: 265–270.

Godron DA (1854) De la fécondation naturelle et artificielle des *Aegilops* par le *Triticum. Ann. Sci. Nat., Bot.*, Sér. 3, 19: 215–222.

Gotsov K and Panayotov I (1972) Natural hybridization of male sterile lines of common wheat × *Ae. cylindrica* Host. *Wheat Inform. Service* 33/34: 20–21.

Griffin WB (1987) Out-crossing in New Zealand wheats measured by occurrence of purple grain. *New Zeal. J. Agric. Res.* 30: 287–290.

Guadagnuolo R, Savova-Bianchi D, and Felber F (2001) Gene flow from wheat (*Triticum aestivum* L.) to jointed goatgrass (*Aegilops cylindrica* Host), as revealed by RAPD and microsatellite markers. *Theor. Appl. Genet.* 103: 1–8.

Guedes-Pinto H, Lima-Brito J, Ribeiro-Carvalho C, and Gustafson JP (2001) Genetic control of crossability of triticale with rye. *Plant Breed.* 120: 27–31.

Gustafson DI, Horak MJ, Rempel CB, Metz SG, Gigax DR, and Hucl P (2005) An empirical model for pollen-mediated gene flow in wheat. *Crop Sci.* 45: 1286–1294.

Hammer K (1980) Vorarbeiten zur monographischen Darstellung von Wildpflanzensortimenten: *Aegilops* L. *Kulturpflanze* 28: 33–180.

Hanf M (1982) *Les adventices d'Europe: Leurs plantules, leurs semences*, trans. L Skawron (BASF: Ludwigshafen, Germany). 497 p.

Hanson BD, Mallory-Smith CA, Shafii B, Thill DC, and Zemetra RS (2005) Pollen-mediated gene flow from blue aleurone wheat to other wheat cultivars. *Crop Sci.* 45: 1610–1617.

Harker KN, Clayton GW, Blackshaw RE, O'Donovan JT, Johnson EN, Gan Y, Holm FA, Sapsford KL, Irvine RB, and van Acker RC (2005) Glyphosate-resistant wheat persistence in western Canadian cropping systems. *Weed Sci.* 53: 846–859.

Harlan JR (1992) *Crops and Man.* 2nd ed. (American Society of Agronomy: Madison, WI, USA). 284 p.

Harper JL, Lovell PH, and Moore KG (1970) The shapes and sizes of seeds. *Annu. Rev. Ecol. Syst.* 1: 327–356.

Harrington JB (1932) Natural crossing in wheat, oats, and barley at Saskatoon, Saskatchewan. *Sci. Agric.* (Ottawa) 12: 470–483.

Hegde SG and Waines JG (2004) Hybridization and introgression between bread wheat and wild and weedy relatives in North America. *Crop Sci.* 44: 1145–1155.

Heyne EG and Smith GS (1967) Wheat breeding. In: Quisenberry KS and Reitz LP (eds.) *Wheat and Wheat Improvement*. Agronomy Monograph 13 (American Society of Agronomy: Madison, WI, USA), pp. 269–306.

Hills MJ, Hall LM, Messenger DF, Graf RJ, Beres BL, and Eudes F (2007) Evaluation of crossability between triticale (×*Triticosecale* Wittmack) and common wheat, durum wheat, and rye. *Environ. Biosafety Res.* 6: 249–257.

Hitchcock AS (1951) [1950 on title page; printed in 1951] *Manual of the Grasses of the United States*, rev. A Chase. 2nd ed. USDA [United States Department of Agriculture] Miscellaneous Publication No. 200 (Government Printing Office: Washington, DC, USA). 1051 p.

Hömmö L and Pulli S (1993) Winterhardiness of some winter wheat (*Triticum aestivum*), rye (*Secale cereale*), triticale (×*Triticosecale*), and winter barley (*Hordeum vulgare*) cultivars tested at six locations in Finland. *Agric. Sci. Finl.* 2: 311–327.

Hucl P (1996) Out-crossing rates for 10 Canadian wheat cultivars. *Can. J. Plant Sci.* 76: 423–427.

Hucl P and Matus-Cádiz MA (2001) Isolation distances for minimizing out-crossing in spring wheat. *Crop Sci.* 41: 1348–1351.

Huygen I, Veeman M, and Lerohl M (2003) Cost implications of alternative GE tolerance levels: Non-genetically modified wheat in western Canada. *AgBioForum* 6: 169–177.

Innes RL and Kerber ER (1994) Resistance to leaf rust and stem rust in *Triticum tauschii* and inheritance in hexaploid wheat of resistance transferred from *T. tauschii*. *Genome* 37: 813–822.

Jacot Y, Ammann K, Rufener Al Mazyad P, Chueca C, Davin J, Gressel J, Loureiro I, Wang H, and Benavente E (2004) Hybridization between wheat and wild relatives, a European Union research programme. In: den Nijs HCM, Bartsch D, and Sweet J (eds.) *Introgression from Genetically Modified Plants into Wild Relatives* (CAB International: Wallingford, UK), pp. 1–17.

Jain SK (1975) Population structure and the effects of breeding system. In: Frankel OH and Hawkes JG (eds.) *Crop Genetic Resources for Today and Tomorrow* (Cambridge University Press: Cambridge, UK), pp 15–36.

Jauhar PP (1995) Morphological and cytological characteristics of some wheat × barley hybrids. *Theor. Appl. Genet.* 90: 872–877.

Jauhar PP and Chibbar RN (1999) Chromosome-mediated and direct gene transfers in wheat. *Genome* 42: 570–583.

Jauhar PP, Riera-Lizarazu O, Dewey WG, Gill BS, Crane CF, and Bennett JH (1991)

Chromosome pairing relationships among the **A, B,** and **D** genomes of bread wheat. *Theor. Appl. Genet.* 82: 441–449.

Jensen NF (1968) Results of a survey on isolation requirements for wheat. *Annu. Wheat Newsl.* 15: 26–28.

Jia SR (2002) Studies on gene flow in China—a review. In: *Proceedings of the 7th International Symposium on the Biosafety of Genetically Modified Organisms (ISBGMO), Beijing, China, October 10–16, 2002* (International Society for Biosafety Research [ISBR], n.p.), pp. 92–96. www.isbr.info/symposia/docs/isb gmo.pdf.

Jiang J, Friebe B, and Gill BS (1994) Recent advances in alien gene transfer in wheat. *Euphytica* 73: 199–212.

Johnston CO and Parker JH (1929) *Aegilops cylindrica* Host: A wheat-field weed in Kansas. *Trans. Kans. Acad. Sci.* 32: 80–84.

Jones HD (2005) Wheat transformation: Current technology and applications to grain development and composition. *J. Cereal Sci.* 41: 137–147.

Jones HD, Wilkinson M, Doherty A, and Wu H (2007) High-throughput *Agrobacterium* transformation of wheat: A tool for functional genomics. In: Buck HT, Nisi JE, and Salomón N (eds.) *Wheat Production in Stressed Environments: Proceedings of the 7th International Wheat Conference, Mar del Plata, Argentina, November 27–December 2, 2005.* Vol. 12 in *Developments in Plant Breeding* (Springer: Dordrecht, The Netherlands), pp. 693–699.

Keeler KH (1989) Can genetically engineered crops become weeds? *Bio/Technol.* 7: 1134–1139.

Keeler KH, Turner CE, and Bolick MR (1996) Movement of crop transgenes into wild plants. In: Duke SO (ed.) *Herbicide Resistant Crops: Agricultural, Environmental, Economic, Regulatory, and Technical Aspects* (CRC Press: Boca Raton, FL, USA), pp. 303–330.

Kertesz C, Matuz J, Proksza J, and Kertesz Z (1995) Comparison of variety maintenance methods in wheat. *Annu. Wheat Newsl.* 41: 102.

Khan MN, Heyne EG, and Arp AL (1973) Pollen distribution and the seed set on *Triticum aestivum* L. *Crop Sci.* 13: 223–226.

Khanna VK (1990) Germination, pollen fertility, and crossability between triticale and wheat and reversion patterns in early segregating generations. *Cereal Res. Commun.* 18: 359–362.

Kihara H (1944) Die Entdeckung der **DD**-Analysatoren beim Weizen. *Agric. Hort.* 19: 889–890.

Kihara H (1954) Considerations on the evolution and distribution of *Aegilops* species based on the analyser method. *Cytologia* 19: 336–357.

Kimber G and Abu Bakar M (1979) Wheat hybrid information systems. *Cereal Res. Commun.* 7: 257–260.

Kimber G and Feldman M (1987) *Wild Wheat: An Introduction.* Special Report 353 (College of Agriculture, University of Missouri-Columbia: Columbia, MO, USA). 142 p.

Kimber G and Sears ER (1987) Evolution in the genus *Triticum* and the origin of cultivated wheat. In: Heyne EG (ed.) *Wheat and Wheat Improvement.* 2nd ed. Agronomy Monograph 13 (American Society of Agronomy, Crop Science Society of America, and Soil Science Society of America: Madison, WI, USA), pp. 154–164.

Kimber G and Zhao YH (1983) The D genome of the Triticeae. *Can. J. Genet. Cytol.* 25: 581–589.

Kirby EJM (2002) Botany of the wheat plant. In: Curtis BC, Rajaram S, and Gómez-Macpherson H (eds.) *Bread Wheat: Improvement and Production.* FAO Plant Production and Protection Series No. 30 (Food and Agriculture Organization of the United Nations [FAO]: Rome, Italy). www.fao.org/DOCREP/006/ Y4011E/y4011e05.htm#bm05.

Knott DR (1960) The inheritance of rust resistance: VI. The transfer of stem rust resistance from *Agropyron elongatum* to common wheat. *Can. J. Plant Sci.* 41: 109–123.

Koba T, Handa T, and Shimada T (1991) Efficient production of wheat-barley hybrids and preferential elimination of barley chromosomes. *Theor. Appl. Genet.* 81: 285–292.

Kobyljanskij VD (1989) *Rye.* Vol. 2, part 1 in *Flora of Cultivated Plants of the U.S.S.R.* (Agropromizdat: Leningrad, Russia) [in Russian]. 367 p.

Koebner RMD, Martin PK, and Anamthawat-Jonnson K (1995) Multiple branching stems in a hybrid between wheat (*Triticum aestivum*) and lymegrass (*Leymus mollis*). *Can. J. Bot.* 73: 1504–1507.

Komatsuzaki M and Endo O (1996) Seed longevity and emergence of volunteer wheat in upland fields. *Weed Res. Japan* 41: 197–204.

Körber-Grohne U (1988) *Nutzpflanzen in Deutschland: Kulturgeschichte und Biologie* (K. Theiss: Stuttgart, Germany). 490 p. .

Kruse A (1969) Intergeneric hybrids between *Triticum aestivum* L. (v. Koga II 2n = 42) and *Avena sativa* L. (v. Stal 2n = 42) with pseudogamous seed formation, In *Royal Veterinary and Agricultural College Yearbook* (Royal Veterinary and Agricultural College: Copenhagen, Denmark), pp. 188–200.

——— (1973) *Hordeum* × *Triticum* hybrids. *Hereditas* 73: 157–161.

Larter EN (1995) Triticale. In: Simmonds S and Smartt J (eds.) *Evolution of Crop Plants.* 2nd ed. (Longman Scientific and Technical: Harlow, UK), pp. 181–183.

Leighty CE (1915) Natural wheat-rye hybrid. *Am. Soc. Agron.* 7: 209–216.

Leighty CE and Sando WJ (1924) The blooming of wheat flowers. *J. Agric. Res.* 27: 231–244.

Lein A (1943) The genetical basis of the crossability between wheat and rye. *Z. Indukt. Abstam.-u-Vererb.* 81: 28–59.

Lelley T (1992) Triticale, still a promise. *Plant Breed.* 109: 1–17.

Lersten NR (1987) Morphology and anatomy of the wheat plant. In: Heyne EG (ed.) *Wheat and Wheat Improvement.* 2nd ed. Agronomy Monograph 13 (American Society of Agronomy, Crop Science Society of America, and Soil Science Society of America: Madison, WI, USA), pp. 33–75.

Lin Y (2001) Risk assessment of *Bar* gene transfer from **B** and **D** genomes of transformed wheat (*Triticum aestivum*) lines to jointed goatgrass (*Aegilops cylindrica*). *J. Anhui Agric. Univ.* 28: 115–118.

Logojan AA and Molnár-Láng M (2000) Production of *Triticum aestivum-Aegilops biuncialis* chromosome additions. *Cereal Res. Commun.* 28: 221–228.

Loureiro I, Escorial MC, García-Baudín JM, and Chueca MC (2006) Evidence of natural hybridization between *Aegilops geniculata* and wheat under field conditions in central Spain. *Environ. Biosafety Res.* 5: 105–109.

———— (2007a) Hybridization between wheat (*Triticum aestivum*) and the wild species *Aegilops geniculata* and *Ae. biuncialis* under experimental field conditions. *Agric. Ecosys. Environ.* 120: 384–390.

Loureiro I, Escorial MC, González Andujar JL, García-Baudin JM, and Chueca MC (2007b) Wheat pollen dispersal under semiarid field conditions: Potential outcrossing with *Triticum aestivum* and *Triticum turgidum*. *Euphytica* 156: 25–37.

Lu AZ, Wang TY, and Wang HB (2002) Gene flow between wheat cultivars itself and its wild relatives. Master's thesis, Graduate School of Chinese Academy of Agricultural Sciences, China [English abstract].

Lu B-R and Bothmer R von (1991) Production and cytogenetic analysis of the intergeneric hybrids between nine *Elymus* species and common wheat. *Euphytica* 58: 81–95.

Lucas H and Jahier J (1988) Phylogenetic relationships in some diploid species of Triticinae: Cytogenetic analysis of interspecific hybrids. *Theor. Appl. Genet.* 75: 498–502.

Lukaszewski AJ and Gustafson JP (1987) Cytogenetics of triticale. In: Janick J (ed.) *Plant Breeding Reviews*, vol. 5 (AVI: New York, NY, USA), pp. 41–94.

Maan SS (1987) Interspecific and intergeneric hybridization in wheat. In: Heyne EG (ed.) *Wheat and Wheat Improvement.* 2nd ed. Agronomy Monograph 13 (American Society of Agronomy, Crop Science Society of America, and Soil Science Society of America: Madison, WI, USA), pp. 453–461.

Mac Key J (2005) Wheat: Its concept, evolution, and taxonomy. In: Royo C, Nachit MM, Di Fonzo N, Araus JL, Pfeiffer WH, and Slafer GA (eds.) *Durum Wheat Breeding: Current Approaches and Future Strategies*, vol. 1 (Food Products Press: Binghamton, NY, USA), pp. 3–61.

Maestra B and Naranjo T (1998) Homoeologous relationships of *Aegilops speltoides* to bread wheat. *Theor. Appl. Genet.* 97: 181–186.

Mallory-Smith CA and Snyder J (1999) Potential for gene flow between wheat (*Triticum aestivum*) and jointed goatgrass (*Aegilops cylindrica*) in the field. In: Lutman PJW (ed.) *Gene Flow and Agriculture: Relevance for Transgenic Crops; Proceedings of a Symposium Held at the University of Keele, Staffordshire, April 12–14, 1999.* BCPC Symposium Proceedings 72 (British Crop Protection Council [BCPC]: Farnham, UK), 165–169.

Mallory-Smith CA, Hansen J, and Zemetra RS (1996) Gene transfer between wheat and *Aegilops cylindrica*. In: Brown H, Cussans GW, Devine MD, Duke SO, Fernández-Quintanilla C, Helweg A, Labrada RE, Landes M, Kudsk P, and Streibig JC (eds.) *Proceedings of the Second International Weed Control Congress, Copenhagen, Denmark, June 25–28, 1996* (Department of Weed Control and Pesticide Ecology: Slagelse, Denmark), pp. 441–445.

Mandy G (1970) *Pflanzenzüchtung—Kurz und Bündig* (Deutscher Landwirtschaftsverlag: Berlin, Germany). 336 p.

Marañon T (1987) Ecología del polimorphismo somático de semillas y la sinaptospermia en *Aegilops neglecta* Req. ex Bertol. *Anales Jard. Bot. Madrid* 44: 97–107.

Martín A, Álvarez JB, Martín LM, Barro F, and Ballesteros JJ (1999) The development of *Tritordeum*: A novel cereal for food processing. *J. Cereal Sci.* 30: 85–95.

Martin JH, Leonard WH, and Stamp DL (1976) *Principles of Field Crop Production* (Macmillan: New York, NY, USA). 1118 p.

Martin TJ (1990) Outcrossing in twelve hard red winter wheat cultivars. *Crop Sci.* 30: 59–62.

Matsuoka Y, Takumi S, and Kawahara T (2007) Natural variation for fertile triploid F1 hybrid formation in allohexaploid wheat speciation. *Theor. Appl. Genet.* 115: 509–518.

Matus-Cádiz MA, Hucl P, Horak MJ, and Blomquist LK (2004) Gene flow in wheat at the field scale. *Crop Sci.* 44: 718–727.

Matus-Cádiz MA, Hucl P, and Dupuis B (2007) Pollen-mediated gene flow in wheat at the commercial scale. *Crop Sci.* 47: 573–581.

Mayfield L (1927) Goatgrass—a weed pest of central Kansas wheat fields. *Kans. Agric. Student* 7: 40–41.

McFadden ES and Sears ER (1946) The origin of *Triticum spelta* and its free-threshing hexaploid relatives. *J. Hered.* 37: 81–89 and 107–116.

Meissner C (1991) Incorporation of disease resistance from the genus *Aegilops* into *Triticum aestivum*. In: Panayotov I and Pavlova S (eds.) *Proceedings of International Symposium, Wheat Breeding Prospects and Future Approaches, Albena,*

Bulgaria, June 4–8, 1990 (Agricultural Academy: General Toshevo, Bulgaria), pp. 392–397.

Meister GK (1921) Natural hybridization of wheat and rye in Russia. *J. Hered.* 12: 467–470.

Mello-Sampayo T (1971) Genetic regulation of meiotic chromosome pairing by chromosome 3D of *Triticum aestivum*. *Nature New Biol.* 230: 22–23.

Miller TE (1987) Systematics and evolution. In: Lupton FGH (ed.) *Wheat Breeding: Its Scientific Basis* (Chapman and Hall: London, UK), pp. 1–30.

Minasyan K (1969) Natural hybridization in wheat under the influence of changed environment: I. Natural wheat-rye amphidiploids. *Genetika SSSR* 5: 17–27.

Miyashita NT, Mori N, and Tsunewaki K (1994) Molecular variation in chloroplast DNA regions in ancestral species of wheat. *Genetics* 137: 883–889.

Morishita DW (1996) Biology of jointed goatgrass. In: Jenks B (ed.) *Pacific Northwest Jointed Goatgrass Conference, Pocatello, Idaho, USA, February 21–29, 1996* (University of Nebraska: Lincoln, NE, USA), pp. 7–9.

Morris R and Sears ER (1967) The cytogenetics of wheat and its relatives. In: Quisenberry KS and Reitz LP (eds.) *Wheat and Wheat Improvement*. Agronomy Monograph 13 (American Society of Agronomy: Madison, WI, USA), pp. 9–87.

Morrison LA, Crémieux LC, and Mallory-Smith CA (2002a) Infestations of jointed goatgrass (*Aegilops cylindrica*) and its hybrids with wheat in Oregon wheat fields. *Weed Sci.* 50: 737–747.

Morrison LA, Riera-Lizarazu O, Crémieux L, and Mallory-Smith CA (2002b) Jointed goatgrass (*Aegilops cylindrica* Host) × wheat (*Triticum aestivum* L.) hybrids: Hybridization dynamics in Oregon wheat fields. *Crop Sci.* 42: 1863–1872.

Mujeeb-Kazi A (2005) Wide crosses for durum wheat improvement. In: Royo C, Nachit MM, Di Fonzo N, Araus JL, Pfeiffer WH, and Slafer GA (eds.) *Durum Wheat Breeding: Current Approaches and Future Strategies*, vol. 1 (Food Products Press: Binghamton, NY, USA), pp. 703–743.

Mujeeb-Kazi A and Hettel GP (eds.) (1995) *Utilizing Wild Grass Biodiversity in Wheat Improvement: 15 Years of Wide Cross Research at CIMMYT*. CIMMYT Research Report No. 2 (International Maize and Wheat Improvement Center [CIMMYT, or Centro Internacional de Mejoramiento de Maíz y Trigo]: México, DF, Mexico). 140 p.

Mujeeb-Kazi A and Kimber G (1985) The production, cytology, and practicality of wide hybrids in the Triticeae. *Cereal Res. Commun.* 13: 111–124.

Mujeeb-Kazi A, Roldan S, and Miranda JL (1984) Intergeneric hybrids of *Triticum aestivum* with *Agropyron* and *Elymus* species. *Cereal Res. Commun.* 12: 75–79.

Mujeeb-Kazi A, Roldan S, Suh DY, Sitch LA, and Farooq S (1987) Production and cytogenetic analysis of hybrids between *Triticum aestivum* and some caespitose *Agropyron* species. *Genome* 29: 537–553.

Mujeeb-Kazi A, Roldan S, Suh DY, Ter-Kuile N, and Farooq S (1989) Production and cytogenetics of *Triticum aestivum* L. hybrids with some rhizomatous *Agropyron* species. *Theor. Appl. Genet.* 77: 162–168.

Müntzing A (1979) *Triticale: Results and Problems*. Advances in Plant Breeding 10 (Paul Parey: Berlin and Hamburg, Germany). 103 p.

Naranjo T and Fernández-Rueda P (1996) Pairing and recombination between individual chromosomes of wheat and rye in hybrids carrying the *ph1b* mutation. *Theor. Appl. Genet.* 93: 242–248.

Nehra NS, Chibbar RN, and Kartha KK (1995) Wheat transformation: Methods and prospects. *Plant Breed. Abst.* 65: 803–808.

Nkongolo KKC, St-Pierre CA, and Comeau A (1991) Effect of parental genotypes, cross direction, and temperature on the crossability of bread wheat with triticale and on the viability of F1 embryos. *Ann. Appl. Biol.* 118: 161–168.

OECD [Organization for Economic Cooperation and Development] (2008) OECD Scheme for the Varietal Certification of Cereal Seed Moving in International Trade: Annex VIII to the Decision. C(2000)146/FINAL. OECD Seed Schemes, February 2008. www.oecd.org/document/0/0,3343,en_2649_33905_1933504_1_1_1_1,00.html.

Oettler G (1985) The effect of wheat cytoplasm on the synthesis of wheat × rye hybrids. *Theor. Appl. Genet.* 71: 467–471.

——— (2005) The fortune of a botanical curiosity—triticale: Past, present and future. *J. Agric. Sci.* 143: 329–346.

Özkan H, Brandolini A, Schäfer-Pregl R, and Salamini F (2002) AFLP analysis of a collection of tetraploid wheats indicates the origin of emmer and hard wheat domestication in southeast Turkey. *Mol. Biol. Evol.* 19: 1797–1801.

Patnaik D and Khurana P (2001) Wheat biotechnology: A minireview. *EJB* [Electronic Journal of Biotechnology] (online) 4(2): 74–102. www.ejbiotechnology.info/content/vol4/issue2/full/4/4.pdf.

Peart MH (1981) Further experiments on the biological significance of the morphology of seed-dispersal units in grasses. *J. Ecol.* 69: 425–436.

Percival J (1921) *The Wheat Plant* (Duckworth: London, UK). 463 p.

Pickett AA (1989) A review of seed dormancy in self-sown wheat and barley. *Plant Var. Seeds* 2: 131–146.

——— (1993) Cereals: Seed shedding, dormancy, and longevity. *Asp. Appl. Biol.* 35: 17–28.

Plourde A, Comeau A, Fedak G, and St-Pierre CA (1989) Intergeneric hybrids of *Triticum aestivum* × *Leymus multicaulis*. *Genome* 32: 282–287.

Poehlmann JM (1959) *Breeding of Field Crops* (Henry Holt: New York, NY, USA). 427 p.

Popova G (1923) Wild species of *Aegilops* and their mass-hybridisation with wheat in Turkestan. *Bull. Appl. Bot.* 13: 475–482.

Prakash S and Singhal NC (2003) Natural cross-pollination in bread wheat (*Triticum aestivum*). *Seed Res.* 31: 22–26.

Rabinovich SV (1998) Importance of wheat-rye translocations for breeding modern cultivars of *Triticum aestivum* L. *Euphytica* 100: 323–340.

Rajhathy T (1960) Continuous spontaneous crosses between *Aegilops cylindrica* and *Triticum aestivum*. *Wheat Info. Syst.* 11: 20.

Raupp WJ, Browder LE, and Gill BS (1983) Leaf rust resistance in *Aegilops squarrosa*, its transfer and expression in common wheat (*Triticum aestivum* L.). *Phytopathol.* 73: 818.

Raupp WJ, Amri A, Hatchett JH, Gill BS, Wilson DL, and Cox TS (1993) Chromosomal location of Hessian fly-resistance genes H22, H23, and H24 derived from *Triticum tauschii* in the D genome of wheat. *J. Hered.* 84: 142–145.

Redden RJ (1977) Natural outcrossing in wheat substitution lines. *Aust. J. Agric. Res.* 28: 763–768.

Riley R and Chapman V (1958) Genetic control of cytologically diploid behaviour of hexaploid wheat. *Nature* 182: 713–715.

—— (1967) The inheritance in wheat of crossability with rye. *Genet. Res.* 9: 259–267.

Riley R, Unrau J, and Chapman V (1958) Evidence on the origin of the B genome of wheat. *J. Hered.* 49: 91–98.

Rimpau W (1891) Kreuzungsprodukte landwirtschaftlicher Kulturpflanzen. *Land. Jahrb.* 20: 335–371.

Sahrawat AK, Becker D, Lutticke S, and Lorz H (2003) Genetic improvement of wheat via alien gene transfer, an assessment. *Plant Sci.* 165: 1147–1168.

Sarkar P and Stebbins GL (1956) Morphological evidence concerning the origin of the B genome in wheat. *Am. J. Bot.* 43: 297–304.

Schneider A, Molnár I, and Molnár-Láng M (2008) Utilisation of *Aegilops* (goatgrass) species to widen the genetic diversity of cultivated wheat. *Euphytica* 163: 1–19.

Schoenenberger N (2005) Genetic and ecological aspects of gene flow from wheat (*Triticum aestivum* L.) to *Aegilops* L. species. PhD dissertation, University of Neuchâtel. http://doc.rero.ch/lm.php?url=1000,40,4,20061003105103-JU/these_SchoenenbergerN.pdf.

Schoenenberger N, Felber F, Savova-Bianchi D, and Guadagnuolo R (2005) Introgression of wheat DNA markers from A, B, and D genomes in early generation progeny of *Aegilops cylindrica* Host × *Triticum aestivum* L. hybrids. *Theor. Appl. Genet.* 111: 1338–1346.

Schoenenberger N, Guadagnuolo R, Savova-Bianchi D, Knüpfer P, and Felber F

(2006) Molecular analysis, cytogenetics, and fertility of introgression lines from transgenic wheat to *Aegilops cylindrica* Host. *Genetics* 174: 2061–2070.

Sears ER (1956) The transfer of leaf-rust resistance from *Aegilops umbellulata* to wheat. *Genetics in Plant Breeding: Report of a Symposium, Biology Dept., Brookhaven National Laboratory, Upton, New York, May 21–23, 1956.* Brookhaven Symposium in Biology 9 (Brookhaven National Laboratory: Upton, NY, USA), pp. 1–22.

——— (1972) Chromosome engineering in wheat. In: Kimber G and Redei GP (eds.) *Stadler Genetics Symposia*, vol. 4 (Agricultural Experiment Station, University of Missouri-Columbia: Columbia, MO, USA), pp. 25–38.

——— (1976) Genetic control of chromosome pairing in wheat. *Annu. Rev. Genet.* 10: 31–51.

——— (1981) Transfer of alien genetic material to wheat. In: Evans LT and Peacock WJ (eds.) *Wheat Science: Today and Tomorrow* (Cambridge University Press: Cambridge, UK), pp. 75–89.

Seefeldt SS, Zemetra R, Young FL, and Jones SS (1998) Production of herbicide-resistant jointed goatgrass (*Aegilops cylindrica*) × wheat (*Triticum aestivum*) hybrids in the field by natural hybridization. *Weed Sci.* 46: 632–634.

Sencer HA and Hawkes JG (1980) On the origin of cultivated rye. *Biol. J. Linn. Soc.* 13: 299–313.

Sharma HC (1995) How wide can a wide cross be? *Euphytica* 82: 43–64.

Sharma HC and Baenziger PS (1986) Production, morphology, and cytogenetic analysis of *Elymus caninus* (*Agropyron caninum*) × *Triticum aestivum* F1 hybrids and backcross 1 derivatives. *Theor. Appl. Genet.* 71: 750–756.

Sharma HC and Gill BS (1983a) Current status of wide hybridization in wheat. *Euphytica* 32: 17–31.

——— (1983b) New hybrids between *Agropyron* and wheat. *Theor. Appl. Genet.* 66: 111–121.

——— (1986) The use of *ph1* gene in direct transfer and search for *Ph*-like genes in polyploid *Aegilops* species. *Z. Pflanzenzucht.* 96: 1–7.

Singh JS and Sharma TR (1997) Morpho-cytogenetics of *Triticum aestivum* L. × *Aegilops speltoides* Tausch hybrids. *Wheat Info. Serv.* 84: 51–52.

Smith DC (1942) Intergeneric hybridization of cereals and other grasses. *J. Agric. Res.* 64: 33–45.

Snape JW, Chapman V, Moss J, Blanchard CE, and Miller TE (1979) The cross-abilities of wheat varieties with *Hordeum bulbosum*. *Heredity* 42: 291–298.

Snyder J, Mallory-Smith CA, Balter S, Hansen J, and Zemetra RS (2000) Seed production on *Triticum aestivum* by *Aegilops cylindrica* hybrids in the field. *Weed Sci.* 48: 588–593.

Spetsov P, Mingeot D, Jacquemin JM, Samardjieva K, and Marinova E (1997) Transfer of powdery mildew resistance from *Aegilops variabilis* into bread wheat. *Euphytica* 93: 49–54.

Stich LA and Snape JW (1987) Factors affecting haploid production in wheat using the *Hordeum bulbosum* system: 1. Genotypic and environmental effects on pollen grain germination, pollen tube growth, and the frequency of fertilization. *Euphytica* 36: 483–496.

Sukopp U and Sukopp H (1993) Das Modell der Einführung und Einbürgerung nicht einheimischer Arten. *Gaia* 5: 268–288.

Tanner DG and Falk DE (1981) The interaction of genetically controlled crossability in wheat and rye. *Can. J. Genet. Cytol.* 23: 27–32.

Ter-Kuile N, Nabors M, and Mujeeb-Kazi A (1988) Callus-culture induced amphiploids of *Triticum aestivum* and *T. turgidum* × *Aegilops variabilis* F1 hybrids: Production, cytogenetics, and practical significance; 80th Annual Meeting of the American Society for Agronomy, Anaheim, CA, USA. *Agron. Abst.* 80: 98.

The TT (1973) Chromosome location of genes conditioning stem rust resistance transferred from diploid to hexaploid wheat. *Nature New Biol.* 241: 256.

Thomas JB, Kaltsikes PJ, and Anderson RG (1981) Relation between wheat-rye crossability and seed set of common wheat after pollination with other species in the Hordeae. *Euphytica* 30: 121–127.

Tikhenko N, Rutten T, Voylokov A, and Houben A (2008) Analysis of hybrid lethality in F1 wheat-rye hybrid embryos. *Euphytica* 159: 367–375.

Tixier MH, Sourdille P, Charmet G, Gay G, Jaby C, Cadalen T, Bernard S, Nicolas P, and Bernard M (1998) Detection of QTLs for crossability in wheat using a doubled-haploid population. *Theor. Appl. Genet.* 97: 1076–1082.

Treu R and Emberlin J (2000) Pollen dispersal in the crops maize (*Zea mays*), oilseed rape (*Brassica napus* ssp. *oleifera*), potatoes (*Solanum tuberosum*), sugar beet (*Beta vulgaris* ssp. *vulgaris*), and wheat (*Triticum aestivum*). Report for the Soil Association from the National Pollen Research Unit (Soil Association: Bristol, UK). 54 p.

Tsegaye S (1996) Estimation of outcrossing rate in landraces of tetraploid wheat (*Triticum turgidum* L.). *Plant Breed.* 115: 195–197.

Tsvelev NN (1984) *Grasses of the Soviet Union*, vol. 1 (A.A. Balkema: Rotterdam, The Netherlands). 568 p.

Ueno K and Itoh H (1997) Cleistogamy in wheat: Genetic control and the effect of environmental conditions. *Cereal Res. Commun.* 25: 185–189.

USDA [United States Department of Agriculture] (2008) 7 CFR [Code of Federal Regulations] § 201.76. Federal Seed Act Regulations: Minimum Land, Isolation, Field, and Seed Standards. Source: 59 FR [Federal Register] 64516, Dec. 14, 1994, as amended at 65 FR 1710, Jan. 11, 2000 (Government Printing Office: Washington, DC, USA), pp. 372–378. www.access.gpo.gov/nara/cfr/waisidx_08/7cfr201_08.html.

van Acker RC, Brûlé-Babel AL, Friesen LF, and Entz MH (2003) GM/non-GM wheat co-existence in Canada: Roundup Ready(r) wheat as a case study. In: Boelt B (ed.) *1st European Conference on the Coexistence of Genetically Modified Crops with Conventional and Organic Crops, Helsingor, Denmark, November 13–14, 2003* (Danish Institute of Agricultural Sciences [DIAS]: Slagelse, Denmark), pp. 60–68.

van Acker RC, Brûlé-Babel AL, and Friesen LF (2004) Intraspecific gene movement can create environmental risk: The example of Roundup Ready(r) wheat in western Canada. In: Breckling B and Verhoeven R (eds.) *Risk, Hazard, Damage: Specification of Criteria to Assess Environmental Impact of Genetically Modified Organisms; Proceedings of the International Symposium of the Ecological Society of Germany, Austria, and Switzerland, Specialist Group on Gene Ecology, Bremen University, Hanover, December 8–9, 2003*. Naturschutz and Biologische Viefalt, vol. 1 (Bundesamt für Naturschutz: Bonn, Germany), pp. 37–47.

van Slageren MW (1994) *Wild Wheats: A Monograph of Aegilops L. and Amblyopyrum (Jaub. & Spach) Eig (Poaceae)*. Wageningen Agricultural University Papers 94–7 (Wageningen Agricultural University: Wageningen, The Netherlands; International Center for Agricultural Research in the Dry Areas [ICARDA]: Aleppo, Syria). 513 p.

Vasil IK (2007) Molecular genetic improvement of cereals: Transgenic wheat (*Triticum aestivum* L.). *Plant Cell Rep.* 26: 1133–1154.

Vasil IK and Vasil V (1992) Advances in cereal protoplast research. *Physiol. Plant.* 85: 279–283.

Vasil V, Redway FA, and Vasil IK (1990) Regeneration of plants from embryogenic suspension culture protoplasts of wheat. *Bio/Technol.* 8: 429–433.

von der Lippe M and Kowarik I (2007) Crop seed spillage along roads: A factor of uncertainty in the containment of GMO. *Ecography* 30: 483–490.

Waines JG and Barnhart D (1992) Biosystematic research in *Aegilops* and *Triticum*. *Hereditas* 116: 207–212.

Waines JG and Hegde SG (2003) Intraspecific gene flow in bread wheat as affected by reproductive biology and pollination ecology of wheat flowers. *Crop Sci.* 43: 451–463.

Wang ZN, Hang A, Hansen J, Burton C, Mallory-Smith CA, and Zemetra RS (2000) Visualization of A- and B-genome chromosomes in wheat (*Triticum aestivum* L.) × jointed goatgrass (*Aegilops cylindrica* Host) backcross progenies. *Genome* 43: 1038–1044.

Wang ZN, Zemetra RS, Hansen J, and Mallory-Smith CA (2001) The fertility of wheat × jointed goatgrass hybrid and its backcross progenies. *Weed Sci.* 49: 340–345.

Wang ZN, Zemetra RS, Hansen J, Hang A, Mallory-Smith CA, and Burton C (2002) Determination of the paternity of wheat (*Triticum aestivum* L.) × jointed goatgrass (*Aegilops cylindrica* Host) BC1 plants by using genomic *in situ* hybridization (GISH) technique. *Crop Sci.* 42: 939–943.

Wicks CL, Felton WL, Murison RD, and Martin RJ (2000) Changes in fallow weed species in continuous wheat in northern New South Wales, 1981–90. *Aust. J. Exp. Agric.* 40: 831–842.

Wiese MV (1991) *Compendium of Wheat Diseases.* 2nd ed. (APS [American Phytopathological Society] Press: St. Paul, MN, USA). 106 p.

Wilson JA (1968) Problems in hybrid wheat breeding. *Euphytica* 17: 13–34.

Wilson WW, Janzen EL, and Dahl BL (2003) Issues in development and adoption of genetically modified (GM) wheats. *AgBioForum* 6: 101–112.

Yamashita K (1980) Origin and dispersion of wheats with special reference to peripheral diversity. *Z. Pflanzenzucht.* 84: 122–132.

Yang YC, Tuleen NA, and Hart GE (1996) Isolation and identification of *Triticum aestivum* L. em[end]. Thell. cv. 'Chinese Spring'-*T. peregrinum* Hackel disomic chromosome-addition lines. *Theor. Appl. Genet.* 92: 591–598.

Ye X, Xu H, Li Z, Du L, and Zhao L (1997) Crossability analysis of reciprocal hybridizations between common wheat (*Triticum aestivum* L.) and white rye (*Secale cereale* L.) and *Aegilops ovata* L. *Sci. Agric. Sin.* 30: 91–93 [English abstract].

Zaharieva M and Monneveux P (2006) Spontaneous hybridization between bread wheat (*Triticum aestivum* L.) and its wild relatives in Europe. *Crop Sci.* 46: 512–527.

Zemetra RS, Hansen J, and Mallory-Smith CA (1998) Potential for gene transfer between wheat (*Triticum aestivum*) and jointed goatgrass (*Aegilops cylindrica*). *Weed Sci.* 46: 313–317.

Zeven AC (1987) Crossability percentages of some bread wheat varieties and lines with rye. *Euphytica* 36: 299–319.

Zhang H, Jia J, Gale MD, and Devos KM (1998) Relationships between the chromosomes of *Aegilops umbellulata* and wheat. *Theor. Appl. Genet.* 96: 69–75.

Zhao YH and Kimber G (1984) New hybrids with D-genome wheat relatives. *Genetics* 106: 509–515.

Zohary D and Feldman M (1962) Hybridization between amphiploids and the evolution of polyploids in the wheat (*Aegilops-Triticum*) group. *Evolution* 16: 44–61.

Zohary D and Hopf M (2000) *Domestication of Plants in the Old World: The Origin and Spread of Cultivated Plants in West Asia, Europe, and the Nile Valley.* 3rd ed. (Oxford University Press: New York, NY, USA). 328 p.

Glossary

....

Accession   A plant sample (seed, tissue, DNA) stored in a **genebank** for conservation or use

Adventitious presence   Accidental or spontaneous occurrence of **GM traits** in a seed sample or in a cultivated crop

Allele   One of multiple alternative states of a **gene**, found at the same location on a chromosome

Allogamy   Reproduction by cross-fertilization

Alloploid   Polyploid of **hybrid** origin where at least one set of chromosomes is derived from an unrelated parent. Depending on the number of **haploid** chromosome sets, alloploids are referred to as allodiploids, allotetraploids, allohexaploids, and so on.

Aneuploid   Individual with a chromosome number that is not an exact multiple of the **haploid** number

Annual   Plant completing its lifecycle and producing seeds within one year or less

Anther   **Pollen**-producing part of the flower

Anthesis   **Dehiscence** of **anthers** when **pollen** is shed

Apomixis   Seed production in the absence of sexual fertilization

Autochory   Fruit and seed dispersal by means of some kind of physical expulsion, often explosively

Autogamy   Reproduction by self-fertilization

Autoploid   Individual or cell with two or more copies of a single **haploid** set

Awn   Bristle-like appendage (common in cereals and grasses)

Backcrossing   Crossing of a **hybrid** with one of its parents

Biodiversity   The totality of living organisms, including their **genetic materials** and the ecosystems they are a part of

Bridging   A mechanism by which **genetic material** can be transferred to sexually noncompatible plants via intermediary **species**. A bridging species is a third-party plant species or **hybrid** that is sexually compatible with at least one of the two incompatible but related species

Bristle   Stiff hair

Bt   = *Bacillus thuringiensis* Berliner. Naturally-occurring soil bacterium that produces toxins that are lethal for certain insect **species**

Center of diversity   Geographic region in which the greatest variability of a crop, including its **wild relatives**, occurs. The highest diversity occurs in the primary center of diversity, which often coincides with the **center of crop origin**. Secondary centers of diversity are regions of lesser diversity that have developed as a result of the subsequent spread of the crop

Center of origin   Geographic region from which a crop originated and where its domestication and cultivation first began

Character   See **trait**

Characterization   Determination of the structural or functional attributes of a plant to distinguish between **accessions**

Chasmogamy   **Pollination** occurring after the flower opens

Cleistogamy   Self-fertilization within a closed flower

Clone   Cell, group of cells, or organism that is descended from and genetically identical to a single common ancestor

Corm   Stem structure with nodes and internodes that facilitates vegetative propagation (for example, in bananas)

Crop wild relative (CWR)   Non-cultivated plant **taxon** that has an indirect use derived from its relatively close genetic relationship to a crop

Cross-pollination   Transfer of **pollen** from the **anther** of the flower of one plant to the **stigma** of a flower of another plant (see also **allogamy**). Synonym: **outcrossing**

Culm   Stem of grasses and sedges

Cultivar (cv.)   Cultivated **variety** of a domesticated crop plant

Cytoplasm   Totality of cell components, excluding the nucleus

Cytotype   Group forming the **polyploid** components of an ecospecies

Dehiscence   Spontaneous rupturing or opening of an **anther** or fruiting structure

Distribution   Geographical area where a **species** grows naturally or has been cultivated by smallholders for many years

Dormancy   A metabolic state in which seeds are incapable of germinating even under favorable conditions, due to structural (mechanical restrictions, seed coat, hormone structures) or external factors (water, light, temperature, gases)

Embryo rescue   Plant **tissue culture** technique that consists of cross-pollinating two distantly related **species** and then culturing the resulting embryo, which would not survive otherwise

Endemic   Autochthonous or indigenous to a geographic region and restricted to it (i.e., not occurring elsewhere)

***Ex situ* conservation**   Literally, conservation "off-site." Conservation of plant material (the entire plant, seed, tissue, or **pollen**) outside its original habitat, such as in a **genebank**, botanical garden, or field collection

F1 hybrid   The first hybrid offspring following cross-fertilization

Feral   Crop plant growing outside of cultivation in semi-natural or wild habitats, such as along seed transport routes. Ferals may become established and form naturalized populations of crops existing outside of cultivation

Gamete   Sexual cells of plants

GE   = Genetic engineering, see **GM**

Gene   The basic unit of heredity in a living organism. A segment of DNA coding for a specific functional expression or **trait** in a living organism

Genebank   Facility where **germplasm** is stored in the form of seeds, **pollen**, or *in vitro* cultures, or, in the case of a field genebank, as plants growing in the field

Gene flow   Transfer of genetic information between individuals, populations, or species. Gene flow can occur naturally, via spontaneous **hybridization** (seed- or **pollen**-mediated), or artificially, using hybridization techniques such as **ovule** culture, **embryo rescue**, or **protoplast fusion**

Gene introgression   See **introgression**

Gene pool (GP)   1. The totality of **germplasm** available for crop improvement. Normally, three gene pools are distinguished: (a) GP-1 usually corresponds with the traditional concept of the biological **species**, that is, it contains **taxa** that cross easily with the crop, regularly producing fertile offspring; (b) GP-2 includes less closely related, usually wild species from which gene transfer to the crop is possible—but difficult—using conventional breeding techniques (manual crossing), due to the hybridization barriers separating biological species; and (c) GP-3 contains species from which gene transfer to the crop requires the use of sophisticated artificial hybridization techniques, such as **embryo rescue** or **somatic hybridization**

2. All **genes** and their different **alleles** that are present in an interbreeding population

Genetic diversity   Variation in a group of individuals, populations, or **species**, due to genetic differences (as opposed to the expression of the same genetic background in different environments)

Genetic resources   Genetic material (seed, tissue, DNA, blood) of plants, animals, and other living organisms

Genome   1. The totality of **genes** and associated DNA characteristic of an organism

2. A set of chromosomes corresponding to the **haploid** number of a given **species**

Genotype   1. The genetic constitution of an organism

2. A group of individuals with the same genetic constitution

Geocarpic   Plant whose fruits develop and mature below the soil surface

Germplasm   A set of **genetic resources** that may be conserved or used

GM = Genetic modification. Artificial transfer, removal, or rearrangement of **genes** into an organism using molecular techniques. Synonyms: **GE, transgenic**

Habitat Place or niche where an organism lives and interacts with both other organisms and the environment

Haploid 1. The basic set (*n*) of chromosomes in a **gamete**

2. A cell or individual having one set of chromosomes

Heterozygous Having different **alleles** at the same **locus** of homologous chromosomes

Hybrid Offspring from parental plants of different **genotypes**, usually **varieties** or species (see also **F1 hybrid**)

In situ conservation Literally, "on-site" conservation. Conservation of plants and animals in their original habitats, such as in the wild or in farmers' fields

Introgression Permanent incorporation of genetic information from one set of differentiated populations (**species**, subspecies, races, and so on) into another

In vitro Literally, "in the glass." Outside the living organism (e.g., in a laboratory tube)

Landraces Farmer-developed **varieties** of crop plants that are adapted to local environmental conditions, often have local names, and are not being improved upon by formal breeding programs. Synonym: **traditional varieties**

Locus Place on a chromosome where a **gene** controlling a given **trait** is located

Male sterile Producing sterile **pollen** or not producing pollen at all

Modern varieties **Cultivars** resulting from formal, institutional, and scientific plant breeding, using modern techniques for improvement and selection

Monoecious Having male and female flowers on the same plant (e.g., maize)

Monophyly Group of organisms (**taxon, species**)—including all descendants—derived from a common ancestor

Myrmecochory Seed dispersal by ants

Naturalized Living in a habitat or region like a native, in spite of not being indigenous

Outcrossing see **cross-pollination**

Ovary Plant part that contains the **ovule**

Ovule Plant part that becomes the seed after fertilization

Parthenocarpy Development of fruit without **pollination** or fertilization, and with fruit produced in this manner usually not containing seeds

PCR = Polymerase Chain Reaction. Method for amplifying a DNA sequence in copious amounts, using a heat-stable enzyme

Perennial Plant living for three years or more

Phenotype 1. Physical, biochemical or external appearance of an organism, determined by complex interactions between its genetic constitution (**genotype**) and the environment

2. Group of organisms with a similar physical or external appearance

Pistil Female plant organ consisting of the **ovary**, the **style**, and the **stigma**

Ploidy The total number (x) of basic sets of chromosomes in a cell or organism

Pollen Grains produced in a plant's **anthers**, containing male **gametes**

Pollen dispersal Movement of **pollen** via a dispersal agent, such as wind, water, insects and other animals, or humans

Pollination Transfer of **pollen** from the male part of a plant flower to the female part of the same flower or plant (**self-pollination**) or to a different plant (**cross-pollination**)

Polyploid Organism with more than two basic sets of chromosomes in its cells. Depending on the number of chromosome sets, polyploids are referred to as triploids ($2n = 3x$), tetraploids ($2n = 4x$), hexaploids ($2n = 6x$), and so on.

Protandry **Anthers** mature and shed **pollen** before the **stigma** of the same flower becomes receptive

Protogyny The **stigma** becomes receptive before the **anthers** of the same flower mature

Protoplast fusion Artificial hybridization technique in which protoplasts are induced to fuse by using an electric field or a solution of polyethylene glycol. Useful for crossing **species** that would not hybridize under natural conditions or through the use of other artificial methods of sexual hybridization

Rachis Main axis of the inflorescence (or spike) of some cereals, which carries the spikelets

Self-pollination Transfer of **pollen** from the **anther** of a flower to the **stigma** of either the same flower or of a flower on the same plant (see also **autogamy**)

Shattering Release of seeds from the plant at maturity, allowing free dispersal. Shattering is frequent in wild and weedy **species**, whereas domesticated species have often been bred to reduce shattering

Somatic cell Any cell forming the body of an organism, as opposed to germline cells

Somatic hybrid **Hybrid** originating from the fusion of **somatic cells**

Species 1. Biology: A group of organisms capable of interbreeding freely with each other but not with members of other species

2. Taxonomy: Subdivision of a genus

Stacked genes Deliberate or accidental incorporation of two or more (**GE**) **genes** encoding different **traits** into the **genome** of an organism.

Stigma Part of the **pistil** that receives the **pollen**

Strain 1. The body of descendants from a common ancestor (a race, stock, line, or breed)

2. An artificial **variety** of a **species** of domestic animal or cultivated plant

Style The stalk between the **ovary** and the **stigma**

Sucker Shoot arising from the base of the mother plant. In banana plants, the mother plant dies after flowering, and the oldest sucker becomes the next mother plant

Taxon Taxonomic element, group, or population, regardless of its level of classification

Tiller Lateral shoot or branch emerging from the base of the principal stem of a plant; common in grasses

Tissue culture Growth of tissues and/or cells in an artificial medium, separate from the organism. Used to propagate plants, produce **clones** of a plant, or for difficult interspecific/intergeneric crosses (**ovary** culture, **ovule** culture, embryo culture)

Traditional varieties see **landraces**

Trait Attribute of the **phenotype** of an organism resulting from the interaction of a **gene** or genes with the environment. Synonym: **character**

Transgenic See **GM (genetic modification)**

Variety 1. Classical botany: Subdivision of a **species**

2. Agriculture: Group of similar plants that, because of their structural features and performance, can be identified from other varieties of the same species. Synonym: **cultivar**

Vegetative propagation Asexual propagation resulting in the formation of a plant which is genetically identical with the mother plant (**clone**)

Volunteer Crop plant growing in a subsequent crop or in a neighboring crop, arising from a seed that escaped from previous crop harvests, either in the crop field or elsewhere on the farm. Volunteers often arise from seeds that remain viable for several years in seed banks in the soil

Weed Plant considered to be a nuisance, usually referring to unwanted plants in manmade settings, in particular in agricultural areas

Wild relative See **crop wild relative**

World Maps Showing the Modeled Likelihood of Introgression

To estimate the likelihood of gene flow throughout the world, global maps were produced that show modeled (i.e., probabilistic) distributions of CWR that are sexually compatible with the crop, linked with (1) global data on crop production areas and (2) the likelihood of introgression between crops and sexually compatible CWR (as determined in this book). As noted in chapter 2, we have made three assumptions regarding introgression: crop and CWR populations are sufficiently close to each other to allow outcrossing; flowering times overlap; and the nature of the respective gene or trait (i.e., its stability, location in the genome, inheritance pattern, and selective value) favors introgression.

The modeling for the global CWR distributions is based on a minimum number (10) of geographical locations where the species under consideration is known to occur. Geographical information (latitude and longitude) was obtained from herbaria or from germplasm observations publicly available in databases, including the Global Biodiversity Information Facility (GBIF at www.gbif.org), and national and international germplasm banks (e.g., SINGER at http://singer.grinfo.net/; GRIN at www.ars-grin.gov/cgi-bin/npgs/html/index.pl; EURISCO at http://eurisco.ecpgr.org/home_page/home.php). Global species distributions were modeled with the MaxEnt algorithm (Phillips et al. 2006), using 19 bioclimatic variables to define the potential geographic range of each species (Andersson et al., in prep.).

Maps were then produced, using ArcGIS 9.0 (ESRI, Redlands, CA, USA), that link the modeled CWR distributions with (1) global data on crop production areas and (2) the five categories for the likelihood of introgression (very low to very high; see chapter 2) between crops and sexually compatible CWR[1]. The maps show the overlap between crop production areas and the modeled—that is, probabilistic—distributions of CWR. Thus the areas seen in the map include both real distributional ranges as well as areas where the species of interest does not actually occur, but where there is a high likelihood that bioclimatic conditions would allow the plant to grow. Note that the maps will be made publicly available via Google Earth(r) on the institutional webpage of Bioversity International (http://bioversity.org/).

It is important to consider two limitations to this methodology. First, modeling will fail to predict the full variation within a species if the observations that are used as the basis for the model inadequately represent the extremes of the species' natural range. Second, modeled distributional ranges may include regions where the species under consideration does actually not occur in the wild. Therefore, if a species is known to occur only in specific region(s) or on certain continent(s), mapping was restricted to those areas. For example, the mapping exercise for maize and for most cassava wild relatives was restricted to the Americas, excluding continents such as Asia and Australia, where the species are not known to occur. Even so, the modeled distributions of some of these wild relatives are broader than the known distributional ranges of highly endemic species that are constrained to very narrow geographic regions.

REFERENCES
....

Andersson MS, Alvarez DF, Jarvis A, Wood S, Hyman G, and de Vicente MC (in prep.) Mapping the global likelihood of gene flow using crop production information and distribution data of wild relatives.

Phillips SJ, Anderson RP, and Schapire RE (2006) Maximum entropy modeling of species geographic distributions. *Ecol. Model.* 190:231–259.

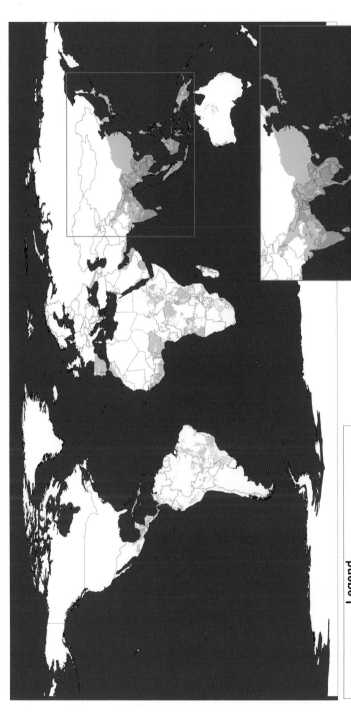

Map 3.1 Likelihood of introgression based on the modeled distributions of cultivated bananas and plantains (*Musa* spp.) and sexually compatible crop wild relatives

Legend

Moderate likelihood
Wild *Musa acuminata, M. balbisiana, M. basjoo, M. cheesmanii**,
M. flaviflora* *M. halabanensis**, *M. itinerans*, *M. laterita**,
M. nagensium* *M. ochracea**, *M. ornata** *M. sanguinea*,
M. schizocarpa, M. sikkimensis*, *M. velutina**

Only crop or only crop wild relatives (i.e., no overlap)

*   Species not mapped due to lack of geographic information

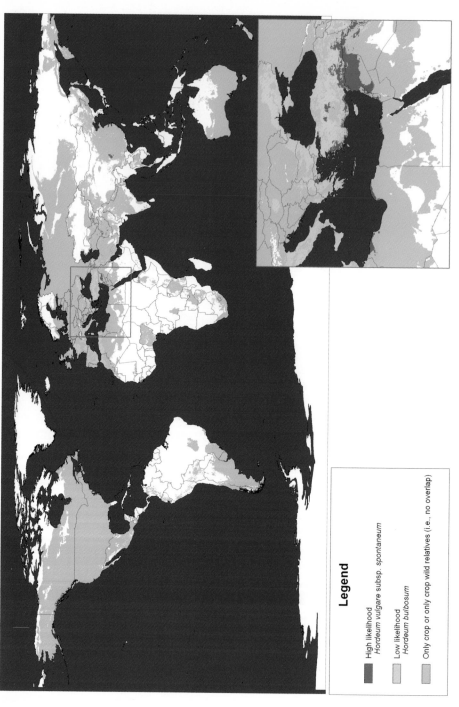

Map 4.1 Likelihood of introgression based on the modeled distributions of domesticated barley (*Hordeum vulgare* L.) and sexually compatible crop wild relatives

Legend

- High likelihood
  Hordeum vulgare subsp. *spontaneum*

- Low likelihood
  Hordeum bulbosum

- Only crop or only crop wild relatives (i.e., no overlap)

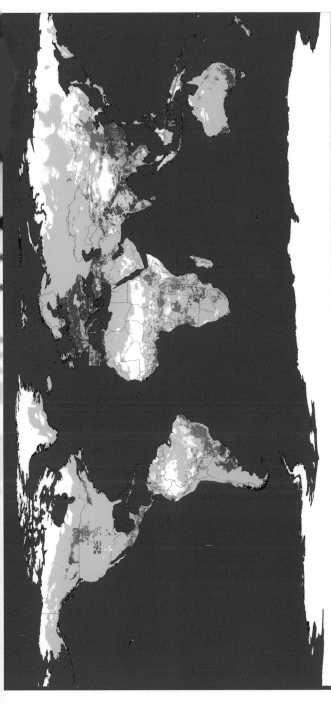

Legend

Very high likelihood
Brassica rapa

High likelihood
Brassica juncea

Moderate likelihood
Brassica oleracea

Low likelihood
Hirschfeldia incana, Raphanus raphanistrum

Only crop or only crop wild relatives (i.e., no overlap)

Map 5.1 Likelihood of introgression based on the modeled distributions of canola (*Brassica napus* L.) and sexually compatible crop wild relatives

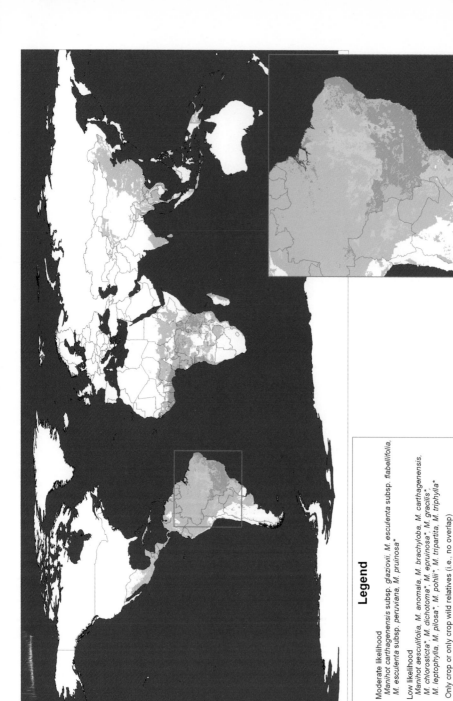

Legend

Moderate likelihood
Manihot carthagenensis subsp. *glaziovii*, *M. esculenta* subsp. *flabellifolia*,
M. esculenta subsp. *peruviana*, *M. pruinosa**

Low likelihood
Manihot aesculifolia, *M. anomala*, *M. brachyloba*, *M. carthagenensis*,
*M. chlorosticta**, *M. dichotoma**, *M. epruinosa**, *M. gracilis**,
M. leptophylla, *M. pilosa**, *M. pohlii**, *M. tripartita*, *M. triphylla**

Only crop or only crop wild relatives (i.e., no overlap)

* Species not mapped due to lack of geographic information

Map 6.1   Likelihood of introgression based on the modeled distributions of cassava (*Manihot esculen* Crant) ally
compatible crop wild relatives

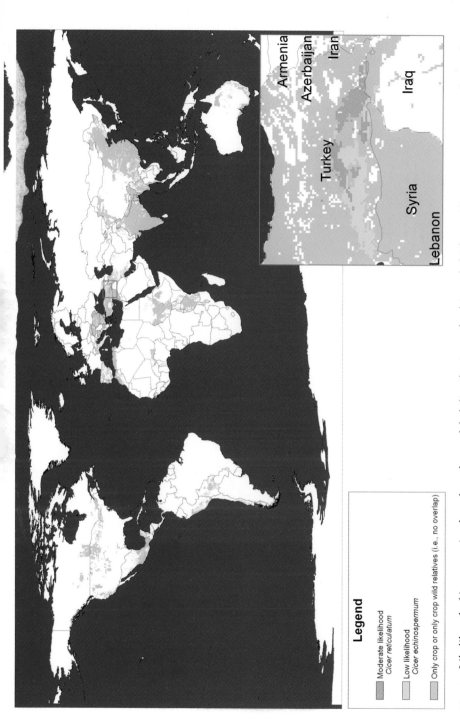

Map 7.1 Likelihood of introgression based on the modeled distributions of cultivated chickpeas (*Cicer arietinum* L.) and sexually compatible crop wild relatives

Legend

Moderate likelihood
Cicer reticulatum

Low likelihood
Cicer echinospermum

Only crop or only crop wild relatives (i.e., no overlap)

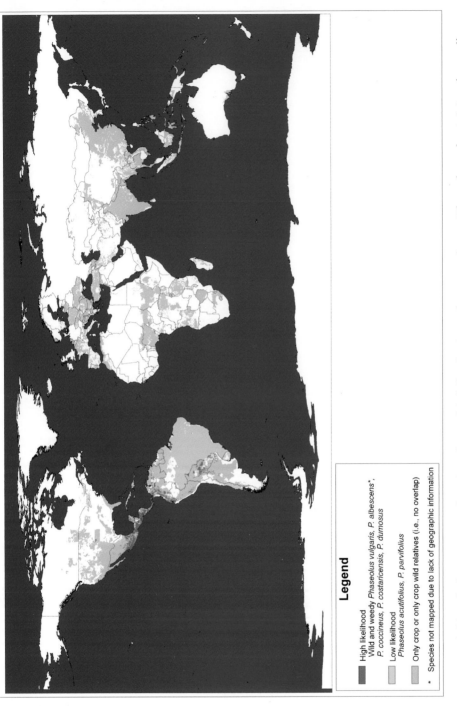

Map 8.1 Likelihood of introgression based on the modeled distributions of common beans (*Phaseolus vulgaris* L.) and sexually compatible crop wild relatives

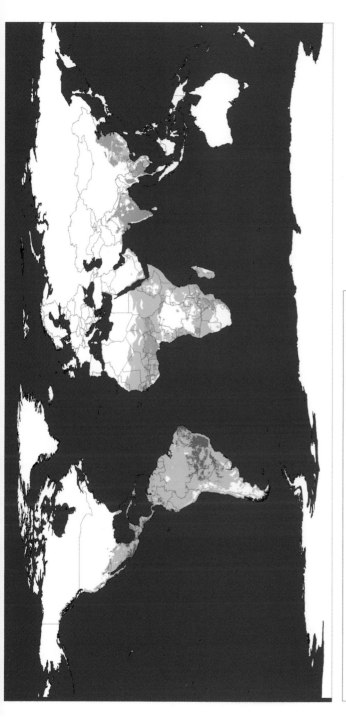

Legend

High likelihood
Wild *Gossypium hirsutum*, wild *G. barbadense*, *G. darwinii**, *G. mustelinum**, *G. tomentosum*

Moderate likelihood
Gossypium anomalum, *G. arboreum*, *G. aridum*, *G. armourianum**, *G. capitis-viridis**, *G. davidsonii**,
G. gossypioides*, *G. harknessii**, *G. herbaceum var. africanum*, *G. klotzschianum**, *G. laxum**, *G. lobatum**,
G. longicalyx*, *G. raimondii**, *G. schwendimanii**, *G. thurberi*, *G. trilobum**, *G. triphyllum*, *G. turneri**

Only crop or only crop wild relatives (i.e., no overlap)

*   Species not mapped due to lack of geographic information

Map 9.1   Likelihood of introgression based on the modeled distributions of cotton (*Gossypium hirsutum* L.) and sexually compatible crop wild relatives

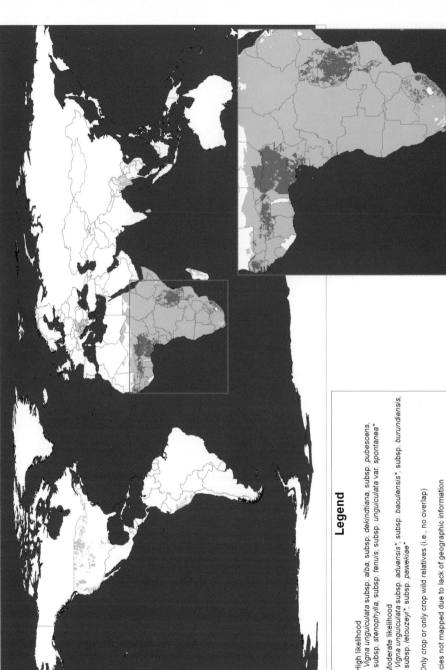

Map 10.1 Likelihood of introgression based on the modeled distributions of cowpeas (*Vigna unguiculata* (L.) Walp.) and sexually compatible crop wild relatives

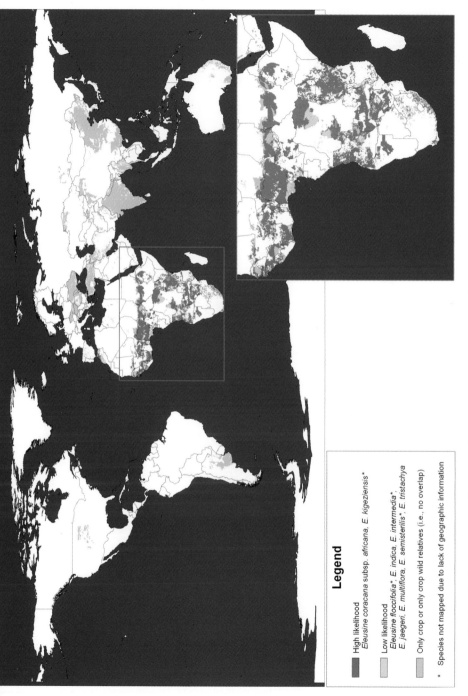

Map 11.1 Likelihood of introgression based on the modeled distributions of finger millet (*Eleusine coracana* (L.) Gaertn.) and sexually compatible crop wild relatives

Legend

High likelihood
Eleusine coracana subsp. *africana, E. kigeziensis**

Low likelihood
Eleusine floccifolia, E. indica, E. intermedia*,
E. jaegeri, E. multiflora, E. semisterilis*, E. tristachya*

Only crop or only crop wild relatives (i.e., no overlap)

* Species not mapped due to lack of geographic information

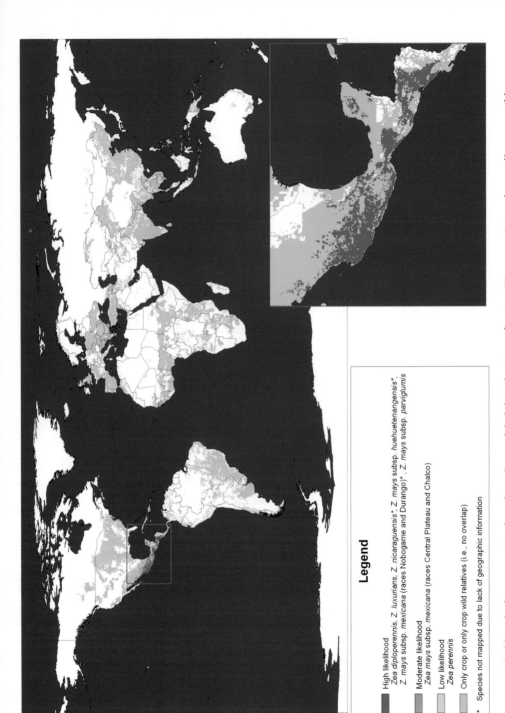

Map 12.1 Likelihood of introgression based on the modeled distributions of maize (*Zea mays* L.) and sexually compatible crop

Legend

High likelihood
Zea diploperennis, Z. luxurians, Z. nicaraguensis*, Z. mays subsp. *huehuetenangensis**
Z. mays subsp. *mexicana* (races Nobogame and Durango)*, *Z. mays* subsp. *parviglumis*

Moderate likelihood
Zea mays subsp. *mexicana* (races Central Plateau and Chalco)

Low likelihood
Zea perennis

Only crop or only crop wild relatives (i.e., no overlap)

* Species not mapped due to lack of geographic information

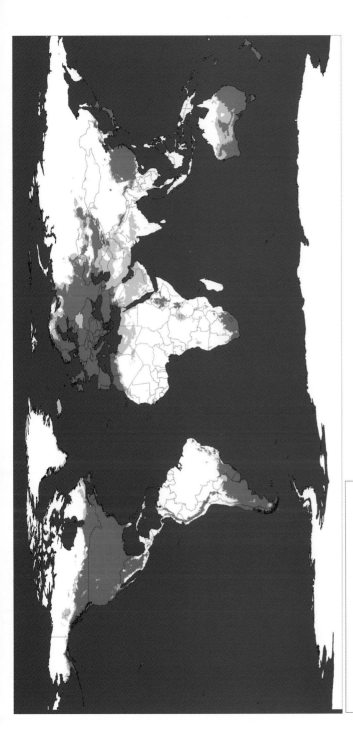

Map 13.1 Likelihood of introgression based on the modeled distributions of oats (*Avena sativa* L.) and sexually compatible crop wild relatives

Legend

High likelihood
Avena fatua, A. ludoviciana, A. occidentalis*, A. sterilis

Only crop or only crop wild relatives (i.e., no overlap)

*   Species not mapped due to lack of geographic information

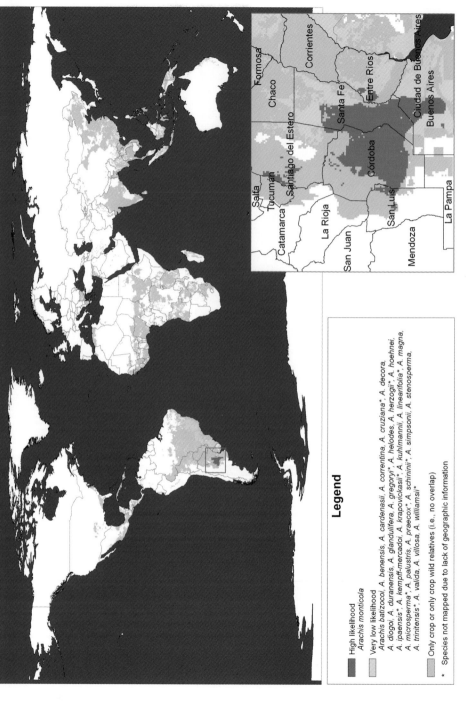

Legend

High likelihood
Arachis monticola

Very low likelihood
Arachis batizocoi, A. benensis, A. cardenasii, A. correntina, A. cruziana, A. decora,
A. diogoi, A. duranensis, A. glandulifera, A. gregoryi, A. helodes, A. herzogi*, A. hoehnei,
A. ipaensis*, A. kempff-mercadoi, A. krapovickasii*, A. kuhlmannii, A. lineariifolia*, A. magna,
A. microsperma*, A. palustris, A. praecox*, A. schininii*, A. simpsonii, A. stenosperma,
A. trinitensis*, A. valida, A. villosa, A. williamsii*

Only crop or only crop wild relatives (i.e., no overlap)

* Species not mapped due to lack of geographic information

Map 14.1 Likelihood of introgression based on the modeled distributions of peanuts (*Arachis hypogaea* L.) and sexually
compatible crop wild relatives

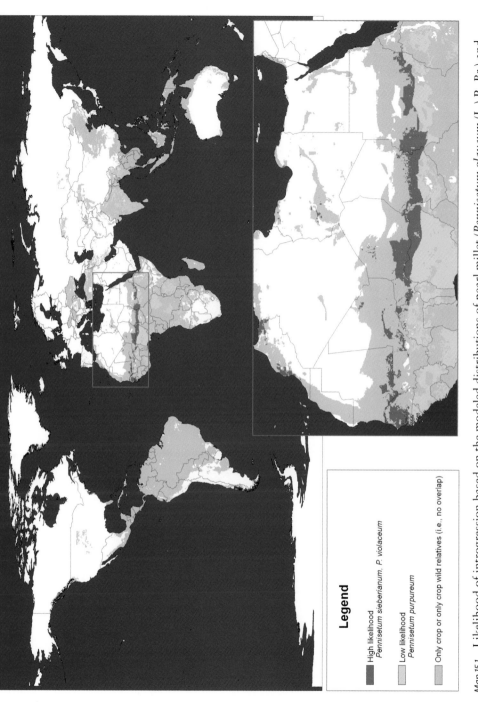

High likelihood
Pennisetum sieberianum, P. violaceum

Low likelihood
Pennisetum purpureum

Only crop or only crop wild relatives (i.e., no overlap)

Map 15.1 Likelihood of introgression based on the modeled distributions of pearl millet (*Pennisetum glaucum* (L.) R. Br.) and sexually compatible crop wild relatives

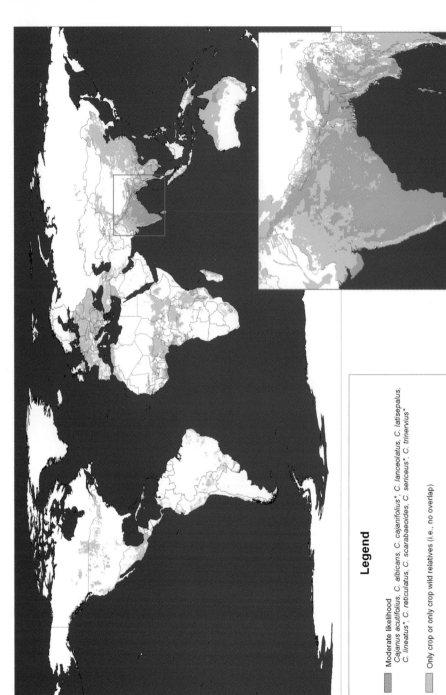

Map 16.1 Likelihood of introgression based on the modeled distributions of pigeonpeas (*Cajanus cajan* (L.) Millsp.) and sexually

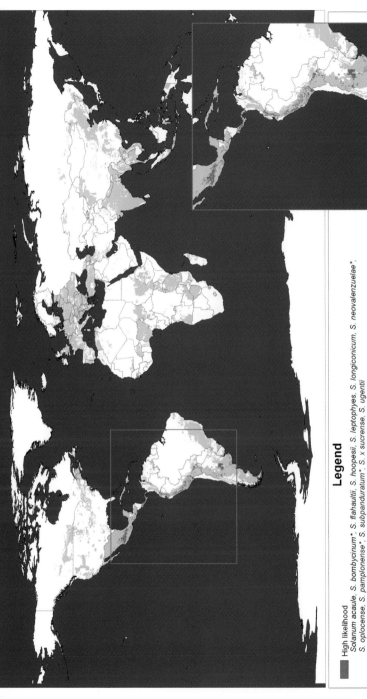

Legend

High likelihood
Solanum acaule, S. bombycinum, S. flahaultii, S. hoopesii, S. leptophyes, S. longiconicum, S. neovalenzuelae*,
S. oplocense, S. pamplonense*, S. subpanduratum*, S. x sucrense, S. ugentii*

Moderate likelihood
*Solanum agrimonifolium, S. boliviense, S. bukasovii, S. colombianum, S. hjertingii, S. infundibuliforme, S. megistacrolobum, S. microdontum, S. oxycarpum, S. paucijugum,
S. pillahuatense, S. raphanifolium, S. stoloniferum, S. tarijense, S. tuquerrense, S. vernei, S. verrucosum, S. vidaurrei, S. violaceimarmoratum*

Only crop or only crop wild relatives (i.e., no overlap)

* Species not mapped due to lack of geographic information, including the following moderate-likelihood species:
*S. acroglossum, S. albornozii, S. amayanum, S. ambosinum, S. anamatophilum, S. ancophilum, S. andreanum, S. anduphilum, S. ayacuchense, S. berthaultii, S. x blanco-galdosii, S. brevicaule, S. buesii,
S. candolleanum, S. cantense, S. chacoense, S. chilliasense, S. chiquidenum, S. chomatophilum, S. coelestipetalum, S. contumazaense, S. x doddsii, S. gandarillasi, S. hastiforme, S. huancabambense,
S. huancavelicae, S. huarochirense, S. incasicum, S. irosinum, S. jalcae, S. kurtzianum, S. laxissimum, S. limbaniense, S. x liitusinum, S. lobbianum, S. marinasense, S. medians, S. multinterruptum,
S. neovavilovii, S. nubicola, S. olmosense, S. opiocense, S. orophilum, S. pampasense, S. paucissectum, S. peloquinianum, S. piurae, S. rhomboideilanceolatum, S. sanctae-rosae, S. sandemanii, S. santolallae,
S. sarasarae, S. sawyeri, S. saxatilis, S. sogarandinum, S. sparsipilum, S. spegazzinii, S. tacnaense, S. tapojense, S. taulisense, S. urubambae, S. venturii, S. yungasense*

Map 17.1 Likelihood of introgression based on the modeled distributions of potatoes (*Solanum tuberosum* L.) and sexually
compatible crop wild relatives

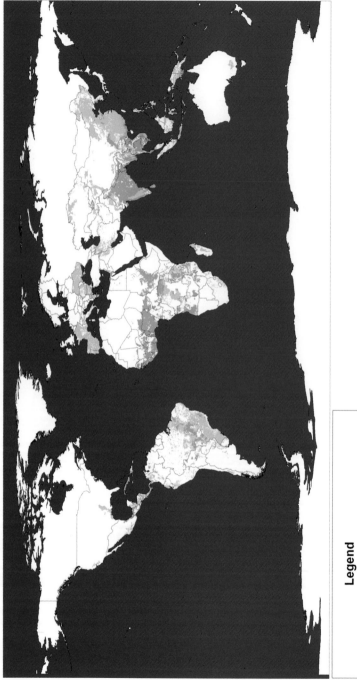

Map 18.1 Likelihood of introgression based on the modeled distributions of Asian cultivated rice (*Oryza sativa* L.) and sexually compatible crop wild relatives

Legend

■ Very high likelihood
Sorghum almum, S. bicolor subsp. *drummondii, S. bicolor* subsp.
verticilliflorum, S. halepense, S. propinquum, shattercane

■ Only crop or only crop wild relatives (i.e., no overlap)

Map 19.1 Likelihood of introgression based on the modeled distributions of sorghum (*Sorghum bicolor* (L.) Moench) and sexually compatible crop wild relatives

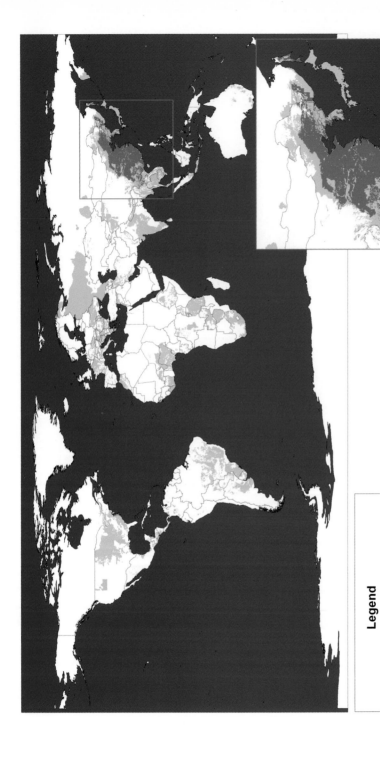

Legend

High likelihood
Glycine soja species complex

Only crop or only crop wild relatives (i.e., no overlap)

Map 20.1 Likelihood of introgression based on the modeled distributions of soybeans (*Glycine max* (L.) Merr.) and sexually compatible crop wild relatives

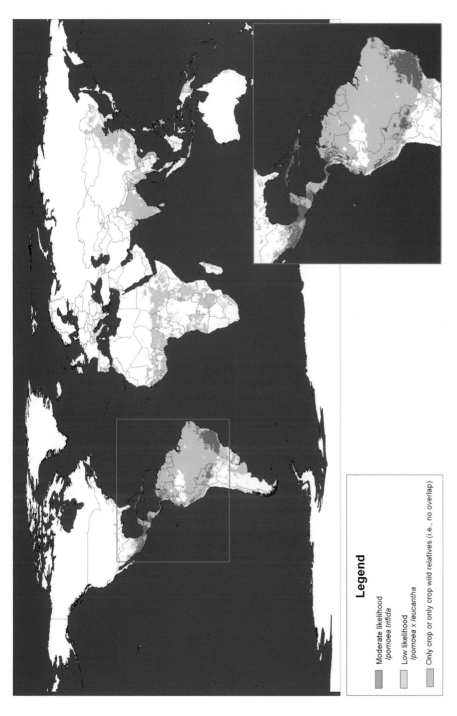

Map 21.1 Likelihood of introgression based on the modeled distributions of sweetpotatoes (*Ipomoea batatas* (L.) Lam.) and sexually compatible crop wild relatives

Legend

Moderate likelihood
Ipomoea trifida

Low likelihood
Ipomoea x leucantha

Only crop or only crop wild relatives (i.e., no overlap)

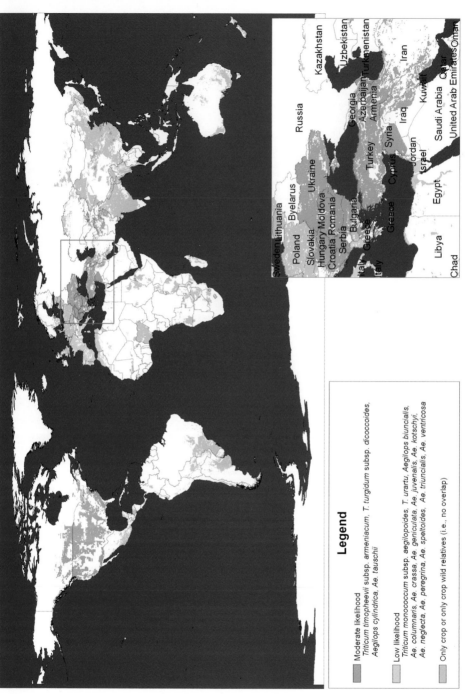

Map 22.1 Likelihood of introgression based on the modeled distributions of bread wheat (*Triticum aestivum* L.) and sexually compatible crop wild relatives.

Index

....

Entries followed by f denote figures; t tables.